A Specialist Periodical Report

Mass Spectrometry
Volume 2

A Review of the Literature Published
between July 1970 and June 1972

Senior Reporter
D. H. Williams, *University Chemical Laboratory, Cambridge*

Reporters
J. H. Bowie, *University of Adelaide*
C. J. W. Brooks, *University of Glasgow*
M. I. Bruce, *Bristol University*
I. Howe, *Cambridge University*
T. J. Mead, *Sheffield University*
B. S. Middleditch, *Baylor College of Medicine, Houston*
H. R. Morris, *Cambridge University*
S. D. Ward, *CEGB, London*
J. M. Wilson, *Manchester University*

ISBN: 0 85186 268 3

© Copyright 1973

The Chemical Society
Burlington House, London, W1V 0BN

Set in Times on Monophoto Filmsetter and printed offset by
J. W. Arrowsmith Ltd., Bristol, England

Made in Great Britain

Foreword

This is the second volume of a series which will appear biennially. Its prime purpose is to summarize the literature which has appeared between July 1970 and June 1972 in an area which can broadly be defined as 'organic mass spectrometry'.

The format which has been used is similar to that adopted in Volume 1. The first chapter deals with ionization techniques and ion analysis, and the second with energetic and kinetic aspects of ion chemistry; the advances of principle which have occurred in these areas have permitted the authors to adopt the style almost of a review article, while still managing to cover the most important literature. In contrast, the vast amount of material to be covered under the reactions of specific functional groups (Chapter 3) and organometallic and co-ordination compounds (Chapter 5) has allowed the authors little choice (within the allotted space) other than to provide an annotated summary of the literature. The material to be covered in the chapters on computerized data acquisition and handling (Chapter 6) and gas chromatography–mass spectrometry (Chapter 7) has permitted the authors to steer a middle course in this respect.

The remaining chapter (that dealing with natural products—Chapter 4) merits a special comment because the person originally assigned to cover this topic unfortunately found himself unable to act only a few weeks before the manuscripts were to be completed. The chapter has therefore been covered at very short notice by four authors to whom I am especially grateful. It is inevitable of course that, within the very short time available to them, these authors will have been unable to cover all the work that should be mentioned. They have asked me to mention that studies of nucleosides are not included in the present report and, in particular, that metabolic studies have not been covered. Indeed, biomedical applications of mass spectrometry (and mass spectrometric studies of drug metabolism in particular) are becoming of increasing importance; special attention could be devoted to these areas in subsequent volumes, with appropriate mention of any studies not covered in the present text.

D. H. Williams

Contents

Chapter 1 Alternative Methods of Ionization and Analysis
By J. M. Wilson

1 General Introduction	1
2 Ionization Potentials and Other Applications of Photoionization	1
3 Chemical Ionization	6
4 Negative Ion Production	14
5 Field Ionization	18
6 Ion Cyclotron Resonance	24

Chapter 2 Kinetic and Energetic Studies of Organic Ions
By I. Howe

1 Introduction	33
2 Energetics of Mass Spectral Processes	34
Sources of Error in Heat of Formation Measurements	34
Formation of Fragments with Excess Energy	37
Ion Structure Evidence from Heats of Formation and Appearance Potentials	38
Bond Energies and Heats of Formation	41
Methods for Determination of Onset Potentials	43
Stevenson's Rule	44
3 The Quasi-Equilibrium Theory	46
Calculations	47
Energy Distributions and Energy States	53
4 Tight Transition States	55
Rearrangement Reactions	55
Anchimeric Assistance	56

5 Metastables	58
Ion Kinetic Energy Spectroscopy	59
Shapes of Metastable Peaks	62
Metastable Ion Abundance Ratios	64
Metastable Transitions in Various Instruments	66
6 Reactions of Isotopically Labelled Species	67
Deuterium Isotope Effects	67
Scrambling Reactions	72
Specific Isotopic Transfer	75
7 Substituent Effects	75
Hammett Correlations	75
Steric Effects	81
8 Collision Processes	82
Ion–Molecule Reactions	83
Collision-induced Metastables	86
9 Determination of Ion Structures	88

Chapter 3 Reactions of Specific Functional Groups
By J. H. Bowie

1 General Introduction	90
2 The Reactions of Specific Functional Groups	93
Hydrocarbons	93
Aliphatic and Acyclic	93
Aromatic	94
Halides	98
Alcohols and Phenols	101
Aldehydes, Ketones, and Quinones	103
Aliphatic	103
Aromatic	106
Acids	108
Esters, Lactones, and Anhydrides	109
Ethers and Peroxides	111
Amines, Amides, Imides, and Related Systems	113
The $\diagdown\mathrm{C}{=}\mathrm{N}{-}$, $\diagdown\mathrm{N}{-}\mathrm{N}\diagup$, $-\mathrm{N}{=}\mathrm{N}-$, $-\mathrm{CN}$, and $-\mathrm{NC}$ Groups	115
The N—O Group	118
Heterocyclic Systems (excluding Sulphur Compounds)	119
Three- and Four-membered Rings	119

Five-membered Rings	120
Six- and Seven-membered Rings	125
Sulphur Compounds	129

3 Scrambling Processes and Skeletal Rearrangements—A Summary   135

4 Negative-ion Mass Spectrometry   137

Chapter 4 Natural Products
By T. J. Mead, H. R. Morris, J. H. Bowie, and I. Howe

1 Introduction   143

2 Lipids   143

3 Amino-acids and Peptides   151
- General   151
- Peptide Sequencing   159
  - Derivatization   159
  - Structural Elucidation   162
- New Methods of Ionization   167

4 Carbohydrates   169

5 Antibiotics   175

6 Isoprenoids   179

7 Steroids   182

8 Alkaloids   186
- Pyrrole and Indole Alkaloids   186
- Pyridine, Quinoline, and Isoquinoline Alkaloids   188
- Miscellaneous Alkaloids   189

9 Flavonoids and Quinones   190

Chapter 5 Organometallic and Co-ordination Compounds
By M. I. Bruce

1 Introduction   193
- Reviews   193
- General Considerations   194

2 Main-group Organometallics — 194
- Group I — 194
- Group II — 194
- Group III — 197
  - Alkyl- and Aryl-borons, *etc.* — 197
  - Boron Heterocycles — 198
  - Boron–Nitrogen Compounds — 198
  - Boronates and other Boron–Oxygen Compounds — 199
  - Other Elements — 201
- Group IV — 202
  - Silicon — 203
  - Silicon Heterocycles — 206
  - Silicon–Nitrogen Compounds — 209
  - Trimethylsilyl Derivatives — 210
  - Germanium — 211
  - Tin — 211
  - Lead — 213
- Group V — 214
  - Phosphorus — 217
  - Arsenic — 221
  - Antimony and Bismuth — 223

3 Transition-metal Organometallics — 223
- Ion–Molecule Reactions — 223
- Metal Carbonyls — 225
- Nitrosyls — 225
- Metal Carbonyl Hydrides and Halides — 226
- Transition-metal Cluster Compounds — 226
- Compounds containing Bonds to Main Group Elements — 227
- Complexes containing Metal–Carbon σ-Bonds — 229
- Hydrocarbon–Metal π-Complexes — 230
  - Olefins, Dienes, and Fluoro-olefins — 230
  - Acetylenes — 230
  - Allyls — 231
  - Cyclobutadiene — 231
  - Cyclopentadienyl — 231
  - Arenes — 233
  - Complexes containing C_7 Hydrocarbons — 234
- Complexes with Donor Ligands — 235
  - Carbene Complexes — 235
  - Nitrogen Ligands — 236
  - Phosphorus Ligands — 237
  - Arsenic Ligands — 237
  - Group VI Ligands — 238
- Ferrocenes — 239

Contents

4 Co-ordination Complexes — 239
- β-Diketonato-complexes — 240
- Schiff-base Complexes — 244
- Sulphur Ligands — 245
- Complexes containing Nitrogen — 246
- Porphyrins — 247
- Carboxylates — 248
- Other Complexes — 248
- Alkoxides, Alkylamides, and Related Compounds — 249

5 Ionization Potential Data — 251
- Table 1: Main-group Organometallics — 251
- Table 2: Transition-metal Organometallics — 252
- Table 3: Co-ordination Compounds — 254

6 Appendix — 255
- Group II — 255
- Group III — 256
  - Carboranes and Related Compounds — 256
- Group IV — 256
  - Silicon — 256
  - Germanium — 257
  - Tin — 257
- Group V — 258
- Carbonyls — 258
- Nitrosyls — 258
- Hydrides, Halides, and Pseudohalides — 259
- Compounds with Bonds to Main Group Elements — 259
- σ-Bonded Complexes — 259
- Acetylene Complexes and σ-Acetylides — 259
- Olefin Complexes — 260
- Allyl Complexes — 260
- Cyclopentadienyl Complexes — 260
- Arene Complexes — 261
- Complexes containing Larger Rings — 261
- Complexes containing Donor Ligands — 261
  - Boron Ligands — 261
  - Carbene and Related Complexes — 261
  - Nitrogen Donors — 261
  - Phosphorus Donors — 262
  - Arsenic, Antimony, and Bismuth Donors — 262
  - Sulphur Donors — 262
- Ferrocenes — 263

Chapter 6 Computerized Data Acquisition and Handling
By S. D. Ward

1 Introduction	264
2 Data Acquisition	265
Electrical Recording	267
System Configurations and Design	267
Accuracy of Mass and Intensity Measurement	273
Photographic Recording	277
System Configurations and Design	277
Accuracy of Mass and Intensity Measurement	279
Reference Compounds for Mass Calibration	279
Other Techniques	279
Metastable Ion Detection	279
Spark Source Mass Spectrometry	280
Time of Flight Mass Spectrometry	280
3 Data Processing	280
Single Spectra and Learning Machine Methods	281
Low Resolution	281
High Resolution	287
Library Searching and Spectrum Comparison Techniques	289
4 Computer Control	293
5 Miscellaneous Applications	296
Gas Chromatography and Mass Spectrometry	296
Ionization and Appearance Potential Measurement	297
Mixture Analysis	299
Computer Programs and General Data Processing	300
6 Future Trends	300

Chapter 7 Gas Chromatography–Mass Spectrometry
By C. J. W. Brooks and B. S. Middleditch

1 General Considerations	302
Introduction	302
Practical Aspects of G.C.–M.S.	302
Selection of Functional Derivatives for G.C.–M.S.	306
Current Trends	309
2 Applications	309
Hydrocarbons	309
Long-chain Compounds	310
Prostaglandins	311

Sphingosine Derivatives	313
Carbohydrates	314
Oxygenated Terpenoids	315
Steroids: (A) Reference Compounds	317
Alcohols	317
Ketones	317
Corticosteroids	318
Steroids: (B) in Biological Material	318
Sterols	318
Bile Acids	319
Hormonal Steroids and Metabolites in the Human	319
Hormonal Steroids of Animals or Plants	321
Amines: (A) Reference Compounds and Reaction Products	321
Amines: (B) in Biological Material	322
Amino-acids and Peptides	323
Drugs and Metabolites: (A) Reference Compounds	324
Drugs and Metabolites: (B) in Biological Material	325
Insect Pheromones and Other Secretions	328
Food Flavours and Aromas	329
Pesticides and Pollutants	331
Organic Geochemistry	332
Miscellaneous	334
Author Index	337

1
Alternative Methods of Ionization and Analysis

BY J. M. WILSON

1 General Introduction

In Volume 1 of this Report, the material covered fell fairly neatly into four categories, Chemical Ionization (CI), Field Ionization (FI), Negative Ion Studies (NI) and Ion Cyclotron Resonance Spectroscopy (ICR). For this article the scope has been broadened to cover Photoionization (PI). In the previous article the emphasis was on the impact of these methods on structural studies, *i.e.* on the structure of the molecules ionized or of the ions produced. These methods are now being extended, and some discussion of energetic and kinetic studies on ions is included here.

There is much more overlap than previously between the sections. The details of the ion–molecule reactions of importance in CI work can be understood better if the primary ions can be produced in well-defined electronic states and within a narrow energy range; this can be achieved by using a vacuum monochromator and photoionizing the molecules. There is also a certain amount of overlap between ICR and CI, and the ease with which the reactions of negative ions can be studied by ICR leads to a degree of overlap between these two sections.

The first section, on electron and photon impact, covers methods of determining ionization and appearance potentials and other applications of photoionization sources. The other sections cover the techniques of FI, CI, NI, and ICR. The advantages of FI and CI in molecular weight determination are becoming more and more obvious. The use of the fragmentation patterns in these modes is still at the development stage. For the sequencing of peptides, CI, FI, and PI have all been claimed to give better results than EI. Of these, FI has the advantage that it is possible to obtain spectra by field desorption of free peptides although, for clarity of spectra and certainty in sequencing, CI appears to be better if suitable derivatives are prepared.

Further developments in ICR spectroscopy have led to more accurate measurements of photodetachment energies of negative ions, and there is now a considerable number of measurements of acidities and basicities in the gas phase. We are now reaching the stage where our knowledge of gas-phase ions can help in our understanding of the behaviour of ions in the condensed phase.

2 Ionization Potentials and Other Applications of Photoionization

The four principal methods of determination of ionization potentials are by observation of absorption spectra, photoionization yields, photoelectron spectra, and electron impact ionization yields. Although the methods differ vastly in the accuracy of results they can produce, even the least accurate, electron impact, is still widely used. The various methods which can be used for accurate measurements of small ions in the gas phase have been described by Herzberg.[1]

The availability of efficient sources of monochromatic light and the ease and accuracy with which ionization potentials and energies of both electronically and vibrationally excited states can be measured, has led to an explosion of publications of results of photoelectron spectra during the past two years. Much of this work is covered in other reviews[2] and will not be considered here. The one advantage of electron impact work was the versatility of the instrument, but even that advantage is now doubtful, since a photoelectron spectrometer can be used to measure ionization potentials of radicals and dissociation energies of ions. Tetrafluorohydrazine decomposed at 225 °C in a silica tube adjacent to the photon source and a photoelectron spectrum of $NF_2^{\cdot}$ was obtained.[3] In a photoelectron spectrometry study of HF and DF, the dissociation limit,of the H—F bond could be observed at the end of the vibrational series of the second IP, and was measurable with greater accuracy than is usually possible by electron impact AP measurements.[4] It is only in small molecules, however, that the vibrational structure can be sufficiently sharp to show up dissociation limits. It is of interest that the first IP as measured in this study was 0.07 eV lower than a previous photoionization measurement.

Lloyd has pointed out that some molecules may have a much lower cross-section for the first IP than for the second. This is particularly true of the Group V trihalides. In $AsCl_3$ there is only a 0.6 eV difference between the two levels as measured by photoelectron spectroscopy,.but the electron impact method gives a result between these two.[5] The same reason is suggested for the 0.7 eV difference between EI and PI measurements of the first IP of 1-methylcyclopentene and methylenecyclopentane; the low cross-section state is lost in the thermal tail of the EI ionization efficiency curve.[6]

This 'thermal tail' is a result of the spread of thermal energy in the electron beam. Tails have been observed in photoionization efficiency curves, but these are ascribed to the presence of thermally excited molecules. Because of this effect the values of $I(Me_4Si)$ and $D(Si-C)$ are only upper limits, but they are 0.3 eV less than the values obtained by electron impact.[7] The thermal effect

---

[1] G. Herzberg, *Quart. Rev.*, 1971, **25**, 201.
[2] S. D. Worley, *Chem. Rev.*, 1971, **71**, 295; D. W. Turner, *Ann. Rev. Phys. Chem.*, 1970, **21**, 107; D. W. Turner, *Phil. Trans. Roy. Soc.*, 1970, **A268**, 7; C. R. Brundle, *Appl. Spectroscopy*, 1971, **25**, 8; D. W. Turner, C. Baker, A. D. Baker, and C. R. Brundle, 'Handbook of 584 Å Spectra', Wiley-Interscience, London, 1970; 'Electronic Structure and Magnetism of Inorganic Compounds', ed. P. Day (Specialist Periodical Reports), The Chemical Society, London, 1972, Vol. 1.
[3] A. B. Cornford, D. C. Frost, F. G. Herring, and C. A. McDowell, *J. Chem. Phys.*, 1971, **54**, 1872.
[4] C. R. Brundle, *Chem. Phys. Letters*, 1970, **7**, 317.
[5] D. R. Lloyd, *Internat. J. Mass Spectrometry Ion Phys.*, 1970, **4**, 500.
[6] M.-Th. Praet, *Org. Mass Spectrometry*, 1970, **4**, Suppl. 65.
[7] G. Di Stefano, *Inorg. Chem.*, 1970, **9**, 1919.

on ionization efficiency curves has been calculated to be as much as 0.1—0.2 eV.[8]

There have been a number of recent attempts to improve the accuracy of ionization and appearance potential measurements by electron impact methods. One group prefers to use an ICR spectrometer for this work, for two reasons: (a) the electron kinetic energy will not be distorted by large electric fields and (b) because of the long residence time of the ions the appearance potential measurements should not be affected by a 'kinetic shift'.[9] The results show a small but reproducible difference in IP between *cis*- and *trans*-pent-2-ene. An on-line acquisition method has been described[10] which involves averaging the ionization efficiency curve over ten scans, followed by mathematical smoothing, followed by correction for the electron energy spread by the EED method.[11] This is a purely mathematical treatment which can give ionization efficiency curves with breaks corresponding to vibrational states. Alternative mathematical methods have been produced involving the second differential of the ionization efficiency curve[12] or a Fourier transform of the first differential.[13] The authors claim that the latter method gives a superior result in terms of clarity of fine structure than does the EED method.

There is a report of success in obtaining good ionization curves using the RPD method which should produce effectively monoenergetic electrons. Breaks are found in the curves which correlate with excited states of neutral products.[14]

An indirect method of determining ionization potentials of radicals has been suggested which relies on the relative abundances of fragment ions in the mass spectra of a series of compounds.[15] In any process such as

$$\begin{bmatrix} R^1 \\ | \\ R^2 \end{bmatrix}^{+} \longrightarrow \begin{array}{l} R^{1+} + R^2 \\ R^1 + R^{2+} \end{array}$$

the abundance of R^{1+} will be greater than that of R^{2+} if $I(R^1) < I(R^2)$. The authors have produced the results shown in Table 1 for cyclohexane derivatives. The ionization potential of the cyclohexyl radical must therefore lie between those of s-butyl and t-butyl, *i.e.* at about 7.2 eV. This is 0.4 eV less than the only measured value.

The importance of photoionization and photoelectron spectroscopy measurements is not only for the measurement of ionization and appearance potentials. There are applications which are more general to mass spectra. It has been suggested that photoelectron spectra can be used as a rough guide to the internal

---

[8] W. A. Chupka, *J. Chem. Phys.*, 1971, **54**, 1936.
[9] M. L. Gross and C. L. Wilkins, *Analyt. Chem.*, 1971, **43**, 1624.
[10] R. A. W. Johnstone, F. A. Mellon, and S. D. Ward, *Internat. J. Mass Spectrometry Ion Physics*, 1970, **5**, 241.
[11] R. E. Winters, J. H. Collins, and W. L. Courchene, *J. Chem. Phys.*, 1966, **45**, 1931.
[12] G. G. Meisels and B. G. Giessner, *Internat. J. Mass Spectrometry Ion Phys.*, 1971, **7**, 489.
[13] B. G. Giessner and G. G. Meisels, *J. Chem. Phys.*, 1971, **55**, 2269.
[14] D. Lewis and W. H. Hammill, *J. Chem. Phys.*, 1970, **52**, 6348.
[15] A. G. Harrison, C. D. Finney, and J. A. Sherk, *Org. Mass Spectrometry*, 1971, **5**, 1313.

Table 1

$C_6H_{11}-R$	$[C_6H_{11}^+]$	$[R^+]$	$I(R\cdot)/eV$
$C_6H_{11}-CHMeEt$	100	16	7.4
$C_6H_{11}-CMe_3$	13	72	6.93

energy distribution of ions produced by electron impact at higher electron energies. This suggestion has been discussed and treated in general with great reservations.[16] The principal reason given is that there are a number of compounds which have fragment ions with appearance potentials which coincide with regions in the photoelectron spectrum where there is zero photoelectron yield. It has also been shown that photoionization and Penning ionization give rise to quite different relative population of the various accessible electronic states of the ion.[17]

The relationship between the photoelectron spectrum and the mass spectrum produced by the same photon beam is one which will obviously require investigation. In the coincidence technique, single-event counting methods are used and only photoelectrons and ions which coincide in time are registered. There are two types of coincidence experiment; it is possible for each mass to measure a kinetic energy spectrum of the electrons produced during its formation; alternatively, for each electron energy level, a mass spectrum can be determined. Experiments of the first type have been described.[18] Eland has described in detail the theoretical limitations of the method[19] and has produced preliminary results from a system designed for experiments of the second type.[20]

In a review of photoionization,[21] Reid has discussed instrumentation, analytical applications, ion efficiency curves, and studies of unimolecular ion decomposition and of ion–molecule reactions. The mass spectra tend to be similar to electron impact spectra in that they are formed by unimolecular decomposition of a series of molecular ions of differing energy content, but a photon source does have the advantages of a well-defined narrow energy range and of low source temperatures (there is no hot filament in the region of the ionization chamber).

PI spectra can be used to detect small differences between stereoisomers which would give identical spectra under electron impact. The *endo*- (1) and *exo*- (2) isomers of tricyclo[3,2,1,0^{2,4}]octane can be differentiated by small intensity differences in spectra obtained using the helium 584 Å line. If the hydrogen Lyman α line is used (10.19 eV) the differences become more marked. The difference between the isomeric ketones (3) and (4) is much more striking.[22]

[16] G. Innorta, S. Torroni, and S. Pignataro, *Org. Mass Spectrometry*, 1972, **6**, 113.
[17] H. Hotop and A. Niehaus, *Internat. J. Mass Spectrometry Ion Phys.*, 1970, **5**, 415.
[18] B. Brehm and E. von Putkammer, *Adv. Mass Spectrometry*, 1968, **4**, 591.
[19] J. H. D. Eland, *Internat. J. Mass Spectrometry Ion Phys.*, 1972, **8**, 143.
[20] J. H. D. Eland and C. J. Danby, *Internat. J. Mass Spectrometry Ion Phys.*, 1972, **8**, 153.
[21] N. W. Reid, *Internat. J. Mass Spectrometry Ion Phys.*, 1971, **6**, 1.
[22] C. E. Brion, J. S. Haywood Farmer, R. E. Pincock, and W. B. Stewart, *Org. Mass Spectrometry*, 1970, **4**, Suppl. 599.

Alternative Methods of Ionization and Analysis

The PI spectra of peptide derivatives are interesting in that, compared with EI spectra, the total number of peaks is reduced, especially in the low-mass region, and the molecular ion and higher-mass ions are more abundant.[23] The sequence

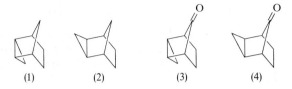

(1) (2) (3) (4)

ions are more often formed by fission of C—C bonds as in (5), but one disadvantage of the method would appear to be that some of the important lower-mass sequence ions may be missing.

$$R-CO-\overset{+}{N}H=\overset{\overset{\displaystyle R^1}{|}}{C}H$$
(5)

Photoionization has certain advantages for the study of ion–molecule reactions, and they are the usual ones of energy resolution and low temperature. Lyman α radiation has been used to study the C_3H_6–C_4D_{10} system,[24] because it can ionize C_3H_6 and not C_4D_{10}. The principle reactions are

$$C_3H_6^{+\cdot} + C_4D_{10} \rightarrow C_3H_6D\cdot + C_4D_9^+$$
$$C_3H_6^{+\cdot} + C_4D_{10} \rightarrow C_6H_6D_2 + C_4D_8^{+\cdot}$$

The second process is much more probable with propylene than with cyclopropane. In spectra of ethanol run using 10.68 eV radiation the only primary ion is $EtOH^{+\cdot}$, and in the absence of other ions it is easier to measure the pressure dependence of the solvated protons produced.[25]

$$EtOH^{+\cdot} \rightarrow Et\overset{+}{O}H_2 \rightarrow (EtOH)_2^+H \rightarrow (EtOH)_3^+H \; etc.$$

Potapov and Sorokin have used a vacuum monochromator to examine the effect of photon energy on ion–molecule reactions.[26] The rate of the reaction

$$MeOD^{+\cdot} + CH_3OD \rightarrow Me\overset{+}{O}DH + \cdot CH_2OD$$

is invariant with photon energy between the first and second ionization potentials of methanol. This is not true of the rate of formation of $Me\overset{+}{O}D_2$ because there

---

[23] V. M. Orlov, Y. M. Varshavsky, and A. A. Kiryushkin, *Org. Mass Spectrometry*, 1972, **6**, 9.
[24] I. Koyano, N. Nakayama, and I. Tanaka, *J. Chem. Phys.*, 1971, **54**, 2384.
[25] W. A. Chupka and M. E. Russell, *J. Phys. Chem.*, 1971, **75**, 3797.
[26] V. K. Potapov and V. V. Sorokin, *Doklady Akad. Nauk S.S.S.R.*, 1970, **192**, 590.

are two reactions involved, and the fragment-ion current will be affected by photon energy.

$$MeOD^{+\cdot} + MeOD \rightarrow Me\overset{+}{O}D_2 + MeO\cdot$$

$$CH_2OD^+ + MeOD \rightarrow Me\overset{+}{O}D_2 + CH_2=O$$

The same authors have shown that the yield of $MeCH\overset{+}{O}H$ from MeCHO does not vary from 10.2—12.5 eV, the reason suggested being that in this region there is no excitation of higher vibrational states of the ground-state ion. The yield of the reaction

$$NH_3^{+\cdot} + NH_3 \rightarrow NH_4^+ + NH_2$$

decreases with increasing photon energy because the excited vibrational states of $NH_3^{+\cdot}$ are accessible and are less reactive.[27]

Some of the most spectacular results of photoion–molecule reactions have come out of the studies on ethane and propane. In the medium-pressure mass spectrum of ethane at 298 K using argon resonance radiation, $(C_2H_6)_2^+$ accounts for 25% of total ionization.[28] At higher temperatures the abundance of this ion decreases, as does that of $C_4H_{11}^+$. In a conventional electron impact source at 498 K, $C_4H_9^+$ is the only abundant C_4 ion observed. In a similar experiment $(C_3H_8)_2^+$ can be obtained from propane.[29]

3 Chemical Ionization

A number of workers in this field have now published experimental details of ion sources used for chemical ionization.[30–32] A quadrupole mass spectrometer has some advantages for this type of application,[33] particularly when used as a g.c.–m.s. combination. The advantage of chemical ionization for gas chromatography work is that the ion source can take all or most of the column effluent, since the source works at higher pressures and faster pumping speeds than usual. A molecular separator should therefore be unnecessary, if the carrier gas is also used as reactant gas. The problem which one faces is to be able to register a complete chromatogram, since so much carrier gas is being ionized. This problem can be solved by using the quadrupole mass spectrometer as a crude mass filter, rejecting all ions below m/e 60. The output will then be the total ionization produced by the sample, with no contribution from the carrier (reactant) gas. Using such a system, Biemann and co-workers could get spectra from 1 µg of fatty acid ester.[34]

[27] V. K. Potapov and V. V. Sorokin, *Doklady Akad. Nauk S.S.S.R.*, 1968, **183**, 386.
[28] S. K. Searles, L. W. Sieck, and P. Ausloos, *J. Chem. Phys.*, 1970, **53**, 849.
[29] L. W. Sieck, S. K. Searles, and P. Ausloos, *J. Chem. Phys.*, 1971, **54**, 91.
[30] D. Beggs, M. L. Vestal, H. M. Fales, and G. W. A. Milne, *Rev. Sci. Instr.*, 1971, **42**, 1578.
[31] A. M. Hogg, *Analyt. Chem.*, 1972, **44**, 227.
[32] J. H. Futrell and L. H. Wojcik, *Rev. Sci. Instr.*, 1971, **42**, 244.
[33] M. S. Story, *Appl. Spectroscopy*, 1971, **25**, 139.
[34] G. P. Arsenault, J. J. Dolhun, and K. Biemann, *Chem. Comm.*, 1970, 1542.

Alternative Methods of Ionization and Analysis

Schoengold and Munson have obtained good CI spectra using about 0.1 μl of sample.[35] They have used both methane and helium as carrier gases and find that only the former gives true CI spectra. The mass spectra obtained in the presence of 1 Torr of helium were very similar to electron impact spectra. Charge exchange to helium produces ions of very high internal energy which undergo drastic decomposition,[36] but the cross-section for charge exchange is an order of magnitude less than for reactions of organic ions, so this is one of the few cases where electron impact ionization of the sample can compete with ion–molecule reactions under these conditions. A more sophisticated combined EI–CI system has been designed which incorporates two sources in a quadrupole mass spectrometer. The electron impact source monitors the gas in the source housing region outside the ionization chamber of the CI source. The system can be programmed to run alternate EI and CI spectra but can also produce simultaneous EI–CI spectra which are claimed to have the advantages of both methods.[37]

Many of the applications of chemical ionization are to compounds which have molecular ions of very low abundance in their electron impact spectra. The electron impact mass spectra of the barbiturates are often unhelpful; the molecular ions are undetectable and the alkyl groups are often eliminated to give identical spectra from different molecules, e.g. (6) and (7). In the chemical ionization spectra of the barbiturates the $(M + 1)^+$ ions have a relative abundance of 46—87% of total ionization and can be used for quantitative analysis of mixtures.[38] Another group of biologically important compounds, the prostaglandins, give characteristic CI spectra. The base peak of PGA$_1$ (8) is $(M - 17)^+$,

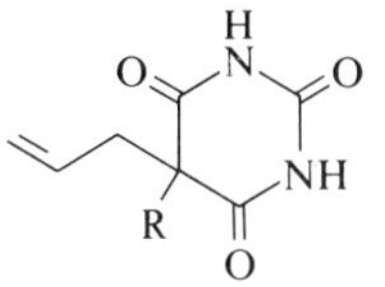

(6) R = CHMe$_2$
(7) R = CH(Me)C$_3$H$_7$

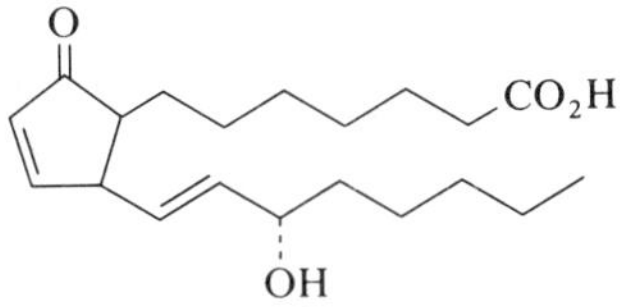

(8)

which is probably formed by dehydration of the protonated molecular ion.[39] More heavily hydroxylated prostaglandins do not form a stable $(M + 1)^+$ ion and are more easily characterized as O-methylated derivatives.

The most successful applications of CI have been to nitrogen compounds, usually because reaction 1 produces a stable cation which does not decompose thermally, whereas in oxygen compounds reaction 3 is sometimes quantitative even at low temperatures. The molecular weight of the steroid aminoglycoside

[35] D. M. Schoengold and B. Munson, *Analyt. Chem.*, 1970, **42**, 1811.
[36] P. Wilmenius and E. Lindholm, *Arkiv Fysik*, 1962, **21**, 97.
[37] G. P. Arsenault, J. J. Dolhun, and K. Biemann, *Analyt. Chem.*, 1971, **43**, 1720.
[38] H. M. Fales, G. W. A. Milne, and T. Axenrod, *Analyt. Chem.*, 1970, **42**, 1432.
[39] D. M. Desiderio and K. Hagele, *Chem. Comm.*, 1971, 1074.

holacurtine (9) was very easily determined by isobutane CI.[40] The more abundant ions in the spectrum can be easily explained: protonation on nitrogen gives the stable $(M + 1)^+$ ion; protonation at oxygen joining the aglycone to the

$$R_3N + CH_5^+ \rightarrow R_3\overset{+}{N}H + CH_4 \qquad (1)$$

$$R-O-R^1 + CH_5^+ \rightarrow R-O\underset{R^1}{\overset{H}{\underset{+}{\diagup}}} + CH_4 \qquad (2)$$

$$R-O\underset{R^1}{\overset{H}{\underset{+}{\diagup}}} \rightarrow R^+ + R^1OH \qquad (3)$$

amino-sugar leads to fission of either C—O bond and production of the ions at m/e 158 and 317. An interesting suggestion is made as to the origin of the ion of m/e 176, *i.e.* that it is formed by protonation of the neutral fragment formed by fission, as in Scheme 1. For this to be true (since the intensity of the peak at

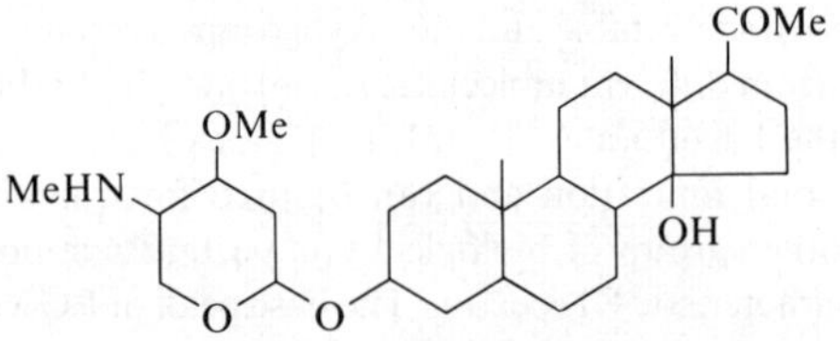

(9)

m/e 176 is of the same order of magnitude as other ions in the spectrum), it would require that the concentration of the neutral species of mass 176 was similar to that of the parent molecule in the ionization chamber, *i.e.* it requires an extremely high ionization efficiency. Since the sensitivity of this method is not much greater than that for electron impact, it is much more probable that the ion of m/e 176 is formed by a hydrogen rearrangement process from $(M + H)^+$.

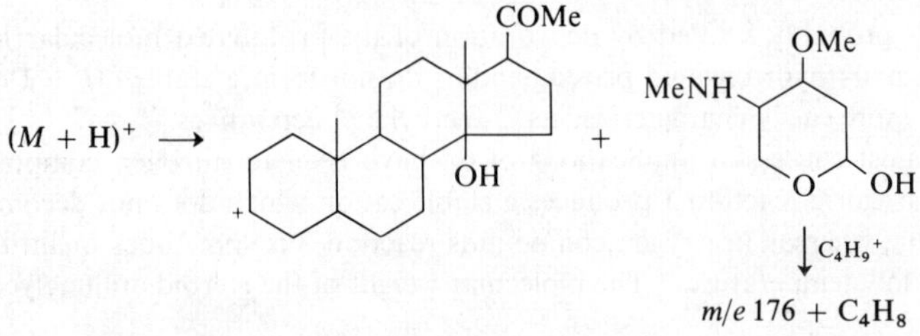

$(M + H)^+ \rightarrow$

$\downarrow C_4H_9^+$

$m/e\ 176 + C_4H_8$

Scheme 1

[40] P. Longevialle, P. Devissagnet, Q. Khuong-Huu, and H. M. Fales, *Compt. rend.*, 1971, **273**, *C*, 1533.

Polytertiary amines such as tetramethylethylenediamine and pentamethyldiethylenetriamine have been analysed successfully by chemical ionization methods.[41] The spectra obtained using N_2 as reactant gas are very similar to electron impact spectra. The principal ions in the N_2 system are N_2^+, N_3^+, and N_4^+, and the first predominates. The most important chemical ionization reaction is charge exchange, and the recombination energy of N_2 (15.75 eV) is only slightly more than the average energy transferred in electron impact experiments. With isobutane, abundant $(M + 1)^+$ ions and very few fragments are observed. With methane the $(M + 1)^+$ ion is still prominent but the $(M - 1)^+$ ion and other fragments are also abundant. Fragmentation can take place either by direct attack on C—C bonds by the reactant ion, or by thermal decomposition of the $(M + 1)^+$ ion, as shown in Scheme 2.

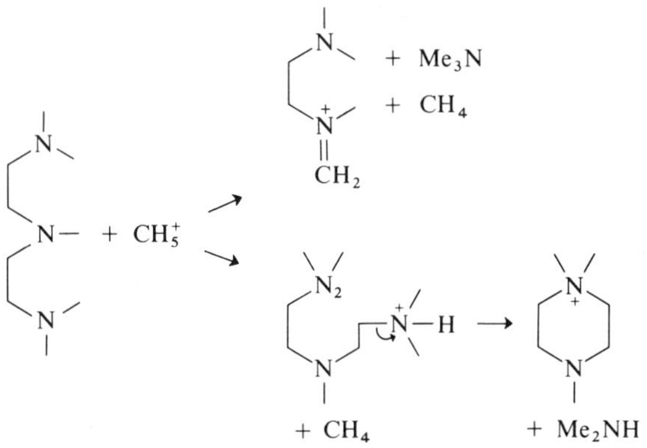

Scheme 2

The free amino-acids all show stable $(M + H)^+$ ions,[42] and fragmentation patterns of the alkyl amino-acids follow Scheme 3. In other amino-acids,

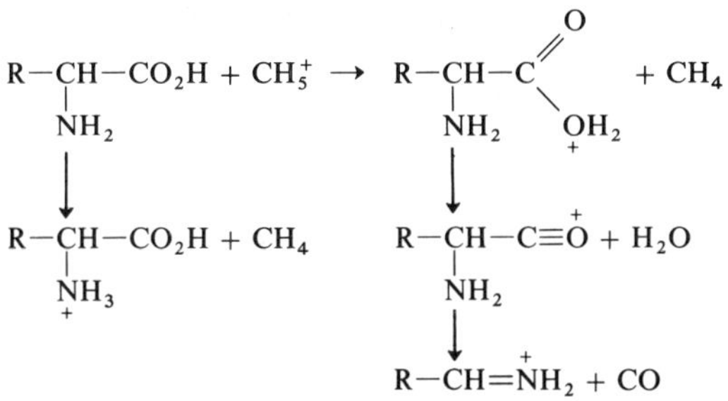

Scheme 3

[41] T. A. Whitney, L. P. Kleemann, and F. H. Field, *Analyt. Chem.*, 1971, **43**, 1048.
[42] G. W. A. Milne, T. Axenrod, and H. M. Fales, *J. Amer. Chem. Soc.*, 1970, **92**, 5170.

neighbouring group effects are important, and ions (10) and (11), it is suggested, are formed by elimination of NH_3 from N-protonated cysteine and β-alanine respectively.

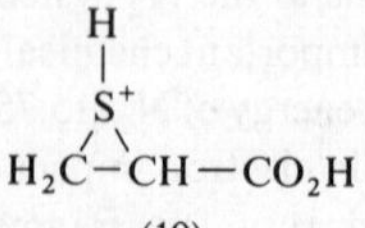

$$\begin{array}{cc} \overset{H}{\underset{\diagup\diagdown}{S^+}} & \overset{+}{O}=C-OH \\ H_2C-CH-CO_2H & \underline{} \\ (10) & (11) \end{array}$$

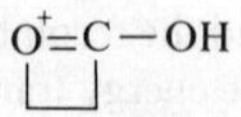

Two groups have described the CI mass spectra of acylpeptide esters;[43,44] the principal features are the ions shown in Scheme 4. The protonated molecular ion is usually prominent and there are two series of 'sequence' ions (12) and (13). The spectra have a greater proportion of ions formed by C—N cleavage in the chain and a smaller proportion due to C—C bond fission in side-chains than is usually found in electron impact spectra. Such spectra are very amenable to a computer-assisted interpretation.[44]

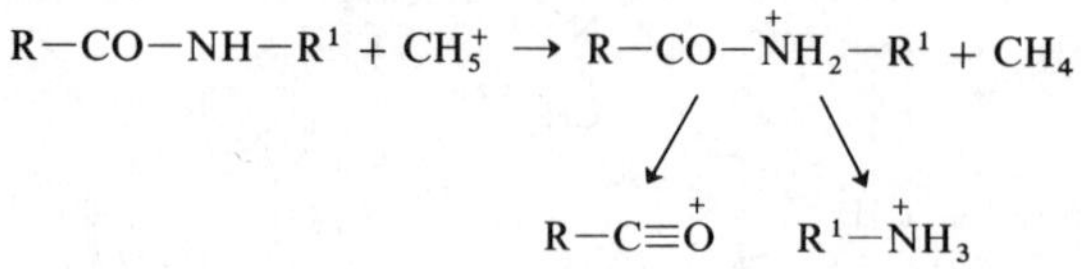

$$R-CO-NH-R^1 + CH_5^+ \rightarrow R-CO-\overset{+}{N}H_2-R^1 + CH_4$$

$$R-\overset{+}{C}\equiv O \qquad R^1-\overset{+}{N}H_3$$
$$(12) \qquad (13)$$

Scheme 4

Another group has examined the chemical ionization spectra of peptides using $(H_2O)_nH^+$ as the reactant ion, instead of CH_5^+ and $C_2H_5^+$. The peptide molecules were bombarded with a resolved ion beam in the collision chamber of a tandem mass spectrometer:[45] as n was increased the abundance of $(M + H)^+$ increased. Although this work produced interesting results, the cost and sophistication of tandem instruments is such that the method is unlikely to be widely applied. The structure of the thyroid hormone releasing factor (14) was based partly on electron impact measurements on various derivatives, but the

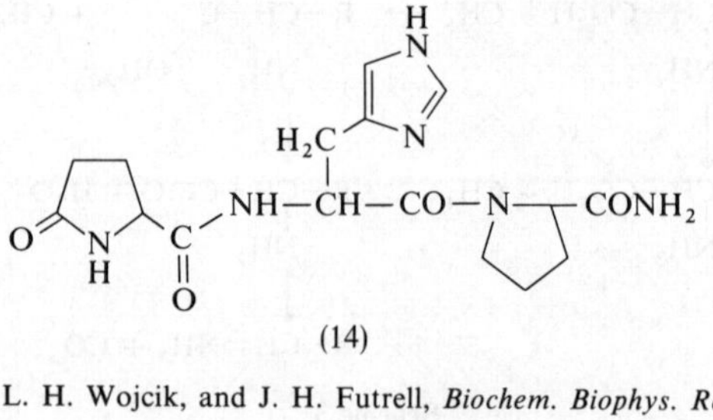

(14)

[43] W. R. Gray, L. H. Wojcik, and J. H. Futrell, *Biochem. Biophys. Res. Comm.*, 1970, **41**, 1111.
[44] A. A. Kiryushkin, H. M. Fales, T. Axenrod, E. J. Gilbert, and G. W. A. Milne, *Org. Mass Spectrometry*, 1971, **5**, 19.
[45] R. J. Beuhler, L. J. Green, and L. Friedman, *J. Amer. Chem. Soc.*, 1971, **93**, 4307.

CI spectrum of the free compound gave a much better correlation with the structure.[46]

It had been reported earlier that simple alcohols with more than six carbon atoms do not have a stable $(M + 1)^+$ ion.[47] Further investigation has shown that although $C_{10}H_{21}OH_2^+$ from n-decanol is unstable, the $(M + 1)^+$ ion from decane-1,10-diol is abundant in its methane CI spectrum and that n-decanol at higher pressures produces a stable ion of formula $(C_{10}H_{21}OH)_2H^+$.[48] The stability of both of these ions is attributed to hydrogen bonding as shown in (15) and (16).

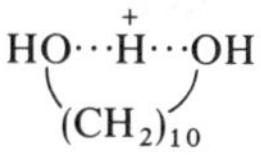

$$\begin{array}{cc} H\overset{+}{O}\cdots H\cdots OH & Me(CH_2)_8\overset{+}{O}\cdots H\cdots O(CH_2)_8Me \\ \diagdown\quad\diagup & |\qquad\qquad| \\ (CH_2)_{10} & H\qquad\qquad H \\ (15) & (16) \end{array}$$

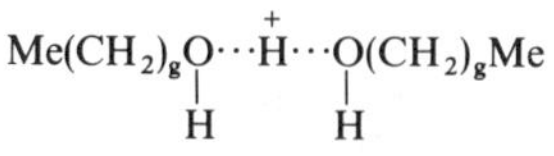

One of the most commonly used arguments in favour of chemical ionization is that it is possible to use a variety of reactant gases, although nearly all of the published work on the method has been done with either methane or isobutane. Michnowicz and Munson have shown that the spectra can also be changed considerably by varying the ion repeller voltage, which affects both ion residence times and ion kinetic energy. Using isobutane and zero repeller voltage, only the $(M + 1)^+$ ion is produced from p-hydroxybenzophenone. Using methane at zero repeller voltage the four ions shown in Scheme 5 are formed. The same

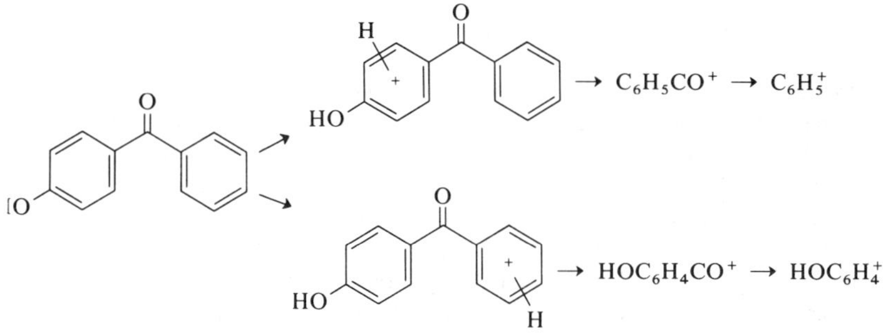

Scheme 5

fragment ions can be observed in greater yield if isobutane is used and the repeller voltage is maintained at +40 V.[49] This may be a useful technique for operators of CI systems. In a study of the methane CI spectra of aryl alkyl ketones it is found that the presence of water, which contributes H_3O^+ as a reactant ion, decreases the amount of fragmentation of the $(M + 1)^+$ ion. The abundance

[46] D. M. Desiderio, R. Burgus, T. F. Dunn, W. Vale, R. Guillemin, and D. N. Ward, *Org. Mass Spectrometry*, 1971, **5**, 221.
[47] F. H. Field, *J. Amer. Chem. Soc.*, 1970, **92**, 2672.
[48] I. Dzidic and J. A. McCloskey, *J. Amer. Chem. Soc.*, 1971, **93**, 4955.
[49] J. Michnowicz and B. Munson, *Org. Mass Spectrometry*, 1970, **4**, Suppl. 481.

of the $(M - 1)^+$ ion increases as the length of the alkyl chain increases, *i.e.* as the number of hydrogen atoms available for abstraction increases.[50]

An ingenious method for distinguishing between primary, secondary, and tertiary alcohols, using mixed reagent gases, has been devised.[51] In a methane–acetone mixture ($CH_4 : C_3H_6O = 20:1$) the principal ions are $C_3H_6\overset{+}{O}H$ and $(C_3H_6O)_2H^+$, and in a similar methane–acetaldehyde mixture the spectrum is mostly CH_3CHOH^+ and $(MeCHO)_2H^+$. The acetaldehyde-derived ions react with primary and secondary alcohols:

$$MeCH=\overset{+}{O}H + Me(CH_2)_5CH_2OH \rightarrow Me(CH_2)_5CH=\overset{+}{O}H + MeCH_2OH$$

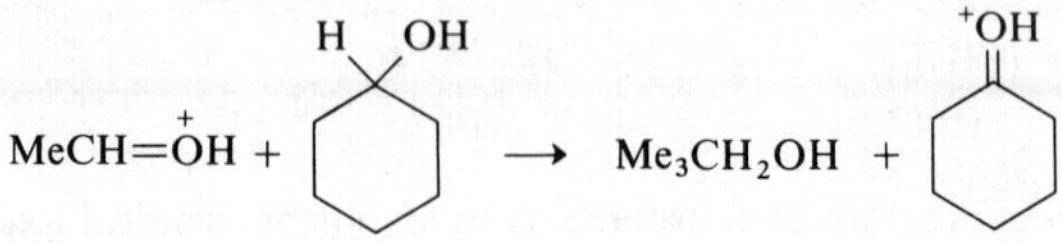

$$MeCH=\overset{+}{O}H + \underset{}{\bigcirc}\!\!\!\!\!{\overset{H\;\;OH}{}} \rightarrow Me_3CH_2OH + \underset{}{\bigcirc}\!\!\!\!\!{\overset{^+OH}{}}$$

The ions from the acetone system will abstract hydride from secondary but not from primary alcohols:

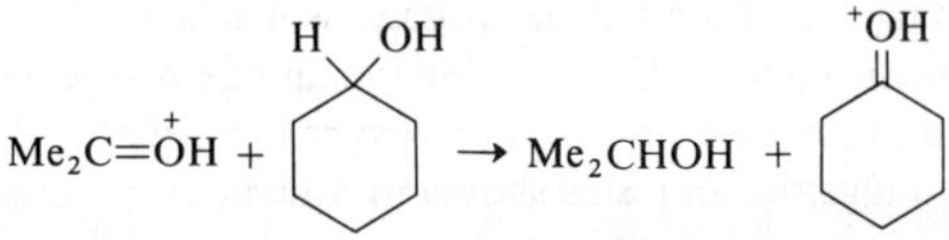

$$Me_2C=\overset{+}{O}H + \underset{}{\bigcirc}\!\!\!\!\!{\overset{H\;\;OH}{}} \rightarrow Me_2CHOH + \underset{}{\bigcirc}\!\!\!\!\!{\overset{^+OH}{}}$$

$$Me_2C=\overset{+}{O}H + Me(CH_2)_5CH_2OH \nrightarrow Me(CH_2)_5CH=\overset{+}{O}H$$

In neither system will the ions abstract hydride from a tertiary alcohol. The same authors have suggested the use of ND_3 as a reactant ion to determine the number of exchangeable protons in amines.[52] Chemical ionization is not generally necessary for such determinations, except in cases where the molecular ion cannot be observed in the electron impact mode.

There has been some interest in the CI mass spectra of inorganic and organometallic compounds. The methane CI spectra of the boron hydrides have been observed:[53] these can be divided into two categories, B_2H_6, B_4H_{10}, and B_5H_{11} which undergo hydride-ion abstraction, and B_5H_9 and B_6H_{10} which are protonated and also react in some condensation reactions, *e.g.*

$$C_2H_5^+ + B_5H_9 \rightarrow CB_5H_{10}^+ + CH_4$$

with production of carborane ions. Some labelling experiments involving ternary mixtures of methane + base + borane have been used to determine the proton affinity of B_5H_9. There is evidence that the following reaction occurs:

$$C_2H_3D_2^+ + B_5H_9 \rightarrow B_5H_9D^+ + C_2H_3D$$

[50] J. Michnowicz and B. Munson, *Org. Mass Spectrometry*, 1972, **6**, 283.
[51] D. F. Hunt and J. F. Ryan, *Tetrahedron Letters*, 1971, 4535.
[52] D. F. Hunt, C. N. McEwen, and R. A. Upham, *Tetrahedron Letters*, 1971, 4539.
[53] J. J. Solomon and R. F. Porter, *J. Amer. Chem. Soc.*, 1972, **94**, 1443.

and that the following does not:

$$B_5HD_9^+ + C_2H_4 \rightarrow C_2H_4D^+ + B_5HD_8$$

It is therefore reasonable to conclude that B_5H_9 has a higher PA than ethylene. From a series of such experiments it has been shown that the value lies between those of H_2O and H_2S, *i.e.* $PA(B_5H_9) = 167 \pm 6$ kcal mol^{-1} (700 ± 25 kJ mol^{-1}). Borazine has a CI mass spectrum in which most of the ionization is due either to protonation or hydride abstraction.[54] The iron tricarbonyl complex of heptafulvene (17) has no ions heavier than $(M - CO)^+$ in its electron impact spectrum. The base peak in its methane CI spectrum corresponds to $(M + 1)^+$.[55]

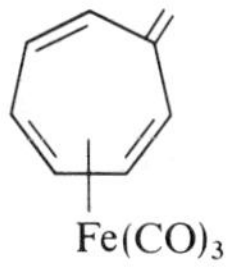

(17)

One of the principal products in the gas-phase radiolysis of toluene in propane is *m*-isopropyltoluene. This can be explained if the C-alkylated cyclohexadienylium ion can rearrange *via* a π-complex prior to loss of a proton, as in Scheme 6.[56] The effect of temperature of CI spectra continues to be studied.

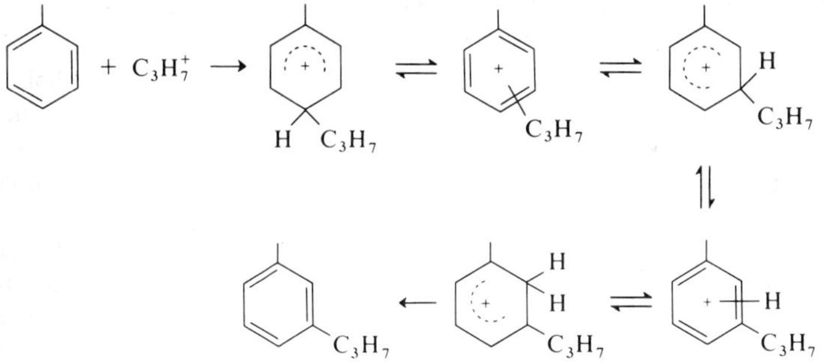

Scheme 6

The bimolecular process involving protonation of benzyl acetate by $C_4H_9^+$ from isobutane is temperature-invariant.[57] The thermal decomposition of protonated tertiary alkyl acetates has been studied in greater detail. The reaction

$$ROAcH^+ \rightarrow R^+ + HOAc$$

[54] R. F. Porter and J. J. Solomon, *J. Amer. Chem. Soc.*, 1971, **93**, 56.
[55] G. T. Rodeheaver, G. C. Farrant, and D. F. Hunt, *J. Organometallic Chem.*, 1971, **30**, C22.
[56] Y. Yamamoto, S. Takamuku, and H. Sakurai, *Bull. Chem. Soc. Japan*, 1972, **45**, 255.
[57] S. Vredenberg, L. Wojcik, and J. H. Futrell, *J. Phys. Chem.*, 1971, **75**, 590.

has been observed for R = Me_3C, Me_2CEt, Me_2CPr, Me_2CBu, and $Me_2C-C_5H_{11}$. As the size of the alkyl group is increased the activation energy for the process remains constant, but the pre-exponential factor increases slightly.[58]

4 Negative Ion Production*

Although there is a considerable ouptut of work on the chemistry of gas-phase negative ions, the method does not appear to be reaching the stage at which it might be routinely used as an analytical method. Much of the work published is exploratory and not aimed at the analytical or structure-determination markets. As in Volume 1, there are a number of observations of molecular negative ions at high electron energies, and further evidence has been produced to suggest that these are formed by attachment of secondary electrons.[59]

As before, the most successful results are found with molecules having low-lying vacant orbitals. Molecular ions are observed in the nitroarenes, which decompose principally by elimination of NO, involving a skeletal rearrangement.[60] In a study of m- and p-dinitrobenzene, the kinetic energy release measured for the elimination of NO from the molecular ion is 0.89 eV for the *para* compound and 0.36 eV for the *meta*.[61] This may be a reflection of the difference in stability between o- and m-nitrophenoxy-anions, the product ions. In the decomposition paths shown in Scheme 7, the p-quinone anion is stable whereas the m-quinone anion-radical decarbonylates. The negative-ion spectra of the nitrophthalic anhydrides have been recorded,[62] and the *ortho*-isomer undergoes a skeletal rearrangement during which it eliminates two molecules of CO_2.

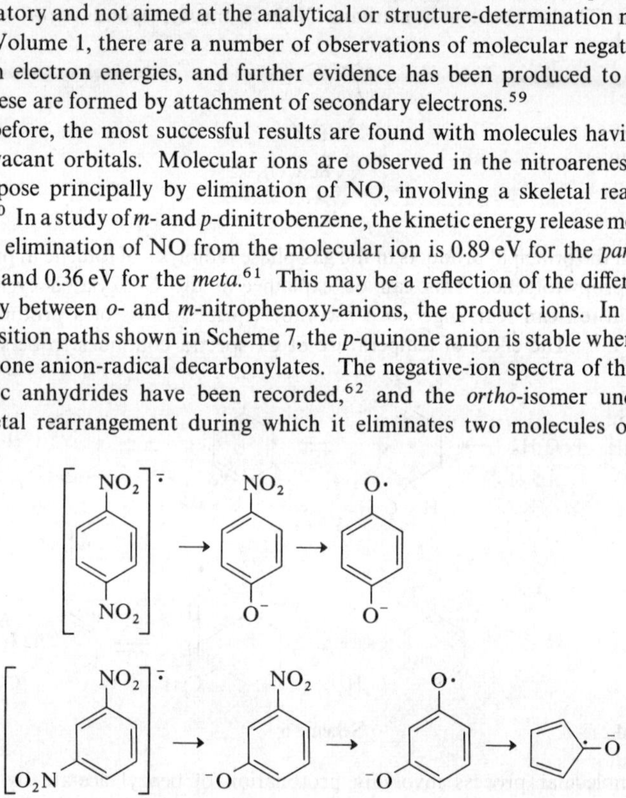

Scheme 7

[58] W. A. Laurie and F. H. Field, *J. Amer. Chem. Soc.*, 1972, **94**, 2913.
[59] T. McAllister, *Chem. Comm.*, 1972, 245.
[60] J. H. Bowie, *Org. Mass Spectrometry*, 1971, **5**, 945.
[61] C. L. Brown and W. P. Weber, *J. Amer. Chem. Soc.*, 1970, **92**, 5775.
[62] T. Blumenthal and J. H. Bowie, *Austral. J. Chem.*, 1971, **24**, 1853.

* For additional discussion of negative ion studies, see the appropriate section of Chapter 3.

The negative-ion mass spectra of nitroaryl esters are claimed to be more characteristic of structure than the positive-ion spectra.[63] The reaction (18) → (19) is observed for a wide variety of R, and the intensity ratios fit a Hammett plot with σ-constants. Hammett correlations have been made for other ions in these spectra with varying success. Nitroaliphatic compounds, as might be predicted,

$$NO_2-\bigcirc-O-CO-\bigcirc-R \qquad NO_2-\bigcirc-O^-$$

(18)   (19)

do not have stable molecular anions, but the NI spectra of $XC(NO_2)_3$, where X = H, F, Cl, Br, I, or Me, are much more characteristic of molecular structure than are the positive-ion spectra.[64] The most prominent ion is NO_2^-, but $C(NO_2)_3^-$ is always abundant and also $CX(NO_2)_2^-$, except when X = H. Esters containing anthraquinone nuclei all give molecular anions of significant abundance,[65] and the fragmentation processes make it possible to distinguish between anthraquinone-1- and -2-carboxylates.

Sulphur compounds have in the past been found to give interesting negative-ion spectra, and the study of them continues. Arylsulphonyl and arysulphinyl compounds often have stable $M^{\bar{\cdot}}$ ions and the principal fragmentation process is fission at sulphur to produce RSO_2^- and RSO^- ions. Aryl thioethers have low-intensity $M^{\bar{\cdot}}$ peaks and the principal fragment is RS^-.[66] 2-Aryl-1,3-dithians, such as (20), have stable $M^{\bar{\cdot}}$ ions and a few abundant fragment ions, due mainly to C—S cleavage with hydrogen rearrangement, in their spectra.[67] Molecular ions are less common in aliphatic sulphur compounds. In DMSO the heaviest negative ion observed is $(M - H)^-$, which is about 7% of the abundance of SO^- which constitutes the base peak.[68]

(20)

Simple nitrogen compounds appear to yield CN^- as the principal negative ion, although alkyl pyridines and pyrazines yield $(M - H)^-$ ions.[69] Alkyl cyanides give intense CN^- peaks and very little ionization at higher mass.[70]

---

[63] J. H. Bowie and B. Nussey, *Org. Mass Spectrometry*, 1972, **6**, 429.
[64] J. T. Larkins, J. M. Nicholson, and F. E. Saalfield, *Org. Mass Spectrometry*, 1971, **5**, 265.
[65] A. C. Ho, J. H. Bowie, and A. Fry, *J. Chem. Soc. (B)*, 1971, 530.
[66] C. Nolde, J. O. Madsen, S.-O. Lawesson, and J. H. Bowie, *Arkiv. Kemi*, 1970, **31**, 481.
[67] J. H. Bowie and P. Y. White, *Org. Mass Spectrometry*, 1972, **6**, 75.
[68] J.-C. Blais, M. Cottin, and B. Gitton, *J. Chim. phys.*, 1970, **67**, 1475.
[69] W. W. Pandler and S. A. Humphrey, *Org. Mass Spectrometry*, 1970, **4**, Suppl. 513.
[70] S. Tsuda, A. Yokohata, and T. Umaba, *Bull. Chem. Soc. Japan*, 1970, **43**, 3383.

This is analogous to the production of halide ions from alkyl halides. Fluorine compounds are the most likely of the halides to yield polyatomic anions in the gas phase; some perfluoroalkanes are reported to have stable molecular anions but unstable molecular cations.[71] Much of the work reported on the negative ions from small halogen-containing molecules is concerned with measurements of the energetics of ion production by dissociative charge transfer. Mathematical deconvolution of ionization efficiency data can reduce the width of resonance peaks by a factor of 2 and reveal low cross-section processes which have previously been hidden in the tail of the main peak.[72] Measurements have been carried out on tetrafluoroethylene,[73] 1,1-difluoroethylene,[74] CF_3OF,[75] alkyl chlorides,[76] SiF_4 and CF_4,[77] BF_3,[78,79] PF_3,[78] BCl_3,[79] SF_6,[80] and small hydrocarbon anions.[81] It is interesting to note that SiF_3^- is more abundant than F^- in the spectrum of SiF_4, whereas CF_3^- is only 7% of the abundance of F^- in CF_4.[77] These are measured at the resonance maxima of the individual peaks rather than at a single electron energy value. The ion F_2^- is common in perfluoro-compounds: MacNeil and Thynne find that their results[78] are consistent with the formation of this ion by both unimolecular and bimolecular processes but Stockdale and co-workers claim that it is produced only by a unimolecular process.[79]

Trichlorovinylmercury(II) chloride and bistrichlorovinylmercury both yield a number of heavy negative ions,[82] as shown in Scheme 8. The ions $C_2Cl_4Hg^-$

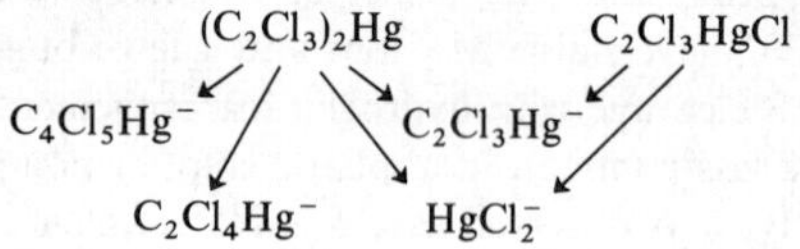

$$C_4Cl_5Hg^- \xleftarrow{} (C_2Cl_3)_2Hg \xrightarrow{} C_2Cl_3HgCl$$
$$\searrow \quad C_2Cl_3Hg^- \quad \swarrow$$
$$C_2Cl_4Hg^- \qquad HgCl_2^-$$

Scheme 8

and $C_2Cl_2^+$ have almost identical appearance potentials and it is suggested that they are formed by the ion-pair process:

$$(C_2Cl_3)_2Hg + e \rightarrow C_2Cl_4Hg^- + C_2Cl_2^+ + e$$

[71] F. B. Dudley, G. H. Cady, and A. L. Crittenden, *Org. Mass Spectrometry*, 1971, **5**, 953.
[72] K. A. G. MacNeil and J. C. J. Thynne, *Internat. J. Mass Spectrometry Ion Phys.*, 1969, **3**, 35; 1970, **4**, 434.
[73] K. A. G. MacNeil and J. C. J. Thynne, *Internat. J. Mass Spectrometry Ion Phys.*, 1970, **5**, 329.
[74] J. C. J. Thynne and K. A. G. MacNeil, *J. Phys. Chem.*, 1971, **75**, 2584.
[75] K. A. G. MacNeil and J. C. J. Thynne, *Internat. J. Mass Spectrometry Ion Phys.*, 1970, **5**, 95.
[76] S. Tsuda, A. Yokohata, and M. Kawai, *Bull. Chem. Soc. Japan*, 1970, **43**, 1649.
[77] K. A. G. MacNeil and J. C. J. Thynne, *Internat. J. Mass Spectrometry Ion Phys.*, 1970, **3**, 455.
[78] K. A. G. MacNeil and J. C. J. Thynne, *J. Phys. Chem.*, 1970, **74**, 2257.
[79] J. A. Stockdale, D. R. Nelson, F. J. Davies, and R. N. Compton, *J. Chem. Phys.*, 1972, **56**, 3336.
[80] B. Lehmann, *Z. Naturforsch.*, 1970, **25a**, 1755.
[81] R. Locht and J. Momigny, *Chem. Phys. Letters*, 1970, **6**, 273.
[82] S. C. Cohen, *Inorg. Nuclear Chem. Letters*, 1970, **6**, 757.

Alternative Methods of Ionization and Analysis 17

The negative-ion spectrum of 2,3-dicarbahexaborane has a base peak corresponding to $(M - 1)^-$ and two principal fragments; their genesis is described as follows:[83]

$$e + C_2B_4H_8 \rightarrow C_2B_4H_7^- + H\cdot$$
$$C_2B_4H_7^- \rightarrow C_2B_4H_5^- + H_2$$
$$C_2B_4H_7^- \rightarrow C_2B_3H_4^- + BH_3$$

The boron hydrides have been examined by ICR and their negative-ion chemistry puts them into two categories.[84] The 'stable' B_nH_{n+4} compounds give mainly the $(M - H)^-$ ion at low energies. The 'unstable' B_nH_{n+6} compounds give mainly $(M - BH_3)^-$ and a series of condensation products $(M + nBH)^-$ where $n = 1$—4.

A number of authors have reported measurements of autodetachment lifetimes of negative ions. Some of these, measured by varying the ion residence time in pulsed conventional mass spectrometers, are very short: 7 µs for $C_{10}H_8^-$ from azulene,[85] 20 µs for ReF_6^-,[86] and 30 µs for $C_6H_4O_2^-$ from benzoquinone.[87] Recent measurements of variation of linewidths of negative-ion resonances with drift voltage in an ICR spectrometer give values of 500 µs for SF_6^- and 200 µs for $C_4F_8^-$ and the authors suggest[88] that other methods give values which are too small.

The study of negative ion–molecule reactions continues both in organic and in inorganic systems. In an investigation[86] of XF_6 molecules it was found that only MoF_6 and ReF_6 attached directly and that SeF_6^-, TrF_6^-, and UF_6^- could only be formed by charge exchange from SF_6^-. Transfer of F^- appears to be a fairly common process as in the following examples:[89]

$$SF_6^- + POF_3 \rightarrow POF_4^- + SF_5$$
$$SF_5^- + POF_3 \rightarrow POF_4^- + SF_4$$

SiF_6^- also transfers F^- to BF_3, SiF_4, PF_5, and PF_3.[90] Examination of the CH_3SH–$ClCN$ system produced a number of reactions including the following:[91]

$$S^- + ClCN \rightarrow SCl + CN^-$$
$$CN^- + CH_3SH \rightarrow SCN^- + CH_4$$

[83] C. L. Brown, K. P. Gross, and T. P. Onak, *Chem. Comm.*, 1972, 68.
[84] R. C. Dunbar, *J. Amer. Chem. Soc.*, 1971, **93**, 4167.
[85] E. L. Chaney, L. G. Christophorou, P. M. Collins, and J. G. Carter, *J. Chem. Phys.*, 1970, **52**, 4413.
[86] J. A. D. Stockdale, R. N. Compton, and H. C. Schweinler, *J. Chem. Phys.*, 1970, **53**, 1502.
[87] L. G. Christophorou, J. G. Carter, and A. A. Christodoulides, *Chem. Phys. Letters*, 1969, **3**, 237.
[88] J. M. S. Henis and C. A. Mabie, *J. Chem. Phys.*, 1970, **53**, 2999.
[89] T. C. Rhyne and J. G. Dillard, *Internat. J. Mass Spectrometry Ion Phys.*, 1971, **5**, 371.
[90] T. C. Rhyne and J. G. Dillard, *Inorg. Chem.*, 1971, **10**, 730.
[91] A. di Domenico, D. K. Sen Sharma, J. L. Franklin, and J. G. Dillard, *J. Chem. Phys.*, 1971, **54**, 4460.

Reactions of O^- with hydrocarbons in most cases are simple proton or hydrogen atom abstractions, but more complex reactions can be observed, *e.g.* with benzene:[92]

$$O^- + C_6H_6 \rightarrow C_6H_4^- + H_2O$$
$$\searrow C_6H_5O^- + H\cdot$$

Working at higher pressures, Kebarle and Yamdagni are producing some very interesting results on the solvation equilibria of negative ions in the gas phase. In the equilibrium

$$(Cl-H-R)^- \rightleftharpoons Cl^- + HR$$

the strength of the hydrogen bond holding the $(Cl-H-R)^-$ ion together increases in the order: H_2O < MeOH < Me_3COH < $CHCl_3$ < PhOH < $MeCO_2H$ < HCO_2H;[93] *i.e.* the order of increasing gas-phase acidity.[94] In a similar series of experiments using the same 'solvent', the anion is changed. In this series the strength of the hydrogen bond in the 'solvated' anion increases as the gas-phase basicity of the ion increases. From the two experiments a rough linear relationship can be derived between the energy of association of the anion $(B\cdots H-R)^-$ and the heterolytic dissociation energies of B—H and RH. This takes the form

$$D(B-HR)^- = 0.2\, D(B^-\!-\!H^+) - 0.134\, D(R^-\!-\!H^+)$$

In another series of similar experiments, the same authors have attempted to obtain some understanding of the behaviour of aprotic solvents.[95] Measurements of equilibria of the type

$$X(MeCN)_n^- \rightleftharpoons X(MeCN)_{n-1}^- + MeCN$$

show that for F^- the enthalpy of solvation for all measured values of n (1—5) is greater than for water as solvent. For Cl^- and Br^- ΔH is greater for MeCN than for water when $n = 0$ or 1, but there is a crossover at higher values of n where water becomes the more strongly associated solvent. For I^- this crossover occurs at even higher values of n. These results are in accord with the solution behaviour of aprotic solvents such as MeCN.

5 Field Ionization

Since the publication of Volume 1, a comprehensive review of field ionization techniques and applications has appeared.[96] Much of the development which has been reported in the past two years has been directed towards improvement in the sensitivity and reproducibility of the FI technique. In a description of a

---

[92] J. A. D. Stockdale, R. N. Compton, and P. W. Reinhardt, *Internat. J. Mass Spectrometry Ion Phys.*, 1970, **4**, 401.
[93] R. Yamdagni and P. Kebarle, *J. Amer. Chem. Soc.*, 1971, **93**, 7139.
[94] J. I. Brauman and L. K. Blair, *J. Amer. Chem. Soc.*, 1968, **90**, 6561.
[95] R. Yamdagni and P. Kebarle, *J. Amer. Chem. Soc.*, 1972, **94**, 2940.
[96] K. Levsen, *Chem.-Ztg.*, 1971, **95**, 725.

g.c.–FI–m.s. combination[97] the necessity for an efficient interface is emphasized (the authors use a membrane separator) because it has been reported that the sensitivity of two wires used in the same source may vary by as much as a factor of 70.[98] Although sensitivity for total ionization may be a factor of 200 less than the EI value, the sensitivity for the molecular ion can often be much greater.

Beckey and his co-workers have produced a considerable amount of detail on the optimum parameters for FI emitters. In one paper they describe the effect on ion current of activation time, heat treatment after activation, ion current during activation, emitter position with respect to the cathode slit, field anode potential, and anode–cathode voltage.[99] In another paper this group discusses the activation of emitters in greater detail.[100] The wire (or blade) is treated at high field gradient with acetone, crotonaldehyde, or benzaldehyde. These compounds polymerize at the emitter surface producing microneedles round the wire, which are responsible for the higher sensitivity. The activating material has much more effect on the sensitivity than the material of the wire. Microneedles are usually grown under conditions of constant field-ion current. Needles grown at higher temperatures have greater thermal stability. It is generally concluded that microneedles grown from benzonitrile at high temperatures (about 900 °C) are the most thermally stable, and the most resistant to chemical attack by molecules in the gas phase. The time required for activation depends upon the diameter of wire used. Although 2.5 μm wires give better sensitivity, 10 μm wires have greater mechanical stability at high temperatures and are to be preferred. Changes in the state of the emitter surface during an FI experiment are reversible. After heating and evacuation of sample the same voltage–current characteristics can be reproduced.[101] Exact details of the method of sample introduction for field desorption experiments have been reported.[102]

An interesting area of research is opening up in the investigation of surface reactions during field ionization. The formation of microneedles at the emitter is a field-induced polymerization, and the use of high temperatures in the activation process causes cross-linking of polymer chains.[103] It is suggested that the field-induced polymerization of acetone takes place *via* the enolate adsorbed *via* oxygen on to the metal surface. If one molecule is field-ionized it will initiate a cationic polymerization.[104] Intermolecular reactions at adsorbed layers can also be observed. Crotonaldehyde on a benzonitrile-activated emitter

[97] J. N. Damico and R. P. Barron, *Analyt. Chem.*, 1971, **43**, 17.
[98] J. C. Tou, L. B. Westover, and E. J. Sutton, *Internat. J. Mass Spectrometry Ion Phys.*, 1969, **3**, 377.
[99] H. D. Beckey, S. Bloching, M. D. Migahed, E. Ochterbeck, and H. R. Schulten, *Internat. J. Mass Spectrometry Ion Phys.*, 1972, **8**, 169.
[100] M. D. Migahed and H. D. Beckey, *Internat. J. Mass Spectrometry Ion Phys.*, 1971, **7**, 1.
[101] I. V. Goldenfeld and V. A. Nazarenko, *Internat. J. Mass Spectrometry Ion Phys.*, 1970, **5**, 197.
[102] H. D. Beckey, A. Heindrichs, and H. U. Winkler, *Internat. J. Mass Spectrometry Ion Phys.*, 1970, **3**, App. 9.
[103] H. D. Beckey and F. W. Rollgen, *Naturwiss.*, 1971, **58**, 23.
[104] F. W. Rollgen and H. D. Beckey, *Surface Sci.*, 1970, **23**, 69.

yields M^+, $(M + H)^+$, $(M - H)^+$, $(2M)^+$, $(2M - H)^+$; in the presence of an excess of benzene vapour all these intensities fall except that of M^+. The others are formed by intermolecular surface reactions and are inhibited by the adsorption of benzene.[105] Ionization due to surface reactions can also be reduced by running the spectra at higher emitter temperatures. Chemisorbed acetone ions are responsible for increased yields of H_3O^+ from water.[106]

It was suggested in Volume 1 that criteria could be applied to peaks in FI spectra which could determine whether they were formed by a simple cleavage process or by a multistep or rearrangement process. The criteria applied are essentially kinetic and rely on the fact that direct fission will be either spontaneous field-dissociation or a thermal dissociation with a very low frequency factor. Skeletal rearrangements are 'slow' and give rise to normal metastable ions but not sharp fragment ions. Direct cleavage reactions should produce sharp fragment ions. Hydrogen rearrangements are in an intermediate category and are likely to give rise to 'fast' metastable ions, or skew fragment peaks, since the decomposition takes place shortly after the beginning of the acceleration process.

Beckey has suggested that in addition to having a relatively low intensity (low R_i) and having a fragment peak tailing to low mass (fast metastable ion), the peak maximum should be shifted to fractionally lower mass.[107] Theoretical calculations of peak shapes for ions with a slightly delayed time of decomposition confirm this 'rearrangement shift'.[108] Displacements have been found, usually less than 0.2 m.u.[109] In the FI mass spectra of a number of propyl esters there is an abundant $C_3H_6^{+}$ ion.[110] This must be a rearrangement peak but does not fit the intensity criterion for such (in this case $R > 0.1$). The peakshape criterion only applies to single-focusing instruments. In this case the kinetic energy distribution of the ions was examined by increasing the emitter voltage and keeping a constant electrostatic analyser voltage in a double-focusing mass spectrometer. All fragments showed some tailing in this experiment but it was more pronounced for the $C_3H_6^{+}$ ion. The authors suggest that since this decomposition is only slightly retarded with respect to others, the molecular ion must be constrained by the field to a conformation similar to that of the transition state (21). It has also been suggested that such six-centre hydrogen rearrangements are exceptions to the general rule.[111]

(21)

[105] M. D. Migahed and H. D. Beckey, *Org. Mass Spectrometry*, 1971, **5**, 453.
[106] F. W. Rollgen and H. D. Beckey, *Surface Sci.*, 1971, **27**, 321.
[107] H. D. Beckey, *Internat. J. Mass Spectrometry Ion Phys.*, 1970, **5**, 182.
[108] K. Levsen and H. D. Beckey, *Internat. J. Mass Spectrometry Ion Phys.*, 1971, **7**, 341.
[109] J. C. Tou, *J. Phys. Chem.*, 1970, **74**, 4596.
[110] P. Schulze and W. J. Richter, *Internat. J. Mass Spectrometry Ion Phys.*, 1971, **6**, 131.
[111] E. M. Chait and F. G. Kitson, *Org. Mass Spectrometry*, 1970, **3**, 533.

Alternative Methods of Ionization and Analysis 21

The use of emitter temperature as another variable parameter has been suggested. Direct cleavage fragments have a much greater temperature dependence than most rearrangements.[112] Again the McLafferty rearrangement appears to be an exception. The difficulty involved in making the distinction between direct cleavage and rearrangement processes is that although the former has a higher frequency factor, there is no real relationship between the activation energies of the two types of reaction. In menthone (22) the activation energy for elimination of C_3H_6 is 0.08 eV whereas the lowest-energy simple fission process is loss of $CH_3^·$, with an activation energy of 0.98 eV. The ratio of the intensities of 'fast' and 'slow' metastables usually increases considerably with temperature for a direct fission and only slightly for a rearrangement. This is not true of menthone because of the very low activation energy for the rearrangement process.[113] The rule is followed in the case of *o*-hydroxybenzyl alcohol, where the *ortho* rearrangement (23) → (24) has a relatively high activation energy.

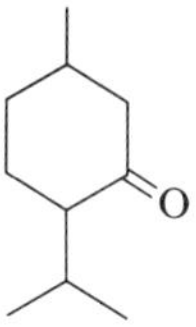

(22) 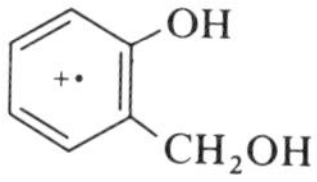   (23) 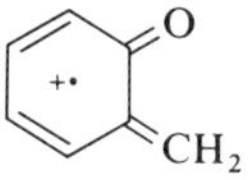   (24)

In a survey of a wide range of types of decomposition, Levsen and Beckey come to the following conclusions: (i) skeletal rearrangement products appear as slow metastables; (ii) a wide selection of hydrogen rearrangements are of the delayed (fast metastable) or slow metastable type; (iii) many direct bond cleavages appear as sharp fragments but there are a number of exceptions, slow direct cleavages, which make the kinetic distinction difficult.[114] There is one exception to the second rule: $C_6H_{12}^{+·}$ from n-hexylbromide appears as a sharp fragment and is considered to be the product of a surface reaction. The ion $C_7H_7^+$ from ethylbenzene appears as a sharp peak at the exact calculated mass: it is therefore concluded that any rearrangement to the tropylium structure is subsequent to fission.

Derrick and Robertson have calculated rate constants for the dissociations

$$Me_4C^{+·} \rightarrow Me_3C^+ + CH_3^·$$
$$Me_3CCH_2CH_2CH_3^{+·} \rightarrow Me_3C^+ + ·CH_2CH_2CH_3$$

from measurements of the linewidths of peaks at varying fields.[115] They claim that each process has a discrete rate constant in the region 10—40 ps. In contrast, Tenschert and Beckey have found a continuous spectrum of rate constants for

[112] H. Knöppel, *Internat. J. Mass Spectrometry Ion Phys.*, 1970, **4**, 97.
[113] K. Levsen and H. D. Beckey, *Internat. J. Mass Spectrometry Ion Phys.*, 1972, **9**, 51.
[114] K. Levsen and H. D. Beckey, *Internat. J. Mass Spectrometry Ion Phys.*, 1972, **9**, 63.
[115] P. J. Derrick and A. J. B. Robertson, *Proc. Roy. Soc.*, 1971, **A324**, 491.

n-butane.[116] They used this to derive an energy distribution curve which shows a fast fall-off above 1 eV. The field ionization spectrum of n-hexanol contains two ions due to two-step processes:

$$C_6H_{14}O^{+\cdot} \rightarrow C_6H_{12}^{+\cdot} \rightarrow C_5H_8^{+\cdot} + CH_4$$
$$C_6H_{14}O^{+\cdot} \rightarrow C_6H_{12}^{+\cdot} \rightarrow C_4H_8^{+\cdot} + C_2H_4$$

Equations have been derived for calculating the apparent mass of metastable ions due to second stage processes.[117]

FI has been used successfully for the analysis of polymers by pyrolysis. The method used was to heat the sample in a gold crucible 5 mm from the emitter. Simple mixtures of monomers and oligomers were obtained from polyisobutane, polystyrene, polypropylene, and polyacrylonitrile.[118] Condensate from cigarette smoke has been analysed by high-resolution FI mass spectrometry.[119] The photo-oligomers of thymine have been analysed by field ionization.[120] The two principal products are (25) and (26). The dimer (25) has a low abundance M^+ ion and intense peaks corresponding to $(M + H)^+$ and fission at a and b. The adduct (26) behaves quite differently with intense peaks due to $M^+, (M - H_2O)^+$, and fragmentation at c. The structure (27) has been assigned to the trimer since the spectrum shows evidence of both types of linkage. The dimeric 1-chloro-2-nitrosocycloalkanes (28; n = 3 or 4) show both monomer and dimer peaks;[121] the monomer peak is sharp and not accompanied by a metastable. This may be a true field dissociation.

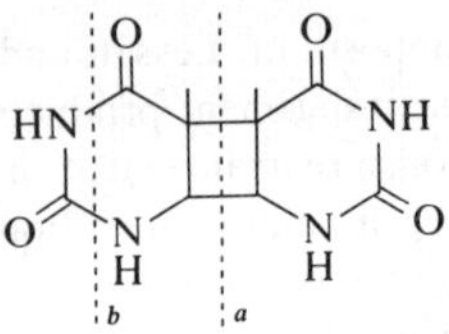

(25) 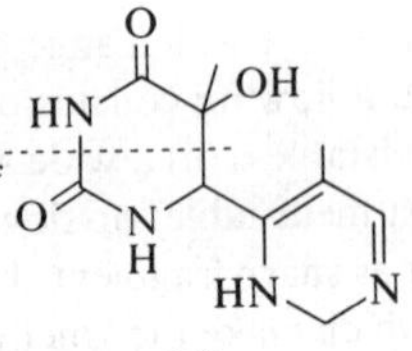   (26)

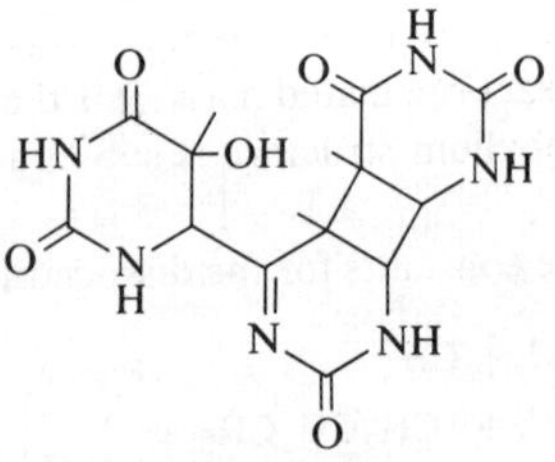

(27) 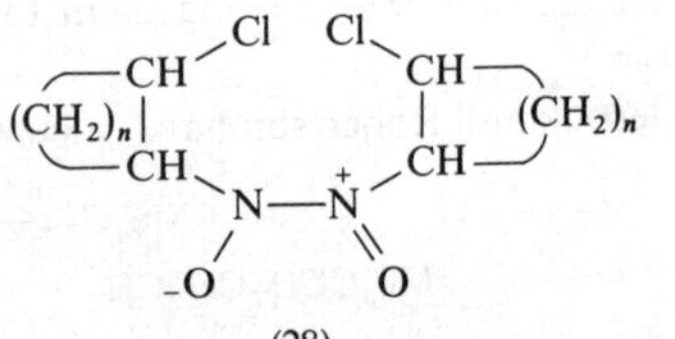   (28)

[116] G. Tenschert and H. D. Beckey, *Internat. J. Mass Spectrometry Ion Phys.*, 1971, **7**, 97.
[117] H. D. Beckey, *Z. Naturforsch.*, 1971, **26a**, 1243.
[118] D. Hummel, K. Rubenacker, H.-J. Dussel, and T. Schweren, *Angew. Chem. Internat. Edn.*, 1971, **10**, 349.
[119] J. B. Forehand and W. F. Kuhn, *Analyt. Chem.*, 1970, **42**, 1839.
[120] C. Fenselau, S. Y. Wang, and P. Brown, *Tetrahedron*, 1970, **26**, 5923.
[121] J. C. Tou and K. Y. Chang, *Org. Mass Spectrometry*, 1970, **3**, 1055.

Field ionization appears to be the best of the available ionization methods for the study of carbohydrates.[122] Free monosaccharides tend to have an abundant $(M + H)^+$ ion and very little M^+. Permethylated derivatives give the most intense $M^{+\cdot}$ peaks. As with electron impact, the spectra of the acetates tend to be dominated by the $MeCO^+$ ion. Permethylated disaccharides give good spectra and show marked intensity differences between 1—4 and 1—6 linked compounds. It would appear that this method is not sufficiently precise to determine linkage positions in trisaccharides, but their molecular weights can be determined. In the spectra of aryl glucosides[123] there are only two major fragments, (29) and (30), which both give linear Hammett plots of intensity ratios.

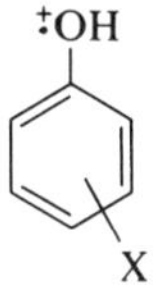

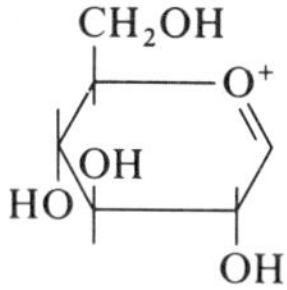

One of the most spectacular applications has been to the cardenolides,[124] of which digitoxin (31) is a typical example. The molecular ion is about 3% of the intensity of the base peak as opposed to 0.003% in the EI spectrum. If the structure (31) is represented as $HO-S3-O-S2-O-S1-O-G$, then the sequence fragments $GOH^{+\cdot}$, $GOS1OH^+$, $GOS1OS2OH^+$ stand out clearly in the FI spectrum, much more so than in the EI spectrum.

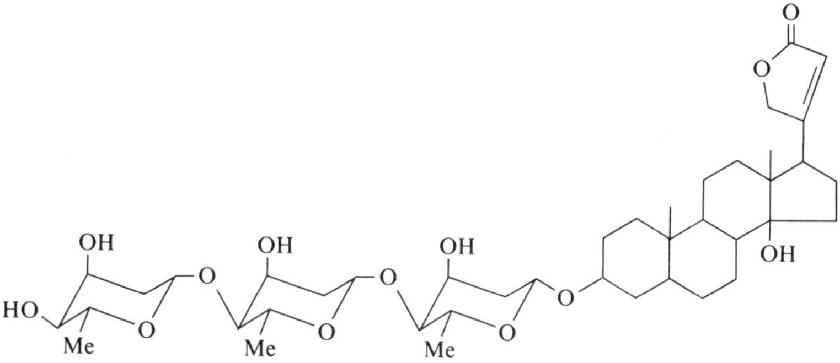

The field ionization mass spectra of derivatized peptides do not show much fragmentation and are inferior to EI and CI spectra for sequence determination.[125]

[122] H. Krone and H. D. Beckey, *Org. Mass Spectrometry*, 1971, **5**, 983.
[123] G. O. Phillips and W. G. Filby, *Chem. Comm.*, 1970, 1269.
[124] P. Brown, F. Bruschweiler, G. R. Pettit, and T. Reichstein, *Org. Mass Spectrometry*, 1971, **5**, 573.
[125] H. U. Winkler and H. D. Beckey, *Biochem. Biophys. Res. Comm.*, 1972, **46**, 391.

The field desorption spectra of free peptides, even of those containing arginine, aspartic acid, and serine residues, show molecular ions or $(M + 1)^+$ ions and fragment ions though not necessarily all the sequence ions required.

6 Ion Cyclotron Resonance

Of the methods described in this chapter in Volume 1, ICR has been developed most in the past two years. There have been several reviews during this time[126] covering studies of the structures of ions, the photochemistry of ions, gas-phase acid–base relationships, measurement of rates of ion–molecule reactions, effect of kinetic energy on reaction rates and general gas-phase ion chemistry.

One of the simplest cases of attempts to find information on the structures of ions is concerned with $C_3H_6^{+\cdot}$. The following reactions are found in cyclopropane–ammonia mixtures:

$$C_3H_6^{+\cdot} + NH_3 \rightarrow CH_4N^+ + C_2H_5^{\cdot}$$
$$C_3H_6^{+\cdot} + NH_3 \rightarrow CH_5N^{+\cdot} + C_2H_4$$

These reactions are not observed when propylene–ammonia mixtures are used. The same reactions are observed with the $C_3H_6^{+\cdot}$ fragment ions from THF and from cyclohexanone but not with the fragment ions from n-propyl chloride, cyclopentane, and pent-1-ene.[127] It is rather facile then to conclude simply that the $C_3H_6^{+\cdot}$ fragments from THF and cyclohexanone are cyclopropane ions, but one can say that they are 'cyclopropane-like' ions, since in the cyclopropane system it is not known whether the ion retains its cyclic structure before it reacts with ammonia. However, there does appear to be a clear-cut distinction between two $C_3H_6^{+\cdot}$ ion structures.

The situation with regard to the structure of $C_3H_7^+$ ions is less clear. In the reaction

$$C_3H_7^+ \rightarrow C_3H_5^+ + H_2$$

there is randomization of all the hydrogen atoms in $C_3H_7^+$ from propane, 1-bromopropane, and 2-bromopropane. This process is, however, 38 kcal mol^{-1} endothermic and it is probable that randomization requires less energy than this. An ICR study of bimolecular reactions of these ions[128] shows that labels in $C_3H_7^+$ from 2-bromopropane retain their identity. This is assumed to be the s-propyl ion. The reactions of labelled C_3H_7 from 1-bromopropane

$$C_3H_5D_2^+ + EtCHO \begin{array}{c} \nearrow EtC\overset{+}{H}OH + C_3H_4D_2 \\ \searrow EtC\overset{+}{H}OD + C_3H_5D \end{array}$$

[126] G. A. Gray, *Adv. Chem. Phys.*, 1971, **19**, 141; J. D. Baldeschwieler and S. S. Woodgate, *Accounts Chem. Res.*, 1971, **4**, 114; G. C. Goode, R. M. O'Malley, and K. R. Jennings, *Nature*, 1970, **227**, 1093; K. H. Leber, *Messtechnik*, 1970, **78**, 109; G. C. Goode, R. M. O'Malley, A. J. Ferrer-Correia, and K. R. Jennings, *Chem. in Britain*, 1971, 12; J. L. Beauchamp, *Ann. Rev. Phys. Chem.*, 1971, **22**, 527.
[127] M. L. Gross and F. W. McLafferty, *J. Amer. Chem. Soc.*, 1971, **93**, 1267.
[128] D. J. McAdoo, F. W. McLafferty, and P. F. Bente, *J. Amer. Chem. Soc.*, 1972, **94**, 2027.

show that the hydrogen atoms have randomized completely. From these results it would appear that there are two $C_3H_7^+$ ions, a s-propyl ion and a rearranging 1-propyl–protonated cyclopropane ion. Another ICR study of the reactivity of $C_3H_7^+$ ions from various sources makes the position less clear.[129] The reaction studied is the alkylation of furan:

$$C_3H_7^+ + C_4H_4O \rightarrow C_7H_{11}O^+$$

The $C_3H_7^+$ fragment ions appear to fall into three groups, (i) very reactive, from propane, 2-bromopropane, and 2-iodopropane, (ii) less reactive, from 1-iodopropane, 1-bromopropane, 2,3-dimethylbutane, n-octane, and 2,3,4-trimethylpentane, and (iii) somewhat less reactive than (ii), from 1-chloropropane, n-pentane, n-hexane, and isopentane. It is not clear whether these differences in reactivity are due to differences in gross structure of the ions or to different vibrational levels of the same ion.

The ion $C_7H_8^{+\cdot}$ produced by elimination of C_3H_6 from the molecular ion of n-butylbenzene has been compared with the molecular ion of toluene. $C_7H_8^{+\cdot}$ from toluene reacts with isopropyl nitrate:[130]

$$C_7H_8^{+\cdot} + C_3H_7ONO_2 \rightarrow C_7H_8NO_2^+ + C_3H_7O\cdot$$

whereas $C_7H_8^{+\cdot}$ from cycloheptatriene does not react. The $C_7H_8^{+\cdot}$ fragment ion from n-butylbenzene does not react in this way. This is a piece of negative evidence, to the effect that this ion does not have the same structure as toluene and the result is not inconsistent with the proposed structure (32).

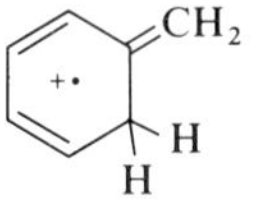

(32)

Styrene is the only olefinic compound which produces a stable dimer ion by ion–molecule reaction with itself:

$$C_8H_8^{+\cdot} + C_8H_8 \rightarrow C_{16}H_{16}^{+\cdot}$$
$$C_{16}H_{16}^{+\cdot} \rightarrow C_{10}H_{10}^{+\cdot} + C_6H_6$$

Neither the dimer $C_{16}H_{16}^{+\cdot}$ nor its fragment $C_{10}H_{10}^{+\cdot}$ is formed from cyclooctatetraene. The results using labelled styrene are shown in Scheme 9. Under

$$PhCD{=}CH_2^{+\cdot} \rightarrow (PhCD{=}CH_2)_2^{+\cdot} \begin{matrix} \nearrow C_{10}H_9D^{+\cdot} \\ \searrow C_{10}H_8D_2^{+\cdot} \end{matrix} \text{ equal amounts}$$

$$PhCH{=}CDH^{+\cdot} \rightarrow (PhCH{=}CDH)^{+\cdot} \rightarrow C_{10}H_8D_2^{+\cdot} + C_6H_6$$

Scheme 9

[129] M. L. Gross, *J. Amer. Chem. Soc.*, 1971, **93**, 253.
[130] M. K. Hoffmann and M. M. Bursey, *Tetrahedron Letters*, 1970, 2539.

low-pressure conditions the molecular ion of 1-phenyltetralin eliminates C_6H_6 to give a fragment $C_{10}H_{10}^{+\cdot}$. Since C_6H_5D is eliminated from the labelled compound (33) this would be consistent with a 1-phenyltetralin structure for the styrene dimer ion.[131]

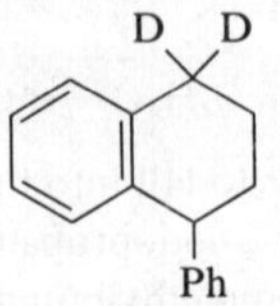

D D

Ph

(33)

Brauman and Smyth have studied the photo-detachment of negative ions using a grating monochromator and a continuously tunable laser. From the threshold energy for the reaction

$$h\nu + PH_2^- \rightarrow PH_2^{\cdot} + e$$

the electron affinity of the radical can be obtained.[132] Values for PH_2 and NH_2 were obtained[133] using a trapped ion cell with ion residence time of 0—500 μs.[134] With the measurement of the electron affinity of $AsH_2^{\cdot}$ [135] it is possible to see the trend in Group V (Table 2).

Table 2

	EA/eV
NH_2	0.74 ± 0.02
PH_2	1.26 ± 0.03
AsH_2	1.27 ± 0.03

Another interesting application of irradiation of ions in ICR has been the photolysis of positive ions.[136] The threshold of the reaction

$$CH_3Cl^+ + h\nu \rightarrow CH_3^+ + Cl\cdot$$

was measured as 3.3 eV. The decomposition of N_2O^+

$$N_2O^+ + h\nu \rightarrow NO^+ + N\cdot$$

took place over the full range of wavelength available and it was not possible to examine the threshold region.

There is an increasing amount of information appearing on acid and base strengths in the gas phase. The relative acidities of a number of amines have been determined.[137] It appears, as with alcohols, that large alkyl groups stabilize

[131] C. L. Wilkins and M. L. Gross, *J. Amer. Chem. Soc.*, 1971, **93**, 895.
[132] K. C. Smyth and J. I. Brauman, *J. Chem. Phys.*, 1972, **56**, 1132.
[133] K. C. Smyth, R. T. McIver, and J. I. Brauman, *J. Chem. Phys.*, 1971, **54**, 2758.
[134] R. T. McIver, *Rev. Sci. Instr.*, 1970, **41**, 555; T. B. McMahon and J. L. Beauchamp, *ibid.*, 1972, **43**, 509.
[135] K. C. Smyth and J. I. Brauman, *J. Chem. Phys.*, 1972, **56**, 4620.
[136] R. C. Dunbar, *J. Amer. Chem. Soc.*, 1971, **93**, 4354.
[137] J. I. Brauman and L. K. Blair, *J. Amer. Chem. Soc.*, 1971, **93**, 3911.

anions. Secondary amines have an enhanced acidity because of the lower N—H bond energy, but a large primary amine may be a stronger acid than a small secondary amine. t-Butylamine and dimethylamine are about equal in acid strength. Measurements of relative basicities of amines[138] show that they follow the same order as in solution, tertiary > secondary > primary. It is also shown that large alkyl groups tend to stabilize positive ions. Both sets of results tend to suggest that the polarizability of an alkyl group is a much more important factor than any permanent inductive effect working in one direction. In the gas phase, pyridine is a stronger base than ammonia and is comparable in basicity to dimethylamine.[139] There is, however, a linear relationship between proton affinity in the gas phase and free energy of protonation in aqueous solution for 4-substituted pyridines.

An ICR study of PH_3 allowed it to be placed in the gas-phase acidity scale $HCN > H_2S > PH_3 > H_2O > NH_3$.[140] The difference between phosphine and ammonia is mainly due to differences in the P—H and N—H bond energies. A more complete comparison of gas-phase acidities of binary hydrides has been carried out and the order is shown in Table 3.[141] The order is consistent with current chemical thinking. Moving to the right along a row, greater electronegativity leads to higher acidity. Moving down a column, compounds become more acidic due to the reduction in bond energy.

Table 3

$$\begin{array}{ccc} CH_4 < NH_3 < H_2O < HF \\ \wedge \quad \wedge \quad \wedge \quad \wedge \\ SiH_4 < PH_3 < H_2S < HCl \\ \wedge \quad \quad \wedge \\ AsH_3 \quad < \quad HBr \\ \wedge \\ HI \end{array}$$

In a study of ICR spectra of N-nitrosamines,[142] a series of relative basicities was determined from the reaction

$$R_2NNOH^{+\cdot} + R_2'NNO \rightarrow R_2'NN\overset{+}{O}H + R_2NNO$$

The formation of the protonated molecular ion from the molecular ion in a cross-section in such mixtures,

$$R_2NNO^{+\cdot} + R_2'NNO \rightarrow R_2'NNOH^+ + \text{neutral product}$$

[138] J. I. Brauman, J. M. Riveros, and L. K. Blair, *J. Amer. Chem. Soc.*, 1971, **93**, 3914.
[139] M. Taagepera, W. G. Henderson, R. T. C. Brownlee, J. L. Beauchamp, D. Holtz, and R. W. Taft, *J. Amer. Chem. Soc.*, 1972, **94**, 1369.
[140] D. Holtz, J. L. Beauchamp, and J. R. Eyler, *J. Amer. Chem. Soc.*, 1970, **92**, 7045.
[141] J. I. Brauman, J. R. Eyler, L. K. Blair, M. J. White, M. B. Comisarow, and K. C. Smyth, *J. Amer. Chem. Soc.*, 1971, **93**, 6360.
[142] H. H. Jaffe, S. Billets, and F. Kaplan, *J. Amer. Chem. Soc.*, 1970, **92**, 6964.

only takes place when R_2^1NNO is more basic than the other component. It is not dependent on the availability of secondary or tertiary C—H bonds in the molecular ion. It is suggested that the molecular ion rearranges, e.g. (34) → (35), and that proton transfer is from oxygen.

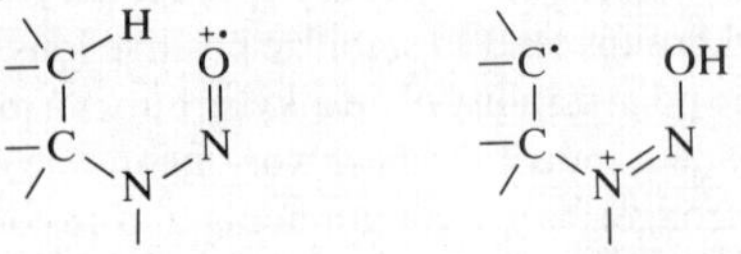

(34)   (35)

The main secondary products from iron pentacarbonyl are $Fe_2(CO)_4^+$ and $Fe_2(CO)_5^+$.[143] The reactions of the primary ions from $Fe(CO)_5$ in binary mixture are more interesting. CO is displaced by NH_3, H_2O, and CH_3F, but not by HCl. The ability to displace a ligand appears to be related to proton affinity (PA) (Table 4).

Table 4	PA/kcal mol^{-1}
HCl	140
CO	143
MeF	151
H$_2$O	164
NH$_3$	207

In the study of ion–molecule reactions, the measurement of absolute rate constants is the aim of a number of workers in the ICR field. To achieve this, expressions have been derived for the measurement of currents of primary and secondary ions both from power absorption peaks and from TIC spectra. These are valid for conditions of low ion density and low conversion.[144] More complete calculations of equations of motion for ions are claimed to give more accurate measured rate constants from power absorption curves.[145] Jenning and co-workers have reported on the factors affecting the sign of a double-resonance signal in ICR.[146] Although this should reflect rates of change in reaction rate with kinetic energy of the ion, the sign of the signal can change as the irradatating field strength is increased, owing to space-charge and 'sweep-out' effects. Such effects can be detected by examination of the total ion current during irradiation.

The variation of rate constant with ion kinetic energy can be measured more precisely using pulsed techniques.[147] In a typical experiment, the electron

[143] M. S. Foster and J. L. Beauchamp, *J. Amer. Chem. Soc.*, 1971, **93**, 4924.
[144] G. C. Goode, R. M. O'Malley, A. J. Ferrer-Correia, R. I. Massey, K. R. Jennings, J. H. Futrell, and P. M. Llewellyn, *Internat. J. Mass Spectrometry Ion Phys.*, 1970, **5**, 393.
[145] W. T. Huntress, *J. Chem. Phys.*, 1971, **55**, 2146.
[146] G. C. Goode, A. J. Ferrer-Correia, and K. R. Jennings, *Internat. J. Mass Spectrometry Ion Phys.*, 1970, **5**, 229.
[147] R. P. Clow and J. H. Futrell, *Internat. J. Mass Spectrometry Ion Phys.*, 1970, **4**, 165; R. T. McIver and R. C. Dunbar, *ibid.*, 1971, **7**, 471.

Alternative Methods of Ionization and Analysis

energy is pulsed from below to above the ionization potential of the molecule being studied to produce a bunch of ions. These are energized by a pulsed RF field with a variable amplitude, which allows kinetic energy dependence of the reaction rate to be measured. Clow and Futrell find that the yield of CH_5^+ drops monotonically with increasing kinetic energy, but there is an increase in the yield of CH_3^+, due to the endothermic process

$$CH_5^+ \rightarrow CH_3^+ + H_2$$

Dunbar has devised an ingenious method for the determination of ion–molecule collision rates.[148] If a 50 μs pulse of ions with resonant frequency ω_c is observed with a marginal oscillator slightly out of tune with ω_c, an interference pattern is produced which is damped by ion–neutral collisions. The damping curve gives a measure of collision rate.

The formation of H_3^+ and its isotopic analogues has been studied in detail.[149] The most interesting case is HD, in which there are two competing processes:

$$HD^+ + HD \rightarrow H_2D^+ + D$$
$$HD^+ + HD \rightarrow HD_2^+ + H$$

The formation of H_2D^+ shows an increase in k with ion kinetic energy followed by a decrease above 1 eV; for HD_2^+, a monotonic decrease is observed.[149] The explanation given is that there are two mechanisms and that at higher kinetic energies a stripping mechanism will be preferred which has a high selectivity for abstraction of H· rather than D·.

$$D_3^+ + C_2H_6 \rightarrow [C_2H_6D^+] + D_2$$

$$[C_2H_6D^+] \begin{array}{c} \nearrow C_2H_5^+ + HD \\ \searrow C_2H_4D^+ + H_2 \end{array}$$

$$C_2H_4D^+ + C_2H_6 \rightarrow C_2H_5^+ + C_2H_5D$$

Scheme 10

In a mixture of D_2 and C_2H_6, the ionic products show that labelling of ethane is taking place by the process shown in Scheme 10. This is presented as an analogy for the Wilzbach labelling of ethane with tritium (Scheme 11).[150]

$$T_2 \rightarrow T^3He^+ + \beta^-$$
$$T^3He^+ + C_2H_6 \rightarrow [C_2H_6T^+] + {}^3He$$
$$[C_2H_6T^+] \rightarrow C_2H_4T^+ + H_2$$
$$C_2H_4T^+ + C_2H_6 \rightarrow C_2H_5^+ + C_2H_5T$$

Scheme 11

[148] R. C. Dunbar, *J. Chem. Phys.*, 1971, **54**, 711.
[149] W. T. Huntress, D. D. Elleman, and M. T. Bowers, *J. Chem. Phys.*, 1971, **55**, 5413.
[150] S. Wexler and L. G. Pobo, *J. Amer. Chem. Soc.*, 1971, **93**, 1327.

The ion–molecule reactions of allene and mixtures of its isotopic analogues are consistent with the formation of an intermediate such as (36), which decomposes preferentially to $C_4H_2D_2^{+\cdot}$ but with a small amount of isotope scrambling.[151] In mixtures of H_2S with ethylene or acetylene, condesation reactions

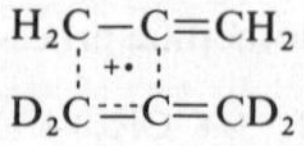

$$H_2C-C=CH_2$$
$${\scriptstyle +\cdot}$$
$$D_2C=C=CD_2$$
(36)

take place to form unstable intermediates which decompose with a certain amount of hydrogen scrambling.[152] The reaction with ethylene is shown in Scheme 12.

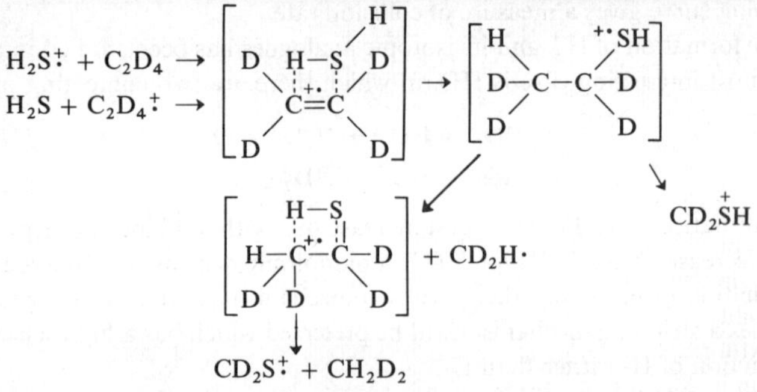

Scheme 12

The ion–molecule chemistry of ammonia involves only two principal primary ions $NH_3^{+\cdot}$ and NH_2^+. Ion ejection techniques and mixtures of ^{14}N, ^{15}N isotopes were used to sort out the reactions.[153] The only reaction of NH_3^+ is proton transfer to NH_3; charge transfer and H· abstraction are not observed. NH_2^+ forms NH_4^+ by proton transfer and $NH_3^{+\cdot}$ by charge transfer.

In methane–ammonia mixtures the condensation reactions involve CH_3^+:

$$CD_3^+ + NH_3 \begin{array}{l} \nearrow CD_2NH_2^{+*} + HD \quad 70\% \\ \searrow CD_3NH^+ + H_2 \quad 30\% \end{array}$$

The ion $CD_2NH_2^+$ is formed in an excited state, and can protonate ammonia. At higher pressures, it is deactivated by collision and the ground state is not a strong enough acid to protonate ammonia.[154] The publication of an investiga-

[151] M. T. Bowers, D. D. Elleman, R. M. O'Malley, and K. R. Jennings, *J. Phys. Chem.*, 1970, **74**, 2583.
[152] S. E. Buttrill, *J. Amer. Chem. Soc.*, 1970, **92**, 3560.
[153] W. T. Huntress, M. M. Mosesman, and D. D. Elleman, *J. Chem. Phys.*, 1971, **54**, 843.
[154] W. T. Huntress and D. D. Elleman, *J. Amer. Chem. Soc.*, 1970, **92**, 3565.

tion of the ICR spectra of butan-2-ol[155] makes possible some generalizations on the ion–molecule chemistry of alcohols. The protonated molecular ion $(M + H)^+$ is produced in abundance, as are the solvated analogues at higher pressures, $(M_2 + H)^+$ and $(M_3 + H)^+$. The reaction of the primary fragment ions varies with size. Smaller fragments tend to protonate the molecule:

$$\text{MeOH} + \text{CH}_2\overset{+}{\text{OH}} \rightarrow \text{Me}\overset{+}{\text{OH}}_2 + \text{CH}_2\text{O}$$

Heavier fragments tend rather to produce dehydration and condensation reactions:

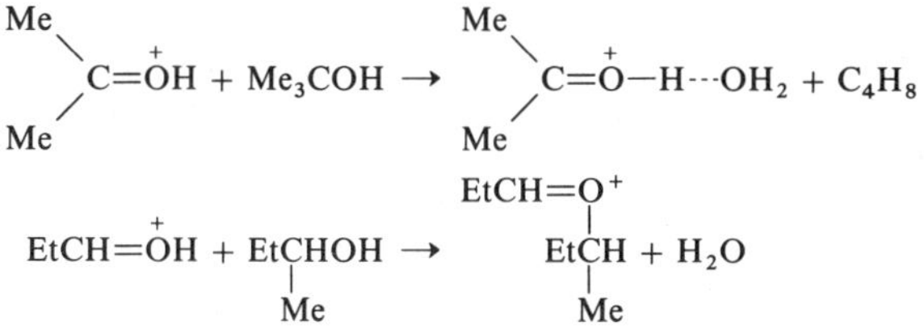

$$\underset{\text{Me}}{\overset{\text{Me}}{\diagdown}}\text{C}=\overset{+}{\text{OH}} + \text{Me}_3\text{COH} \rightarrow \underset{\text{Me}}{\overset{\text{Me}}{\diagdown}}\text{C}=\overset{+}{\text{O}}-\text{H}\cdots\text{OH}_2 + \text{C}_4\text{H}_8$$

$$\text{EtCH}=\overset{+}{\text{OH}} + \underset{\underset{\text{Me}}{|}}{\text{EtCHOH}} \rightarrow \underset{\underset{\text{Me}}{|}}{\overset{\overset{\text{EtCH}=\text{O}^+}{|}}{\text{EtCH}}} + \text{H}_2\text{O}$$

In smaller fragments, the lifetime is limited and decomposition takes place preferably along the hydrogen bond of the intermediate. Larger fragments have more vibrational degrees of freedom, and hence longer lifetimes and a greater possibility of alternative reactions.

Mixtures of hydrogen and fluoromethane yield the MeFH$^+$ ion which undergoes an interesting reaction with both N$_2$ and CO in which the latter molecules behave as nucleophiles:[156]

$$\text{N}_2 + \text{MeFH}^+ \rightarrow \text{MeN}_2^+ + \text{HF}$$
$$\text{CO} + \text{MeFH}^+ \rightarrow \text{MeCO}^+ + \text{HF}$$

Bursey and his collaborators are evolving a 'chemical ionization' method using an ICR spectrometer. They use the reactant gas MeCOCOMe which gives the ions MeCO$^+$, (MeCO)$_2^+$ and (MeCO)$_3^+$. When oxygen and nitrogen compounds are added to this system, abundant $(M + \text{MeCO})^+$ ions are observed.[156] The use of a mixed labelling system shows that the mechanism of the process is:[157]

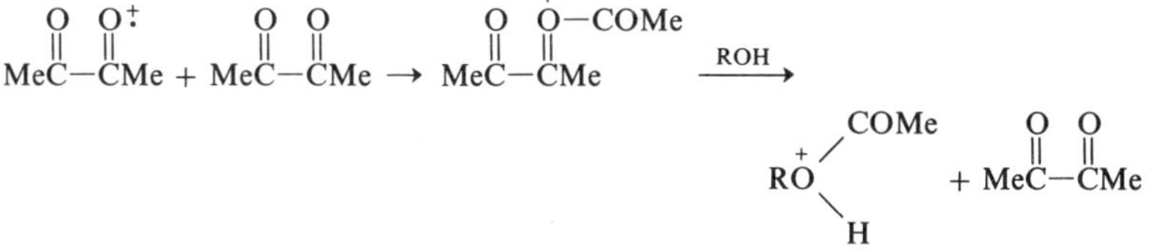

$$\underset{\text{MeC}-\text{CMe}}{\overset{\text{O} \ \ \ \text{O}^{+\cdot}}{\| \ \ \ \|}} + \underset{\text{MeC}-\text{CMe}}{\overset{\text{O} \ \ \ \text{O}}{\| \ \ \ \|}} \rightarrow \underset{\text{MeC}-\text{CMe}}{\overset{\text{O} \ \ \ \overset{+}{\text{O}}-\text{COMe}}{\| \ \ \ \|}} \xrightarrow{\text{ROH}}$$

$$\underset{\text{H}}{\overset{\text{COMe}}{\overset{+}{\text{RO}} \diagup}} + \underset{\text{MeC}-\text{CMe}}{\overset{\text{O} \ \ \ \text{O}}{\| \ \ \ \|}}$$

[155] J. L. Beauchamp and M. C. Caserio, *J. Amer. Chem. Soc.*, 1972, **94**, 2638.
[156] M. M. Bursey, T. A. Elwood, M. K. Hoffman, T. A. Lehman, and J. M. Tesarek, *Analyt. Chem.*, 1970, **42**, 1370.
[157] M. K. Hoffman, T. A. Elwood, T. A. Lehman, and M. M. Bursey, *Tetrahedron Letters*, 1970, 4021.

The reactions in biacetyl–alcohol mixtures include a condensation–dehydration process which is considered to proceed *via* the intermediate (37).[158] The acetylation of phenols by this gas-phase system can be sterically hindered by *ortho*-groups and is completely inhibited in (38).[159]

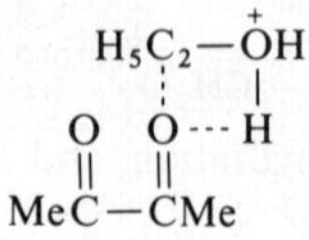

$$\underset{(37)}{\begin{array}{c}H_5C_2-\overset{+}{O}H\\|\\O\quad O\cdots H\\\|\quad\|\\MeC-CMe\end{array}}\qquad\underset{(38)}{Me_2CH-C_6H_3(OH)-CHMe_2}$$

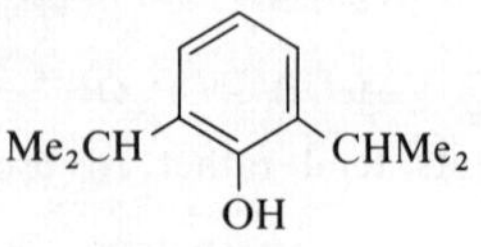

ICR spectra have been used to distinguish between some of the C_5H_{10} isomers.[160] The isomers examined were pent-1-ene, 3-methylbut-1-ene, 2-methylbut-1-ene, *cis*-pent-2-ene, *trans*-pent-2-ene, 2-methylbut-2-ene, and cyclopentane. The first five of these isomers all reacted with ionized buta-1,3-diene to give a series of condensation ions, $C_8H_{13}^+$, $C_7H_{12}^{+\cdot}$, $C_7H_{11}^+$, $C_6H_{10}^{+\cdot}$, $C_6H_9^+$, and each isomer gave different relative yields of these. The reaction of 2-methylbut-2-ene with $C_4H_6^{+\cdot}$ is charge exchange, and this is unique among the isomers. Cyclopentane was also unique in that it was reported to be unreactive. This is an interesting result since the EI mass spectra of all isomers are similar, and it is something of an achievement to distinguish between *cis*- and *trans*-olefins in a mass spectrometer.

---

[158] T. A. Lehman, T. A. Elwood, M. K. Hoffman, and M. M. Bursey, *J. Chem. Soc.* (*B*), 1970, 1717.
[159] S. A. Benezra and M. M. Bursey, *J. Amer. Chem. Soc.*, 1972, **94**, 1024.
[160] M. L. Gross, P.-H. Lin, and S. J. Franklin, *Analyt. Chem.*, 1972, **44**, 974.

2
Kinetic and Energetic Studies of Organic Ions

BY I. HOWE

1 Introduction

The corresponding chapter in Volume 1 of this series concluded with the expectation that much of the ground-work laid during the few years until mid-1970 would be consolidated with rigorous application of the mass spectral theories developed during that period. This expectation has largely been fulfilled in the literature covering the two years until mid-1972. Erroneous and ill-chosen methods for ion-structure elucidation in the mass spectrometer have largely disappeared. Possible pitfalls have often been recognized and in some instances assessments of discrepancies in mass spectral measurements (*e.g.* the kinetic shift) have been attempted. Most important, many authors have realized that there is far more information about ion structures and decomposition mechanisms to be gleaned from a mass spectrum than is obtainable merely from the normal ion abundances in the 70 eV spectrum. Some of this extra information is very easily obtainable (*e.g.* abundances and shapes of metastable peaks, low eV spectra, appearance potential measurements) on most mass spectrometers, and the use of other instruments or modifications (*e.g.* ICR, FI, collision-induced metastables) widens the scope still further. The mass spectra of carefully chosen isotopically labelled derivatives often supply useful information.

The research reported in this chapter covers the methods for elucidation of organic ion structures and reaction mechanisms in the mass spectrometer, usually with recourse to the quasi-equilibrium theory (QET). Papers are also reported which incidentally illustrate the application of kinetic and energetic principles to organic mass spectrometry. The reader is referred to Volume 1 of this series[1] or other recent publications for general and background discussion[2] and review articles.[3-6]

There has inevitably been a shift in emphasis in the coverage of particular topics compared with Volume 1. Heat of formation measurements have grown

[1] I. Howe, in 'Mass Spectrometry', ed. D. H. Williams (Specialist Periodical Reports), The Chemical Society, London, 1971, Vol. 1, Chap. 2, and references cited therein.
[2] D. H. Williams and I. Howe, 'Principles of Organic Mass Spectrometry', McGraw-Hill, London, 1972.
[3] D. H. Williams, 'Advances in Mass Spectrometry', The Institute of Petroleum, 1971, Vol. 5, p. 569.
[4] D. C. DeJongh, *Analyt. Chem.*, 1970, **42**, 169R.
[5] A. L. Burlingame and G. A. Johanson, *Analyt. Chem.*, 1972, **44**, 341R.
[6] K. R. Jennings, in 'Mass Spectrometry—Techniques and Applications', ed. G. W. A. Milne, Wiley-Interscience, 1971, pp. 419—458.

in number and energetic aspects of mass spectral processes have been investigated more extensively. There has been an increase in the more theoretical applications of the QET to large organic ions. Collision processes involving organic ions have been more frequently studied. On the other hand, the frequency of substituent effect publications has decreased. This chapter has been divided for clarity into seven sections of literature coverage (Sections 2—8) with a concluding summary (Section 9). In some cases the division is rather arbitrary, since a paper might equally well be discussed under several different headings.

2 Energetics of Mass Spectral Processes

Throughout this section and the rest of the chapter, kilojoules (abbreviated as kJ) are employed as units of energy in place of kilocalories. Since the majority of publications have used kilocalories, direct conversion into kilojoules has been carried out in these cases. Marginal losses in accuracy attending these conversions are inevitable.

Sources of Error in Heat of Formation Measurements.—The comparison of heats of formation of organic ions in the mass spectrometer is the most widely used criterion for identity or non-identity of ion structures. Despite the extensive use of this method, several energy terms in the heat of formation calculations are often ignored, namely: (i) the kinetic shift, (ii) the competitive shift, (iii) the thermal shift, and (iv) the energy of activation for the reverse mass spectral (recombination) reaction. This method therefore incorporates potential pitfalls[1,7] and it is apparent from the literature that care should be exercised in its application.

For the reaction

$$M + e \rightarrow X^+ + Y + 2e$$

the appearance potential, $AP(X^+)$, of X^+ may be expressed as shown in equation 1. The term E is included to allow for possible formation of either X^+ and/or Y with excess energy at the threshold for reaction.

$$AP(X^+) = \Delta H_f(X^+) + \Delta H_f(Y) - \Delta H_f(M) + E \qquad (1)$$

The kinetic shift arises because an ion must be formed with a unimolecular rate constant greater than $\sim 10^6$ s^{-1} to be detected in the ion source. The kinetic shift is the excess energy (above the minimum energy for reaction) necessary to bring about decomposition with this rate constant. The consequence of the kinetic shift is that AP measurements for daughter ions produced in the source (and hence their heats of formation) will be too high. Since metastable decompositions occur from precursor ions having lower minimum energy than those decomposing in the source, a significant difference (Δ) between the AP's of metastable and daughter ions should be indicative of a kinetic shift. Furthermore, the use of the AP for the metastable should give a heat of formation closer to the true value. This is illustrated in Figure 1.

[7] A. G. Harrison, in 'Topics in Organic Mass Spectrometry', ed. A. L. Burlingame, Wiley-Interscience, New York, 1970.

Kinetic and Energetic Studies of Organic Ions

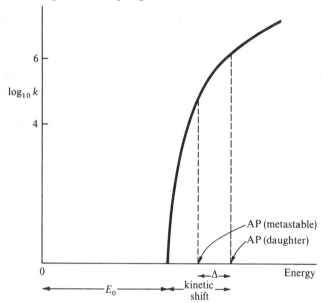

Figure 1 A $\log k$ versus E curve, indicating approximate AP's for metastable and daughter ions

The factor Δ has been measured for a number of fragments from organic molecular ions[8] and found to vary from 0.1 to 1.0 eV. One of the largest Δ values was found for the $C_4H_4^{+\cdot}$ ion formed from the benzene molecular ion, consistent with previous theoretical considerations.[9] An approximate correlation was found between Δ and the intensity of the metastable peak, and it is suggested[8] that the full kinetic shift for processes giving rise to intense metastable peaks is substantially larger than the measurable part.

The metastable transition for which AP's were measured in this publication[8] occurred in the first drift region, where the signal-to-noise ratios were sufficiently high. The ion-accelerating voltage was kept constant to maintain source-tuning conditions and the electrostatic analyser (ESA) voltage was reduced to transmit metastable ions, using the standard modification (metastable defocuser).

The kinetic shift has been estimated for a series of 1,2-diphenylethanes (1).[10] Realistic k vs. E curves for formation of $C_7H_7^+$ and $C_7H_6X^+$ are computed by

$$X\!-\!\!\!\bigcirc\!\!\!-CH_2CH_2\!\!-\!\!\!\bigcirc$$
(1)

[8] R. D. Hickling and K. R. Jennings, *Org. Mass Spectrometry*, 1970, **3**, 1499.
[9] M. Vestal, in 'Fundamental Processes in Radiation Chemistry', ed. P. Ausloos, Wiley-Interscience, New York, 1968, Ch. 2.
[10] F. W. McLafferty, T. Wachs, C. Lifshitz, G. Innorta, and P. Irving, *J. Amer. Chem. Soc.*, 1970, **92**, 6867.

enumerating the exact number of states, and the kinetic shifts are as high as 1.2 eV. It is noted[10] that reactions with tight transition states (restricted internal rotation, e.g. rearrangements) will exhibit larger kinetic shifts than those having loose complexes (i.e. direct cleavage reactions) owing to the slower increase of rate constant k with internal energy E in the former case. The authors also note the occurrence of the 'competitive shift' which may arise where two k vs. E curves cross. The reaction which has the higher AP (curve B in Figure 2) may need to proceed at a rate considerably faster than 10^6 s^{-1} to compete with the other reaction (A) sufficiently to be observed in the source.

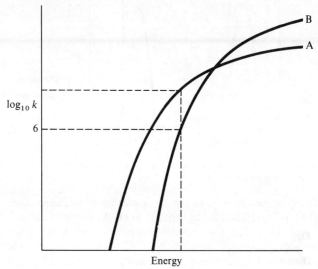

Figure 2 *Rate curves for competing reactions, illustrating the competitive shift*

Another factor should be recognized which may compromise IP and AP measurements. The presence, in molecules, of thermal energy before ionization above 0 K means that, neglecting other factors, less energy than the true critical potential would be required to produce the ion at threshold. It has been shown that the vibrational and rotational energies of moderately complex molecules, even at room temperature, are considerably greater than the uncertainties often quoted for AP's. Chupka has investigated the effect of thermal energy on the shape of photo-ionization efficiency curves of some alkanes.[11] This energy discrepancy in onset potential measurements has been designated the 'thermal shift', and of course acts in the opposite direction to the kinetic shift. For n-hexane the thermal shift between ionization potentials determined at 28 and 150 °C was found to be 0.16 eV.[11] Good agreement was found between calculated and observed thermal shift, which increased with molecular size, as predicted. It

[11] W. A. Chupka, *J. Chem. Phys.*, 1971, **54**, 1936.

Kinetic and Energetic Studies of Organic Ions

was shown that, within experimental error, the thermal energy is fully effective in the dissociation of the parent ions, consistent with one of the hypotheses of the QET.

It is evident that, in a large molecule, the thermal energy is significant relative to the average internal energy gained upon electron impact with 70 eV electrons. The distributions of thermal energy for 1,2-diphenylethane (1; X = H) have been calculated at 75 and 200 °C and the average thermal energies are approximately 0.3 and 0.7 eV respectively (see Figure 3).[10]

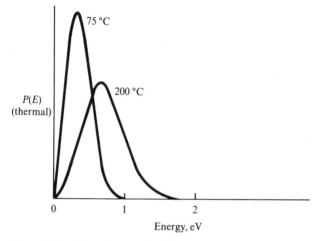

Figure 3 *The thermal energy distribution for 1,2-diphenylethane calculated at 75 and 200 °C*
(Reproduced by permission from *J. Amer. Chem. Soc.*, 1970, **92**, 6867)

Formation of Fragments with Excess Energy.—Even at the minimum energy for unimolecular decomposition, the product ion and neutral species may be formed with excess internal and/or kinetic energy* owing to an activation energy for the recombination reaction (see the term E in equation 1).

The partition of excess internal energy of the fragmenting species between internal and kinetic modes in unimolecular decompositions has been investigated for the reaction (2) → (3) in some substituted benzaldoxime methyl ethers.[12] The total excess energy (including any possible kinetic shift) was calculated (see equation 2) from the appearance potential of the $M^+ - $ HCN ion, with the additional assumption that this ion (3) corresponds to the substituted anisole molecular ion. The kinetic energy released in the transitions (2) → (3) (measured from metastable peak widths) tends to increase, both absolutely and as a fraction

---

[12] R. G. Cooks, D. W. Setser, K. R. Jennings, and S. Jones, *Internat. J. Mass Spectrometry Ion Phys.*, 1971, **7**, 493.

* For other examples of kinetic energy release in metastable transitions, see Section 5.

of the available excess energy, with increasing electron-donating power of the
p-substituent. The minimum percentage of excess energy released as kinetic

$$\underset{X}{\underset{(2)}{\underset{\diagdown}{\diagup}}}\text{—CH=N—OMe}\quad\rceil^{\ddagger} \longrightarrow \underset{X}{\underset{(3)}{\underset{\diagdown}{\diagup}}}\text{—OMe}\quad\rceil^{\ddagger} + \text{HCN}$$

energy was 10% for the *p*-CN substituent, and the maximum was 59% for the
p-OMe substituent.

$$E_{\text{excess}} = \text{AP}(M^+ - \text{HCN}) + \Delta H_\text{f}(\text{OE}) - \text{IP}(\text{A}) - \Delta H_\text{f}(\text{A}) - \Delta H_\text{f}(\text{HCN}) \quad (2)$$

where A = anisole, OE = oxime methyl ether.

The distribution of excess internal energy between internal and kinetic energy of the products has also been investigated for dissociation of negative ions in the mass spectrometer.[13] For the dissociative electron capture process described in equation 3, the thermochemical relationship described in equation 4 may be written. EA(A) is the electron affinity of neutral A and E^* is the excess energy

$$M + e \longrightarrow A^- + B \quad (3)$$

$$\text{AP}(A^-) = \Delta H_\text{f}(A) - \text{EA}(A) + \Delta H_\text{f}(B) - \Delta H_\text{f}(M) + E^* \quad (4)$$

of the system. E^* was measured from the relevant AP in cases where all the other terms in equation 4 were known.[13] All cases reported involved the dissociation of small (<12 atoms) organic and inorganic ions. The translational energies (E_t) released in the dissociations were measured from metastable peak widths at half height. It was found that the approximate relationship described by equation 5 existed between E_t and E^*, where N is the number of oscillators

$$E_t = E^*/\alpha N \quad (5)$$

and α is an empirical correction factor for the effective number of oscillators. The mean value of α was found to be 0.42, close to the 0.44 determined similarly for positive ions.[14] From this approximate correlation it is possible to obtain estimates for E^* from measurement of E_t, and hence unknown electron affinities and bond dissociation energies in negative ions can be estimated (by application of equation 4). In connection with energy release in unimolecular ionic fragmentations, it is of note that formulae for calculating average translational and rotational energies of the fragments have been presented.[15]

Ion Structure Evidence from Heats of Formation and Appearance Potentials.—
Measurements of heats of formation have continued to be used to yield structural information about hydrocarbon ions at their formation. For example, the appearance potentials of the fragment ions in the spectra of the C_6H_{10} isomers (4—8) have been used to calculate heats of formation (employing equation 1 and

[13] J. J. DeCorpo, D. A. Bafus, and J. L. Franklin, *J. Chem. Phys.*, 1971, **54**, 1592.
[14] M. A. Haney and J. L. Franklin, *J. Chem. Phys.*, 1968, **48**, 4093.
[15] C. E. Klots, *Z. Naturforsch.*, 1972, **27a**, 553.

neglecting E).[16] All of the spectra contain $C_4H_6^{+\cdot}$ ions formed by loss of C_2H_4 in one step from the molecular ion and the respective heats of formation calculated for this ion from compounds (4)—(8) are 1132, 1148, 1091, 1138, and

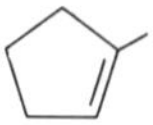

(4) 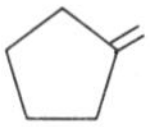 (5) 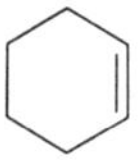 (6)   Me(CH$_2$)$_3$C≡CH   Me(CH$_2$)$_2$C≡CMe
                        (7)   (8)

1111 kJ mol^{-1}. In the case of (6), the $C_4H_6^{+\cdot}$ ion has long been considered to arise *via* the retro-Diels–Alder reaction. The heat of formation would appear to be consistent with a buta-1,2-diene structure ($\Delta H_f = 1086$ kJ mol^{-1}, from its molecular ion) rather than buta-1,3-diene ($\Delta H_f = 989$ kJ mol^{-1}).[16] However, the calculated heats of formation for the $C_4H_6^{+\cdot}$ fragment ions should be regarded as maximum values, owing to possible excess energy terms (see above).

Appearance potential measurements in the mass spectra of norbornene (9) and nortricyclene (10) have yielded some related ΔH_f results.[17] The retro-Diels–Alder elimination of C_2H_4 from the molecular ion is an important process in both cases, yielding the base peak ($C_5H_6^{+\cdot}$, m/e 66) in the case of (9). The calculated ΔH_f value for $C_5H_6^{+\cdot}$ from (9) and (10) corresponds exactly (958 kJ mol^{-1}) to that calculated for the cyclopentadiene ion (11) from its IP. However, as the authors point out, uncertainty in ΔH_f values of the two neutrals (9) and (10)

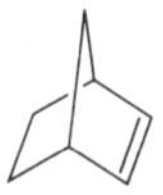

(9) 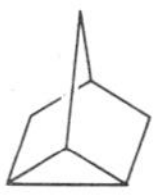 (10)  (11) 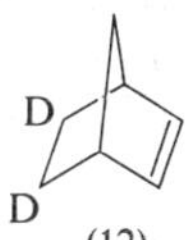 (12)

leads to uncertainty in the calculated value for $C_5H_6^{+\cdot}$ from these two sources. For example, the ΔH_f value for $C_5H_6^{+\cdot}$ from (9) could be up to 38 kJ mol^{-1} greater. Owing to possible excess energy terms, this value would still be consistent with the elimination of C_2H_4 from the molecular ion of (9) to yield (11). Deuterium labelling results confirm this conclusion: the molecular ion of (12) eliminates only $C_2H_2D_2$, even in metastable transitions.[17]

A suggestion has been made[18] concerning the structure of the $C_3H_3^+$ ion formed by fragmentation in the mass spectrometer. Previous mass spectral evidence had suggested that two stable structures might exist for this ion, with a difference in stability of 46 ± 17 kJ mol^{-1}. It was further observed[18] that the $C_3H_5^+ \rightarrow C_3H_3^+ + H_2$ reaction was accompanied by a metastable peak which was a composite of two individual peaks (see also pp. 63, 64) with a difference in kinetic energy release of 59 kJ mol^{-1}. This is also consistent with the formation of two

[16] M.-Th. Praet, *Org. Mass Spectrometry*, 1970, **4**, 65.
[17] J. L. Holmes and D. McGillivray, *Org. Mass Spectrometry*, 1971, **5**, 1349.
[18] P. Goldberg, J. A. Hopkinson, A. Mathias, and A. E. Williams, *Org. Mass Spectrometry*, 1970, **3**, 1009.

different $C_3H_3^+$ ions (possibly cyclic and linear) having energy differences within the limits defined by the heat of formation differences above.

Heats of formation of ions determined from IP's of the corresponding neutral (even-electron species or radical) are not susceptible to some of the inaccuracies attending AP measurements.[2,7] Nevertheless, IP's determined using a conventional electron impact source are generally higher than the adiabatic IP by 0.1—0.7 eV, the error becoming larger with unfavourable Franck–Condon factors in the threshold ionization region.[19] Lossing has continued his determination of reliable IP's of radicals by use of an energy-resolved electron beam.[19] Vinyl (C_2H_3), allyl (C_3H_5), and benzyl ($C_6H_5CH_2$) radicals were generated pyrolytically and their IP's determined. By application of equation 6 it is possible either (i) to determine $\Delta H_f(R^+)$ from knowledge of $\Delta H_f(R\cdot)$ or (ii) to determine $\Delta H_f(R\cdot)$ by adopting a value for $\Delta H_f(R^+)$ from AP methods. For vinyl, allyl, and benzyl cations, the calculated heats of formation were respectively 1113, 946, and 891 kJ mol^{-1}.

$$IP(R\cdot) = \Delta H_f(R^+) - \Delta H_f(R\cdot) \qquad (6)$$

More AP results have been reported concerning the long-standing benzyl vs. tropylium problem.[20] AP's were measured for the $M^+ - CH_3$ ions in the mass spectra of m- and p-substituted ethylbenzenes (13). It is argued that if the $M^+ - CH_3$ ions formed at threshold have the benzyl structure (14), with ring-orientation of the substituent retained, then the difference in AP between m- and p-isomers is expected to be close to the difference between the IP's of the corresponding benzyl radicals. It is concluded from the data that (14) is formed when X = NH_2, OMe, CN, or F, whereas (15) is formed when X = Me or OH, although the experimental error in the onset potential determinations does allow some leeway for the conclusions. It should be generally noted that if (14) is involved for the m-substituent, it does not follow that (14) [and not (15)] should be formed from the p-substituted compound, and vice versa.

X—⟨⟩—CH$_2$Me   X—⟨⟩—CH$_2^+$ ⟨⟩—X

(13)   (14)   (15)

The mass spectrum of the methiodide (16) of cinnoline has the highest mass peak at the mass of the cation (m/e 145) with an AP of 6.5 eV.[21] This very low AP is contrasted with that of cinnoline itself (9.15 eV), and it is concluded that the m/e 145 ion arises via ionization of the volatile free radical (17), itself formed in a pyrolytic process.

Appearance potential measurements have been suggested[22] as a method for determination of ring conformations. If a given fragment ion (X^+) from two

[19] F. P. Lossing, *Canad. J. Chem.*, 1971, **49**, 357.
[20] D. A. Lightner, S. Majeti, and G. B. Quistad, *Tetrahedron Letters*, 1970, **44**, 3857.
[21] G. Hvistendhal and K. Undheim, *Tetrahedron*, 1972, **28**, 1737.
[22] K. Pihlaja and J. Jalonen, *Org. Mass Spectrometry*, 1971, **5**, 1737.

isomeric molecular ions can be assumed to have identical ΔH_f values by both routes, then it may be argued (see equation 1) that the difference (Δ) between the

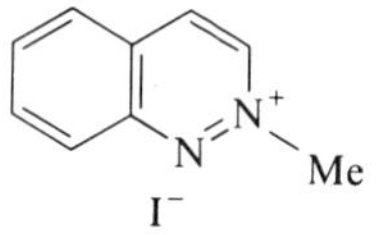

(16) 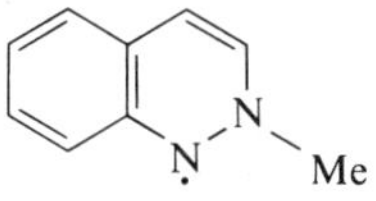   (17)

two AP's of X^+ will be equal to the difference in the heats of formation of the neutral molecules. For example, the $M^+ - CH_3$ ions from (18) and (19) have AP's 9.593 ± 0.006 and 9.448 ± 0.008 respectively, giving a Δ value of 0.145 ± 0.013 eV (14.0 ± 1.3 kJ mol^{-1}). It is argued that the less stable isomer [*i.e.* (19)] has the lower AP owing to the release of energy of non-bonded interactions in formation of (20). The Δ value obtained in fact corresponds closely to the difference (14.0 kJ mol^{-1}) between the ΔH_f values for neutral (18) and (19). Other examples

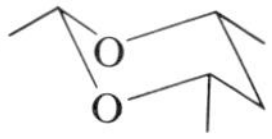

(18) 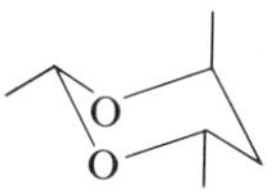   (19) 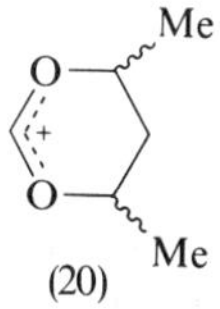   (20)

are cited and it is concluded that AP measurements may be a valuable tool in conformational analysis. However, it should be emphasized that very accurate AP measurements are required and the limits of error claimed in this publication[22] (as little as ±0.002 eV) are much narrower than those normally quoted for onset potentials determined by the semi-log plot with only one internal standard.

Bond Energies and Heats of Formation.—By use of AP data from the spectra of some organosilanes and by the use of an empirical bond-energy scheme, heats of formation have been determined for the species R, R^+, R_2, and RSiMe$_3$ (R = Ph$_n$Me$_{3-n}$Si).[23] The authors neglect excess energies of the fragments, which is probably a valid assumption for the direct bond cleavages involved. Particularly significant are the results concerning the Si—Si bond dissociation energies in the disilyls (21). For Me$_3$SiSiMe$_3$ the Si—Si bond dissociation

$(R^1R^2R^3Si)_2$

(21) R's = Me or Ph

energy is 280 ± 8 kJ mol^{-1}, and *increases* on replacement of methyl by phenyl to 377 ± 25 kJ mol^{-1} in Ph$_3$SiSiPh$_3$. This trend is in marked contrast to the ethane series where the central C—C bond dissociation energy decreases from 368 kJ mol^{-1} in C$_2$H$_6$ to 50 kJ mol^{-1} in Ph$_3$CCPh$_3$. There are several contributing explanations for this result:[23] (i) steric crowding is less in $Ph_3SiSiPh_3$ compared with Ph_3CCPh_3 (the Si—Si distance is 0.23 nm compared with the C—C distance of 0.16 nm), (ii) as aryl groups draw out the p-character in the Si orbitals, they must increase the s-character of the Si—Si bond, thereby strengthening it, and (iii) whereas $Ph_3C\cdot$ may gain stability by assuming a planar configuration, the $Ph_3Si\cdot$ radical probably retains a pyramidal configuration.

The C—N bond energies in compounds (22) and (23) have been determined using equation 7, where $D(X-Y)$ is the X—Y bond dissociation energy.[24] The

$$\begin{array}{cc} \underset{\underset{(22)}{NH_2\ \ NH_2}}{CH_2CH_2CH_2} & \underset{\underset{(23)}{NH_2\ NH_2}}{MeCH-CH_2} \end{array}$$

equation neglects excess energies. It was found that the primary C—N bond energy in (22) was 1.0 eV (97 kJ mol^{-1}) greater than the secondary C—N bond energy in (23).

$$AP(X^+) = D(X-Y) + IP(X\cdot) \qquad (7)$$

Other recent studies of ionic heats of formation by mass spectrometry have included the following: various forms of the molecular ion of HNCO,[25] the $C_4H_8OS^{+\cdot}$ ion formed *via* direct ionization of (24) and *via* fragmentation,[26] and other ions formed from sulphur–oxygen heterocycles.[26,27] Other bond-energy studies by mass spectrometry include: energies of bonds broken in the ionic fragmentation of stereoisomers of the D-homoestrane series,[28] bond energies

(24)

in $MeCF_3$ and other fluoroethanes,[29] Si—O dissociation energies in some siloxanes,[30] bond dissociation energies in phenylboranes,[31] and the C—F bond energy in C_2F_4 from negative-ion work.[32]

[23] J. M. Gaidis, P. R. Briggs, and T. W. Shannon, *J. Phys. Chem.*, 1971, **75**, 974.
[24] M. Hertzberg, G. White, R. S. Olfky, and F. E. Saalfeld, *J. Phys. Chem.*, 1972, **76**, 60.
[25] D. J. Bogan and C. W. Hand, *J. Phys. Chem.*, 1971, **75**, 1532.
[26] G. Condé-Caprace and J. E. Collin, *Org. Mass Spectrometry*, 1972, **6**, 341.
[27] G. Condé-Caprace and J. E. Collin, *Org. Mass Spectrometry*, 1972, **6**, 415.
[28] V. I. Zaretskii, V. L. Sadovskaya, N. S. Wulfson, V. F. Sizoy, and V. G. Mermison, *Org. Mass Spectrometry*, 1971, **5**, 1179.
[29] J. M. Simmie and E. Tschuikow-Roux, *Internat. J. Mass Spectrometry Ion Phys.*, 1971, **7**, 41.
[30] J. Borossay, B. Csákvári, and L. Szepes, *Internat. J. Mass Spectrometry Ion Phys.*, 1971, **7**, 47.
[31] R. H. Cragg, D. A. Gallagher, J. P. N. Husband, G. Lawson, and J. F. J. Todd, *Chem. Comm.*, 1970, 1562.
[32] J. C. J. Thynne and K. A. G. MacNeil, *Internat. J. Mass Spectrometry Ion Phys.*, 1970, **5**, 329.

Kinetic and Energetic Studies of Organic Ions 43

Methods for Determination of Onset Potentials.—Several methods for determination of IP's and AP's are currently in use, and in the literature covered by these reports the semi-logarithmic-plot method[33] and the extrapolated-voltage-difference (EVD) method (Warren's method)[34] have been used with approximately equal frequency. Minor modifications of the above techniques have been employed and several other methods have been used, depending on the accuracy required and the specialized instrumentation available. Various substances (e.g. Ar, Kr, Xe, CS_2, benzene, and toluene, whose IP's are accurately known) have been used as energy standards. Widely differing errors in measured onset potentials have been reported.

The semi-log method[33] requires that the log of the ion current (normalized to 50 eV) for the sample and standard should be parallel (or nearly parallel) over one or two orders of magnitude of ion current intensity.[2] In a typical example[28] linearity was reported over the range from 10^{-4} to 10^{-2} of the 50 eV intensity, with an error of ± 0.1 eV and with Ar and Xe as standards. The curves were not parallel in some cases (especially for fragment ions) and it was found necessary to complement the results by using Warren's EVD technique.[34] At least two standards would seem to be necessary in onset potential measurements, preferably with IP's close to that being measured. Using the semi-log plot, the literature indicates that accuracies of much better than 0.02 eV should be regarded as unlikely, even with the use of two standards. Accuracies better than 0.05 eV have been reported[16,28] using Warren's method.

The semi-log method has been used for the determination of the apparent IP of binary mixtures of the isomeric compounds (25) and (26).[35] Parallel plots were found for different compositions in the gas phase, but no linear relationship between recorded IP and composition was found. The IP of (25) was determined as 6.8 eV, 0.8 eV lower than that of (26). The reason for the lower IP of (25) is considered to be the stable molecular ion (27).

$^+NMe_3$ / O^- NMe_2 / OMe $^+NMe_3$ / $O^\cdot$

(25)   (26)   (27)

A rapid method (see, for example, ref. 10) for onset potential measurements may be employed, known as Kiser's energy compensation technique. The onset potential is taken as that electron voltage at which the abundance of the ion is 0.1% of its 50 eV value, referred to the 0.1% value of a standard. This method in effect gives the same result as that obtained with the semi-log plot when the curves are parallel, the difference being that only one point on the curve is measured in the 0.1% method. Another rapid method has been described[36]

[33] F. P. Lossing, A. W. Tickner, and W. A. Bryce, *J. Chem. Phys.*, 1951, **19**, 1254.
[34] J. Warren, *Nature*, 1950, **165**, 810.
[35] G. Hvistendahl and K. Undheim, *Org. Mass Spectrometry*, 1972, **6**, 217.
[36] G. D. Flesch and H. J. Svec, *Internat. J. Mass Spectrometry Ion Phys.*, 1972, **9**, 106.

which defines a unique point on the low-energy portion of the ionization efficiency curve. This point is defined such that a 1 eV increase in electron-beam energy produces a ten-fold increase in ion intensity. The two points for the unknown and calibrant are then compared.

The critical-slope method of Honig[37] has also been used for IP measurements.[10,35] In one case[35] the results were identical with those found by the semi-log plot and in the other case[10] a precision of 0.1 eV was reported.

There are a number of other graphical methods that have been used in the recent literature for onset potential determinations from impact with electrons having non-homogeneous energy, but space only permits brief reference here. The method of linear extrapolation has been employed[24,30] and also the method of vanishing currents.[30] AP's of product ions from dissociative electron-capture processes have been determined on a time-of-flight mass spectrometer[13] by the energy-distribution-difference (EDD) method of Winters et al.[38] Errors down to 0.05 eV have been claimed for the EDD method.[39] The determination of AP's from second differential ionization efficiency curves has been reported and possible errors have been discussed in terms of the threshold law.[40] The deconvolution of ionization efficiency curves by the EDD technique and by Fourier transforms has been examined with a view to improving AP measurements.[41]

Special methods have been developed which utilize a narrow electron-energy spread. For example, AP's have been obtained using a two-stage double-hemispherical electron-energy selector attached to a quadrupole mass analyser.[17] New IP's for hydrocarbon radicals have been reported using a monoenergetic electron beam with a width at half-height of about 0.07 eV.[19]

In concluding this sub-section, it should be emphasized that even if the experimental errors in onset potential measurements are small (<0.1 eV), contributions from other discrepancy factors may be large (see pp. 34—37).

Stevenson's Rule.—In the fragmentation of alkanes in the mass spectrometer, Stevenson noted that the positive charge remained on the more substituted fragment,[42] i.e. the one with the lower IP. This principle may be extended to all mass spectral fragmentations, with the generalization that in the dissociation of the ion $AB^{+\cdot}$ the positive charge will remain on the fragment (A or B) of lower IP.

This principle is demonstrated in Figure 4, which illustrates the energy changes involved in production of A^+ and B^+ from the molecule AB.[2] Reverse activation energies and excess energies are neglected. In Figure 4, the formation of A^+ has the lower energy requirement and A· has the lower IP. It should be noted that the rule does not predict which of all the possible fragmentation reactions

---

[37] R. E. Honig, *J. Chem. Phys.*, 1948, **16**, 105.
[38] R. E. Winters, J. H. Collins, and W. L. Courchene, *J. Chem. Phys.*, 1966, **45**, 1931.
[39] C. Lageot, *Org. Mass Spectrometry*, 1971, **5**, 839, 845.
[40] G. G. Meisels and B. G. Giessner, *Internat. J. Mass Spectrometry Ion Phys.*, 1971, **7**, 489.
[41] B. G. Giessner and G. G. Meisels, *J. Chem. Phys.*, 1971, **55**, 2269.
[42] D. P. Stevenson, *Discuss. Faraday Soc.*, 1951, **10**, 35.

Kinetic and Energetic Studies of Organic Ions 45

should be the most prevalent, but indicates in which direction the charge will go for scission of a given bond.

Harrison et al.[43] have illustrated the application of this rule with a large number of examples. In all cases where the radical IP's differ by more than

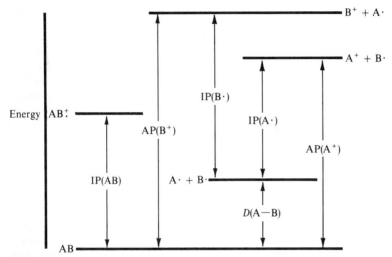

Figure 4 *The energy changes involved in the ionization and dissociation of a molecule* AB

0.3 eV the species of lower IP dominates in the mass spectrum. For example, the relative abundances of the ions CMe_3^+ and CH_2OH^+ in the spectrum of (28) are 100 and 7.4% respectively and the IP's of the corresponding radicals are 6.93 and 7.6 eV. In the spectrum of (29) however, CMe_3^+ and $MeCH(OH)^+$ have relative abundances of 100 and 79% respectively and the IP of $MeCH(OH)\cdot$ is 6.9 eV.

$$\text{Me}-\underset{\underset{\text{Me}}{|}}{\overset{\overset{\text{Me}}{|}}{\text{C}}}-\text{CH}_2\text{OH} \qquad \text{Me}-\underset{\underset{\text{Me}}{|}}{\overset{\overset{\text{Me}}{|}}{\text{C}}}-\text{CH}\underset{\text{OH}}{\overset{\text{Me}}{\diagup}}$$

(28) (29)

Apparent contradictions to the rule have been found in the spectra of some oxime ethers.[44] The molecular ion (30) forms (31) rather than (32) in a metastable process. One explanation is that either or both of the reactions (30) → (31) and (30) → (32) do not form the ions or neutrals with the structures indicated.

---

[43] A. G. Harrison, C. D. Finney, and J. A. Sherk, *Org. Mass Spectrometry*, 1971, **5**, 1313.
[44] R. G. Cooks and A. G. Varvoglis, *Org. Mass Spectrometry*, 1971, **5**, 687.

Charge localization arguments have been applied to the spectra of aminoketones.[45] The spectrum of (33) provides an instance of the McLafferty rearrangement in which the charge remains on the olefinic fragment [see

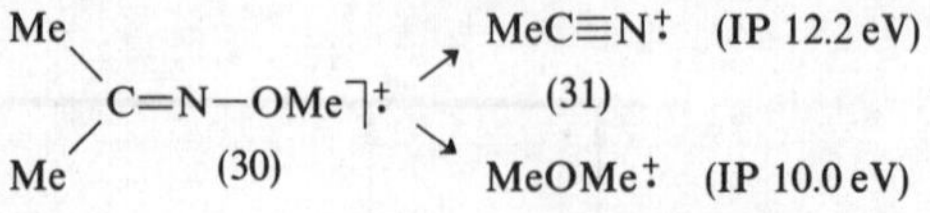

$$\begin{matrix} \text{Me} \\ \diagdown \\ \diagup \\ \text{Me} \end{matrix} \text{C=N--OMe} \Big]^{\ddagger} \diagup^{\diagdown} \begin{matrix} \text{MeC}{\equiv}\text{N}^{\ddagger} & (\text{IP } 12.2 \text{ eV}) \\ (31) \\ \text{MeOMe}^{\ddagger} & (\text{IP } 10.0 \text{ eV}) \\ (32) \end{matrix}$$
(30)

(33) → (34)]. The enol would be expected to have the higher IP in this case. It is argued that this reaction is triggered by an electron-deficient carbonyl group, but this suggestion would not seem to be supported by the observation that the rearrangement persists at ionizing voltages below the IP of the carbonyl group.

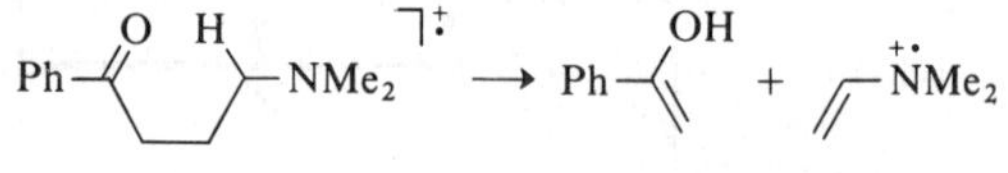

$$\text{Ph}{-}\underset{(33)}{\overset{\text{O}\quad\text{H}}{\diagup\diagdown\diagup\diagdown}}{-}\text{NMe}_2 \Bigg]^{\ddagger} \longrightarrow \text{Ph}{-}\underset{(34)}{\overset{\text{OH}}{\diagup\diagdown}} + \overset{+\cdot}{\diagup\diagup}{-}\text{NMe}_2$$

Charge localization arguments have also been applied to the spectra of some substituted thioureas,[46] and substituted purine and pyrimidine nucleoside analogues.[47] Arguments against a charge localization approach have been forwarded, with evidence from the fragmentation of methionine and selenomethionine.[48] It should be noted in conclusion that charge localization and QET approaches are not necessarily incompatible.[1]

3 The Quasi-Equilibrium Theory

Many of the publications reported in this chapter employ the QET to interpret mass spectral features. The QET has been used to varying degrees of complexity. For example, there are publications (largely in the physical chemistry journals) which deal with the more theoretical aspects of the QET and its development.[15,49—52] The mathematics of such papers are often complex and largely beyond the scope of this chapter, although mass spectral calculations have been performed for large organic ions using the more exact rate equations and

[45] P. J. Wagner, *Org. Mass Spectrometry*, 1970, **3**, 1307.
[46] M. A. Baldwin, A. M. Kirkien, A. G. Loudon, and A. Maccoll, *Org. Mass Spectrometry*, 1970, **4**, 81.
[47] W. F. A. Grose, T. A. Eggelte, and N. M. M. Nibbering, *Org. Mass Spectrometry*, 1971, **5**, 833.
[48] T. W. Bentley, R. A. W. Johnstone, and F. A. Mellon, *J. Chem. Soc. (B)*, 1971, 1800.
[49] K. H. Lau and S. H. Lin, *J. Phys. Chem.*, 1971, **75**, 981.
[50] C. E. Klots, *J. Phys. Chem.*, 1971, **75**, 1526.
[51] R. D. Finney and G. G. Hall, *Internat. J. Mass Spectrometry Ion Phys.*, 1970, **4**, 489.
[52] P. F. Knewstubb, *Internat. J. Mass Spectrometry Ion Phys.*, 1971, **6**, 217, 229.

are reported below. In contrast, there are the large numbers of publications which merely employ general consequences of the QET to explain mass spectral features. Such publications are widely reported in other sections of this chapter (in particular, see Sections 4—7). In between these extremes are those publications employing QET calculations which are simplified or modified, while preserving correct general principles. This section deals with this last-mentioned class of papers. In addition, publications are covered dealing with energy distributions and energy states of ions in the mass spectrometer.

Calculations.—The differences between high- and low-voltage spectra in terms of the QET have been frequently expressed in the recent literature. In particular, the property has often been employed that rearrangement reactions tend to predominate over direct cleavage reactions as the electron-beam energy is decreased (see also Section 4). This principle may be expressed neatly[53] in terms of the simplified equation of the QET (equation 8), but publications[10,54] in the recent literature have emphasized that the more exact form of the QET can be applied to large organic ions.

$$k = v\left(\frac{E - E_0}{E}\right)^{s-1} \qquad (8)$$

The molecular ion of the dimethylthiocarbamate (35) forms abundant ions (36) and (37), the latter involving rearrangement.[54,55] As predicted by equation 8, (37) increases dramatically compared with ion (36) in low-energy electron impact.

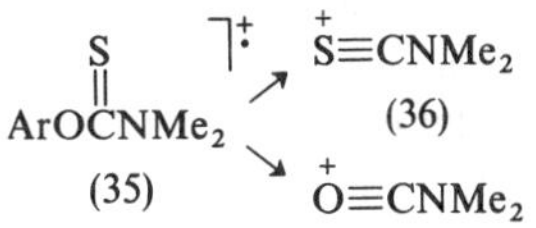

$$\begin{array}{c} S \\ \| \\ ArOCNMe_2 \\ (35) \end{array} \begin{array}{c} \overset{+}{\rceil} \\ \nearrow \\ \searrow \end{array} \begin{array}{c} \overset{+}{S \equiv CNMe_2} \\ (36) \\ \overset{+}{O \equiv CNMe_2} \\ (37) \end{array}$$

According to the QET, the rate constant for dissociation of an isolated ion with internal energy E can be expressed as shown in equation 9,[54] where h is Planck's constant and $W(E)$ is the number of states of the ion with energy less than and equal to E. The symbol $\neq$ refers to the activated complex.

$$k(E) = \frac{1}{h} \frac{W^{\ne}(E - E_0)}{dW(E)/dE} \qquad (9)$$

If rearrangement and direct cleavage reactions occur with respective rate constants k_r and k_d, then

$$\frac{k_r(E)}{k_d(E)} = \frac{W_r^{\ne}(E - E_0^r)}{W_d^{\ne}(E - E_0^d)} \qquad (10)$$

[53] D. H. Williams and R. G. Cooks, *Chem. Comm.*, 1968, 663.
[54] J. C. Tou, *J. Phys. Chem.*, 1971, **75**, 1903.
[55] J. C. Tou and R. M. Rodia, *Org. Mass Spectrometry*, 1972, **6**, 493.

The activated complex of a rearrangement reaction involves new bond formation and hence some vibrational frequencies will increase and some internal rotations will be stopped. This type of activated complex is called a tight complex. In a direct cleavage reaction, the activated complex (known as a loose complex) involves stretching of a bond along the reaction co-ordinate. Some vibrational frequencies will decrease and certain torsional and skeletal vibrations might change to internal rotations in the activated complex.[54]

Experimental data frequently show that $E_0^r < E_0^d$, and therefore $k_r/k_d > 1$ at low energy (above E_0^r). However, $W_r^{\ddagger}$ increases much more slowly than $W_d^{\ddagger}$ because the energies of the vibrational and rotational states are, on average, lower for the loose complex. Hence the excess energy above E_0 can be distributed over more extra states (compared with the tight complex) as E is increased.

At high energy, $k_r/k_d \to W_r^{\ddagger}(E)/W_d^{\ddagger}(E)$. Hence $k_r/k_d < 1$, since $W_r^{\ddagger}(E)$ is always less than $W_d^{\ddagger}(E)$. That is, the tight complex has a lower frequency factor than the loose complex,[54] as employed in the applications of equation 8.

The situation at high and low energy is shown diagrammatically in Figure 5 (see refs. 10 and 54). For simplicity, the high- and low-energy distributions are assumed to have shapes H and L respectively. When $P(E)$ shifts from high to low energy, the distribution of the rate ratio k_r/k_d shifts to the region of higher values, and increasing numbers of rearrangement ions are produced compared with those formed by direct cleavage.

McLafferty et al. have employed the RRKM theory to account for some substituent effects in 1,2-diphenylethanes (1).[10] An exact enumeration of states was employed (similarly to equation 9). The procedure of estimating frequency factors (see equation 8) is replaced in effect by estimating the changes in relative number of vibrational and rotational states in passing from the active molecular ion to the activated complex. In rearrangement reactions rotational states are frozen out in the activated complex, so that the right configuration for reaction may be attained. In the decomposition of (38) to $H_2NC_7H_6^+$, a relatively tight activated complex for a single-bond cleavage is suggested,[10] since the reaction appears to have a slower rise of k with E than the corresponding benzylic cleavage in 1,2-diphenylethane itself. In the transition state (39) the orbital

H_2N—⟨⟩—CH_2—CH_2—⟨⟩]‡ → $H_2\overset{+}{N}$=⟨⟩=$CH_2 \cdots \dot{C}H_2$—⟨⟩

(38)   (39)

↓

$H_2NC_7H_6^+$

overlap which is necessary to minimize the activation energy freezes out the N—C and one C—C rotation which are possible in the active molecular ion. This example falls within a general scheme which requires a slower rise of k with E for reactions involving a restricted geometry in the transition state.

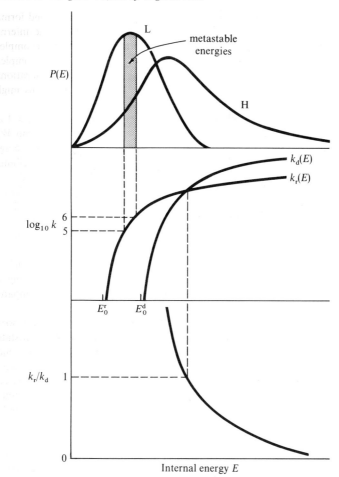

Figure 5 *Rate and energy curves for rearrangement and direct cleavage reactions*

It is well established that equation 8 yields too shallow a rise of k with E near threshold. This fact should not compromise sound qualitative conclusions, and it is possible to obtain semi-quantitative data by modifying the equation.[56] For example, in the mass spectrum of methyl salicylate (40), the increase with increasing electron energy in the abundance ratio $[M^+ - \text{OMe}]:[M^+ - \text{MeOH}]$ would be explained on the basis of a lower frequency factor and activation energy for the rearrangement reaction.

[56] A. N. H. Yeo and D. H. Williams, *J. Amer. Chem. Soc.*, 1970, **92**, 3984.

Yeo and Williams have put this argument on a semi-quantitative basis by assuming realistic values for energy distributions and rate functions and computing ion abundances using these functions.[56] Equation 8 is modified to give

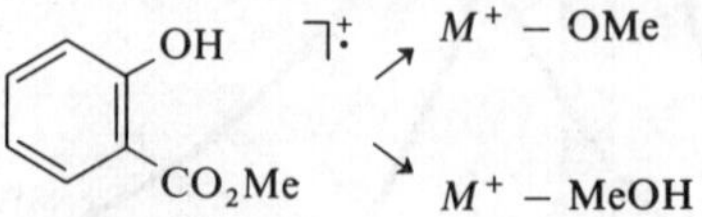

$$\underset{(40)}{\underset{CO_2Me}{\bigodot}OH} \rceil^{\ddagger} \quad \nearrow M^+ - OMe \\ \searrow M^+ - MeOH$$

equation 11. The parameter x is varied from 0.2 at threshold to 0.5 at the maximum internal energy of M^+. This modification has the effect of producing a sharper rise of k with E near threshold. If a v value of 4×10^{13} s^{-1} is adopted for the simple cleavage reaction and 6×10^{11} s^{-1} for the rearrangement, good agreement is found between observed and calculated spectra for electron-beam energies between 12 and 20 eV. Other similar examples are reported.[56]

$$k = v\left(\frac{E - E_0}{E}\right)^{(s-1)x} \qquad (11)$$

Benezra and Bursey have employed a simplified QET equation to investigate *ortho*-effects upon the elimination of CH$_2$CO from halogenated phenyl acetates (41) and acetanilides (42).[57] By adjustment of parameters to give a best fit with observed spectra at a series of electron-beam energies, values of v were compared for keten loss from the *o*- and *p*-compounds for each of the series (41) and (42).

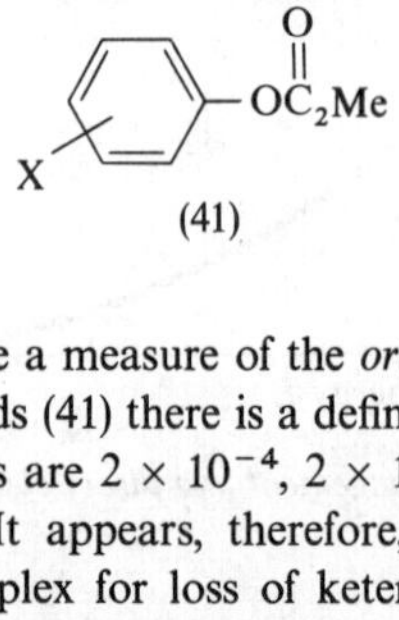

$$\underset{(41)}{X\text{-}\bigodot\text{-}OC_2Me} \qquad \underset{(42)}{X\text{-}\bigodot\text{-}NHCMe}$$
(with C=O groups)

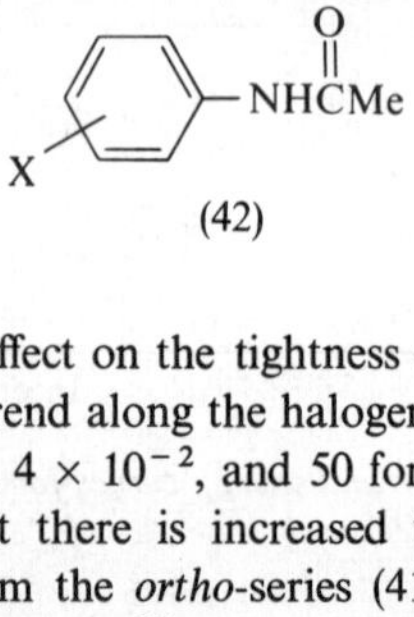

v_o/v_p should be a measure of the *ortho*-effect on the tightness of the complex.[57] For compounds (41) there is a definite trend along the halogen series. At 20 eV the v_o/v_p ratios are 2×10^{-4}, 2×10^{-3}, 4×10^{-2}, and 50 for F, Cl, Br, and I, respectively. It appears, therefore, that there is increased tightening in the activated complex for loss of keten from the *ortho*-series (41) as the halogen becomes more electronegative. It is suggested[57] that the *ortho*-effect in series (41) is predominantly a solvation effect or something akin to this, not a steric effect. For dihalogenosubstituted phenyl acetates similar *ortho*-effects were found.[58] For the acetanilides (42) the situation is complicated and the *ortho*-effect appears to incorporate other factors.[57]

Determination of frequency factors for loss of OMe from the ionized methyl ethers (43)—(45) to produce the C$_7$H$_7^+$ ion suggests that the reaction is a direct

---

[57] S. A. Benezra and M. M. Bursey, *J. Chem. Soc. (B)*, 1971, 1515.
[58] S. A. Benezra and M. M. Bursey, *Z. Naturforsch.*, 1972, **27a**, 670.

Kinetic and Energetic Studies of Organic Ions   51

cleavage in all cases and that the molecular ions have retained structural integrity prior to dissociation.[59] $C_7H_7^+$ ions have been studied kinetically in a field ionization source and these ions are formed in the benzyl structure for very fast decompositions.[60]

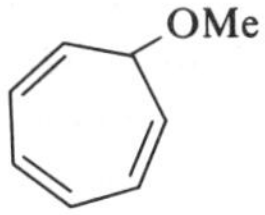

    (43) 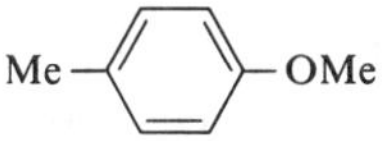   (44) 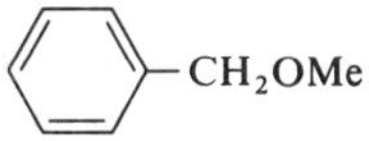   (45)

(43) cycloheptatriene–OMe; (44) Me–C$_6$H$_4$–OMe; (45) C$_6$H$_5$–CH$_2$OMe

In many mass-spectra consecutive reactions occur, especially at high electron-beam energies, and this situation has been treated using the modified QET equation 11.[61] Mass spectra of seven aromatic compounds were used as the models, in which the molecular ions each undergo two consecutive reactions $M^{+\cdot} \rightarrow A^+ \rightarrow B^+$ [e.g. (46) → (47) → (48)]. The following factors must be considered:[61] (i) the energy distribution of the molecular ion, (ii) the k vs. E

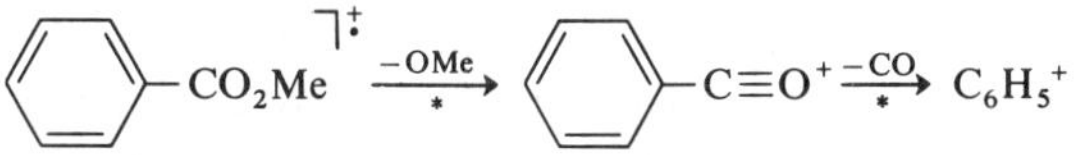

$$[C_6H_5\text{-}CO_2Me]^{+\cdot} \xrightarrow{-OMe^\cdot} [C_6H_5\text{-}C{\equiv}O]^+ \xrightarrow{-CO} C_6H_5^+$$

    (46)   (47)   (48)

curves applicable to the decomposition of $M^{+\cdot}$ and A^+, and (iii) the partition of internal energy between A^+ and the neutral $(M - A)$. Acceptable approximations were adopted for these three requirements and good agreement was found between observed and calculated spectra, including metastable peak intensities. The authors point out that the calculated spectra are relatively insensitive to frequency factors adopted, and compounds undergoing competing reactions[56] should be studied for estimations of frequency factors. The mass spectrum of thiacyclobutane (49) at 70 eV has been calculated using the QET.[62] The results

(49) thietane (four-membered ring with S)

indicate that the calculated ion abundances are not very sensitive to changes in mechanism and, unless reaction pathways are clearly marked by metastables, caution should be exercised in the interpretation of such calculations.

The QET has been applied for the first time to negative ions in the mass spectrometer.[63] The dissociation rate constants, as a function of internal energy of the parent negative ions (SF_6^- and a number of fluorocarbon anions) were computed for the processes observed and used to estimate kinetic shifts and appearance

---

[59] M. K. Hoffmann and M. M. Bursey, *Chem. Comm.*, 1971, 824.
[60] K. Levsen and H. D. Beckey, *Internat. J. Mass Spectrometry Ion Phys.*, 1972, **9**, 63.
[61] A. N. H. Yeo and D. H. Williams, *Org. Mass Spectrometry*, 1971, **5**, 135.
[62] J. R. Gilbert and A. J. Stace, *Org. Mass Spectrometry*, 1971, **5**, 1119.
[63] C. Lifshitz, A. M. Peers, R. Grajower, and M. Weiss, *J. Chem. Phys.*, 1970, **53**, 4605.

potentials. The results (*e.g.* metastable intensities and temperature effects) were in agreement with the QET.

According to the QET, the rate constant for decomposition forms a continuous function of internal energy. This has often been verified, *e.g.* breakdown curves (graphs of average rate constants as a function of time) have been derived for a series of field-ionized molecules.[64] In accordance with the QET, the shape of the breakdown curves can be explained on the basis of a continuous spectrum of rate constants. Decay curves of product-ion intensity *vs.* lifetime have been obtained for 20 metastable transitions using a cycloidal mass spectrometer.[65] The curves provide evidence for a continuum of rate constants for gaseous unimolecular reactions. The observed and calculated peak shapes (using the QET) of fragment ions caused by rearrangement reactions in a field ionization source have been found to agree qualitatively.[66]

The first unambiguous direct measurement of $k(E)$ functions has been reported using the technique of charge exchange ionization.[67,68] Ions from the benzonitrile, benzene, and thiophen molecules were produced in well-defined states of excitation by charge exchange with Xe^+, Kr^+, Ar^+, CO^+, and N_2^+. The rate constants k for unimolecular decomposition were found to be monotonically increasing functions of the excitation energy. In the case of the benzene molecular ion, the reactions forming the $C_6H_5^+$ and $C_4H_4^{+\cdot}$ daughter ions were found not to be in competition, in contradiction with the basic tenet of the QET. However, this particular contradiction has been critically discussed.[15] Charge exchange-induced cleavage of some *N*-phenylbenzamides, phenyl benzoates, and benzophenones to produce benzoyl ions has also led to the suggestion of isolated states in the molecular ion, or the occurrence of radiative transitions.[69] The photon-emission argument was invoked as a possible explanation for high-energy molecular ions which fail to decompose.[70]

It is evident from some of the publications reported in this section that a general word of caution should be sounded about QET calculations upon polyatomic organic ions. Some of the parameters incorporated into the calculations are of an empirical nature. Consequently there often remains a lot of leeway for adjustment of parameters to achieve agreement between calculation and observation. Data taken at multiple electron energies can help to alleviate this problem. Too little is known about the energy states and structures of molecular ions. In calculations employing exact enumerations of states to evaluate rate curves, arbitrary assumptions are necessary about the vibrational and rotational states of the organic ion and the activated complex. In addition, there is no reliable method available at this time for obtaining electron-impact

[64] G. Tenschert and H. D. Beckey, *Internat. J. Mass Spectrometry Ion Phys.*, 1971, **7**, 97.
[65] R. B. Fairweather and F. W. McLafferty, *Org. Mass Spectrometry*, 1970, **4**, 221.
[66] K. Levsen and H. D. Beckey, *Internat. J. Mass Spectrometry Ion Phys.*, 1971, **7**, 341.
[67] B. Andlauer and C. Ottinger, *J. Chem. Phys.*, 1971, **55**, 1471.
[68] B. Andlauer and C. Ottinger, *Z. Naturforsch.*, 1972, **27a**, 293.
[69] J. Turk and R. H. Shapiro, *Org. Mass Spectrometry*, 1972, **6**, 189.
[70] J. Turk and R. H. Shapiro, *Org. Mass Spectrometry*, 1971, **5**, 1373.

energy distributions (see below) for polyatomic ions without resort to empirical ion abundances. Nevertheless, valuable contributions to the understanding of mass spectra have been made by some of the QET calculations reported above.

Energy Distributions and Energy States.—The electron-impact energy distributions employed for molecular ions in the QET calculations reported above are in most cases semi-empirical. A considerable number of such distributions have been employed, some of which are smoothed and others discontinuous. Other techniques have been used which have some theoretical basis: (i) the second derivative of the electron-impact ionization efficiency curve and (ii) the first derivative of the photon-impact total ionization efficiency curve. The threshold laws which validate these methods are only approximations. Attempts have been made to construct semi-empirical energy distributions containing fine structure.[10]

The compounds under study were substituted 1,2-diphenylethanes (1).[10] Photoelectron spectra were modified, using the threshold law, thermal energy convolution, and empirical observation of ion abundances. The distributions incorporate electronic fine structure which has been used to interpret temperature effects on metastable ion abundances.[10] Although useful information on ionic states can be obtained from photoelectron spectra, it has been suggested that too crude an approximation to the electron-impact energy distribution of molecular ions is obtained on occasions by this method.[71] In passing, it should be noted that the energy distribution for field-ionized molecules has been estimated.[72]

The application of molecular orbital theories to mass spectral processes continues to be strictly limited. Too little is known about the structures and energy states of organic ions produced by electron and photon impact. However, a molecular orbital interpretation has been used to rationalize the spectra of some N-, O-, and S-containing five-membered heterocycles.[73] For example,

$$\begin{array}{c} \text{Ph}\underset{0.37}{\diagdown}\text{O} \\ \text{C} \quad \text{X}_{0.25} \\ 0.63 \mid \quad \text{C—O} \\ \diagup \text{N} \diagdown \diagup_{0.64} \\ \text{Ph} \quad 0.47 \quad \text{N} \end{array} \Biggr]^{\pm} \longrightarrow \begin{array}{c} \text{Ph}\underset{0.83}{\diagdown}\text{O} \\ \text{C} \\ 0.43 \\ \text{N} \quad \text{C}\underset{0.53}{\overset{0.79}{\diagup}}\text{O} \\ \text{Ph} \quad 0.65 \quad \text{N} \end{array} \Biggr]^{\pm} \longrightarrow \text{PhC} \equiv \text{O}^{+}$$

(50)   (51)

the dominant ring-cleavage process in the mass spectrum of 4,5-diphenylisosydnone (50) is the formation of the benzoyl ion. This cleavage shows a strong metastable peak and is thus predicted[73] to be a ground-state process in the doublet molecular ion. The ground-state bond orders suggest that the 1,5 bond should be the first to cleave [see (50) → (51)]. The subsequent formation of

[71] G. Innorta, S. Torroni, and S. Pignataro, *Org. Mass Spectrometry*, 1972, **6**, 113.
[72] K. Levsen and H. D. Beckey, *Internat. J. Mass Spectrometry Ion Phys.*, 1972, **9**, 51.
[73] R. C. Dougherty, R. L. Foltz, and L. B. Kier, *Tetrahedron*, 1970, **26**, 1989.

PhCO⁺ is suggested by bond-order arguments and by the stability of the benzoyl cation.

It had previously been suggested[74] that the molecular ion of isocyanic acid (HNCO) undergoes a slow doublet → quartet spin-forbidden predissociation. This suggestion accounts for the occurrence of a metastable peak for formation of HCO⁺ from this small molecular ion. It has now been proposed on the basis of HNDO calculations that a slow transition from an open chain to a cyclic configuration is possible in HNCO⁺˙ and can account for the observed spectrum and metastable peak.[25]

Several authors have drawn analogies between mass spectral and photochemical processes and commented upon the electronic states involved. Correlations between the Norrish Type II photochemical process and the mass spectral McLafferty rearrangement have been investigated for a number of aromatic ketones.[75] Generally the McLafferty rearrangement occurs even when the Norrish Type II process does not, but a low quantum yield for the photochemical reaction usually indicates a weak ion for the mass spectral rearrangement. It is suggested[75] that states corresponding to ionization of the carbonyl function are populated in the molecular ion, even though the ionization potentials suggest that the lowest electronic state does not correspond to ionization of the carbonyl group. For example, acetone and anthracene have respective ionization potentials of 9.7 and 7.2 eV, but the McLafferty rearrangement in 2-butyrylanthracene (52)

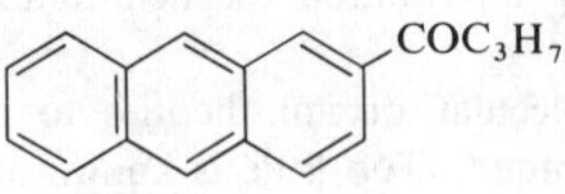

(52) — 2-butyrylanthracene with COC_3H_7 substituent

occurs with an intensity of 6% of the molecular ion intensity at 70 eV. However, the fact that the photochemical rearrangement can occur in ketones demonstrates that charge localization on carbonyl is not a prerequisite for rearrangement. Among other comparisons made with mass spectral processes have been thermolytic fragmentations in benzotriazoles[76] and photochemical rearrangements in isoxazoles.[77]

Qualitative correlations have been attempted for temperature effects in the mass spectra of eight methyl ketones.[78] Simplified equations have been derived relating the daughter/parent ion ratio to the average energy of the molecular ion. Despite some very tentative simplifications, such as incorporation of average energies into the simplified QET equation 8 and the use of derivations

[74] C. G. Rowland, J. H. D. Eland, and C. J. Danby, *Chem. Comm.*, 1968, 1535.
[75] M. M. Bursey, D. G. Whitten, M. T. McCall, W. E. Punch, M. K. Hoffmann, and S. E. Benezra, *Org. Mass Spectrometry*, 1970, **4**, 157.
[76] M. Ohashi, K. Tsujimoto, A. Yoshino, and T. Yonezawa, *Org. Mass Spectrometry*, 1970, **4**, 203.
[77] T. Nishiwaki, *Org. Mass Spectrometry*, 1971, **5**, 123.
[78] J. Julien and J. M. Pechine, *Org. Mass Spectrometry*, 1970, **4**, 325.

from a steady-state approximation, correlations are obtained which fit the derivations.

4 Tight Transition States

In this section, only those publications are reported in which kinetic or energetic evidence for a rearrangement or anchimeric assistance is produced. The reader is referred to Chapter 3 for a full coverage of rearrangement reactions. Sections 5 and 6 of this chapter also contain reports of rearrangement reactions in metastable ions.

Rearrangement Reactions.—It has now been firmly established that differences between the kinetic properties of rearrangement and direct cleavage reactions can be illustrated by differences between high-and low-voltage spectra (see pp. 47—50) and also by metastable ion abundances.[1] These properties have been substantiated for eight aromatic compounds.[79] In all cases the abundance of the rearrangement reaction from the molecular ion increases relative to that of the direct cleavage reaction as the electron-beam energy is lowered. In some cases the effect is dramatic. For example, the molecular ion of diphenyl ether (53) eliminates competitively the neutrals CO and OC_6H_5. At 70 eV the

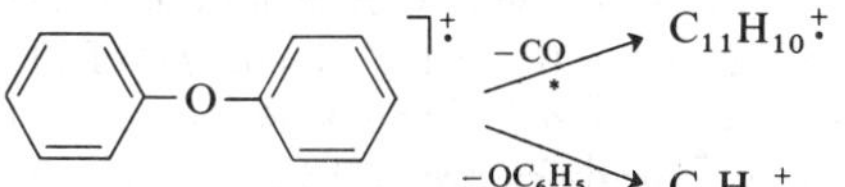

$$\text{(53)} \xrightarrow{-CO} C_{11}H_{10}^{+\cdot}$$
$$\xrightarrow{-OC_6H_5} C_6H_5^+$$

$[C_{11}H_{10}^{+\cdot}]:[C_6H_5^+]$ ratio is 0.82, increasing to 112 at 12 eV. Metastable abundance ratios are also in accord with predictions. In simple terms, the metastable abundance ratio represents the limiting ratio of normal ion abundances as the beam energy is lowered towards threshold. Accordingly, the ratio m^* (rearrangement):m^* (cleavage) is high in all eight cases reported,[79] and in three of these cases no unimolecular metastable for the cleavage reaction is detectable [e.g. for (53)].

Ethylene ketals containing no aromatic groups yield no parent ions in their mass spectra, giving daughter ions by cleavage next to the ketal function. A series of hydroxyethylene ketals showed this behaviour but, in addition, loss of ethylene glycol $(CH_2OH)_2$ occurred in some cases [e.g. from (54)] via a double hydrogen rearrangement.[80] This rearrangement increased in importance at

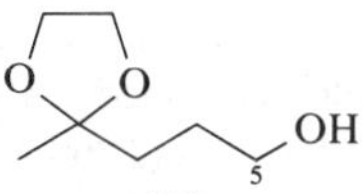

(54)

---

[79] P. Brown, *Org. Mass Spectrometry*, 1970, **3**, 1175.
[80] J. R. Dias and C. Djerassi, *Org. Mass Spectrometry*, 1972, **6**, 385.

low eV. Deuterium labelling established that C-5 is not lost in the rearrangement process from (54).

Because ions spend such a short time ($< 10^{-11}$ s) in a field ionization (FI) source, a reaction having a rate constant less than about 10^{10} s^{-1} will not occur in such a source to a sufficient extent to be observed as a normal daughter ion. It is therefore a consequence of the QET that rearrangement reactions (having low frequency factors) have little chance of being observed as daughter ions in FI spectra, particularly as the molecular ions do not have appreciable excess energy to elevate the rate constant to values greater than 10^{10} s^{-1}. These arguments have again been extensively applied in the literature covered by these reports,[60,66,72,81,82] and FI and EI spectra have been compared to extract information about rearrangements. Chapter 1 gives a full treatment of FI spectroscopy.

Although the above rule is useful and generally applicable, it appears that there are many known instances of 'fast' hydrogen rearrangements as evidenced by abundant FI peaks.[60,81] Elimination of RCO_2H from some aliphatic carboxylic esters can take place in a very short time interval (10^{-12}—10^{-11} s) and in some cases the resulting $C_nH_{2n}^{+\cdot}$ peak is the base peak in the spectrum.[82]

Simple AP and IP measurements can sometimes provide evidence for rearrangement reactions. Possible ring-expansions have been investigated, by means of IP and AP values, in the molecular ions of halogenotoluenes.[83] For example, the differences AP − IP for the elimination of Cl from the molecular ions of chlorobenzene (55) and o-, m-, and p-chlorotoluenes (56) are respectively 3.6, 2.8, 2.9, and 2.7 eV. The difference (~0.8 eV) between the activation energy for (55) and the three isomeric chlorotoluenes (56) is too great to be a substituent effect of the methyl group. A ring-expansion (56) → (57) is therefore suggested, followed by Cl loss with a relatively low activation energy.

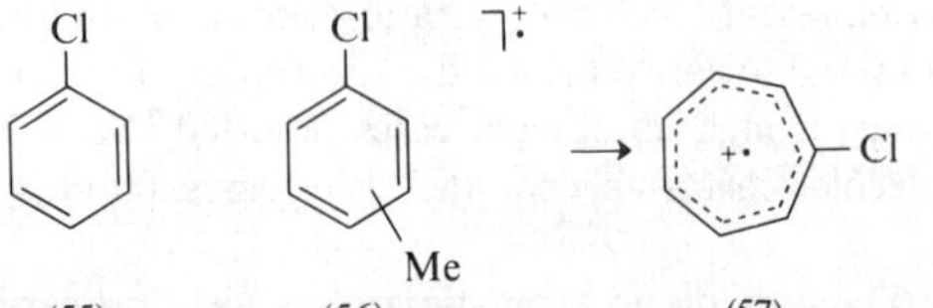

(55)   (56)   (57)

Anchimeric Assistance.—The kinetic properties of rearrangements may also be used to identify reactions which may be formulated as direct cleavages, but which involve anchimeric assistance by another group. The spectra of a series of ring-substituted 5-bromo-2-phenylpent-2-enes (58) have been obtained at several beam energies.[84] In all cases the abundance of the M^+ − Br ion increases relative to that of M^+ − CH_2Br as the electron energy is lowered. This is consistent with a lower frequency factor and tighter transition state for bromine

[81] H. D. Beckey, *Internat. J. Mass Spectrometry Ion Phys.*, 1970, **5**, 182.
[82] P. Schulze and W. J. Richter, *Internat. J. Mass Spectrometry Ion Phys.*, 1971, **6**, 131.
[83] A. N. H. Yeo and D. H. Williams, *Chem. Comm.*, 1970, 886.
[84] K. B. Tomer, J. Turk, and R. H. Shapiro, *Org. Mass Spectrometry*, 1972, **6**, 235.

Kinetic and Energetic Studies of Organic Ions 57

elimination compared with loss of CH_2Br. Anchimeric assistance in the expulsion of Br is therefore strongly indicated, and the fact that Br loss yields the more abundant metastable strengthens the case. As far as the mechanism for the anchimeric assistance is concerned it is proposed that,[84] with substituents on the ring capable of electron donation, the transition state for Br loss resembles a bicyclic ion such as (59), whereas with substituents which destabilize a positive charge the loss of Br proceeds *via* the cyclopropylmethyl-like transition state (60).

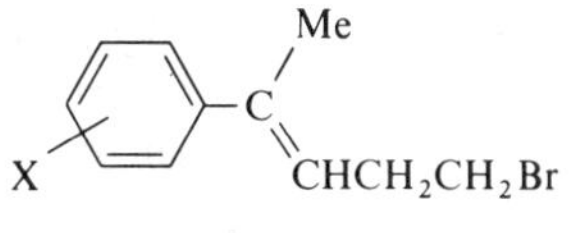

(58) 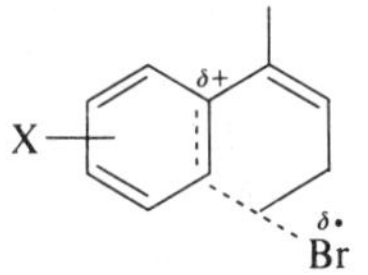   (59) 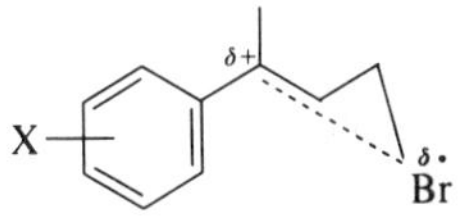   (60)

Similar kinetic data for other allylic and homallylic bromides, along with deuterium labelling and IP and AP data, suggest that Br elimination takes place *via* cyclic transition states with the aid of π-electron density.[84]

Other evidence for anchimeric assistance in the expulsion of bromine is forthcoming from the mass spectra of some 1,2-dibromo-compounds.[85] Competing 'reference' direct cleavage reactions were not available, thus precluding a study of relative daughter-ion abundances at different electron energies. However, relative molecular-ion abundances in a given pair of *cis–trans* isomers were sufficiently different to suggest anchimeric assistance. For example, 1,2-*cis*- and -*trans*-dibromocyclopentanes yield the same breakdown pattern, but the molecular ion in the *cis*-isomer carries 23 times the fraction of the total ion current at 70 eV as that carried by the *trans*-isomer. This suggests that Br is eliminated with the lower activation energy from the *trans*-molecular ion, presumably *via* a transition state (61).

Convincing evidence for aryl participation has been forthcoming from the spectra of a large number of 1-, 4-, and 6-substituted azulenes.[86] The anchimeric assistance occurs only in the fragmentation reactions of the 4-substituted azulenes (62). Some of the 70 eV spectra provide concrete evidence for such participation, without recourse to confirmatory data, such as metastable ion

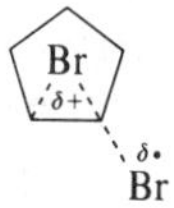

(61) 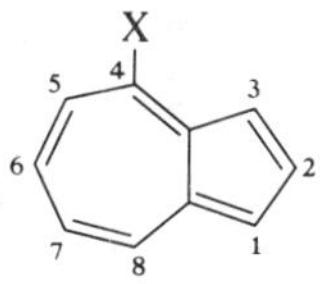   (62)

[85] J. M. Péchiné, *Org. Mass Spectrometry*, 1971, **5**, 705.
[86] R. G. Cooks, N. L. Wolfe, J. R. Curtis, H. E. Petty, and R. N. McDonald, *J. Org. Chem.*, 1970, **35**, 4048.

abundances, low eV spectra and appearance potentials (all of which are in fact provided). When X = CH_2CH_2OTs, for example, the M^+ − Ts peak is less than 0.5% of the base peak for the 1-isomer, is absent for the 6-isomer, but forms the base peak in the 4-isomer (63). In the isomeric 1-naphthyl derivative (65), the M^+ − Ts peak is of negligible intensity. The δ-cleavage reaction in the 4-isomer is therefore assisted by the azulyl ring and the reaction (63) → (64)

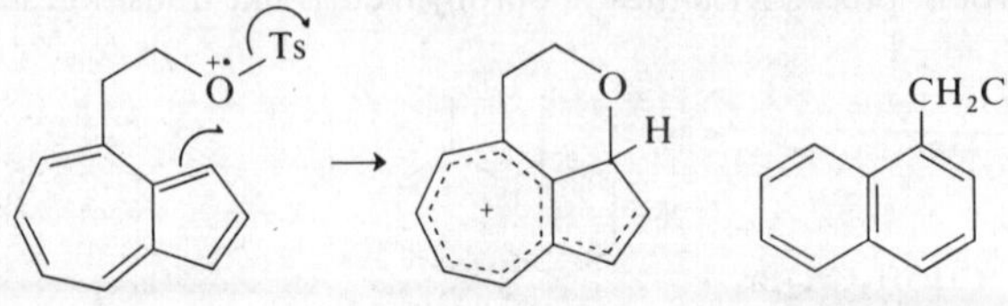

(63)   (64)   (65)

is suggested, forming the stable tropylium-like ion. It has been suggested from some IP and AP data that charge localization on the side-chain is a necessary criterion for the assisted reaction.[86] The evidence comes from the AP's for the product ions, which are similar to the ionization energy of the side-chain function (as gauged from the corresponding monosubstituted aliphatic compound).

5 Metastables

There are several reasons for the extensive use of metastable transitions in the investigation of mass spectral processes: (i) unlike reactions occurring in the ion source, metastable transitions occur from a narrow range of ion energies corresponding to rate constants for decomposition of 10^4—10^6 s^{-1}, (ii) the occurrence, absence, or relative abundance of metastable transitions for different processes gives valuable information on relative activation energies, (iii) information about rearrangement reactions may be obtained by appropriate comparison of metastable and normal ion abundances, (iv) competing metastable abundance ratios are a valuable criterion for identity or non-identity of ion structures, (v) since metastable transitions uniquely define a reaction, it is much simpler (and often the only way) to investigate scrambling reactions in metastables rather than from decompositions in the source, (vi) the metastable defocusing technique has enabled metastables to be uniquely and sensitively identified, (vii) it is convenient to investigate collision-induced decompositions on the metastable mode, and (viii) isotope effects are largest in metastable transitions.

This section is not intended to be exhaustive as far as studies of metastable transitions are concerned, and extensive reference may be found to their use in other sections (see also refs. 2, 3, and 6 for critical discussions). Rather, in this section, the progress in ion-kinetic-energy spectroscopy is reported and publications concerned with energetic and kinetic properties of metastables are reviewed (e.g. kinetic energy release and metastable lifetimes). Identification of ion structures from metastable abundance ratios are reported. Collision-induced metastables are covered in Section 8.

Kinetic and Energetic Studies of Organic Ions 59

Ion Kinetic Energy (IKE) Spectroscopy.—For the reaction $m_1^+ \rightarrow m_2^+$ in the metastable drift regions, the m_2^+ ions possess only a fraction m_2/m_1 of the kinetic energy of the main ion beam. Therefore, because of decompositions in the first drift region, the ions entering the electrostatic analyser (ESA) of a double-focusing instrument will have a spectrum of kinetic energies, varying downwards from the energy of the normal ion beam (neglecting those metastable decompositions of doubly charged ions which produce ions of kinetic energy above that of the ion beam). If an electron-multiplier detector is placed at the energy-resolving (β) slit at the exit of the ESA and the electrostatic voltage is continuously varied, a spectrum of ion kinetic energies is obtained. Beynon et al. have developed this technique.[87] An IKE spectrum is, in effect, a display of all reactions occurring with rate constants $\sim 10^5$ s^{-1} in the first metastable drift region. The spectra are usually complex and contain peaks of widely differing shapes and intensity.

Some of the advantages and weaknesses of IKE spectroscopy have been revealed by a detailed study on nonan-4-one (66) and deuteriated analogues.[88]

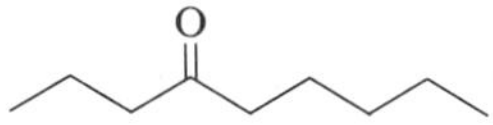

(66)

Detection of transitions is possible for which no metastable is observed in the normal mass spectrum. In favourable cases (*i.e.* the absence of overlapping peaks) the sensitivity of the method can exceed that of the defocusing technique and is not limited to ratios of $m_1 : m_2 < 2$. However, the width of an IKE peak sometimes makes assignments of parent and daughter mass numbers difficult, but use of the defocusing technique in conjunction can overcome this.[88] If the positions of the magnetic and electric sectors can be reversed, then one particular ion can be selected by the magnetic sector and its metastable decompositions examined by sweeping the ESA voltage.

The structure of the McLafferty rearrangement ion from a series of alkyl phenyl ketones has been examined using IKE spectroscopy.[89] The m/e 120 ion was generated by direct ionization of acetophenone (67) and *via* fragmentation of the molecular ion of a series of alkyl phenyl ketones (see (68) $\rightarrow$ (69)). In all

PhCOMe]$^{+\cdot}$ (68) $\rightarrow$ (69)

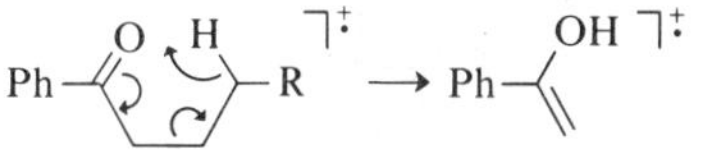

(67)   (68)   (69)

[87] J. H. Beynon, R. M. Caprioli, W. E. Baitinger, and J. W. Amy, *Internat. J. Mass Spectrometry Ion Phys.*, 1969, **3**, 313.
[88] G. Eadon, C. Djerassi, J. H. Beynon, and R. M. Caprioli, *Org. Mass Spectrometry*, 1971, **5**, 917.
[89] J. H. Beynon, R. M. Caprioli, and T. W. Shannon, *Org. Mass Spectrometry*, 1971, **5**, 967.

cases the m/e 120 ion eliminates a CH_3 radical in a metastable transition and the energy released in this process has been measured from the peak width in the IKE spectrum, and confirmed by ordinary metastable defocusing. Kinetic energy releases as low as 10^{-6} eV can be detected by IKE spectroscopy and energy releases as low as 2×10^{-4} eV have actually been measured in organic spectra.[90] The energy released should be characteristic of the ion structure, and therefore distinction should be possible between the keto-structure (67) and the enol structure (69). For m/e 120 ions generated from (68) the energy release (from the IKE scan) was 59 ± 8 meV, whereas for (67) the release was only 7.6 meV. It is concluded that the molecular ion of acetophenone has the keto-structure (67), and that the product ion (69) from the single McLafferty rearrangement has the enol structure but nevertheless loses CH_3 in a metastable transition.[89] These conclusions are comparable with those found for the $C_3H_6O^{+\cdot}$ ion from the unimolecular[91] and ICR[91,92] mass spectrometry of alkanones. It was found that the ion has an enolic structure (70), whether formed *via* a one-step or a two-step McLafferty rearrangement.

$$\begin{bmatrix} & OH & \\ & | & \\ & C & \\ H_2C & \diagup \diagdown & Me \end{bmatrix}^{+\cdot}$$

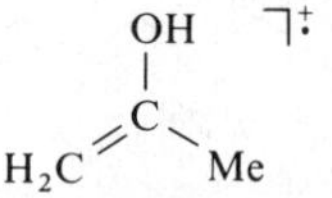

(70)

Measurement of IKE spectra has been reported using a double-focusing instrument of Mattauch–Herzog geometry.[93] It is shown that the IKE spectrum in most cases gives a unique fingerprint and may therefore be used as a structural probe to distinguish between isomers having very similar mass spectra. If the IKE spectra are different, the two compounds have different structures.[93] The differences between the IKE spectra of isomers have also been emphasized for pyrimidine, pyrazine, and pyridazine[94] and for the phenylenediamines.[95] However, the diagnostic usefulness of IKE spectroscopy in identifying isomers must be regarded as strictly limited. For example, the IKE spectra of the six isomeric methyl indoles (71) are all similar, except that for the 1-isomer an

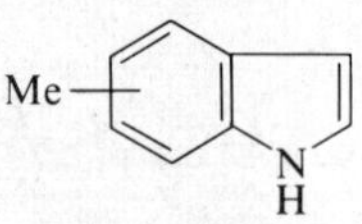

(71) 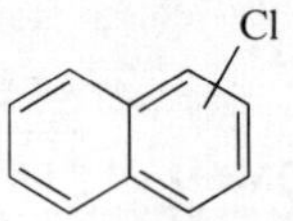   (72)

[90] J. H. Beynon, R. M. Caprioli, and T. Ast, *Org. Mass Spectrometry*, 1971, **5**, 229.
[91] F. W. McLafferty, D. J. McAdoo, J. S. Smith, and R. A. Kornfeld, *J. Amer. Chem. Soc.*, 1971, **93**, 3720.
[92] G. Eadon, J. Diekman, and C. Djerassi, *J. Amer. Chem. Soc.*, 1970, **92**, 6205.
[93] E. M. Chait and W. B. Askew, *Org. Mass Spectrometry*, 1971, **5**, 147.
[94] J. H. Beynon, R. M. Caprioli, and T. Ast, *Org. Mass Spectrometry*, 1972, **6**, 273.
[95] R. M. Caprioli, J. H. Beynon, and T. Ast, *Org. Mass Spectrometry*, 1971, **5**, 417.

additional peak arose corresponding to loss of CH_3 from the molecular ion.[96] The IKE spectra for the two isomeric chloronaphthalenes (72) are identical,[97] indicating that those molecular ions decomposing with rate constants $\sim 10^5 \text{ s}^{-1}$ have the same structure and internal energy distribution, whether formed from the 1- or the 2-isomer. Similar conclusions were reached for seven isomeric dichloronaphthalenes, suggesting that a randomization reaction is occurring. The use of IKE spectra as a criterion for ion structure identity is tantamount to the use of competing metastable ion abundance ratios (see below).

IKE spectra reported so far in this section have been obtained by scanning the ESA voltage from zero to E (where E is the voltage corresponding to transmission of the main ion beam). Since doubly charged ions after acceleration possess twice the kinetic energy of singly charged ions, their metastable decompositions may be detected by scanning the electric sector voltage over the range E to $2E$.

The IKE spectra of the three isomeric phenylenediamines (73) are identical except for small differences in intensity.[95] In particular, dish-topped peaks, resulting from cleavage of doubly charged ions into two single charged ions, are observed. The major transition corresponds to the reaction $M^{2+} \rightarrow 80^+ + 28^+$, and there is evidence to suggest that the m/e 28 ion is H_2CN^+. The energy released in this transition is about 2.6 eV for all three isomers. If it is assumed that most, or all, of this large kinetic energy release results from electrostatic repulsion between two localized positive charges in the molecular ion, then the charge separation corresponds to 5.7 Å (0.57 nm). It appears from the similarity in the IKE spectra and kinetic energy released that the doubly charged molecular ions surviving to the metastable drift region have the same structure for all three isomers. Either the p-phenylenediamine structure or the seven-membered structure (74) is a reasonable suggestion consistent with a charge separation of 0.57 nm.

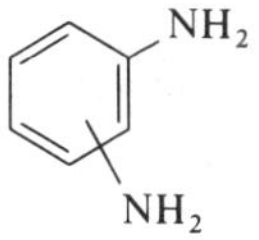

(73) 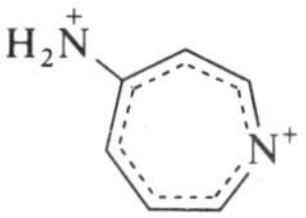   (74)

From the kinetic energy released in the transition $M^{2+} \rightarrow C_3H_2N^+ + H_2CN^+$ in the IKE spectra of pyrazine and pyrimidine, it was concluded that the doubly charged molecular ion has an open-chain structure, with the charges located four atoms apart.[94] In the case of the benzene doubly charged molecular ion, a charge separation of ~ 0.62 nm was indicated, consistent with a straight-chain ion.[98] (The diameter of the benzene ring is 0.29 nm.) IKE studies have

[96] S. Safe, W. D. Jameison, and O. Hutzinger, *Org. Mass Spectrometry*, 1972, **6**, 33.
[97] S. Safe, O. Hutzinger, and M. Cook, *J.C.S. Chem. Comm.*, 1972, 260.
[98] J. H. Beynon, R. M. Caprioli, W. E. Baitinger, and J. W. Amy, *Org. Mass Spectrometry*, 1970, **3**, 963.

also been carried out upon the doubly charged toluene molecular ion.[99] Kinetic energy released upon the fragmentation of some doubly-charged diatomic molecules has been measured by IKE spectroscopy.[100] The results agree closely with theoretical potential energy diagrams.

Shapes of Metastable Peaks.—The decomposition of a doubly charged ion with kinetic energy release is an intuitively reasonable event, as discussed above. Many instances are also known of kinetic energy release in decomposition of single charged ions in the mass spectrometer to yield broad metastable peaks. This section reports on some of the recent literature in this field.

There have been several publications incorporating calculations of metastable peak shapes (appropriate to different instrument geometries) which would be anticipated when kinetic energy is released. Beynon et al. have carried out a detailed theoretical study of the shapes of metastable peaks resulting from decompositions in front of the magnetic sector.[101] Such metastable peaks can be observed by IKE spectroscopy in the normal way by varying the ESA voltage. Figure 6 shows the peak shape computed for the reaction $139^+ \rightarrow 109^+ + 30^+$ in o-nitrophenol. The release of kinetic energy is assumed to be 0.75 eV. Figure 6(a) shows the peak shape at infinite collector slit length, whereas Figure 6(b)

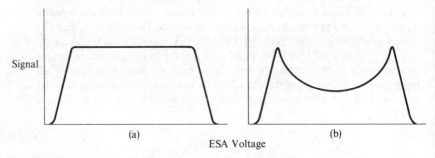

Figure 6 *Computed metastable peak shapes when the collector slit length is* (a) *infinite* (b) 0.2 in

shows that the peak becomes dish-shaped if a collector slit length of only 0.2 inches is assumed. The tendency towards a dish-shape as the slit-length is decreased is due to a discrimination effect against ions which dissociate in the direction of the collector slit (see also refs. 2, 6, 102, and 103 for similar discussions).

[99] T. Ast, J. H. Beynon, and R. G. Cooks, *Org. Mass Spectrometry*, 1972, **6**, 741.
[100] J. H. Beynon, R. M. Caprioli, and J. W. Richardson, *J. Amer. Chem. Soc.*, 1971, **93**, 1852.
[101] J. H. Beynon, A. E. Fontaine, and G. R. Lester, *Internat. J. Mass Spectrometry Ion Phys.*, 1972, **8**, 341.
[102] C. Reichert, R. E. Fraas, and R. W. Kiser, *Internat. J. Mass Spectrometry Ion Phys.*, 1970, **5**, 457.
[103] C. G. Rowland, *Internat. J. Mass Spectrometry Ion Phys.*, 1971, **6**, 155.

Kinetic and Energetic Studies of Organic Ions

The possibility of isomerization between the molecular ions of anthracene (75), phenanthrene (76), and diphenylacetylene (77) prior to decomposition has been investigated by measurement of the kinetic energy released in the

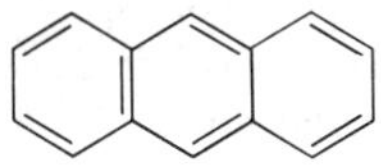

(75) 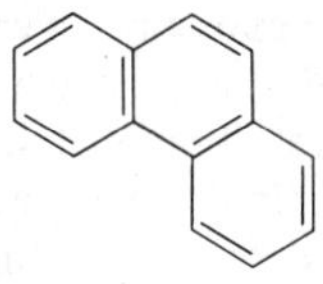 (76) 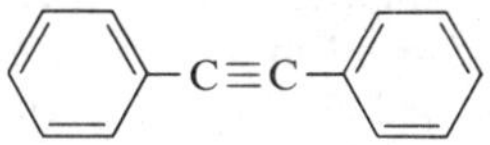 (77)

elimination of C_2H_2.[104] The respective kinetic energy releases are 0.35 ± 0.03, 0.42 ± 0.02, and 0.38 ± 0.02 eV. These data, along with the observation of similar energy requirements for $M^+ - C_2H_2$ formation from the three compounds, suggest that the molecular ions of (75)—(77) have rearranged to similar structures, or a mixture of structures, prior to low-energy dissociation to $M^+ - C_2H_2$.

It should be borne in mind that differences in metastable peak shapes may be used as a criterion for non-identity of ion structures. For example, the $C_3H_8N^+$ ions (m/e 58) formed from various amines all eliminate C_2H_4 in a metastable transition and the metastable peaks have either a gaussian shape or are flat-topped, with a kinetic energy release of 0.4 eV.[105] This indicates that the two groups of ions have different structures at energies appropriate to metastable transitions, since the reaction must be taking place over two different potential surfaces. Without exception, kinetic energy release occurs only from those $C_3H_8N^+$ ions formed with either of the primary amino-structures (79) or (80).

$\overset{+}{C}H_2NHCH_2Me$ $\underset{Me}{\overset{Me}{>}}\overset{+}{C}-NH_2$ $MeCH_2\overset{+}{C}HNH_2$

(78)   (79)   (80)

$Me\overset{+}{C}HNHMe$ $\underset{Me}{\overset{Me}{>}}N-CH_2^+$

(81)   (82)

All other structures [(78), (81), and (82), containing secondary or tertiary amino-groups] yield a gaussian metastable peak for C_2H_4 loss. Competing metastable abundance ratios further support the assignments (see below).

An unusual case of fine structure in a metastable peak has been reported for the collision-induced reaction $H_3^+ \rightarrow H^+ + H_2$.[106] Since the H^+ product

---

[104] C. G. Rowland, *Internat. J. Mass Spectrometry Ion Phys.*, 1971, 7, 79.
[105] N. A. Uccella, I. Howe, and D. H. Williams, *J. Chem. Soc. (B)*, 1971, 1933.
[106] R. G. Cooks and J. H. Beynon, *Chem. Comm.*, 1971, 1282.

ion cannot incorporate vibrational energy, any excess energy must be distributed between kinetic energy of the products and vibrational energy of H_2. The metastable peak appears to exhibit fine structure and it is argued that H_2 is formed in a number of vibrational states. The centre of the peak corresponds to product formation with minimum kinetic energy release and therefore maximum vibrational energy of H_2. Each notch on the peak corresponds to H_2 formed in a successively lower vibrational state. The energy increments observed are consistent with formation of the vibrational states $v = 14, 13, 12,$ and 11 of H_2. A composite metastable peak (consisting of narrow and broad components) is also reported[106] for the reaction $H_2^+ \rightarrow H^+ + H$. Transitions from the ground and first electronically excited state of H_2^+ account for the results.

Metastable Ion Abundance Ratios.—One of the most successful techniques used as a criterion for identity or non-identity of ion structures is the ratio of competing metastable ion abundances. When two competing reactions from the same ion both give reasonably abundant metastable ions, the metastable abundance ratio, $[m_1^*] : [m_2^*]$, may be employed to characterize the ion structure.

However, it has been shown by QET calculations and by experiment that $[m_1^*] : [m_2^*]$ for two competing reactions is a function of the internal energy distribution of the ions generating the metastables.[107] The ratio is therefore a function of the electron-beam energy, and where the precursor ion A^+ is itself a fragment ion from a primary process, the ratio will depend upon the mode of formation of A^+. These conclusions arise because the energy band giving rise to metastables is not infinitesimal and varies from about 0.2 eV for reactions with high frequency factors ($\sim 10^{13}$ s^{-1}) up to several eV for reactions with low frequency factors ($\sim 10^8$ s^{-1}).[107] The authors emphasize that relatively small changes in $[m_1^*] : [m_2^*]$ (say a factor of up to 5) do not constitute reliable evidence for different structures. The occurrence of two (or more) identical decompositions in similar ratios from two precursors is good evidence that the two precursors are decomposing from the same structure (or mixture of structures).

It has experimentally been shown[79,107] that $[m_1^*] : [m_2^*]$ is a function of electron-beam energy. For example, $[m_1^*] : [m_2^*]$ for decompositions from (83) increases

$$\text{(83)} \quad \langle \text{Ph} \rangle^{+\cdot}-C_4H_9 \quad \overset{m_1^*}{\underset{m_2^*}{\rightleftarrows}} \quad \begin{array}{l} C_7H_8^{+\cdot} \\ C_7H_7^+ \end{array}$$

smoothly from 6.2 at an electron energy of 70 eV to 14 at 11 eV.[79] Such a variation is acceptable within the framework of the QET and reactions from isolated electronic states are not therefore indicated. The variation would have to be considerably more drastic for such a postulate.

[107] A. N. H. Yeo and D. H. Williams, *J. Amer. Chem. Soc.*, 1971, **93**, 395.

Nevertheless, the criterion of competing metastable transitions has been employed for structural investigations of ions decomposing with rate constants around 10^5 s^{-1}. Five groups of structurally distinct $C_3H_8N^+$ ions (78)—(82) have been generated, and their decomposing structures have been characterized by this method.[105] Three groups of ions are found: (79 and 80) decompose from one structure, (81) and (82) decompose from another structure, and ion (78) constitutes the third group. Extensive ^{2}H- and ^{13}C-labelling reveals information on the mechanisms of decomposition and, in view of the consistency with which the different $C_3H_8N^+$ ions fall into their respective groups, it should be possible to identify structural entities in unknown amines in terms of their metastable decompositions from $C_3H_8N^+$ (both abundances and peak shapes).[105]

The criterion of metastable abundance ratios has also been applied to $C_nH_{2n-1}^+$ and $C_nH_{2n-3}^+$ ions[108] and some hydrocarbon and oxygen-containing C_7 and C_8 ions.[109] The $C_3H_5O^+$ ions generated from (84)—(87) were found to isomerize to common structures.[110] The $C_3H_5O^+$ ions generated from methyl propionate remain structurally distinct (presumably as the propionyl ion MeCH$_2$CO$^+$). A novel proposal is made about metastable rearrangements, to explain the

(84)   (85)   (86)   (87)

loss of CO from the $C_3H_5O^+$ ions (A^+) generated from (84)—(87). Competition between ethylene loss and rearrangement to $C_2H_5CO^+$ is proposed for the A^+ ion. Loss of CO then occurs more rapidly with a lower activation energy from $C_2H_5CO^+$ than the reaction for reversion to A^+ (otherwise the $C_2H_5CO^+$ ion would be seen to lose ethylene in a metastable transition). Therefore, when isomeric precursor ions A^+ and B^+ undergo several metastable transitions, one of which is common to both precursor ions, it is possible that the common transition is occurring from the same precursor ion structure.[110]

There has been some discussion about the structure of the $M^+ - CH_3$ ion from stilbene (88).[111–113] The corresponding $C_{13}H_9^+$ ion ($M^+ - H$) from fluorene (89) appears from its very different metastable transitions to have a different structure,[113] and the $C_{13}H_9^+$ ($M^+ - H$) ion from phenalene (90) has yet another structure at these low energies for metastables.[114]

[108] M. A. Shaw, R. Westwood, and D. H. Williams, *J. Chem. Soc.* (*B*), 1970, 1773.
[109] A. J. Dale, W. D. Weringa, and D. H. Williams, *Org. Mass Spectrometry*, 1972, **6**, 501.
[110] T. J. Mead and D. H. Williams, *J. Chem. Soc.* (*B*), 1971, 1654.
[111] J. H. Bowie and P. Y. White, *Austral. J. Chem.*, 1971, **24**, 205.
[112] J. H. Bowie and P. Y. White, *Org. Mass Spectrometry*, 1972, **6**, 135.
[113] H. Güsten, L. Klasinc, J. Marsel, and D. Milivojević, *Org. Mass Spectrometry*, 1972, **6**, 175.
[114] J. H. Bowie and T. K. Bradshaw, *Austral. J. Chem.*, 1970, **23**, 1431.

As indicated previously,[1] metastable : daughter ion ratios should not be used as a criterion for ion structure identity. Slightly different ratios in the spectra

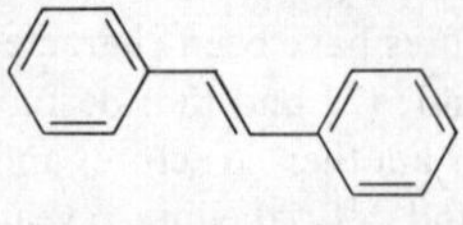

(88) 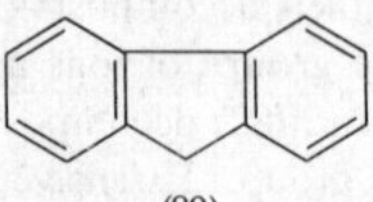   (89) 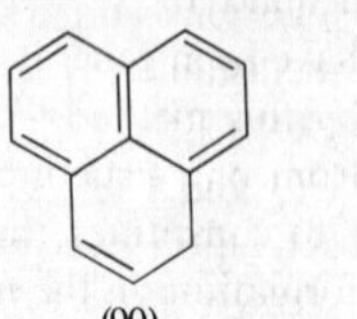   (90)

of some naphthoquinones are therefore not necessarily inconsistent with identical ion structures at energies appropriate to metastable transitions.[115]

Metastables may be used simply to exclude identical ion structures. The $NO_2C_7H_6^+$ ions generated from the *m*- and *p*-isomers of (91) undergo different

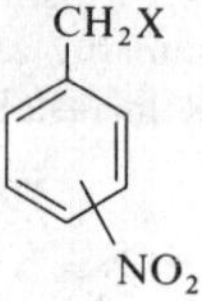

(91) X = Me or Br

metastable transitions (NO loss for the *p*-isomer and NO_2 loss for the *m*-isomer).[116] Hence both of the $NO_2C_7H_6^+$ ions cannot have the nitrotropylium structure at 'metastable energies' and it is likely that both exist in the benzyl form.

Metastable Transitions in Various Instruments.* —A large number of ordinary and consecutive metastable transitions have been measured from various hydrocarbons.[117] A detection sensitivity was reported of 2×10^{-8} of the total ionization, which is estimated to be sufficient to detect metastable transitions for any breakdown path of the molecular ion.

An instrument modification has been described for observation of consecutive metastable transitions[118] (see $m_1^+ \to m_3^+$, below). The metastable peak in the normal spectrum for the process $m_2^+ \to m_3^+$ is first brought to focus; the overall consecutive metastable transition may then be observed by raising the accelerating voltage by the factor m_1/m_2.

$$m_1^+ \xrightarrow{*} m_2^+ \xrightarrow{*} m_3^+$$

[115] T. A. Elwood, K. H. Dudley, J. M. Tesarek, P. F. Rogerson, and M. M. Bursey, *Org. Mass Spectrometry*, 1970, **3**, 841.
[116] R. Westwood, D. H. Williams, and A. N. H. Yeo, *Org. Mass Spectrometry*, 1970, **3**, 1485.
[117] U. Löhle and C. Ottinger, *Internat. J. Mass Spectrometry Ion Phys.*, 1970, **5**, 265.
[118] L. P. Hills and J. H. Futrell, *Org. Mass Spectrometry*, 1971, **5**, 1019.

* See also Collision Processes, Section 8.

A metastable peak has been observed in the spectrum of p-chlorophenol (92) for the overall reaction $M^+ \rightarrow C_5H_5^+ + COCl$.[119] Since this reaction requires an extensive rearrangement if it occurs in one step, the possibility of consecutive metastable processes has been investigated. In the first place, ΔH_f values for CO, Cl, and COCl indicate that it is energetically more favourable to form CO and Cl as separate fragments in the overall reaction (neglecting reverse activation energies). Using the accepted techniques for observation of consecutive metastable transitions, it was found that the overall process (92) $\rightarrow$ (93) $\rightarrow$ (94) takes place in the metastable drift regions. Hence the mechanism

$$Cl-\langle\text{+·}\rangle-OH \rightarrow C_6H_5O^+ + Cl \rightarrow C_5H_5^+ + Cl + CO$$
$$(92) \qquad\qquad (93) \qquad\qquad (94)$$

involving ring-chlorine rearrangement does not necessarily exist. In terms of the QET, it is argued that the rate of the reaction (93) $\rightarrow$ (94) must increase rapidly with energy close to threshold, since the decomposing $C_6H_5O^+$ ions possess low internal energy.

Field ionization and its applications are covered in Chapter 1, but the position of FI metastables in the spectrum of ion lifetimes should be briefly mentioned here. The use of a sharp-edge FI source enables metastable ions with mean lifetimes in the range 3×10^{-13}—10^{-10} s to be studied by peak-broadening techniques and those in the range 10^{-9}—10^{-4} s to be studied by varying the ion transit time from the first slit electrode of the source to the entrance slit of the analyser.[120]

6 Reactions of Isotopically Labelled Species

Deuterium Isotope Effects.—An increasing number of deuterium isotope effects on mass spectral reactions have been reported in the recent literature, and in some cases mechanistic inferences have been made. The simultaneous occurrence of H/D scrambling (see below and also Chapter 3) creates a complication, but this may be surmounted if several different deuteriated samples are employed.

Studies on the loss of H and D from the singly charged molecular ions of the partially deuteriated toluenes (95)—(97) have revealed that (i) the extent of H/D scrambling prior to reaction increases with decreasing internal energy and (ii) the primary deuterium isotope effect for the reaction also increases with decreasing internal energy.[121] Both these conclusions are consistent with the QET, and the variation of isotope effect with energy is consistent with the variation of the isotope effect with temperature in neutral chemistry. The simplest example of the isotope effect calculation for compounds (95)—(97) is found for the low-energy metastable reactions. The metastable abundance

---
[119] J. C. Tou, *J. Phys. Chem.*, 1970, **74**, 3076.
[120] B. W. Viney, *Internat. J. Mass Spectrometry Ion Phys.*, 1972, **8**, 417.
[121] I. Howe and F. W. McLafferty, *J. Amer. Chem. Soc.*, 1971, **93**, 99.

ratios ($[m_H^*]:[m_D^*]$) for the reactions involving loss of H and D are respectively 4.68 ± 0.09, 1.68 ± 0.09, and 4.77 ± 0.15 for compounds (95), (96), and (97). After correcting for the number of H and D atoms in the respective molecules,

(95) 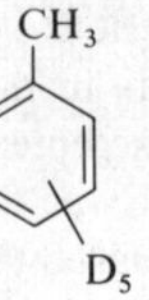 (96) 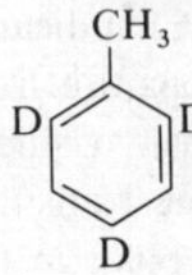 (97)

it becomes evident that an isotope effect of 2.8 is operating in favour of H loss, with a 100% degree of H/D scrambling. A similar conclusion was reached independently[122] and the slightly larger 'isotope factor' calculated (3.5) is probably attributable to the sampling of slightly longer lifetimes. Studies on the deuteriated cycloheptatrienes (98) and (99) have revealed similar variations

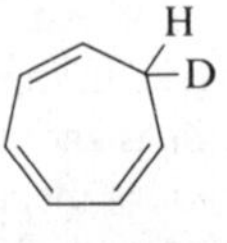

(98) 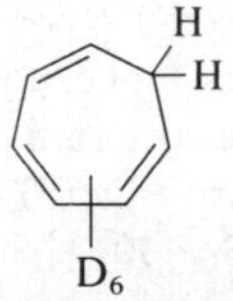 (99)

of isotope effect with internal energy to the toluene case.[121] This is consistent with (but does not prove) isomerization of the toluene molecular ion to a cycloheptatriene structure prior to decomposition. The higher-energy studies on the toluene molecular ion indicate a slight preference for loss of α-hydrogens, consistent with partial decomposition from a skeletally unrearranged structure (cf. also ref. 60).

It has been shown from deuterium labelling[121] and from double ^{13}C labelling in toluene[123] that the $C_7H_7^+$ ion is completely scrambled prior to acytelene loss, consistent with the tropylium structure. In the spectrum of the π-cycloheptatrienyl complex (100), incomplete hydrogen randomization has been reported prior to $C_7H_7^+$ formation.[124]

Isotope effects have also been evaluated for loss of H and D from doubly charged toluene molecular ions, using one of Beynon's elegant metastable

$(C_7H_8)Cr(CO)_3$
(100)

[122] J. H. Beynon, J. E. Corn, W. E. Baitinger, R. M. Caprioli, and R. A. Benkeser, *Org. Mass Spectrometry*, 1970, **3**, 1371.
[123] A. S. Siegel, *J. Amer. Chem. Soc.*, 1970, **92**, 5277.
[124] F. E. Tibbetts, M. M. Bursey, W. F. Little, B. R. Willeford, S. A. Benezra, M. K. Hoffman, and P. W. Jennings, *Org. Mass Spectrometry*, 1972, **6**, 475.

techniques.[125] The decompositions are followed with the electric sector set at $2E$ to transmit the ions formed from charge exchange between the doubly charged ions and neutral benzene in the first drift region (see also Section 8). An isotope effect of about 1.3 operates, favouring H loss from the doubly charged molecular ion. Side-chain hydrogens are preferentially eliminated. For loss of H_2, HD, and D_2, however, randomization appears to be complete.

Deuterium isotope effects upon the loss of hydrogen (atom or molecule) are of course to be anticipated, and other examples quoted in the recent literature include the molecular ion of pyrazole (101)[126] and the benzyloxy-cation formed from benzyl nitrate.[127] In the latter example,[127] an isotope effect of 1.8 was determined from peak heights for loss of H(D) in the source at 10 eV from $C_6H_5CHDO^+$. The α-D_2 ion, $C_6H_5CD_2O^+$, eliminated exclusively a D atom.

Variations predicted by the QET in the isotope effect and extent of scrambling with internal energy have been observed for the loss of water from the molecular ions of nicotinic and isonicotinic acid [(102) and (103), respectively].[128] From the mass spectra of specifically deuteriated analogues of (103) it was shown that

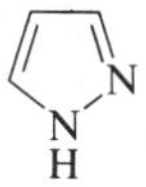

(101) 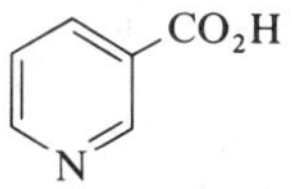   (102) 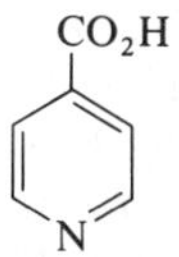   (103)

the isotope effect operating on the loss of water from the molecular ion increases with decreasing internal energy, being 1.6, 2.0, and 2.3 in the ion source and first and second drift regions, respectively. Partial exchange occurs between the hydroxylic- and 3-hydrogens, and the extent of exchange also depends on the internal energy of the molecular ion, increasing from $55 \pm 2\%$ prior to decomposition in the source to 100% in the second drift region. This variation in the degree of scrambling with energy is analogous to that found for deuteriated benzoic acids.[129] Incidentally, the occurrence of an isotope effect for water elimination is contrary to the assertion that this elimination under electron impact is always attended by an isotope effect near unity.[130]

Further QET calculations have been carried out on small molecular ions to evaluate the magnitude of the primary deuterium isotope effect. The loss of H (or D) from the molecular ion of [α-^{2}H]ethanol (104) is amenable to such

CH_3CHDOH
(104)

[125] T. Ast, J. H. Beynon, and R. G. Cooks, *J. Amer. Chem. Soc.*, 1972, **94**, 1834.
[126] J. van Thuijl, K. J. Klebe, and J. J. van Houte, *Org. Mass Spectrometry*, 1971, **5**, 1101.
[127] P. J. Smith, *Canad. J. Chem.*, 1971, **49**, 333.
[128] R. Neeter and N. M. M. Nibbering, *Org. Mass Spectrometry*, 1971, **5**, 735.
[129] I. Howe and F. W. McLafferty, *J. Amer. Chem. Soc.*, 1970, **92**, 3797.
[130] M. M. Green, J. G. McGrew, and J. M. Moldowan, *J. Amer. Chem. Soc.*, 1971, **93**, 6700.

calculations[131] since the spectra of deuteriated analogues show that the loss of a hydrogen atom occurs from the α-position largely without scrambling. The calculations[131] were carried out for daughter ions formed in the source, using an improved form of the QET. The vibrational frequencies of the ethanol molecular ion were assumed to be the same as those for neutral ethanol and it was calculated that the activation energy for α-C—D bond rupture is almost 0.05 eV higher than that for α-C—H rupture. This difference in activation energies is sufficient to yield a calculated $[M^+ - H]:[M^+ - D]$ ratio of 3.0—3.2 (the exact ratio depending on the energy distribution employed for M^+), which compares favourably with the observed ratio of 3.57. The ratio for metastables would be expected to be much higher, since k vs. E curves rise sharply at low energies for small molecules. The mass spectra of deuteriated methanols have been investigated in the range 11—35 eV.[132] The isotope effects for the various reactions were measured and QET calculations were found to give satisfactory agreement with the data when half the theoretical number of oscillators were employed.

Isotope effects have been investigated for all the possible deuteriated methanes.[133] Calculations indicate that the activation energy for D loss from the molecular ion is 0.08 eV higher than that for H loss. In this small molecule (rapid rise of k with E, see equation 8) this activation energy difference is sufficiently large to cause no observation of D loss for unimolecular metastable decompositions of partially deuteriated methanes. A metastable for D loss from CD_4^{+} was observed. In the collision-induced spectra, an isotope effect was measurable and was found to be greater than that for formation of normal ions in the source. This isotope-effect order (unimolecular metastables > collision-induced metastables > decompositions in the source) is consistent with observations on toluene[121] and provides qualitative information about the average internal energy added by the collision-induced reaction.

Reports of primary deuterium isotope effects upon hydrogen-transfer reactions in the mass spectrometer have continued to increase. For example, a significant isotope effect, operating against deuterium, occurs in the water loss from the molecular ion of o-toluic acid (105).[134] Once again scrambling (between methyl and carboxy-groups) appeared to increase with increasing lifetime. Isotope effects have also been reported for metastable decompositions involving hydrogen transfer from the ions (79)—(82).[105]

In the course of investigation of the $C_3H_6O^{+}$ ion from different sources, it has been noted[91] that there is an isotope effect upon the loss of methyl from this ion when (106) is the parent ion. The isotope effect was detected by comparing the product and precursor ion intensities for the above reaction, employing different deuteriated analogues of (106). Only in the case of deuterium substitu-

---

[131] M. Corval, *Bull. Soc. chim. France*, 1970, 2871.
[132] M. Corval and P. Masclet, *Org. Mass Spectrometry*, 1972, **6**, 511.
[133] L. P. Hills, M. L. Vestal, and J. H. Futrell, *J. Chem. Phys.*, 1971, **54**, 3834.
[134] M. J. Lacey, C. G. MacDonald, and J. S. Shannon, *Org. Mass Spectrometry*, 1971, **5**, 1391.

Kinetic and Energetic Studies of Organic Ions

tion on oxygen is the reaction slowed. Therefore, loss of methyl from the enol ion (70) occurs *via* hydrogen transfer from oxygen in the rate-determining step.

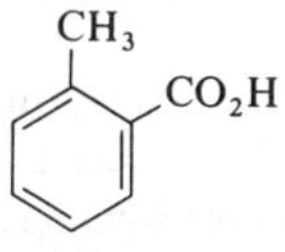

(105) 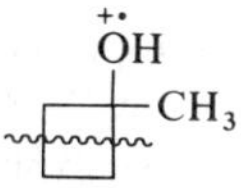 (106)

When a primary deuterium isotope effect occurs in the mass spectrometer upon a reaction not involving elimination of a hydrogen-containing neutral, it is sometimes possible to use a reference metastable to investigate the isotope effect. For example, the molecular ion of *p*-bromophenol (107) competitively eliminates Br and CO in metastable transitions, and it is observed that the metastable ratio $m^*[M^+ - Br] : m^*[M^+ - CO]$ increases by a factor of 3 after substitution of deuterium on the phenolic oxygen.[135] Hence there is a deuterium isotope effect on the rate of CO loss from (107), assuming no such effect on Br loss. This isotope effect may be employed to yield mechanistic

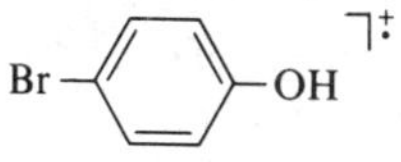

(107) 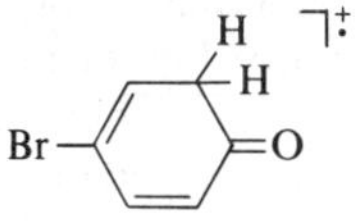 (108) 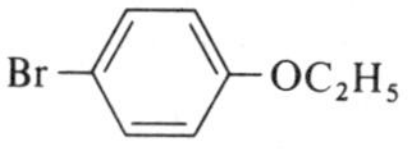 (109)

information on the elimination of ethylene from the molecular ion of *p*-bromophenetole (109). The resulting ion, at formation, may have either structure (107) or (108), but investigation of the metastable ratio for loss of Br and CO from the M^+ − ethylene ion from (109) and (110) reveals that there is an isotope effect on CO loss. This confirms that hydrogen transfer has occurred *via* a four-membered transition state to yield (107), since direct formation of (108) would not present an opportunity for an isotope effect on further CO loss. Similarly, elimination of CH_2CO from the molecular ion of *p*-chloroacetanilide (111) occurs *via* a four-membered transition state with hydrogen transfer to oxygen.[136] This is consistent with results based on steric effects (see Section 7).

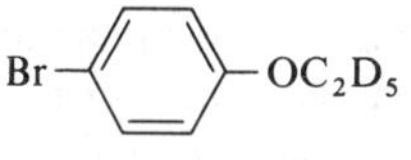

(110) 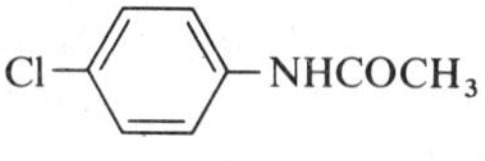 (111)

[135] I. Howe and D. H. Williams, *Chem. Comm.*, 1971, 1195.
[136] N. A. Uccella, I. Howe, and D. H. Williams, *Org. Mass Spectrometry*, 1972, **6**, 229.

It is possible that the magnitude of the kinetic deuterium isotope effect might be employed to distinguish between concerted and stepwise processes in the rearrangement of organic ions, but even in neutral chemistry there remain some doubts about the applicability of the deuterium isotope effect to this particular problem.[137]

Secondary deuterium isotope effects have been noted in the decomposition of a number of $C_3H_6O^{+\cdot}$ ions.[91] Replacement of H by D on a methyl group makes loss of that methyl less favoured. An approximate secondary deuterium isotope effect of 1.10 is calculated.

Scrambling Reactions.—The large volume of publications concerning the mass spectra of isotopically labelled organic molecules over the past few years has revealed that hydrogen and carbon scrambling may be extensive on the time scale of the mass spectrometer. Many of these scrambling reactions possess activation energies below that for an elimination reaction and it is a common situation that, where scrambling occurs prior to decomposition, it is more extensive in longer-lifetime ions (e.g. metastables) compared with shorter-lifetime ions (e.g. decompositions in the source). The spectra of compounds (95)—(99), (102), (103), and (105) all show this effect. A scrambling reaction is just another rearrangement in which the precursor and product ions have the same elemental composition. Such rearrangements would be expected to have tight transition states (low frequency factor) and, according to the QET, compete less effectively with other reactions at high energy (see, for example, Figure 5). If the scrambling reaction has a higher activation energy E_0 than a high frequency factor decomposition reaction, then it will not be expected to compete at any energy.

Scrambling reactions are covered fully in Chapter 3, but some examples (confined to kinetic studies) are reported below. In calculations of statistical distributions in scrambling reactions, formula 12 might be found useful. This formula gives the number of ways of choosing n atoms from p atoms.

$$\frac{p!}{(p-n)!n!} \tag{12}$$

Extensive ^{2}H-, ^{13}C-, and ^{15}N-labelling has revealed a variation of the degree of scrambling with internal energy in the parent ion of some benzyl cyanides prior to HCN elimination.[138] For (112) itself it is revealed from reactions in the source that the high-energy molecular ions lose hydrogen cyanide containing exclusively the original cyano-group after almost complete hydrogen randomization. In the low-energy metastable ions the complete hydrogen scrambling persisted, but surprisingly 22% of the C atom lost in the HCN originated from the α-position. It therefore appears that two different scrambling processes are competing with HCN loss from the molecular ion. The hydrogen rearrangement competes successfully at all energies but the carbon randomization reaction

---

[137] H. Kwart and M. C. Latimore, *J. Amer. Chem. Soc.*, 1971, **93**, 3770.
[138] T. A. Molenaar-Langeveld, N. M. M. Nibbering, and Th. J. deBoer, *Org. Mass Spectrometry*, 1971, **5**, 725.

between the α- and β-positions appears to have a similar activation energy to that for HCN loss and therefore proceeds at a similar rate in low-energy ions. This carbon scrambling reaction presumably involves a very tight transition state and fails to compete effectively with HCN loss (itself a rearrangement) at high energies. From labelling studies on the three positional isomers of (113) it is found[138] that the hydrogen cyanide eliminated at all energies from the molecular ion contains primarily the side-chain cyano-group, thus indicating that few or none of the molecular ions have rearranged to the seven-membered-ring isomer (114). However, hydrogen scrambling is extensive, but cannot proceed *via* reversible isomerization to (114).

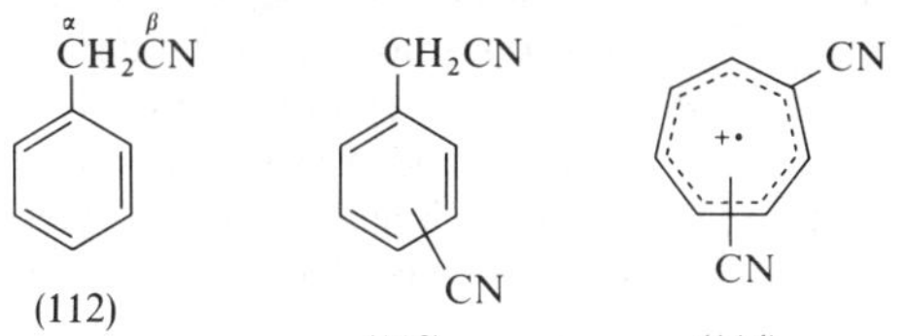

$\overset{\alpha}{C}H_2\overset{\beta}{C}N$ CH₂CN CN

(112) (113) (114)

The dependence of atom scrambling on internal energy and hence lifetime can also be demonstrated by varying the repeller voltage.[139] In the mass spectra of labelled thiophens an increase in repeller voltage (which decreases the lifetime in the source) brought about a decrease in scrambled product in the source, as predicted by the QET.

It has been shown from the spectra of suitably deuterated derivatives that the ion (115) eliminates ethylene containing hydrogens from the ethyl group, with increasing incorporation of methylene hydrogens as the electron energy is lowered.[140] This is consistent with the reversible hydrogen transfer (115) ⇌ (116) prior to ethylene loss.

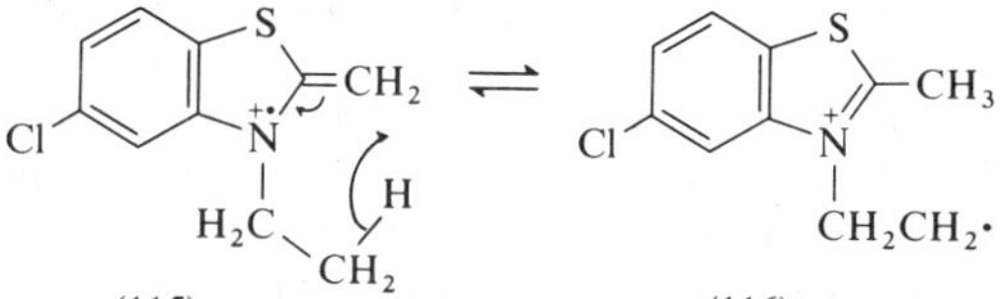

(115) (116)

It is apparent therefore that randomization reactions are most extensive at low energies (other instances are found for ketone molecular ions[141] and $C_8H_8^{+\cdot}$ and $C_8H_9^+$ ions[142]). At first glance, therefore, the reverse effect in the loss of

[139] F. de Jong, H. J. M. Sinnige, and M. J. Janssen, *Org. Mass Spectrometry*, 1970, **3**, 1539.
[140] P. R. Briggs, T. W. Shannon, and P. Vouros, *Org. Mass Spectrometry*, 1971, **5**, 545.
[141] A. N. H. Yeo, *Chem. Comm.*, 1970, 987.
[142] A. Venema, N. M. M. Nibbering, and Th. J. deBoer, *Org. Mass Spectrometry*, 1970, **3**, 1589.

C_4H_8 from the parent ion of phenyl n-butyl ether [see (117) → (118)] is surprising.[143] For this reaction in the source at 70 eV the hydrogen transferred comes from all four positions of the butyl chain, but largely from the 3-position in metastable transitions. This increased specificity with a decrease in internal

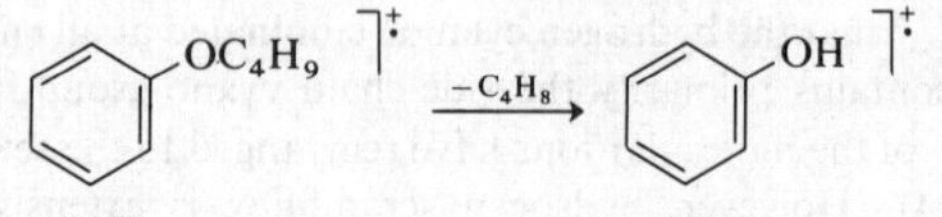

(117) $\xrightarrow{-C_4H_8}$ (118)

energy is an uncommon situation and it is suggested that different specific reactions are occurring, involving transition states of different sizes.[143]

Studies of the IKE spectra (metastable transitions) of the doubly labelled ethyl benzoate (119) have helped to determine the breakdown mechanism of the $M^+ - C_2D_4$ ion (120).[144] This ion eliminates ^{18}OD, ^{18}OH, OD, and OH in the IKE spectrum. Hydroxyl radicals containing H are lost in almost twice the

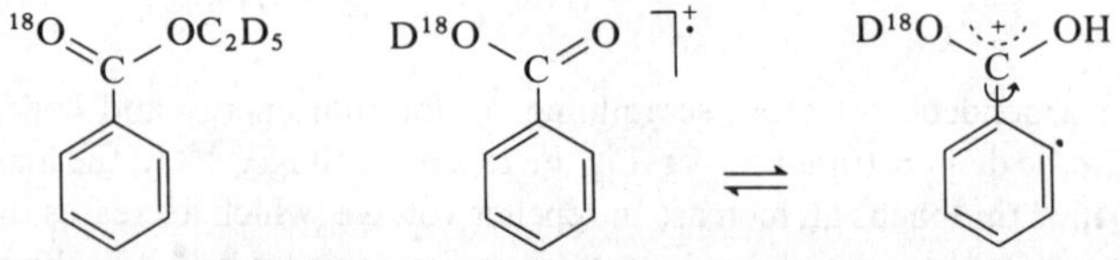

(119)   (120) ⇌ (121)

abundance of those containing D, consistent with equilibration between *ortho-* and carboxyl hydrogens.[145] However, the two oxygen atoms have not become completely equivalent with respect to the available H and D atoms. It appears that scrambling between *ortho-* and carboxyl hydrogens occurs *via* reactions such as (120) ⇌ (121) followed by revolutions about the Ar—C bond. To explain the results it is calculated that about eight such half-revolutions have occurred.[144]

In general it should be noted that a scrambling reaction is more likely to be observed the higher the activation energy for unimolecular decomposition, since the scrambling reaction may then compete more effectively (see, for example, Figure 5). Many instances are known of hydrogen scrambling over only part of an ion (e.g. stilbene,[112,146] cyclohexane-1,2-diol,[147] and $C_3H_8N^+$ ions[105]) and the presence of heteroatoms often increases the barrier to randomization processes.[105,110]

[143] A. N. H. Yeo and C. Djerassi, *J. Amer. Chem. Soc.*, 1972, **94**, 482.
[144] R. H. Shapiro, K. B. Tomer, R. M. Caprioli, and J. H. Beynon, *Org. Mass Spectrometry*, 1970, **3**, 1333.
[145] R. H. Shapiro, K. B. Tomer, J. H. Beynon, and R. M. Caprioli, *Org. Mass Spectrometry*, 1970, **3**, 1593.
[146] H. Güsten, L. Klasinc, J. Marsel, and D. Milivojević, *Org. Mass Spectrometry*, 1971, **5**, 357.
[147] F. Benoit and J. L. Holmes, *Canad. J. Chem.*, 1971, **49**, 1161.

Specific Isotopic Transfer.—It should be borne in mind that frequent examples are reported of specific reactions in isotopic species (see Chapter 3). Such specific reactions give valuable information about ion structures and mechanisms. For example the molecular ion of thionyl[1-^{13}C]aniline (122) eliminates CO in a low-energy process with complete retention of ^{13}C.[148] The ion (123) is

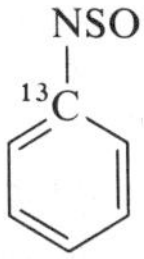

(122) 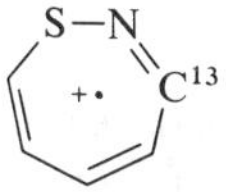   (123) 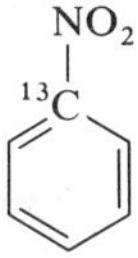   (124)

suggested for the M^+ − CO structure because further loss of CS retains all the label. The spectrum of nitro[1-^{13}C]benzene (124) demonstrates that the NO_2 substituent does not shift around the ring prior to elimination of NO from a rearranged molecular ion.[149] The M^+ − NO ion eliminates exclusively ^{13}CO.

7 Substituent Effects

There has been a shift of emphasis in the literature covering substituent effects in organic mass spectrometry since the publication of Volume 1 of these reports.[1] The period 1968—70 saw considerable activity in this field, with authors defining the effects of substituents on ion abundances in terms of the QET. This period of interest in substituent effects benefted organic mass spectrometry generally, since the knowledge of factors influencing ion abundances from relatively large organic molecules was necessarily increased beyond the previously somewhat superficial level. With the effects of substituents on ion abundances thus qualitatively defined, authors over the period 1970—72 have been cautious about making mechanistic inferences from substituent-effect data. The factors influencing ion abundances are fairly complex and additional data (*e.g.* IP and AP measurements, variable eV studies) are now frequently employed in substituent effect studies, in order to extract mechanistic information.

Hammett Correlations.—Kinetic derivations have been attempted in order to explain mass spectral substituent effects in general and the spectra of substituted benzophenones (125) in particular.[150] A combination of previous methods is partly employed and the factors influencing ion abundances are discussed. The authors concentrate on the relative abundances of the ions $YC_6H_4CO^+$ and $C_6H_5CO^+$, formed in competition from the molecular ion (125): a plot of $\log[YC_6H_4CO^+]/[C_6H_5CO^+]$ against σ^+ yields a straight line. It is argued that the daughter ion ratio $[A_1^+]:[A_2^+]$ is a measure of relative rate constants for their formation. The rate constants thus obtained, however, are weighted

---

[148] A. S. Siegel, *Org. Mass Spectrometry*, 1970, **3**, 875.
[149] F. Benoit and J. L. Holmes, *Chem. Comm.*, 1970, 1031.
[150] N. Einolf and B. Munson, *Org. Mass Spectrometry*, 1971, **5**, 397.

average values which depend upon the distribution of energies produced by electron impact in the molecular ion[150] (see also Figure 5). The variation of $[A_1^+]:[A_2^+]$ with electron-beam energy has been investigated. At high eV (*i.e.*

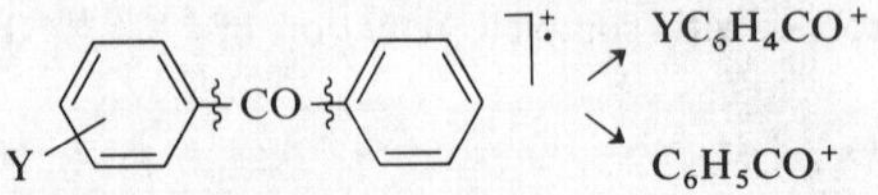

$$Y-C_6H_4-CO-C_6H_5 \quad]^{\ddagger} \quad \rightarrow YC_6H_4CO^+$$
$$\searrow C_6H_5CO^+$$
(125)

above 30 eV) the ratio becomes independent of eV. As the authors point out, it does not follow that this ratio at high electron energies corresponds to the ratio of frequency factors. Equation 13 gives an expression for the variation of k_1/k_2 with the specific energy of the molecular ion. $[A_1^+]:[A_2^+]$ is obtained by summing over the whole energy distribution, which always contains a substantial contribution from lower-energy reactions (see Figure 5). Therefore the final term in equation 13 has a substantial effect on $[A_1^+]:[A_2^+]$ at all electron-beam energies and some preliminary charge-exchange results confirm this. For example, molecular ions generated from (125; Y = p-NO$_2$) with a specific energy of 15.8 eV from charge exchange with Ar$^+$ yield a $[NO_2C_6H_4CO^+]:$ $[C_6H_5CO^+]$ ratio larger (0.23) than the value (0.13) obtained from 70 eV electrons, and of course much larger than the low eV value. This result indicates the considerable contribution of the low-energy fraction of the energy distribution in electron-impact spectra. In passing, it should be noted that substituent effects in benzophenones have been examined by chemical ionization mass spectrometry.[151]

$$\log k_1/k_2 = \log v_1/v_2 + (s-1)\log\left(\frac{E - E_1^0}{E - E_2^0}\right) \quad (13)$$

Other correlations between daughter ion ratios and σ constants have been reported in the recent literature. $\text{Log}[A_1^+]/[A_2^+]$ for daughter ions from the substituted ethylenediamines (126) correlate well with Taft's σ^0 constant.[152]

The mass spectra of a series of substituted polyenic compounds (127) have been measured and the ratio $[PhCO^+]:[M^+]$ has been correlated with the substituent constant σ.[153] Transmission of substituent influence through the chain of conjugated double bonds is discussed in terms of the earlier kinetic

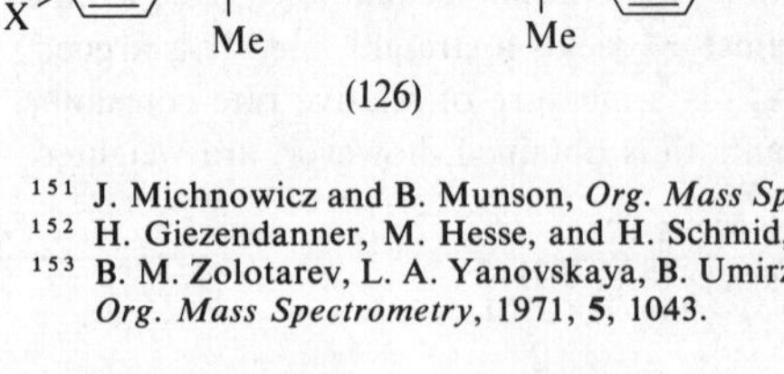

$$X-C_6H_4-N(Me)-CH_2-CH_2-N(Me)-C_6H_5 \quad]^{\ddagger}$$
(126)

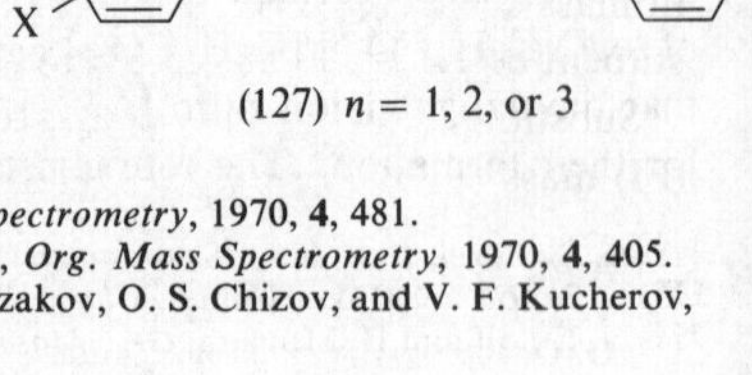

$$X-C_6H_4-(CH=CH)_n-CO-C_6H_5$$
(127) $n = 1, 2,$ or 3

---

[151] J. Michnowicz and B. Munson, *Org. Mass Spectrometry*, 1970, **4**, 481.
[152] H. Giezendanner, M. Hesse, and H. Schmid, *Org. Mass Spectrometry*, 1970, **4**, 405.
[153] B. M. Zolotarev, L. A. Yanovskaya, B. Umirzakov, O. S. Chizov, and V. F. Kucherov, *Org. Mass Spectrometry*, 1971, **5**, 1043.

Kinetic and Energetic Studies of Organic Ions 77

approach to substituent effects. The transmission coefficients are calculated as $\pi' = 0.72$, 0.25, and 0.11 for $n = 1$, 2, and 3, respectively. However, it should be emphasized that there are factors operating which will influence daughter : parent ion ratios whether there is a saturated or unsaturated system separating the two rings. In particular there is a substituent effect on the IP and the fraction of ions with insufficient energy to decompose. Nevertheless, the Hammett correlations are close to linearity and to this extent the results may be used predictively.

The effect of substituent on both the rearrangement ($YC_6H_4NH_2^{+\cdot}$) and direct cleavage ($MeCO^+$) product ions from the molecular ions of substituted acetanilides (128) has been investigated.[154] The rearrangement ion shows the expected

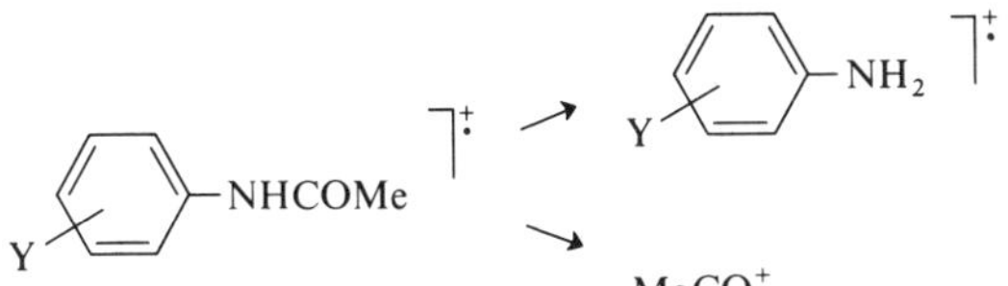

<chemical structure: Y-C6H4-NHCOMe (128) rearranging to Y-C6H4-NH2 radical cation, and MeCO+>

kinetic properties for all substituents, becoming more prevalent relative to the direct cleavage reaction at lower energies (see also Section 4). Hammett plots were made for daughter : parent ion ratios in both these reactions. Good correlations were obtained with small, but positive, ρ values. The authors rightly point out that this correlation largely arises because of the correlation of IP with σ^+, and it is unwise to attach too much importance to the correlation of $\log Z/Z_0$ with σ^+.

Similarly, a reduction in Z values by electron-donating substituents in the formation of $p\text{-}Y\cdot C_6H_4CH_2O^+$ ions from p-substituted benzyl nitrates is probably due largely to the effect of the substituent on the IP[127] (see below for quantitative correlations of this effect). In the spectra of some substituted aryl esters (129), a similar substituent effect was observed on the ratio $[RCH_2CO^+]$:

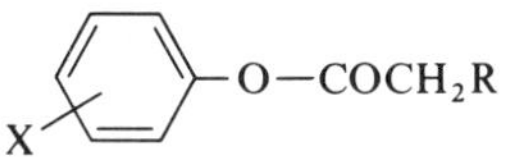

<chemical structure: X-C6H4-O-COCH2R>

(129) R = H, CH_3, F, or Cl

$[M^+]$.[155] It is worth repeating at this stage that the majority of Hammett plots in mass spectrometry give positive ρ values, owing partly to the effect of substituent on both the IP and ions with insufficient energy to decompose.

Substituent effects have been evaluated for the first time in field ionization (FI) mass spectrometry.[156] Owing to the short ion lifetimes in an FI source,

[154] A. A. Gamble, J. R. Gilbert, and J. G. Tillett, *J. Chem. Soc. (B)*, 1970, 1231.
[155] V. J. Feil and J. M. Sugihara, *Org. Mass Spectrometry*, 1972, **6**, 265.
[156] G. O. Phillips, W. G. Filby, and W. L. Mead, *Chem. Comm.*, 1970, 1269.

production and measurement of daughter ions can be a problem, but in the FI mass spectra of the substituted aryl-O-glucosides (130) daughter ions were obtained having up to 5% of the molecular ion intensity. Log Z/Z_0 plots were made for the daughter ions (131) and (132), formed by rearrangement and direct

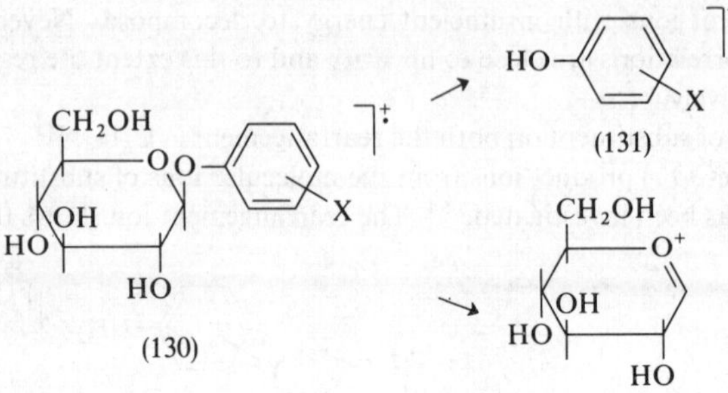

cleavage respectively. There was some scatter, but both plots gave positive ρ values.

Other substituent-effect studies on positive-ion abundances have employed substituted phenyl azides,[157] aroylhydrazones,[158] and 5-bromo-2-phenyl-2-pentenes (58).[84] The charge-localization concept has been invoked to explain the McLafferty rearrangement from the molecular ions of some substituted phenyl butyrophenones (133), and was found not to give a completely satisfactory explanation of the data.[159]

It would appear that measurement of the substituent effect on the approximate activation energy AP–IP might be profitable for mechanistic investigations. IP's and AP's have been determined for competing reactions (loss of CH_3 and CH_2O) from the molecular ion of substituted anisoles (134).[160] Values of AP–IP are found not to differ by more than 0.2 eV for m- and p-X pairs with the same substituent. This result implies either common rearranged molecular

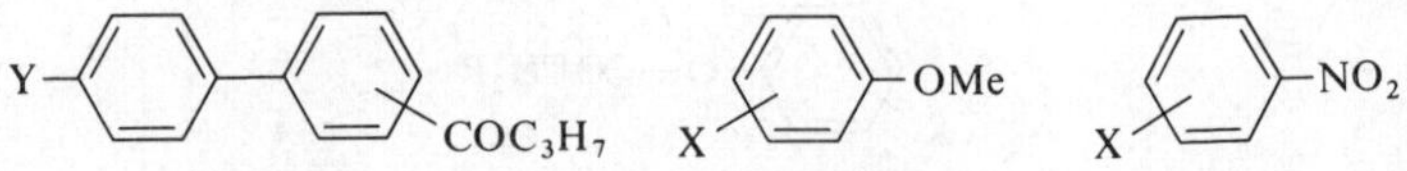

ions or similar activation energies in unrearranged ions.[160] Values of AP–IP were also found to be similar for m- and p-isomers undergoing the $M^+ \rightarrow M^+ - NO_2$ reaction in substituted nitrobenzenes (135).[161] However, the author

[157] R. A. Abramovitch, E. P. Kyba, and E. F. V. Scriven, *J. Org. Chem.*, 1971, **36**, 3796.
[158] D. G. I. Kingston, H. P. Tannenbaum, G. B. Baker, J. R. Dimmock, and W. G. Taylor, *J. Chem. Soc.* (*C*), 1970, 2574.
[159] D. A. Lightner and F. S. Steinberg, *Canad. J. Chem.*, 1971, **49**, 660.
[160] P. Brown, *Org. Mass Spectrometry*, 1970, **4**, 519.
[161] P. Brown, *Org. Mass Spectrometry*, 1970, **4**, 533.

indicates that it is not possible to be conclusive from these results about positional scrambling in the molecular ion. Correlations of AP and IP with σ constants are also presented and the disadvantages and limitations of AP–IP measurements are discussed[160,161] (see Section 2 for discussion of errors in energy measurements such as these, and also ref. 10 for discussion of the kinetic shift in substituent-effect reactions).

A useful empirical expression has been employed to calculate the vertical IP's of disubstituted benzenes.[162] It has previously been noted[163] that a good correlation exists between IP and σ^+ for the substituent in monosubstituted benzenes. In order that the best possible predictions may be made in the case of disubstituted benzenes, a new set of Hammett constants (σ_{IP}^+) have been defined[162] in terms of the vertical IP's obtained by photoelectron spectroscopy. The empirical equation 14 is then employed to calculate the ionization potential $I_{1,2}$ of a disubstituted benzene. The σ values are σ_{IP} for the respective substituents.

$$I_{1,2} = 1.11(\sigma_1 + \sigma_2) \pm 0.81(\sigma_1\sigma_2) + 9.28 \qquad (14)$$

In a series of disubstituted benzenes where one of the substituents is constant, equation 14 reduces to equation 15, in which k and k' are constants. Hence

$$I_{1,2} = k\sigma_2 + k' \qquad (15)$$

the IP's of such a series should correlate with σ_{IP}, and this is largely borne out in practice.

The IP's of a limited number of disubstituted derivatives of diphenylmethane (136) and diphenylethane (137) have been measured (or values taken from the literature) in order to evaluate whether only non-classical transannular conjugation effects exist in these neutral molecules.[164] Since the IP corresponds to removal of an electron from the highest occupied molecular orbital, and if it is assumed as a starting hypothesis that (136) and (137) are equivalent to two isolated aromatic rings not interacting at all, then: (i) if $X = NH_2$ and Y changes

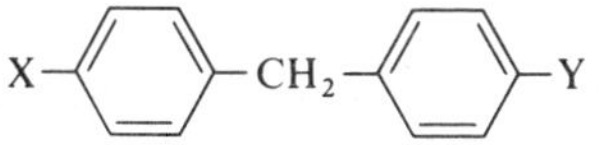

X—⟨⟩—CH$_2$—⟨⟩—Y   X—⟨⟩—CH$_2$CH$_2$—⟨⟩—Y

(136) 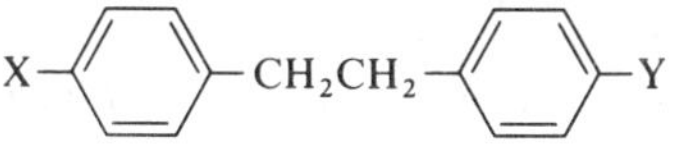   (137)

and is more electron-withdrawing, the energy required to remove the least firmly bound electron from these systems should be equal to the value for the NH_2—C_6H_4 moiety; (ii) if $X = H$ and Y is variable, the IP should increase on going from $Y = NH_2$ to $Y = H$, and for electron-withdrawing substituents the electron should then be extracted from the unsubstituted ring. In an IP vs. σ_p^+ plot there should be a break at $Y = H$; (iii) if $X = NO_2$ and Y is variable, the

[162] T. W. Bentley and R. A. W. Johnstone, *J. Chem. Soc. (B)*, 1971, 263.
[163] G. F. Crable and G. L. Kearns, *J. Phys. Chem.*, 1962, **66**, 436.
[164] S. Pignataro, V. Mancini, G. Innorta, and G. Distefano, *Z. Naturforsch.*, 1972, **27a**, 534.

electron should be removed from the Y-substituted ring. The results largely fit this model, indicating that the 'homoconjugative effect' is negligible in this particular system.[164]

The first observations of substituent effects in negative-ion mass spectrometry have been reported.[165] The compounds studied were various series of substituted nitroaryl benzoates (138) and aryl nitrobenzoates (139). The intention of the study was to determine whether linear correlations of log Z/Z_0 with σ, or of functions proposed by Harrison,[166] were obtainable, and if so, to extract any mechanistic information. Some very good correlations were obtained, for example between log Z/Z_0 and σ (positive ρ) for the formation of m/e 166 from (140). It was concluded from this and other effects that (i) valence isomerism

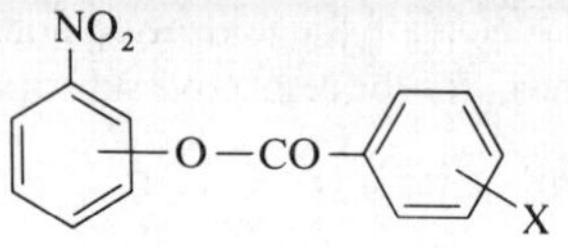

(138)

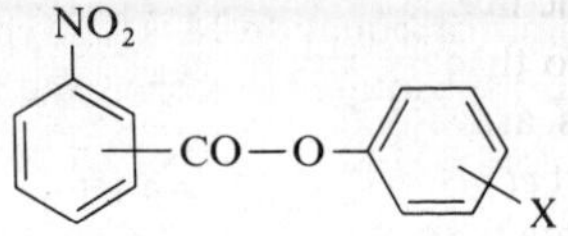

(139)

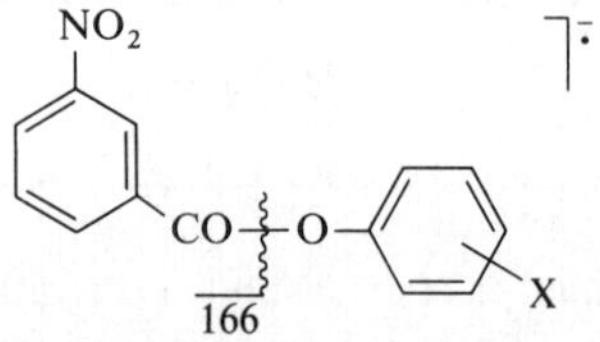

(140)

does not accompany the fragmentations of the molecular anions and (ii) the ρ values should not be used to define the nature of the transition states for the reactions.[165] Other substituent-effect studies in negative-ion mass spectrometry are awaited with interest and, in particular, AP measurements should reveal whether similar effects are influencing the ion abundances as were documented for positive-ion substituent effects.

In an extensive publication, McLafferty et al.[10] have shown that the effect of substituents on ion abundances in the spectra of 1,2-diphenylethanes (1) is consistent with the predictions of the QET, as evidenced by rate constants calculated utilizing the RRKM theory with exact enumeration of states. A number of contributory factors have been re-emphasized and identified, and correlated with σ^+ or σ_p^+. These factors, which influence the relative ion abundances of M^+, $XC_7H_6^+$, and $C_7H_7^+$, may be divided into those influencing the rate curves [$k(E)$ vs. E] for the competing reactions and those influencing the energy distribution [$P(E)$ vs. E, see also Section 3].

Among the factors influencing the $k(E)$ functions are:[10] (i) the activation energies for the two reactions, which depend on (a) the IP's of the radicals

[165] J. H. Bowie and B. Nussey, Org. Mass Spectrometry, 1972, 6, 429.
[166] M. S. Chin and A. G. Harrison, Org. Mass Spectrometry, 1969, 2, 1073.

Kinetic and Energetic Studies of Organic Ions 81

$YC_7H_6^+$ and $C_7H_7^+$, (b) the dissociation energy of the central C—C bond in the neutral, and (c) the reverse activation energy; (ii) the number of active states in the molecular ion, which depends on (a) the vibrational degrees of freedom, and (b) the geometry of the activated complex (cf. the frequency factor employed in the simplified equation 8).

Among the factors influencing the $P(E)$ curve are: (i) the energy states of the molecule; (a) the IP of (1) is correlated with σ^+, (b) splitting of states is increased by electron-donating substituents, (c) additional states arise from the presence of non-bonding electrons; (ii) relative populations of the energy states which vary substantially with the substituent X.

Steric Effects.—It was seen in the first volume of these reports[1] that the concept of steric inhibition in mass spectral processes appears to be significant, with the proviso that reductions in daughter ion intensities on substitution by blocking groups are not due to competition reactions involving the blocking group. Gilbert et al. in a series of simple experiments[167,168] have used the steric effect to elucidate the mechanisms of hydrogen rearrangement reactions in butylbenzenes [(141) → (142)], acetanilides [(143) → (144)], and phenyl acetates [(145) → (146)].

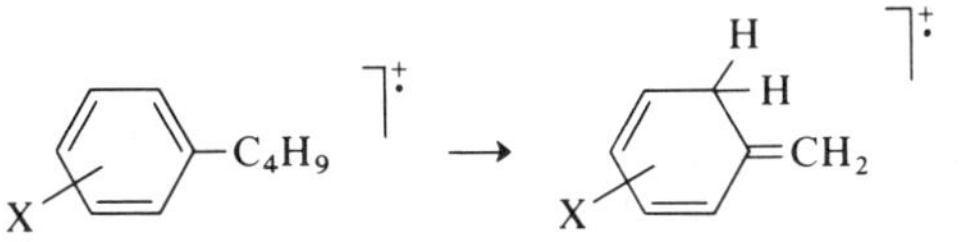

$$X\text{-}C_6H_4\text{-}C_4H_9\ ]^{+\cdot} \quad \rightarrow \quad X\text{-}C_6H_4\text{=}CH_2\ (\text{+2H})\ ]^{+\cdot}$$

(141)   (142)

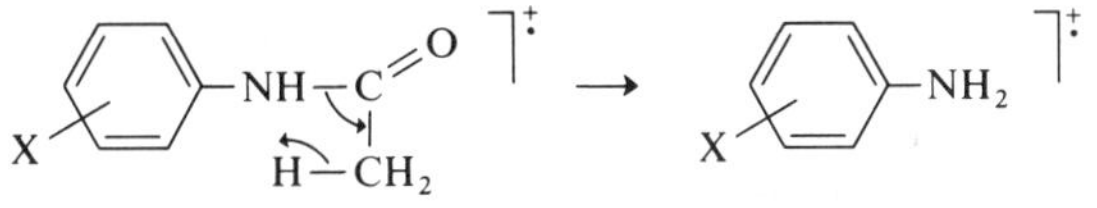

$$X\text{-}C_6H_4\text{-}NH\text{-}C(=O)\text{-}CH_2\text{-}H\ ]^{+\cdot} \quad \rightarrow \quad X\text{-}C_6H_4\text{-}NH_2\ ]^{+\cdot}$$

(143)   (144)

$$X\text{-}C_6H_4\text{-}O\text{-}C(=O)\text{-}CH_2\text{-}H\ ]^{+\cdot} \quad \rightarrow \quad X\text{-}C_6H_4\text{-}OH\ ]^{+\cdot}$$

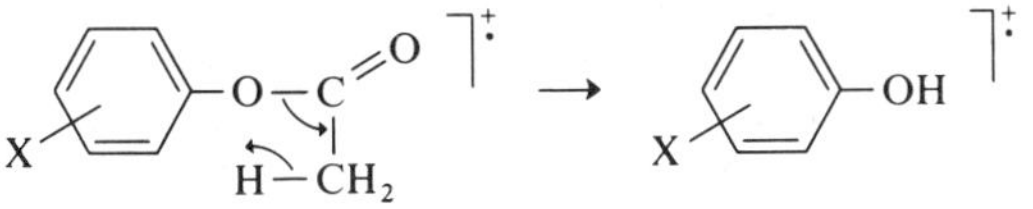

(145)   (146)

Daughter : parent ion ratios were measured for the unsubstituted case (X = H) and for various methyl- and dimethyl-substituted derivatives. For example, for the butylbenzenes (141) at 70 eV $[M^+ - C_3H_6]:[M^+]$ was 1.51 (X = H), 0.71 (X = m-Me), 0.11 (X = o-Me), and 0.00 (X = 2,6-dimethyl). Thus it appears that a steric effect is operating and the hydrogen transfer in formation of (142) takes place via a six-membered-ring transition state and not via a four-membered-

---

[167] A. A. Gamble, J. R. Gilbert, and J. G. Tillett, *Org. Mass Spectrometry*, 1970, **3**, 1223.
[168] A. A. Gamble, J. R. Gilbert, and J. G. Tillett, *Org. Mass Spectrometry*, 1971, **5**, 1093.

ring transition state to form the substituted toluene. In a related ICR study[169] it was found that $C_7H_8^{+\cdot}$ ions from butylbenzene differ in reactivity from those from toluene, probably because of different structure.

For the rearrangements from (143) and (145), however, the *o*-methyl groups had little inhibiting effect on the elimination of CH_2CO, which indicates that the four-membered cyclic transition states provide the preferred pathway, as shown.[167,168] In the case of the acetanilides [(143) → (144)] the same conclusion was reached from isotope effect studies.[136] The molecular ions of enol acetates also appear to lose keten *via* a four-membered transition state.[170] It has been shown by deuterium labelling that keten is eliminated in a four-membered transition state from the molecular ion of methylacetylsalicylate (147).[171] As a general rule, it appears that the hydrogen is transferred in the above rearrangements to a vacant *p*-orbital of a heteroatom (if present) or alternatively *via* a six-membered transition state to a vacant π-orbital of the ring (if no heteroatom is available).

Steric effects have been investigated for some substituted benzophenones. It appears from ion abundance data that there is little steric inhibition of resonance of the dimethylamino-group by the bromo-substituents flanking it [see (148)] on the formation of the benzoyl ion.[172]

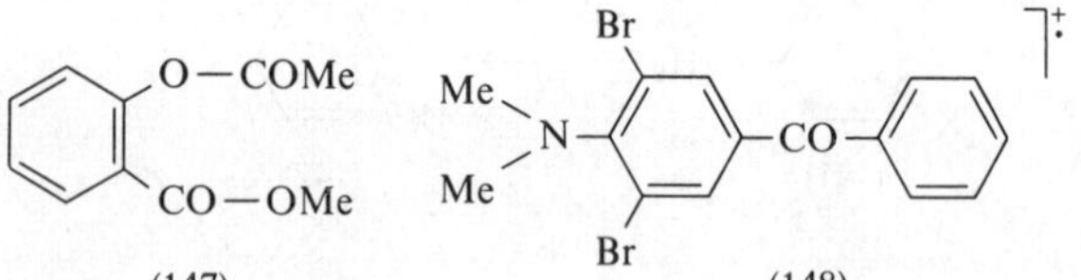

(147)   (148)

8 Collision Processes

Two important methods for the investigation of ion–molecule reactions in the gas phase are chemical ionization (CI) and ion cyclotron resonance (ICR) mass spectrometry. Work in these two areas has increased rapidly in the last few years and the recent literature is covered in Chapter 1. The literature covered in this section is confined largely to a few energetic and kinetic studies of ion–molecule reactions. The reader is referred to publications in the CI[173–175] and the ICR[176,177] fields which are concerned with kinetic interpretation of data. ICR is a particularly useful tool for structure elucidation of low-energy

[169] M. M. Bursey, M. K. Hoffman, and S. A. Benezra, *Chem. Comm.*, 1971, 1417.
[170] H. Nakata and A. Tatematsu, *Org. Mass Spectrometry*, 1970, **4**, 211.
[171] H. Nakata and A. Tatematsu, *Org. Mass Spectrometry*, 1971, **5**, 1343.
[172] M. M. Bursey and C. E. Twine, *J. Org. Chem.*, 1971, **36**, 137.
[173] D. P. Beggs and F. H. Field, *J. Amer. Chem. Soc.*, 1971, **93**, 1567, 1576.
[174] S. Vredenberg, L. Wojcik, and J. H. Futrell, *J. Phys. Chem.*, 1971, **75**, 590.
[175] G. Sroka, C. Chang, and G. G. Meisels, *J. Amer. Chem. Soc.*, 1972, **94**, 1052.
[176] G. C. Goode, R. M. O'Malley, A. J. Ferrer-Correia, R. I. Massey, and K. R. Jennings, *Internat. J. Mass Spectrometry Ion Phys.*, 1970, **5**, 393.
[177] T. McAllister, *Internat. J. Mass Spectrometry Ion Phys.*, 1972, **8**, 162.

ions.[178—181] Collision-induced metastable reactions of organic ions (another expanding topic of research) are covered in this section. Brief reference to charge-exchange reactions may be found in Sections 3 and 7.

Ion–Molecule Reactions.—High-pressure ion sources have proved suitable for the investigation of ion–molecule reactions of organic species, free from solvent interaction. Both the kinetics and energetics of such processes have been studied and some of the publications are reported below. Extensive reviews have appeared on the topic.[182,183]

It is worth recalling a few simple kinetic properties of ion–molecule reactions. First it should be noted that most ion–molecule reactions proceed with little or no energy of activation (see also, the reverse activation energy for unimolecular decompositions, Section 2). It is informative to relate the rate of an ion–molecule reaction to the theoretical collision frequency; ideally it is calculated that a rate constant of about 10^{-9} cm^3 mol^{-1} s^{-1} should result if every collision is reactive. In practice, many ion–molecule reactions in the gas phase exhibit this maximum rate constant and are usually appreciably faster than reactions between neutral species. Examples of rate constants for ion–molecule reactions may be found in many recent publications (see for example refs. 184—186). Clearly it will not always be the case that all collisions are reactive. For example, a simple proton transfer or hydrogen abstraction is likely to occur with higher probability than a condensation reaction where more intimate contact between ion and neutral is required.[186] Contrast, for example, reaction (149) ($k = 1.28 \times 10^{-9}$ cm^3 mol^{-1} s^{-1}) with reaction (150) ($k = 5.5 \times 10^{-11}$ cm^3 mol^{-1} s^{-1}).[186]

$$CH_3F^{+\cdot} + CH_3F \rightarrow CH_3FH^+ + CH_2F\cdot \qquad (149)$$
$$CH_3Br^{+\cdot} + CH_3Br \rightarrow (CH_3)_2Br^+ + Br\cdot \qquad (150)$$

The large cross-section for ion–molecule reactions is mainly due to (i) long-range attractive forces and (ii) the formation of a collision complex of relatively long lifetime, with possible orbiting interaction between the two species. The cross-section varies with the relative kinetic energy of the reacting species.

Ion–molecule reactions have been investigated for C_2F_6[187] and C_3F_8;[188] the most abundant ions in the high-pressure spectrum of C_2F_6 are $C_2F_5^+$ and CF_3^+. For ionization with 70 eV electrons these two species have as much as

---

[178] C. L. Wilkins and M. L. Gross, *J. Amer. Chem. Soc.*, 1971, **93**, 895.
[179] M. L. Gross and F. W. McLafferty, *J. Amer. Chem. Soc.*, 1971, **93**, 1267.
[180] S. Wexler and L. G. Pobo, *J. Amer. Chem. Soc.*, 1971, **93**, 1327.
[181] T. A. Lehman, T. A. Elwood, J. T. Bursey, M. M. Bursey, and J. L. Beauchamp, *J. Amer. Chem. Soc.*, 1971, **93**, 2108.
[182] E. W. McDaniel, V. Cermak, A. Dalgarno, E. E. Ferguson, and L. Friedman, 'Ion–Molecule Reactions', Wiley-Interscience, New York, 1970.
[183] J. E. Parker and R. S. Lehrle, *Internat. J. Mass Spectrometry Ion Phys.*, 1971, **7**, 421.
[184] O. P. Strausz, W. K. Duholke, and H. E. Gunning, *J. Amer. Chem. Soc.*, 1970, **92**, 4128.
[185] M. S. Foster and J. L. Beauchamp, *J. Amer. Chem. Soc.*, 1972, **94**, 2425.
[186] J. L. Beauchamp, D. Holtz, S. D. Woodgate, and S. L. Patt, *J. Amer. Chem. Soc.*, 1972, **94**, 2798.
[187] R. E. Marcotte and T. O. Tiernan, *J. Chem. Phys.*, 1971, **54**, 3385.
[188] T. Su, L. Kevan, and T. O. Tiernan, *J. Chem. Phys.*, 1971, **54**, 4871.

2.4 and 2.9 eV respectively of internal energy, which accounts for the endothermic reactions that they undergo. In general, the bimolecular rate constants in this system are over two orders of magnitude slower than those typically found for ion–molecule reactions in hydrocarbon systems [e.g. reaction (151) for which

$$CF^+ + C_2F_6 \rightarrow CF_3^+ + C_2F_4 \qquad (151)$$

$k = 1.3 \times 10^{-11}$ cm^3 mol^{-1} s^{-1}]. This suggests that sizeable activation energies are involved; reactions observed include F and F_2 transfer.

In the corresponding high-pressure C_3F_8 system,[188] the most common reaction is F^- transfer which has a typical rate constant of 5×10^{-11} cm^3 mol^{-1} s^{-1}. Such rate constants are generally more than an order of magnitude slower than those for analogous H^- transfers in alkanes. Kinetic energy dependence of reaction cross-section was also determined. The negative ion–molecule reactions of C_3F_8 have also been investigated.[189] The main ion is F^- with smaller amounts of CF_3^- and $C_2F_5^-$. The collision-induced dissociations (152) and (153) have rate constants of 3×10^{-12} and 8×10^{-12} cm^3 mol^{-1} s^{-1} respectively.

$$CF_3^- + M \rightarrow F^- + CF_2 + M \qquad (152)$$
$$C_2F_5^- + M \rightarrow F^- + C_2F_4 + M \qquad (153)$$

Hydrocarbon ion–molecule reactions have proved to be a popular research topic. The reactions in mixtures of CH_4 and CD_4 at pressures up to about 0.65 Torr have been investigated in the ion source of a quadrupole mass filter.[190] At pressures above 0.2 Torr of methane, the two principal ions in the mass spectrum are CH_5^+ and $C_2H_5^+$. In the mixture of CH_4 and CD_4, kinetic plots and measurements of isotopic distributions in the ethyl ions reveal that randomization of hydrogens occurs in the collision complex [see (154)] between methyl ions and methane prior to formation of the ethyl ion.

The principal reaction leading to methanium ion (CH_5^+) formation above 0.15 Torr is described by reaction (155). In contrast to the case reported above [see (154)] any further exchange of hydrogens [other than that described in

$$CH_3^+ + CD_4 \rightarrow [C_2H_3D_4^+] \begin{array}{l} \nearrow C_2H_3D_2^+ + D_2 \\ \rightarrow C_2H_2D_3^+ + HD \\ \searrow C_2HD_4^+ + H_2 \end{array}$$
$$(154)$$

$$CH_4^{+\cdot} + CD_4 \rightarrow CH_4D^+ + CD_3 \qquad (155)$$

(155)] is not detectable. Investigations of hydrogen scrambling in the high-pressure ethylene system have also been carried out.[191] The condensation of $C_2H_4^+$ with C_2D_4 to produce $C_3(H,D)_5^+$ ions proceeds with equilibration of the hydrogens, but here a long-lived $C_4(H,D)_8$ ion is not unreasonable on simple valence considerations.

[189] T. Su, L. Kevan, and T. O. Tiernan, *J. Phys. Chem.*, 1971, **75**, 2534.
[190] S.-L. Chong and J. L. Franklin, *J. Chem. Phys.*, 1971, **55**, 641.
[191] W. T. Huntress, *J. Chem. Phys.*, 1972, **56**, 5111.

The high-pressure (0.1—1.0 Torr) mass spectra of methane have also been determined at temperatures between 77 and 300 K.[192] An important feature of this publication is the dominance at low temperatures of hydrocarbon ions with a high H : C ratio whose structures must deviate considerably from conventional ideas about valence [see reactions (156)—(158)]. The values of ΔH for reactions

$$CH_5^+ + CH_4 \rightarrow C_2H_9^+ \quad (156)$$
$$C_2H_5^+ + CH_4 \rightarrow C_3H_9^+ \quad (157)$$
$$C_2H_9^+ + CH_4 \rightarrow C_3H_{13}^+ \quad (158)$$

(156)—(158) are -17.3, -10.0, and -6.2 kJ mol^{-1} respectively. Thus the complexes are bound by weak chemical forces. An interesting feature of the observations was the occurrence of second drift region collision-induced metastables for the reverse reactions (156)—(158). These metastable peaks were all very sharp, indicating very little kinetic energy release.

The high-pressure mass spectrometry (from impact with 10.0 eV and 11.6—11.8 eV photons) of cyclohexene, cyclopentene, and the three isomeric methylcyclopentenes (159)—(161) have been investigated[193] (see Section 2 for the

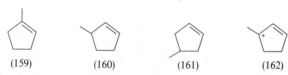

(159)　　　　(160)　　　　(161)　　　　(162)

unimolecular reactions undergone by C_6H_{10} molecular ions). Bimolecular rate constants were determined for the various reactions (H transfer, H_2 transfer, and condensation) undergone between the parent ion and the parent molecule. Some of the rate constants obtained may be rationalized in terms of the ion structures. For example, the hydrogen-transfer reaction (163) is important

$$C_6H_{10}^{+\cdot} + C_6H_{10} \rightarrow C_6H_9^+ + C_6H_{11}^{\cdot} \quad (163)$$

only for isomers (160) and (161) and is faster for (160) ($k = 2.8 \times 10^{-10} \text{ cm}^3 \text{ mol}^{-1} \text{ s}^{-1}$). In both these cases, there is an opportunity for extraction of a tertiary hydrogen from the molecular ion, to form a tertiary carbonium ion. In the case of (160), the product ion is allylic (162), thus conferring extra stability. Incidentally, the sum of all the rate constants for the reactions undergone between parent molecule and ion is remarkably constant ($\sim 5 \times 10^{-10} \text{ cm}^3 \text{ mol}^{-1} \text{ s}^{-1}$) for all five compounds studied, so presumably the same high proportion of collisions is active in all cases. Studies of ion–molecule reactions of the parent ions of propylene and cyclopropane with $n\text{-}C_4D_{10}$ have yielded kinetic information on hydrogen-transfer reactions.[194]

An unusual termolecular reaction has been investigated between $C_3H_5^+$ (produced from electron impact on four different molecules) and C_2D_4 to

[192] F. H. Field and D. P. Beggs, *J. Amer. Chem. Soc.*, 1971, **93**, 1585.
[193] R. Lesclaux, S. Searles, L. W. Sieck, and P. Ausloos, *J. Chem. Phys.*, 1971, **54**, 3411.
[194] I. Koyano, N. Nakayama, and I. Tanaka, *J. Chem. Phys.*, 1971, **54**, 2384.

produce $C_5H_5D_4^+$.[195] The reaction is second order in C_2D_4 [see (164)]. The extra molecule of ethylene is necessary to stabilize the rapidly formed collision

$$C_3H_5^+ + 2C_2D_4 \rightarrow C_5H_5D_4^+ + C_2D_4 \qquad (164)$$

complex. The high termolecular rate constant explains the high yield of $C_5H_9^+$ in the ethylene high-pressure mass spectrum.

Collision-induced Metastables.—Collision-induced dissociation may be defined as the conversion of kinetic energy into internal excitation energy of the reactant ion sufficient to lead to its fragmentation.[183] A considerable increase in internal energy may occur when organic ions of high kinetic energy (e.g. 8 keV) collide with neutral molecules. These higher-energy ions subsequently undergo decompositions which may be frequently observed in the metastable drift regions as collision-induced metastables.

Instrumental modifications have been described for studies of collision-induced fragmentations by introducing any collision gas at a controlled pressure into the first drift region, preceding the electric sector.[196,197] The target gas can be introduced simply, without adversely affecting the normal mode of operation of the mass spectrometer.

The first drift region has been used in this way for sensitive measurements of collision-induced metastable decompositions, using the defocused mode.[197] Argon and helium have been used as target gases, and under controlled pressure conditions an optimum pressure for observation of collision-induced metastables may be determined. The effect of increasing pressure on the 'collision-induced spectrum' of decompositions from a given ion is similar to that of increasing the electron-beam energy on the normal spectrum. There is an increase in the prevalence of decompositions requiring more energy. The AP's (from the normal spectrum) of the fragments formed by collision-induced dissociation confirm that several eV may be added by collision. For example, four abundant collision-induced metastables are observed from the molecular ion of acetophenone (165),

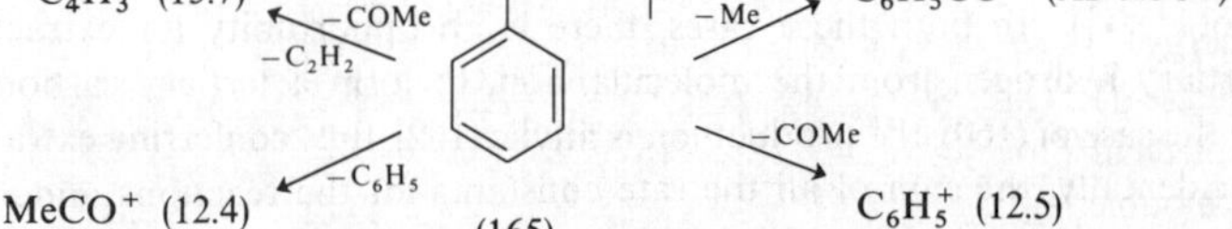

$$C_4H_3^+ \;(15.7) \xleftarrow[-C_2H_2]{-COMe} \text{COMe-}C_6H_5^{+\cdot} \xrightarrow{-Me} C_6H_5CO^+ \;(AP\;9.6\;eV)$$

$$MeCO^+ \;(12.4) \xleftarrow{-C_6H_5} (165) \xrightarrow{-COMe} C_6H_5^+ \;(12.5)$$

and one of these reactions ($C_4H_3^+$ formation) has an energy requirement 6.1 eV above that for the lowest-energy unimolecular reaction.[197] However, loss of total ion intensity occurs with increasing pressure owing to collisional scattering. The fact that the molecular ion intensity decreases with increasing pressure

[195] A. G. Harrison and A. A. Herod, *Canad. J. Chem.*, 1970, **48**, 3549.
[196] J. H. Beynon, R. M. Caprioli, and T. Ast, *Internat. J. Mass Spectrometry Ion Phys.*, 1971, **7**, 88.
[197] F. W. McLafferty, R. A. Kornfeld, S.-C. Tsai, I. Howe, and P. F. Bente, submitted for publication.

according to equation 16 is consistent with occurrence of the bimolecular reaction (166).

$$\log [M^+] = -kP \qquad (16)$$

$$M^+ + A \rightarrow M^{+*} + A \qquad (166)$$
$$\downarrow$$
$$\text{products}$$

The optimum pressure for observation of collision-induced metastables has been found to be about 10^{-4} Torr for a number of cases studied,[197] although the optimum pressure varies with such factors as the target gas, the length of the drift tube, and the kinetic energy of the incident ion. In the RMH-2, having a long drift tube of 590 mm, the optimum pressure is $\sim 10^{-5}$ Torr.[196] That collision-induced metastable ions decompose from higher average internal energies than unimolecular metastables is illustrated by a change in the deuterium isotope effect for loss of a hydrogen atom from the scrambled toluene molecular ion. This isotope effect in the first drift region decreases from 2.80 ± 0.06 to 2.33 ± 0.08 on introduction of argon at a pressure of 2×10^{-4} Torr, in accord with an increase in the internal energy.[121]

The conversion of part of the kinetic energy of the incident ion directly (on collision with a neutral) into internal energy has been established for the reaction (167) → (168) from the dimethyl ether molecular ion.[198] The unimolecular

$$\overset{+}{\text{MeOMe}} \rightarrow \overset{+}{\text{MeO}}=CH_2 + H\cdot$$
$$(167) \qquad\qquad (168)$$

reaction (167) → (168) at a pressure of 2×10^{-7} Torr gives a very narrow metastable peak in the first drift region. As air is admitted to the region, a collision-induced peak appears which is broader and shifted towards higher mass. This mass shift corresponds to a loss on average of 3.6 eV of kinetic energy from the molecular ion (167) after collision and prior to fragmentation. This loss of kinetic energy must correspond in magnitude to the internal energy transferred to the reactant, assuming that the neutral does not carry off excess energy. This excess internal energy is reflected in the greater kinetic energy release (150 meV) in the collision-induced process compared with the unimolecular process (4 meV).[198]

Double ionization of singly charged ions by collision with neutrals[199] according to the charge-transfer reaction (169) has now been utilized in the first drift region to generate a new type of 'collision-induced mass spectrum'.[200] If the collection of singly charged ions formed in the usual way in an electron-impact source are then passed through a pressurized first drift region ($p = 10^{-6}$—10^{-4} Torr), the doubly charged ions formed by reaction (169) can be separated from the singly charged ions by operating the electric sector at half the normal

[198] J. H. Beynon, M. Bertrand, E. G. Jones, and R. G. Cooks, *J.C.S. Chem. Comm.*, 1972, 341.
[199] K. R. Jennings, *Internat. J. Mass Spectrometry Ion Phys.*, 1968, **1**, 227.
[200] R. G. Cooks, J. H. Beynon, and T. Ast, *J. Amer. Chem. Soc.*, 1972, **94**, 1004.

voltage. By scanning the magnet in the usual way, a new type of mass spectrum is obtained (the $E/2$ mass spectrum). For aniline, the $E/2$ spectrum is markedly different from the normal spectrum. At optimum pressures ($\sim 5 \times 10^{-5}$ Torr) the beam intensity is about 10^{-3} of that of the main beam of singly charged ions. As discussed above[198] a slight shift in the mass position of the metastable peak maximum should occur after collision (owing to loss of kinetic energy). In the case of benzene and aniline for the charge-stripping reaction (169), a loss of

$$A^+ + N \rightarrow A^{2+} + N + e^- \qquad (169)$$

16 ± 2 eV was found.[200] This is close to the differences between the first and second IP's of these molecules.

Another variation of this technique has been described.[99,125,201,202] It is possible to observe a 'doubly charged ion' mass spectrum by doubling the electric sector voltage so that only the ions formed by the charge-exchange reaction (170) in the first drift region are observed. The spectra obtained give

$$M^{2+} + N \rightarrow M^+ + N^+ \qquad (170)$$

some reflection of the distribution of doubly charged ions in the first drift region. These '2E spectra' have been obtained for benzene, phenol, diphenyl ether, and the isomeric dihydroxybenzenes,[201] for toluene,[99] and for various hydrocarbons.[202] The significance of differences between these and the normal singly charged ion spectra has been discussed.

9 Determination of Ion Structures

It can be seen from the preceding sections of this chapter that new methods have become available and old ones more established for determination of ion structures and their decomposition mechanisms. Some or all of these methods may incorporate pitfalls and it is important to evaluate the reliability of each method and decide whether we would expect the different approaches to give the same or different answers. It should be noted that an ion structure may be a function of its lifetime and therefore of its energy. Figure 7 summarizes the ion lifetimes appropriate to different techniques used. A summary of methods used for ion structure and decomposition mechanism determinations follows below:

(i) Heat of formation measurements. Ideally, identical heats of formation should indicate identical structures at formation of the ions, but there are potential sources of error in heat of formation determinations. (ii) QET calculations. These have been used to gain information about transition states of reactions, particularly by evaluation of frequency factors, and in a few cases by application of the more exact theory. It should always be ascertained that invalid assumptions are not obscured by complex mathematics. (iii) Kinetic properties of rearrangements. The variation of relative daughter ion abundances with electron energy has been used to confirm rearrangements, or suggest anchimeric assistance.

[201] J. H. Beynon, A. Mathias, and A. E. Williams, *Org. Mass Spectrometry*, 1971, **5**, 303.
[202] T. Ast, J. H. Beynon, and R. G. Cooks, *Org. Mass Spectrometry*, 1972, **6**, 749.

Kinetic and Energetic Studies of Organic Ions

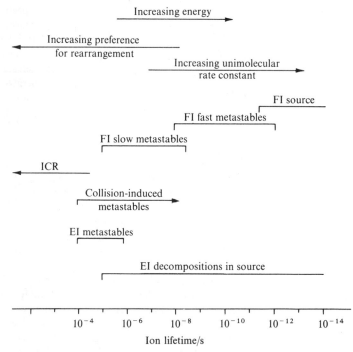

Figure 7 *Ion lifetimes appropriate to different mass spectral techniques*

(iv) Metastable ion abundance ratios. This criterion compares ion structures from different sources at energies just above the lowest activation energy for decomposition. The applicability of the method is limited by the requirement for two or more competing reactions. IKE spectroscopy provides an alternative and sensitive method provided that individual peaks can be uniquely located. (v) Shapes of metastable peaks. These provide a simple criterion for the non-identity of ion structures where shapes are different. Valuable information about the structures of doubly charged ions has been forthcoming from the kinetic energy released in their metastable transitions. (vi) Isotopic labelling. This gives information on scrambling processes and on decomposition mechanisms. A high degree of scrambling usually obscures conclusions about decomposition mechanisms. Deuterium isotope effects have been used for reaction mechanism determination. (vii) Substituent effects. The applications of Hammett equations are strictly limited. Steric effects appear to be a simple and effective tool for mechanism evaluation. (viii) Ion–molecule reactions. ICR spectroscopy is ideal for obtaining information about low-energy ions. Kinetic studies in CI spectroscopy have yielded thermodynamic information. Deuterium labelling and bimolecular rate-constant evaluations have yielded information about collision complexes.

3
Reactions of Specific Functional Groups

BY J. H. BOWIE

1 General Introduction

The material contained in this chapter follows the general format adopted in the previous report.[1] A functional-group approach is used, which allows both a summary of mass spectral processes for the general reader, and a review of progress for the specialist. Some overlap between chapters occurred in the 1970 report, primarily because of the difficulty of dividing mass spectrometry into specific areas. Some overlap is again inevitable, especially between certain sections of Chapters 1—3. The growth of publications in mass spectrometry continues, and over two thousand papers were scanned for this section. The material chosen for review to some extent reflects the interests of the reviewer and, consequently, some important contributions may have been omitted.

The literature prior to 1967 is still best covered by the text of Budzikiewicz, Djerassi, and Williams.[2] The literature since 1967 has been listed by Kiser and Sullivan (1968),[3] De Jongh (1970),[4] Burlingame and Johanson* (1972),[5] and the Chemical Society Specialist Report (1971).[6] The current literature is summarized by the Mass Spectrometry Bulletin,[7]† but references from the less common journals may be more than a year out of date. A computer library of literature references, based on the Mass Spectrometry Bulletin, has been described by Biemann.[8] Many books,[9—21] reviews on a variety of topics,[22—37] and reference data[38—45] have been published since 1970.

---

[1] J. H. Bowie, in 'Mass Spectrometry', ed. D. H. Williams (Specialist Periodical Reports), The Chemical Society, London, 1971, Vol. 1, Chapter 3.
[2] H. Budzikiewicz, C. Djerassi, and D. H. Williams, 'Mass Spectrometry of Organic Compounds', Holden-Day, San Francisco, 1967.
[3] R. W. Kiser and R. E. Sullivan, *Analyt. Chem.*, 1968, **40**, 273R.
[4] D. C. De Jongh, *Analyt. Chem.*, 1970, **42**, 169R.
[5] A. L. Burlingame and G. A. Johanson, *Analyt. Chem.*, 1972, **44**, 337R.
[6] 'Mass Spectrometry', ed. D. H. Williams (Specialist Periodical Reports), The Chemical Society, London, 1971, Vol. 1.
[7] Mass Spectrometry Bulletin, Mass Spectrometry Data Centre, A.W.R.E., Aldermaston, Berkshire, England.
[8] H. S. Hertz, D. A. Evans, and K. Biemann, *Org. Mass Spectrometry*, 1970, **4**, 453.
[9] 'Topics in Organic Mass Spectrometry', ed. A. L. Burlingame, Wiley-Interscience, New York, 1970.

* This review does not include material from the journals *Organic Mass Spectrometry* and *International Journal of Mass Spectrometry and Ion Physics*.
† The Mass Spectrometry Bulletin can now be obtained on magnetic tape.

Reactions of Specific Functional Groups 91

The past two years have seen a period of consolidation in the knowledge of the fragmentation of functional groups, with most major advances occurring in related areas. Basic fragmentations and rearrangement processes continue to be investigated using labelling techniques, while cleavage processes and ion structures have also been studied widely using metastable characteristics, thermodynamic data and ion cyclotron resonance techniques (Chapters 1 and 2). The Ion Kinetic Energy method (IKE; Chapter 2) has been applied to a variety of systems, and there have been many achievements in alternative methods of ionization (Chapter 1), including negative-ion mass spectrometry. A technique has been described by Beynon[46,47] which allows the measurement of 'doubly charged ion' mass spectra by the observation of singly charged ions produced from charge-exchange reactions of doubly charged ions. The acquisition of data by computer, and the computer-aided interpretation of spectra, continues to be a major area of research (Chapter 6). A working knowledge of metastable

[10] 'Recent Developments in Mass Spectrometry', ed. K. Ogata and T. Hayakawa (Proceedings of the International Conference, Kyoto, Japan, 1969), University Park Press, Baltimore, London, and Tokyo, 1970.
[11] J. W. Robinson, 'Mass Spectrometry in Undergraduate Instrumental Analysis', Marcel Dekker, New York, 1970.
[12] 'Dynamic Mass Spectrometry', ed. D. Price and J. E. Williams, Heyden, London, 1970, Vol. 1.
[13] 'Mass Spectrometry, 1970', ed. G. W. A. Milne, Wiley, New York, 1970.
[14] C. E. Melton, 'Principles of Mass Spectrometry and Negative Ions', Marcel Dekker, New York, 1970.
[15] M. von Ardenne, K. Steinfelder, and R. Tümmler, 'Electronenanlagerungs-Massenspektrographie Organischen Substanen', Springer-Verlag, Berlin, 1971.
[16] 'Dynamic Mass Spectrometry', ed. D. Price, Heyden, London, 1971, Vol. 2.
[17] 'Mass Spectrometry: Techniques and Applications', ed. G. W. A. Milne, Wiley-Interscience, New York, 1971.
[18] Q. N. Porter and J. Baldas, 'Mass Spectrometry of Heterocyclic Compounds', Wiley-Interscience, New York, London, Sydney, and Toronto, 1971.
[19] 'Recent Topics in Mass Spectrometry', ed. R. I. Reed, Gordon and Breach, London, New York, and Paris, 1971.
[20] S. R. Schrader, 'Introductory Mass Spectrometry', Allyn and Bacon, Boston, 1971.
[21] 'Advances in Mass Spectrometry', ed. A. Quayle, The Institute of Petroleum, London, 1971, Vol. 5.
[22] R. T. Alpin, *Ann. Reports (B)*, 1969, **66**, 5.
[23] J. H. Bowie, 'Mass Spectrometry of Carbonyl Compounds', in 'The Chemistry of the Carbonyl Group', ed. J. Zabicky, Interscience, London, 1970, Vol. 2.
[24] D. J. Curran and S. Siggia, 'Mass Spectrometry of the $C-N$ Double Bond', in 'The Chemistry of the $C-N$ Double Bond', ed. S. Patai, Wiley-Interscience, London, 1970.
[25] A. G. Loudon and A. Maccoll, 'Mass Spectrometry of the Double Bond', in 'The Chemistry of Alkenes', ed. J. Zabicky, Wiley-Interscience, London, 1970.
[26] T. W. Bentley and R. A. W. Johnstone, 'Mechanism and Structure in Mass Spectrometry. A Comparison with Other Processes', in 'Advances in Physical Organic Chemistry', Academic Press, London and New York, 1970, **8**, 151.
[27] R. T. Aplin, *Ann. Reports (B)*, 1970, **67**, 7.
[28] G. Spiteller and M. Spiteller, 'Decomposition Reactions with Electrons', *Chimia*, 1970, **24**, 298.
[29] R. E. Honig, 'Mass Spectrometric Techniques', in 'Modern Analytical Techniques for Metals and Alloys', ed. R. F. Bunshah, Wiley-Interscience, New York, 1970, Vol. 3, Part 2, Ch. 12.
[30] A. L. Burlingame, D. H. Smith, T. D. Merren, and R. W. Olsen, 'Real Time High-Resolution Mass Spectrometry', in 'Progress in Analytical Chemistry', Plenum Press, New York and London, 1970, **4**, 17.

ions and the 'metastable-defocusing technique'[48-51] is essential for those using mass spectrometry as a structural aid, and the excellent review[52] by Jennings on metastable transitions is strongly recommended. A method for the computerized acquisition and reduction of data on metastables produced by the defocusing technique has been described by McLafferty.[53] The work on theoretical aspects of organic mass spectrometry seems to have lost some of its momentum and direction, but a considerable achievement has been the calculation of the mass spectra of a range of organic compounds.[54,55]

It has been shown[56] that the influence of substituents on the fragmentation of aliphatic compounds decreases in the following order: NMe_2 > NH_2 > ethylene ketal > $NHCO_2Me$ > I > MeO > MeS > C=O > SH > CO_2Me > Br > Cl > OH > CH_2OH > CO_2H. A new method for the calculation of formulae from exact-mass data has been reported,[57] as has a method for deriving the mass spectrum of a fully deuteriated compound from the spectra of partially deuteriated analogues.[58]

[31] P. W. Harland, K. A. G. MacNeil, and J. C. J. Thynne, 'Studies of Negative-Ion Formation at Low-Electron Energies', in 'Dynamic Mass Spectrometry', Heyden, London, and Sadler Research Laboratories, Inc., Philadelphia Pa., 1970, Vol. 1, p. 122.
[32] A. L. Burlingame, 'Topics in Organic Mass Spectrometry', in 'Advances in Analytical Chemistry and Instrumentation', Wiley-Interscience, London, New York, Sydney, and Toronto, 1970, Vol. 8.
[33] J. I. Brauman, 'Mass Spectrometry', in 'Spectral Analysis—Methods and Techniques', ed. J. A. Blackburn, Marcel Dekker, New York, 1970, Vol. 7, p. 235.
[34] S. Yamanaka and E. Abe, 'Mass Spectra of Heterocyclic Compounds', *Kagaku No Ryaki, Zokan*, 1970, **92**, 97 (*Chem. Abs.*, 1970, **73**, 301).
[35] R. A. Jones, 'Physicochemical Properties of Pyrroles', in 'Advances in Heterocyclic Chemistry', ed. A. R. Katritzky and A. J. Boulton, Academic Press, New York and London, 1970, Vol. 11, p. 383.
[36] G. Spiteller, 'Mass Spectrometry of Heterocyclic Compounds', in 'Physical Methods of Heterocyclic Chemistry', ed. A. R. Katritzky, Academic Press, London, 1971, Vol. 3, p. 224.
[37] A. E. Williams and H. E. Stagg, *Analyst*, 1971, **96**, 1.
[38] J. L. Franklin, J. G. Dielard, H. M. Rosenstock, J. T. Herron, K. Draxe, and F. H. Field, 'Ionisation Potentials, Appearance Potentials, and Heats of Formation of Gaseous Positive Ions', National Standard Ref. Data System, Nat. Bureau of Standards, 26, Superintendent of Documents, U.S. Govt. Printing Office, Washington PC 20402, 1969.
[39] G. Ege, 'Numerical Tables for Mass Spectrometry and Elemental Analysis', Verlag Chemie Publishers, Weinheim/Bergstr., 1970.
[40] 'Atlas of Mass Spectral Data', ed. E. Stenhagen, S. Abrahamsson, and F. W. McLafferty, Interscience, New York, 1970, Vols. 1—3.
[41] Mass Spectrometry Data Centre, A.W.R.E., Aldermaston, Reading, 1970, Vols. 1 and 2.
[42] R. Burks, J. S. Littler, and R. L. Cleaver, 'Tables for Use in High-Resolution Mass Spectrometry', Heyden, New York and Rheine, with Sadler Res. Labs. Inc. Philadelphia, 1970.
[43] J. H. Beynon, R. M. Caprioli, A. W. Kundert, and R. B. Spencer, 'Tables of Ion Energies for Metastable Transitions in Mass Spectrometry', Elsevier, Amsterdam, 1970.
[44] A. Cornu and R. Massot, 'Second Supplement to Compilation of Mass-Spectral Data', Heyden, London, and Presses Univ. de France, 1971.
[45] A. Tatematsu and T. Tsuchiya, 'Structure Indexed Literature of Organic Mass Spectra 1968', Academic Press, Japan, 1971.
[46] J. H. Beynon, R. M. Caprioli, W. E. Baitinger, and J. W. Amy, *Org. Mass Spectrometry*, 1970, **3**, 455.

The use of mass spectrometers in atmospheric extraterrestrial research has been described for the Explorer 31 Satellite[59] and for the Apollo 15 and 16 lunar missions.[60] The latter type employs a Nier-type system, scans from 12 to 66 a.m.u., and weighs only 11 kg. The lunar results are still being evaluated, but already the applications of mass spectrometers beyond the moon are being considered,[61] and a miniature double-focusing mass spectrometer of Mattauch–Herzog geometry, weighing 4 kg, is planned for the 1976 Mars probe.[62]

2 The Reactions of Specific Functional Groups

Hydrocarbons.—*Aliphatic and Acyclic.* The analysis of the fragmentation of hydrocarbons by an examination of decompositions in the first field-free region of the mass spectrometer has been described by several groups,[63,64] and it has been shown that the metastable transitions in the mass spectrum of methane may be used to calibrate a low-mass scale.[65] A study of competing metastable transitions indicates that the ions $C_4H_8^{+\cdot}$ and $C_6H_{12}^{+\cdot}$, when produced from different substrates, decompose through common intermediates.[66] The spectra of *cis*- and *trans*-1,2-dimethylcyclohexane,[67] alkylcyclanes,[68] decalin,[69] substituted adamantanes,[70–72] homoadamantane derivatives[73] and bicyclo[3,3,1]-nonanes[74] have been reported. Deuterium-labelling studies demonstrate that a large amount of hydrogen migration (or scrambling) accompanies or precedes the fragmentation of the homoadamantane molecular ion.[73]

---

[47] J. H. Beynon, A. Mathias, and A. E. Williams, *Org. Mass. Spectrometry*, 1971, **5**, 303.
[48] M. Barber and R. M. Elliot, 12th Annual Conference on Mass Spectrometry and Allied Topics, ASTME14, Montreal, 1964.
[49] K. R. Jennings, *J. Chem. Phys.*, 1965, **43**, 4176.
[50] M. Barber, W. A. Wolstenholme, and K. R. Jennings, *Nature*, 1967, **43**, 1832, and references cited therein.
[51] A. H. Struck and H. W. Major, paper presented to the ASTME14 Meeting, Dallas, Texas, 1969.
[52] K. R. Jennings, 'Some Aspects of Metastable Transitions', in 'Mass Spectrometry: Techniques and Applications', ed. G. W. A. Milne, Wiley-Interscience, New York, 1971, 419.
[53] J. E. Coutant and F. W. McLafferty, *Internat. J. Mass Spectrometry Ion Phys.*, 1972, **8**, 323.
[54] A. N. H. Yeo and D. H. Williams, *J. Amer. Chem. Soc.*, 1970, **92**, 3984.
[55] A. N. H. Yeo and D. H. Williams, *Org. Mass Spectrometry*, 1971, **5**, 135.
[56] G. Remberg and G. Spiteller, *Chem. Ber.*, 1971, **103**, 3640.
[57] C. P. Moss, *Org. Mass Spectrometry*, 1971, **5**, 353.
[58] A. N. H. Yeo, *Chem. Comm.*, 1970, 988.
[59] J. H. Hoffman, *Internat. J. Mass Spectrometry Ion Phys.*, 1970, **4**, 315.
[60] J. H. Hoffman, *Internat. J. Mass Spectrometry Ion Phys.*, 1972, **8**, 403.
[61] L. F. Herzog, *Internat. J. Mass Spectrometry Ion Phys.*, 1970, **4**, 337.
[62] A. O. Nier and J. L. Hayden, *Internat. J. Mass Spectrometry Ion Phys.*, 1971, **6**, 339.
[63] A. Herlan, *Org. Mass Spectrometry*, 1970, **4**, 425.
[64] P. Goldberg, J. A. Hopkinson, A. Mathias, and A. E. Williams, *Org. Mass Spectrometry*, 1970, **3**, 1009.
[65] J. H. Beynon, R. M. Caprioli, W. E. Baitinger, and J. W. Amy, *Org. Mass Spectrometry*, 1970, **3**, 479.
[66] G. A. Smith and D. H. Williams, *J. Chem. Soc. (B)*, 1970, 1529.
[67] A. Lageot, *Org. Mass Spectrometry*, 1971, **5**, 839.

The spectra of ethylene[75] and of various straight-chain olefins[76,77] have been discussed. Molecular ions corresponding to (1) decompose by the ubiquitous McLafferty rearrangement, but in general isomerization precedes fragmentation.[77] Methylenecycloalkanes,[78] 1-methylcyclopentene and four isomers,[79] norbornene,[80] bicyclo[2,2,2]oct-2-ene derivatives[81] and mono- and di-alkylacetylenes[82] have been studied. Metastable and thermodynamic data show that the norbornene (2) and nortricyclene (3) molecular ions do not fragment through a common intermediate.[80] The elimination of C_2H_4 from (3) involves C-1 and C-7 (C-4 and C-7).[80] The molecular ion of 7-methylcyclohepta-1,3,5-triene eliminates a methyl radical which originates from both the 7-methyl substituent and the cycloheptatriene ring.[83] The spectrum of (4) exhibits the following ions: $(M-{}^{13}CH_3^{\cdot}):(M-Me^{\cdot}) = 77:23$ (ions formed in ion source); 51:49 (metastable ions).[83]

$$\begin{array}{c} R \\ \diagdown \\ C=C \\ \diagup \quad \diagdown \\ H \quad (CH_2)_3R \end{array} \Bigg]^{\ddagger}$$

(1)   (2)   (3)   (4)

Aromatic. Two groups[84,85] have shown that the C and H scrambling which occurs in the benzene molecular ion (cf. Vol. 1, pp. 53, 97) involves both valence isomerism and independent H scrambling. The daughter ions produced by the losses of C_2H_2 and its labelled analogues from (5) show that 70% of the scrambling involves valence isomerism together with the breakage and reformation of C—H bonds, but that 30% of the scrambling occurs without C—H bond cleavage.[84]

[68] R. I. Reed, *Rev. Port. Quim.*, 1970, **12**, 16.
[69] M. Yamamoto and K. Hirota, *Mass Spectrometry (Japan)*, 1970, **17**, 653.
[70] N. F. Karpenko, O. S. Chizhov, S. S. Novikov, A. G. Yurchenko, and G. I. Danilenko, *Zhur. org. Khim.*, 1971, **7**, 416.
[71] N. F. Karpenko, O. S. Chizhov, S. S. Novikov, A. G. Yurchenko, V. D. Sukhoverkhov, and N. A. Smirnova, *Izvest. Akad. Nauk S.S.S.R., Ser. khim.*, 1971, 439.
[72] R. L. Greene, W. A. Kleschick, and G. H. Wahl, *Tetrahedron Letters*, 1971, 4577.
[73] A. G. Yurchenko, F. N. Stepanov, S. S. Isaeva, B. M. Zolotarev, V. I. Kadentsev, and O. S. Chizhov, *Org. Mass Spectrometry*, 1970, **3**, 1401.
[74] J. K. MacLeod, M. Vegar, and R. J. Wells in ref. 10, p. 1197.
[75] Y. Masuda, *Mass Spectrometry (Japan)*, 1971, **19**, 158.
[76] D. G. Earnshaw, F. G. Doolittle, and A. W. Decora, *Org. Mass Spectrometry*, 1971, **5**, 801.
[77] K. K. Mayer and C. Djerassi, *Org. Mass Spectrometry*, 1971, **5**, 817.
[78] A. Z. Shizhmamedbekova, F. A. Aslanov, M. M. Gadzhiev, T. E. Galamova, and F. N. Akhmedova, *Doklady Akad. Nauk Azerb. S.S.R.*, 1970, **26**, 34.
[79] M.-Th. Praet, *Org. Mass Spectrometry*, 1970, **4**, 65.
[80] J. L. Holmes and D. McGillivray, *Org. Mass Spectrometry*, 1971, **5**, 1349.
[81] C. M. Cimarusti and J. Wolinsky, *J. Org. Chem.*, 1971, **36**, 1871.
[82] H. Luftmann and G. Spiteller, *Org. Mass Spectrometry*, 1971, **5**, 1073.
[83] A. Venema, N. M. M. Nibbering, and Th. J. de Boer, *Tetrahedron Letters*, 1971, 2141.
[84] W. O. Perry, J. H. Beynon, W. E. Baitinger, J. W. Amy, R. M. Caprioli, R. N. Renaud, L. C. Leitch, and S. Meyerson, *J. Amer. Chem. Soc.*, 1970, **92**, 7236.
[85] R. J. Dickinson and D. H. Williams, *J. Chem. Soc. (B)*, 1971, 249.

Reactions of Specific Functional Groups

The ratio of the metastable transitions for the losses of C_2H_2, C_2HD, $C^{13}CH_2$, and $C^{13}CHD$ from (6) proves that the decomposition of the lower-energy molecular ions of benzene is preceded or accompanied by complete carbon

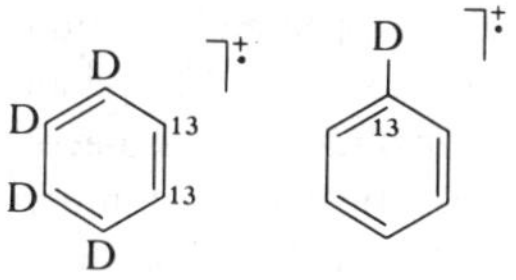

(5)   (6)

and hydrogen randomization with statistical cleavage and reformation of C—H bonds.[85] An examination[86] of the decompositions of the doubly charged benzene molecular ion has demonstrated that the decomposing ions $C_4H_4^{2+}$ and C_6H^{2+} have linear structures, a conclusion reached earlier[87] for the molecular ion $C_6H_6^{2+}$. The IKE spectra of benzene and other aromatic systems have been investigated.[88]

The problem of the structure of the ion $C_7H_7^+$ derived from toluene, and of its carbon and hydrogen scrambling processes, has been the subject of much debate (Vol. 1, p. 97). Previous work has not distinguished between scrambling processes which proceed by either (a) a series of benzyl–tropylium ion interconversions, or (b) insertion of the methyl carbon atom randomly between adjacent carbon atoms of the ring. If the latter process is operative, loss of a hydrogen atom from the [2,6-$^{13}C_2$]toluene molecular ion (7) should produce only (8) and (9), but if the former process occurs, (8)—(10) should be formed.[89] Siegel has shown that decomposition of the ion $C_5{}^{13}C_2H_7^+$ indicates a completely random distribution of carbon, thus favouring a series of benzyl–tropylium ion

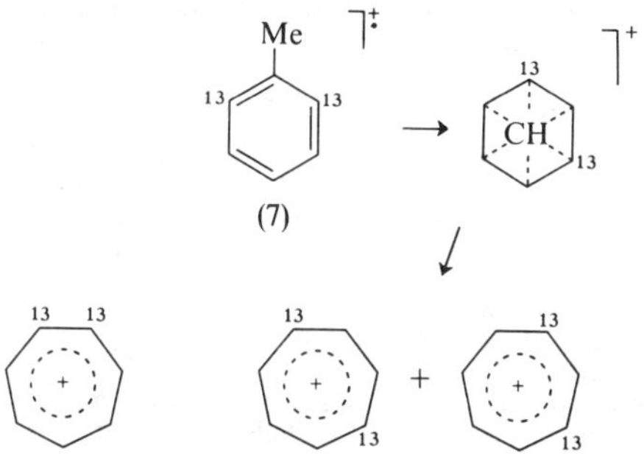

(10)   (8)   (9)

[86] J. H. Beynon, R. M. Caprioli, W. E. Baitinger, and J. W. Amy, *Org. Mass Spectrometry*, 1970, **3**, 963.
[87] J. H. Beynon and A. E. Fontaine, *Chem. Comm.*, 1969, 723.
[88] J. H. Beynon, R. M. Caprioli, W. E. Baitinger, and J. W. Amy, *Org. Mass Spectrometry*, 1970, **3**, 455.
[89] A. S. Siegel, *J. Amer. Chem. Soc.*, 1970, **92**, 5277.

isomerizations.[89] The competitive losses of H and D from the methyl group and phenyl ring of the singly charged[89-91] and doubly charged[92] molecular ions of deuteriated toluenes have been reported. The problem of the formation of benzyl or tropylium species from ethylbenzenes has been studied,[93,94] a classification of alkylbenzenes has appeared,[95] various aspects of the spectra of substituted benzenes have been considered,[96-98] and the spectra of phenylcyclopropanes[99,100] and phenylcyclohexadiene isomers[101] have been listed.

Several groups[102-105] have investigated tetralin derivatives. The molecular ion (11) eliminates C_2H_4, $C_2H_2D_2$, and C_2D_4 in the ratio 1 : 2 : 1.[104] A small amount of H/D scrambling occurs, but the above result indicates that the mechanism for the loss of ethylene from the tetralin molecular ion is more complex than the expected retro-Diels–Alder process. Electron-introducing groups placed at the 5-position of tetralin retard the fragmentation of the B ring [see (11)] relative to tetralin, whereas electron-withdrawing groups slightly enhance fragmentation.[103] The molecular ion (12) eliminates C_6H_5D and C_6H_6 in the ratio 20 : 1.[105,106]

(11)   (12) $Ph^{13}CD_2Ph$ (13)

Metastable transitions in the spectrum of (13) show that the loss of a methyl radical from the molecular ion of diphenylmethane occurs with almost complete carbon and hydrogen scrambling, accompanied by statistical C—H bond cleavage and reformation.[107] The loss of a methyl radical from the diphenylmethyl cation is analogous, occurring with complete scrambling.[107] Other

[90] J. H. Beynon, J. E. Corn, W. E. Baitinger, R. M. Caprioli, and R. A. Benkeser, *Org. Mass Spectrometry*, 1970, 3, 1371.
[91] I. Howe and F. W. McLafferty, *J. Amer. Chem. Soc.*, 1971, **93**, 99.
[92] T. Ast, J. H. Beynon, and R. G. Cooks, *J. Amer. Chem. Soc.*, 1972, **94**, 1834.
[93] D. A. Lightner, S. Majeti, and G. B. Quistad, *Tetrahedron Letters*, 1970, 3857.
[94] H. E. Audier, G. Bouchoux, and M. Fetizon, *Compt. rend.*, 1971, **272**, C, 953.
[95] R. G. Gillis, *Org. Mass Spectrometry*, 1971, **5**, 79.
[96] R. H. Wiley, *J. Polymer Sci.*, 1970, **8**, 792.
[97] A. A. Polyakova, I. M. Lukanshenko, R. V. Popanova, S. A. Rang, and O. I. Eizen, *Zhur. org. Khim.*, 1971, 7, 98.
[98] T. W. Bentley and R. A. W. Johnstone, *J. Chem. Soc. (B)*, 1971, 263.
[99] K. A. Chochua, O. S. Chizhov, Yu. S. Shabarov, and N. A. Kazbulatova, *Zhur. org. Khim.*, 1971, 7, 2024.
[100] K. A. Chochua, O. S. Chizhov, and Yu. S. Shabarov, *Zhur. org. Khim.*, 1971, **7**, 2032.
[101] A. M. Braun, *Org. Mass Spectrometry*, 1970, **3**, 1479.
[102] A. G. Loudon, A. Maccoll, and S. K. Wong, *J. Chem. Soc. (B)*, 1970, 1727.
[103] D. A. Lightner and F. S. Steinberg, *Org. Mass Spectrometry*, 1970, **3**, 1095.
[104] H.-F. Grützmacher and M. Puschmann, *Chem. Ber.*, 1971, **104**, 2079.
[105] M. L. Gross, C. L. Wilkins, and T. G. Regulski, *Org. Mass Spectrometry*, 1971, **5**, 99.
[106] C. L. Wilkins and M. L. Gross, *J. Amer. Chem. Soc.*, 1971, **93**, 895.
[107] J. H. Bowie, P. Y. White, and T. K. Bradshaw, *J. C. S. Perkin II*, in press.

aspects of the spectra of diarylmethanes have been considered,[108—111] including the proximity effects which occur in the spectra of *o*-substituted diphenylmethanes,[108,109] and the correlations between ionization potentials and σ^+ values.[111] 1,2-Diphenylethanes have also been described.[111,112] The metastable transitions of 1-phenyl[1,2-$^{13}C_2$]ethylene, measured at 15 eV in the second field-free region, show that the loss of the elements of acetylene from the styrene molecular ion occurs with complete carbon scrambling.[113] It was previously suggested that the molecular ions of styrene and cyclo-octatetraene may fragment through a common intermediate;[114a] ICR mass spectrometry of both systems shows that the molecular ions of the two isomers undergo different ion–molecule reactions,[114b] but this observation does not of course preclude the possibility that the decomposing ions decompose to a common intermediate. The fragmentations of the stilbene molecular ion are still inciting comment[115—118] and it has been shown[115] that the loss of C_6H_6 is accompanied by little hydrogen scrambling and no carbon scrambling between the olefinic and phenyl carbons, and that the major loss originates as indicated in (14).

The phenylacetylene molecular ion is completely scrambled with respect to hydrogen during the elimination of C_2H_2;[119] the triply charged molecular ion of biphenyl eliminates a methyl cation;[120] bridged biphenyls,[121] methylnaphthalenes,[122] and binaphthyls[123] have been discussed; the ion $C_{12}H_8^{+}$ produced from anthracene, phenanthrene, and diphenylacetylene is formed with similar energy requirements.[124]

[108] W. M. Scott, M. E. Wachs, C. Steelink, and J. D. Fitzpatrick, *Org. Mass Spectrometry*, 1970, **3**, 657.
[109] S. Meyerson and E. K. Fields, *Org. Mass Spectrometry*, 1971, **5**, 1227.
[110] H. Budzikiewicz, J. Rullkötter, and H. M. Schiebel, *Org. Mass Spectrometry*, 1972, **6**, 251.
[111] S. Pignataro, V. Mancini, G. Innorta, and G. Distefano, *Z. Naturforsch.*, 1972, **27a**, 534.
[112] J. F. Manville and P. J. Smith, *Org. Mass Spectrometry*, 1971, **5**, 95.
[113] A. Venema, N. M. M. Nibbering, and Th. J. de Boer, *Org. Mass Spectrometry*, 1970, **3**, 1589.
[114] (a) J. L. Franklin and S. R. Carroll, *J. Amer. Chem. Soc.*, 1969, **91**, 5940; (b) C. L. Wilkins and M. L. Gross, *J. Amer. Chem. Soc.*, 1971, **93**, 895.
[115] J. H. Bowie and P. Y. White, *Austral. J. Chem.*, 1971, **24**, 205.
[116] H. Güsten, L. Klasinc, J. Marsel, and D. Milivojević, *Org. Mass Spectrometry*, 1971, **5**, 357.
[117] J. H. Bowie and P. Y. White, *Org. Mass. Spectrometry*, 1972, **6**, 135.
[118] H. Güsten, L. Klasinc, J. Marsel, and D. Milivojević, *Org. Mass Spectrometry*, 1972, **6**, 175.
[119] S. Safe, *J. Chem. Soc. (B)*, 1971, 962.
[120] J. H. Beynon, R. M. Caprioli, W. E. Baitinger, and J. W. Amy, *Org. Mass. Spectrometry*, 1970, **3**, 661.
[121] A. Maquestiau, Y. van Havenbeke, and F. Delalieu, *Org. Mass Spectrometry*, 1971, **5**, 1015.
[122] R. V. Popanova, I. M. Lukashenko, A. A. Polyakova, S. A. Rang, and O. I. Eizen, *Zhur. org. Khim.*, 1971, **7**, 2038.
[123] M. H. Harris, A. G. Loudon, and R. Z. Mazengo, *Org. Mass Spectrometry*, 1971, **5**, 1123.
[124] C. G. Rowland, *Internat. J. Mass Spectrometry Ion Phys.*, 1971, **7**, 79.

An interesting correlation between thermochemistry and mass spectrometry has been observed for the methylphenanthrenes (15) and (16).[125] Pyrolysis of these compounds yields fluorene, and their spectra contain the fluorenyl cation ($C_{13}H_9^+$) which is produced by rearrangement.[125] The spectra of benzononatrienes[126] and hexabenzocoronene[127] have been recorded, while that of tetramethano-o-tetraphenylene (17) contains the ion $(M - 4H\cdot)^{2+}$ as base peak.[128]

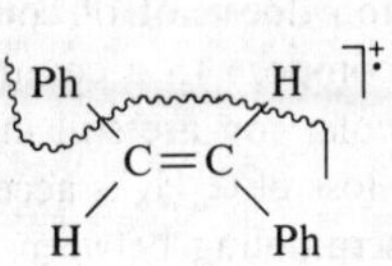

(14)

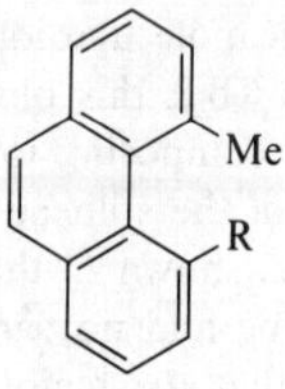

(15) R = H
(16) R = Me

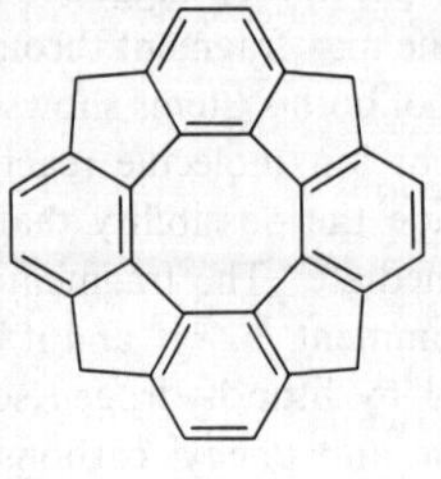

(17)

Halides.—The losses of CH_4 and C_2H_4 from $C_4H_9^+$ ions have been studied using labelled (^{13}C and ^2H) butyl cations produced from isomeric butyl halides.[129] This study has shown that whereas all the carbon atoms in the ion $C_4H_9^+$ are equivalent after 10^{-7} s, statistical distribution of hydrogen is only attained for metastable transitions (*i.e.* after 10^{-5}—10^{-6} s), and a protonated cyclobutane has been proposed as an intermediate in the fragmentation pathway.[129] The ions $C_4H_7^+$, $C_5H_9^+$, and $C_6H_9^+$, when produced from a variety of bromocompounds, isomerize to common intermediates before decomposition.[130] The spectra of 1,1,1-trifluorethane,[131] 1,2-dichlorobutane,[132] 1,2-dibromobutane,[133] 1,2-dibromopentane,[133] 1,2-dibromocyclohexane,[133] dichlorocyclohexanes,[134] chlorocyclohexenes,[134] dichlorocycloheptanes,[134] cyclic fluoroethers,[135] and cyclic polyfluorinated alkenyl alcohols[136] have been reported. Stereoselective hydrogen atom abstraction is noted during the loss of C_2H_4Cl

[125] R. C. Dougherty, H. E. Bertorello, and M. M. de Bertorello, *Org. Mass Spectrometry*, 1971, **5**, 1321.
[126] R. R. Fraser and A. Juraydini, *Canad. J. Chem.*, 1970, **48**, 3591.
[127] R. I. Reed and A. Tennent, *Org. Mass Spectrometry*, 1971, **5**, 619.
[128] D. Hellwinkel and G. Reiff, *Angew. Chem. Internat. Edn.*, 1970, **9**, 527.
[129] R. Liardon and T. Gäumann, *Helv. Chim. Acta*, 1971, **54**, 1968.
[130] M. A. Shaw, R. Westwood, and D. H. Williams, *J. Chem. Soc. (B)*, 1970, 1773.
[131] J. M. Simmie and E. Tschiukov-Roux, *Internat. J. Mass Spectrometry Ion Phys.*, 1971, **7**, 41.
[132] D. S. Ashton, J. M. Tedder, and J. C. Walton, *J. Chem. Soc. (B)*, 1970, 1775.
[133] J. M. Péchiné, *Org. Mass Spectrometry*, 1971, **5**, 705.
[134] F. Gaymard, A. Vermeglio, A. Cambon, and R. Guidj, *Bull. Soc. chim. France*, 1971, 2238.
[135] J. L. Cotter, *Org. Mass Spectrometry*, 1972, **6**, 345.
[136] S. P. Levine, C. D. Bertino, J. D. Park, and R. H. Shapiro, *Org. Mass Spectrometry*, 1970, **4**, 461.

Reactions of Specific Functional Groups

from the *exo*-2-norbornyl chloride molecular ion [see (18)],[137] and chlorine migration occurs during the decomposition of (19).[138]

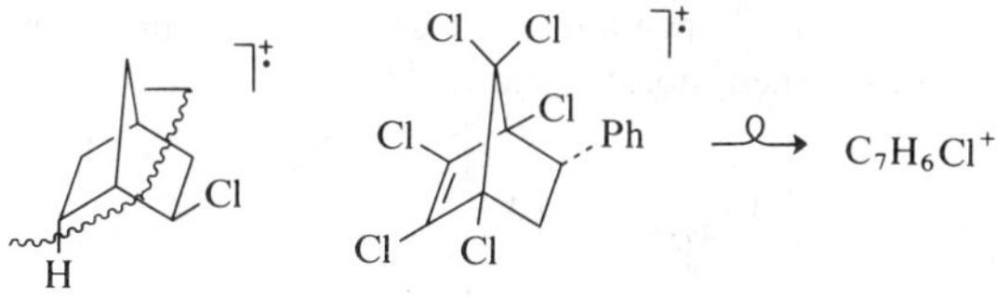

(18) (19) → $C_7H_6Cl^+$

The ions $C_9H_9^+$, $C_{10}H_{11}^+$, $C_{11}H_{13}^+$, and $C_{12}H_{15}^+$ are produced by the loss of bromine atoms from the molecular ions of the cyclo-octatetraenyl bromides (20; $n = 1$—4).[139] All the atoms in the first three ions have lost positional identity prior to fragmentation in the first field-free region of the mass spectrometer, while the last, $C_{12}H_{15}^+$ is only partially scrambled. Examination of the spectrum of (21) shows that the decompositions of $C_{11}H_{13}^+$ may be interpreted in terms of a decomposing ion in which some, but not all, of the C—H bonds have broken and reformed. Each ion may be produced from a variety of substrates, and decomposition occurs through common intermediates.[139]

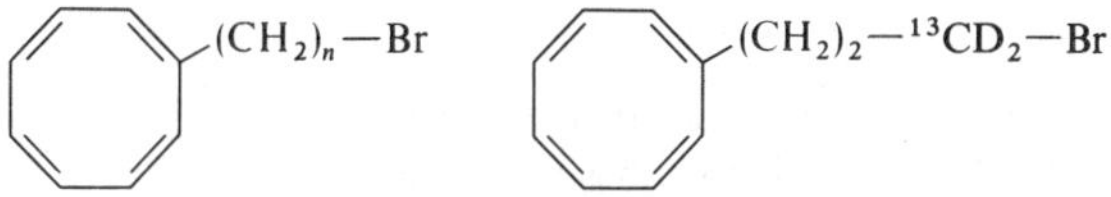

(20) $(CH_2)_n$—Br   (21) $(CH_2)_2$—$^{13}CD_2$—Br

Loss of an iodine atom from the iodobenzene molecular ion produces the ion $C_6H_5^+$ in which all atoms scramble prior to the loss of C_2H_2.[85] The molecular ions of fluoro- and chloro-toluenes, but not of bromo- and iodo-toluenes, undergo ring-expansion prior to fragmentation.[140] Molecular-ion isomerization has been proposed to explain the similarities between the decompositions $M - Cl\cdot$ in a series of *m*- and *p*-R-chlorobenzenes (R = NH_2, Me, H, F, Cl, CF_3, or CN).[141] Aromatic perchlorocarbons may be used as reference standards for mass measurements.[142] The molecular ions of (22) and (23) do not isomerize to the hexachlorobenzene molecular ion prior to fragmentation,[143] but the spectra of hexakistrifluoromethylbenzene and its isomers (24)—(26) are very

---

[137] J. L. Holmes and D. McGillivray, *Org. Mass Spectrometry*, 1971, **5**, 1339.
[138] D. S. Weinberg, C. Stafford, and C. G. Cardenas, *J. Org. Chem.*, 1971, **36**, 1893.
[139] J. H. Bowie, G. E. Gream, and M. Mular, *Austral. J. Chem.*, 1972, **25**, 1107.
[140] A. N. H. Yeo and D. H. Williams, *Chem. Comm.*, 1970, 886.
[141] P. Brown, *Org. Mass Spectrometry*, 1970, **3**, 639.
[142] M. E. Frerburger, B. Mason Hughes, L. Spialter, and T. O. Tiernan, *Org. Mass Spectrometry*, 1971, **5**, 885.
[143] I. Agranat, R. M. J. Loewenstein, and E. D. Bergmann, *Org. Mass Spectrometry*, 1971, **5**, 289.

similar.[144] Several reports[145−147] of fluorobenzenes have appeared, and of particular interest are the pentafluorophenyl nitrogen compounds,[146] where the loss of the nitrogen substituent competes with fragmentation of the fluorogroups. For example, both pentafluoroaniline and decafluorophenylaniline lose HF and HCN from their molecular ions.[146]

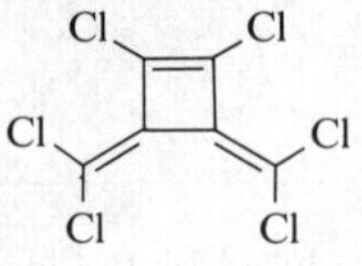

(22) 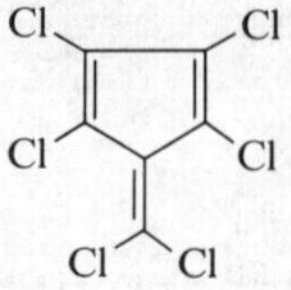 (23) 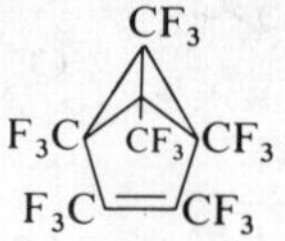 (24)

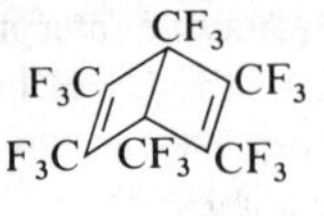

(25) 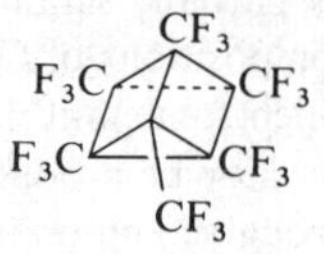 (26) 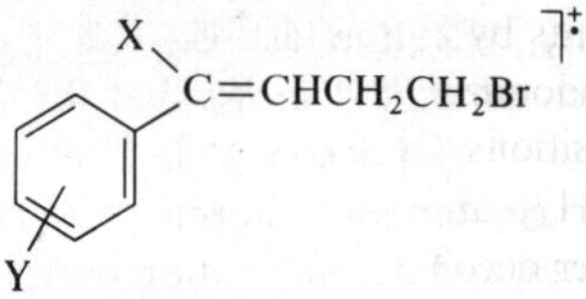 (27)

A major decomposition mode in the spectrum of 1-phenylethylbromide is $M - Br\cdot - C_2H_2$. Labelling studies (^{13}C) show that, at 70 eV, the ion $C_8H_9^+$ ($M - Br\cdot$) eliminates C_2H_2 from the side-chain, but that carbon scrambling occurs at lower energies.[113] Bromine expulsion from (27) is affected by the π-electron density of the aryl ring.[148] Hydrogen scrambling has been noted for monosubstituted diphenyls,[149] chlorine scrambling occurs for chloronaphthalenes,[150] and recent results are not inconsistent with randomization of chlorine between the phenyl rings of polychlorinated diphenyls.[151,152] Chlorinated aromatic fungicides,[153] fluoroglucitols,[154] and halogenoacetylenes[155] have been studied.

[144] M. G. Barlow, R. N. Haszeldine, and R. Hubbard, *J. Chem. Soc.* (*C*), 1970, 1232.
[145] L. D. Smithson, A. K. Bhattacharya, and C. Tamborski, *Org. Mass Spectrometry*, 1970, **4**, 1.
[146] G. F. Lanthier and J. M. Miller, *Org. Mass Spectrometry*, 1971, **6**, 89.
[147] J. Hitzke, F. Peter, and J. Guion, *Org. Mass Spectrometry*, 1972, **6**, 349.
[148] K. B. Tomer, J. Turk, and R. H. Shapiro, *Org. Mass Spectrometry*, 1972, **6**, 235.
[149] S. Safe, *Org. Mass Spectrometry*, 1971, **5**, 1221.
[150] S. Safe, O. Hutzinger, and M. Cook, *Chem. Comm.*, 1972, 260.
[151] S. Safe and O. Hutzinger, *Chem. Comm.*, 1971, 446.
[152] S. Safe and O. Hutzinger, *Chem. Comm.*, 1972, 686.
[153] O. Hutzinger, W. D. Jamieson, and S. Safe, *J. Assoc. Offic. Analyt. Chemists*, 1971, **54**, 178.
[154] J. Adamson, A. D. Barford, E. M. Bessell, A. B. Foster, M. Jarman, and J. H. Westwood, *Org. Mass Spectrometry*, 1971, **5**, 865.
[155] E. Kloster-Jensen, C. Pascual, and J. Vogt, *Helv. Chem. Acta*, 1970, **53**, 2109.

Alcohols and Phenols.—Isotope effects have been calculated for various decompositions from the molecular ions of deuteriated methanol[156] and ethanol.[157] Conflicting results have been obtained concerning the properties of the ion $C_3H_7O^+$ produced by loss of a methyl radical from the 2-hydroxy-2-methylpropane molecular ion.[158,159] The ion $C_3H_7O^+$ eliminates C_2H_4 to produce CH_2OH^+, and high-resolution measurements on the product ions CH_2OH^+ and $^{13}CH_2OH^+$ in the spectra of (28) and (29) led Siegel[158] to suggest that the

$$\begin{array}{cc} \text{Me} & \text{Me} \\ | & | \\ \text{Me}-^{13}\text{C}-\text{OH} & ^{13}\text{CH}_3-\text{C}-\text{OH} \\ | & | \\ \text{Me} & \text{Me} \\ (28) & (29) \end{array}$$

three carbon atoms in $C_3H_7O^+$ are completely equivalent. Similar measurements by Harrison[159] using (28) indicated that only a small amount of carbon randomization occurs. The situation has been clarified for low-energy decompositions, by metastable defocusing of (28) for the processes $[C_2{}^{13}CH_7O^+ - C_2H_4]$ and $[C_2{}^{13}CH_7O^+ - C^{13}CH_4]$, and these measurements show that after decomposition, the central carbon atom is retained to an extent of approximately 90%.[159] The carbon atoms are therefore not equivalent, but the spectra of analogous deuteriated compounds show that almost complete hydrogen randomization precedes the loss of C_2H_4 from $C_3H_7O^+$, and a protonated hydroxypropane cation is proposed to rationalize these observations.[159] Deuteriated n-pentyl alcohol has been studied.[160]

Further reports on cyclohexanols have appeared,[160,161] and the spectra of deuteriated cyclohexanols show that the hydrogen attached to oxygen undergoes partial scrambling with the 2, 3, 5, and 6 ring hydrogens during the formation of major ions.[161] The losses of water from the *cis*- and *trans*-4-isopropylcyclohexanol molecular ions show a high degree of stereoselectivity [see (30) and (31)].[162] Adamantan-2-ol[163] and bicyclo[2,2,1]heptan-2-ols[164] fragment by initial elimination of water, and the retro-Diels–Alder process is the basic cleavage of bicyclo[2,2,1]octyl alcohols.[165] The distance between two hydroxy-groups must be less than 3.5 Å if the loss of water is to occur from these substituents.[166] The major breakdown of *trans*-cyclopentane-1,2-diol is

---

[156] M. Corval and P. Masclet, *Org. Mass Spectrometry*, 1972, **6**, 511.
[157] M. Corval, *Bull. Soc. chim. France*, 1970, 2871.
[158] A. S. Siegel, *Org. Mass Spectrometry*, 1970, **3**, 1417.
[159] C. W. Tsang and A. G. Harrison, *Org. Mass Spectrometry*, 1971, **5**, 877.
[160] M. M. Green, R. J. Cook, J. M. Schwab, and R. B. Roy, *J. Amer. Chem. Soc.*, 1970, **92**, 3076.
[161] R. H. Shapiro, S. P. Levine, and A. M. Duffield, *Org. Mass Spectrometry*, 1971, **5**, 383.
[162] M. M. Green and R. B. Roy, *J. Amer. Chem. Soc.*, 1970, **92**, 6368.
[163] J. W. Greidanus, *Canad. J. Chem.*, 1971, **49**, 3210.
[164] K. Humski and L. Klasinc, *J. Org. Chem.*, 1971, **36**, 3057.
[165] H. Kwart and T. A. Blazer, *J. Org. Chem.*, 1970, **35**, 2726.
[166] C. C. Fenselau and C. H. Robinson, *J. Amer. Chem. Soc.*, 1971, **93**, 3070.

illustrated in (32);[167] metastable defocusing studies show that the loss of water from the cyclohexane-1,2-diol molecular ion involves prior scrambling of the hydrogens depicted in (33),[168] and the spectra of bicyclic[2,2,1]heptanediols[169] and various αδ-diols[170] have been recorded.

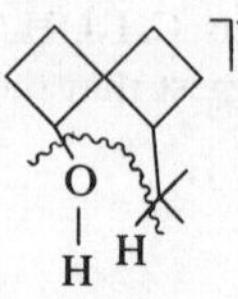

(30) 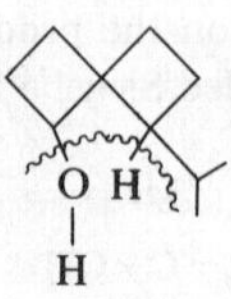   (31) 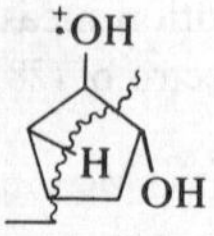   (32) 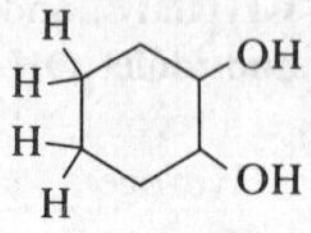   (33)

The base peak in the spectrum of (34) is an ion $C_9H_9^+$, which is produced both by the simple cleavage reaction A, and by a rearrangement process B which involves the specific incorporation of the two α-hydrogens into $C_9H_9^+$.[139] The molecular ion of [1-^{13}C]phenol specifically eliminates ^{13}CO.[85] The loss of water from the 6-phenylhexanol molecular ion specifically involves a benzylic (C-6) hydrogen (35), whereas the analogous elimination from 5-phenylpentanol involves hydrogens at both the 4- and 5-positions.[171] The main cleavage of phenylcyclopropylcarbinol is shown in (36).[172] A study of consecutive and competing metastable transitions from chloro- and bromo-phenol molecular ions demonstrate that ring-halogen rearrangement is not necessarily a prerequisite for fragmentation.[173] The oxygen atoms of the m-hydroxybenzyl alcohol molecular ion (37) retain their positional identity, demonstrating that ring-expansion to a tropylium species (38) does not occur.[174] Various phloroglucinol derivatives have been studied.[175]

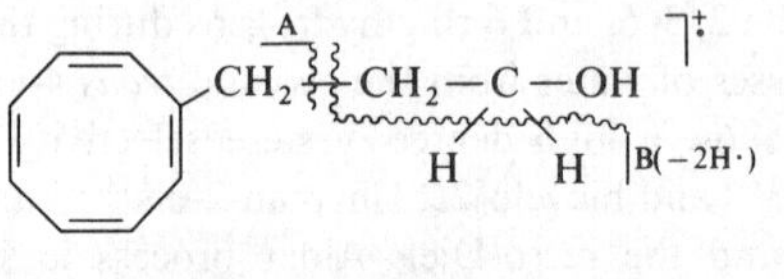

(34) 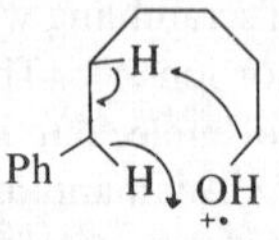   (35) 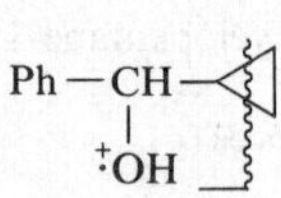   (36) 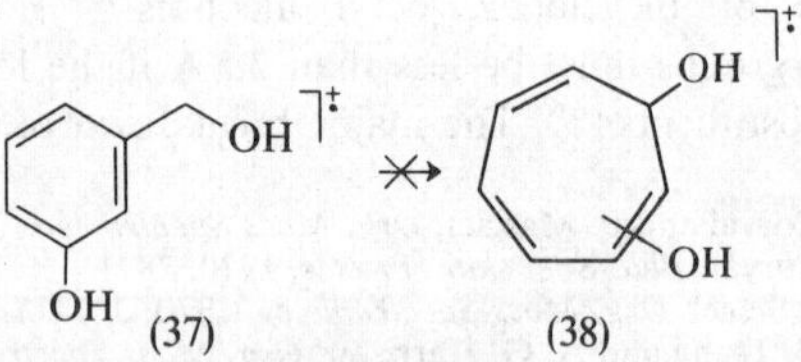   (37)   (38)

[167] G. A. Singy and A. Buchs, *Helv. Chim. Acta*, 1971, **54**, 537.
[168] F. Benoit and J. L. Holmes, *Canad. J. Chem.*, 1971, **49**, 1161.
[169] H. F. Grützmacher and K. H. Fechner, *Tetrahedron*, 1971, **27**, 5011.
[170] Gy. Horváth and J. Kuszmann, *Org. Mass Spectrometry*, 1972, **6**, 447.
[171] S. Meyerson and L. C. Leitch, *J. Amer. Chem. Soc.*, 1971, **93**, 2244.
[172] K. Sisido, M. Tanouti, and K. Utimoto, *Tetrahedron Letters*, 1970, 2209.
[173] J. C. Tou, *J. Phys. Chem.*, 1971, **74**, 3076.
[174] T. A. Molenaar-Langeveld and N. M. M. Nibbering, *Tetrahedron*, 1972, **28**, 1043.
[175] S. J. Shaw and P. V. R. Shannon, *Org. Mass Spectrometry*, 1970, **3**, 941.

Aldehydes, Ketones, and Quinones.—*Aliphatic.* Further work[176,177] with C_3—C_9 n-alkanals usefully complements earlier work in this field (Vol. 1, p. 104). Structure elucidation of aliphatic aldehydes may be accomplished by the determination of the mass spectra of their alkenyl ether derivatives.[178] The mass spectra of aldehydes and ketones (together with peaks due to SO and SO_2) may be obtained by introduction of their bisulphite adducts into the mass spectrometer using the direct probe.[179] The spectra of aldehyde esters have been recorded.[180]

Hydrogen randomization occurs in the high-energy molecular ion of undecan-2-one prior to loss of a methyl radical by α-cleavage.[181] The activation energies for cleavages of alkyl ketones have been calculated,[182] the spectra of bridgehead acetone derivatives have been discussed,[183] and the IKE spectrum of nonan-4-one has been measured.[184] Various aspects of the McLafferty

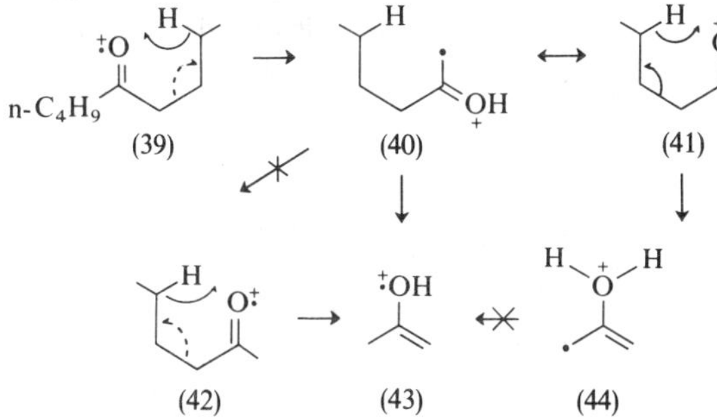

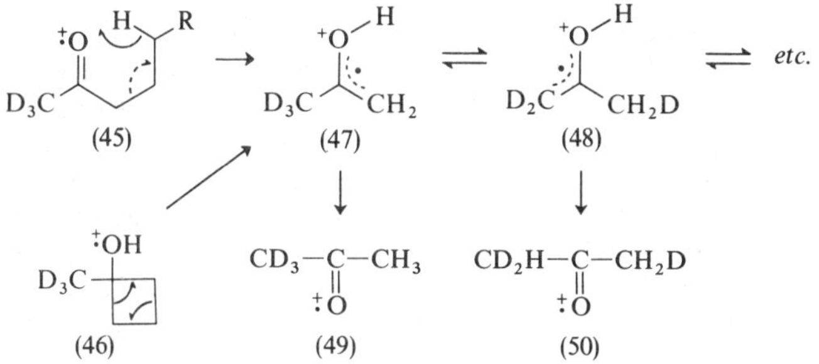

[176] A. G. Harrison, *Org. Mass Spectrometry*, 1970, **3**, 549.
[177] S. Meyerson, C. Fenselau, J. L. Young, W. R. Landis, E. Selke, and L. C. Leitch, *Org. Mass Spectrometry*, 1970, **3**, 689.
[178] P. E. Manni, W. G. Andrus, and J. N. Wells, *Analyt. Chem.*, 1971, **43**, 265.
[179] C. Fenselau, J. L. Young, and W. R. Landis, *Org. Mass Spectrometry*, 1970, **3**, 1085.
[180] A. C. Noble and W. W. Nawar, *J. Agric. Food Chem.*, 1971, **19**, 1039.
[181] A. N. H. Yeo, *Chem. Comm.*, 1970, 987.
[182] J. Jullien and J. M. Pechine, *Org. Mass Spectrometry*, 1970, **4**, 325.
[183] R. R. Sauers, M. Gorodetsky, J. A. Whittle, and C. K. Hu, *J. Amer. Chem. Soc.*, 1971, **93**, 5520.
[184] G. Eadon, C. Djerassi, J. H. Beynon, and R. M. Caprioli, *Org. Mass Spectrometry*, 1971, **5**, 917.

rearrangement of simple ketones continue to excite interest. ICR studies[185] of the reactions of the double McLafferty ion produced from (39) show that a substantial part of this ion exists in the enol form (43), that (43) is formed directly from (40) and not through either the intermediacy of an oxonium species (44), or by the enol–keto process (40) → (42) → (43). In contrast to the ICR experiments, McLafferty[186,187] has shown from a study of the metastable transitions of the ion $C_3H_4D_3O^+$ produced by the McLafferty rearrangement of (45) and cleavage of the cyclobutanol (46), that at least a portion of this ion decomposes through the keto-form by normal α-cleavage reactions. Initial formation of the enol (47) followed by hydrogen rearrangements of the type (47) → (48), accompanied by conversion to the corresponding keto tautomers (49) and (50), may account for the losses of variously labelled methyl radicals from $C_3H_4D_3O^+$.[186,187] The possibility that the product ion from the McLafferty rearrangement may correspond to a cyclobutanol species has been discounted, as the rearrangement ion from (51) does not suffer the decompositions expected for (52).[188] The McLafferty rearrangement has been investigated using Field-Ion mass spectometry.[189]

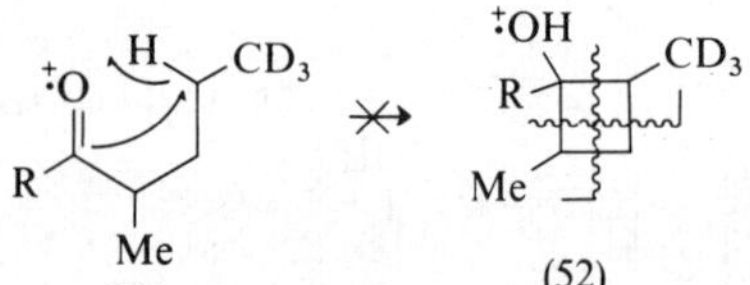

The effect of non-conjugated double bonds on the McLafferty rearrangement has been studied,[190] and it has been shown that double-bond migration is favoured to the δε-positions, resulting in a McLafferty rearrangement which leaves the charge on the hydrocarbon fragment [e.g. (53) → (54)]. Attention has also been drawn to the extensive hydrogen migration and bond lability which occurs when isoprenyl ketones are subjected to electron impact.[191] The α-cleavage reaction is a major decomposition of αβ-unsaturated ketones, but the reaction (55) → (56) is also observed.[192] The molecular ions of alkyl acetylenic ketones undergo the α-cleavage reaction and also eliminate CO.[193] Further

---

[185] C. Eadon, J. Diekman, and C. Djerassi, *J. Amer. Chem. Soc.*, 1970, **92**, 6205.
[186] D. J. McAdoo, F. W. McLafferty, and J. S. Smith, *J. Amer. Chem. Soc.*, 1970, **92**, 6343.
[187] F. W. McLafferty, D. J. McAdoo, J. S. Smith, and R. Kornfeld, *J. Amer. Chem. Soc.*, 1971, **93**, 3720.
[188] A. F. Gerrard, R. L. Hale, R. Liedke, W. H. Faue, and C. A. Brown, *Org. Mass Spectrometry*, 1970, **3**, 683.
[189] E. M. Chait and F. G. Kitson, *Org. Mass Spectrometry*, 1970, **3**, 533.
[190] J. R. Dias, Y. M. Sheikh, and C. Djerassi, *J. Amer. Chem. Soc.*, 1972, **94**, 473.
[191] U. T. Bhalerao and H. Rapoport, *J. Amer. Chem. Soc.*, 1971, **93**, 105.
[192] Y. M. Sheikh, A. M. Duffield, and C. Djerassi, *Org. Mass Spectrometry*, 1970, **4**, 273.
[193] R. T. Aplin and R. Mestres, *Org. Mass Spectrometry*, 1970, **3**, 1067.

Reactions of Specific Functional Groups 105

aspects of the spectra of β-diketones[194,195] and fluorinated β-diketones[196] have appeared. The basic cleavages noted for the ε-diketone system are the α-cleavage process and the McLafferty rearrangement.[197] The elimination of C_2H_4 occurs from (57) and related species, and it has been suggested that rearrangement to a cyclobutanol system (57) → (58) may account for the elimination.[197]

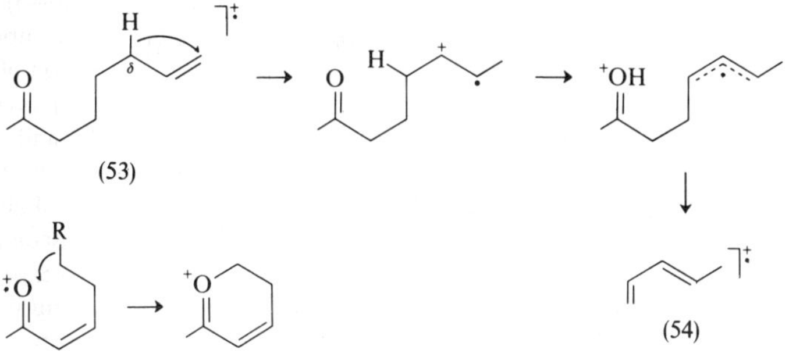

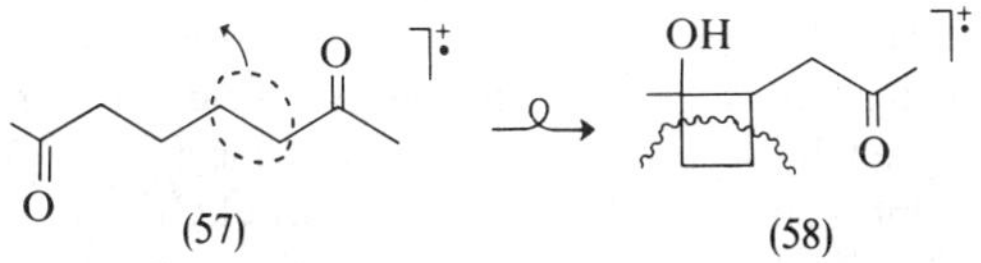

Cyclobutanols are formed on photolysis of C_{11}—C_{16} cyclic ketones, but are not formed upon electron impact.[198] γ-Hydrogen transfer occurs to the carbonyl group of cyclic ketones when the number of carbon atoms is greater than ten [(59) → (60) + (61)].[198] Large differences are observed between abundances of ions in the spectra of cis- and trans-bicyclo[3,3,0]octan-3-ones.[199] Cyclic ketones fused to a cyclobutane ring,[200] 7-methyloctan-4-one,[201] spiroalkanones with five- or six-membered rings,[202,203] further aspects of polycyclic ketones,[204,205] and 5,5-dimethylcyclohexane-1,3-diones[206] have been reported.

[194] K. S. Patel, K. L. Rinehart, and J. C. Bailar, *Org. Mass Spectrometry*, 1970, **3**, 1239.
[195] M. E. Rennekamp, J. V. Paukstelis, and R. G. Cooks, *Tetrahedron*, 1971, **27**, 4407.
[196] M. Rubesch, A. L. Clobes, M. L. Morris, and R. D. Koob, *Org. Mass Spectrometry*, 1971, **5**, 237.
[197] J. Kossanyi and J. K. Mogto, *Org. Mass Spectrometry*, 1970, **3**, 721.
[198] K. H. Schulte-Elte, B. Willhalm, A. F. Thomas, M. Stoll, and G. Ohloff, *Helv. Chim. Acta*, 1971, **54**, 1759.
[199] R. Granger, J.-P. Vidal, J.-P. Girard, and J.-P. Chapat, *Compt. rend.*, 1970, **270**, C, 2023.
[200] P. Singh, *Tetrahedron Letters*, 1971, 1071.
[201] G. Eadon and C. Djerassi, *J. Amer. Chem. Soc.*, 1970, **92**, 3084.
[202] G. D. Christiansen and D. A. Lightner, *J. Org. Chem.*, 1971, **36**, 948.
[203] W. D. Weringa, *Org. Mass Spectrometry*, 1971, **5**, 1055.
[204] J. Deutsch and A. Mandelbaum, *Org. Mass Spectrometry*, 1971, **5**, 53.
[205] J. Deutsch and A. Mandelbaum, *J. Chem. Soc. (B)*, 1971, 886.
[206] G. Ozolins, A. Strakov, and D. Brutane, *Latv. P.S.R. Zinat. Akad. Vestis. Kim. Ser.*, 1971, 150.

The decompositions of cyclic cyclopentane-1,3-diones occur from both *keto* and *enol* forms of the molecular ion, with the basic cleavage occurring as indicated in (62).[207]

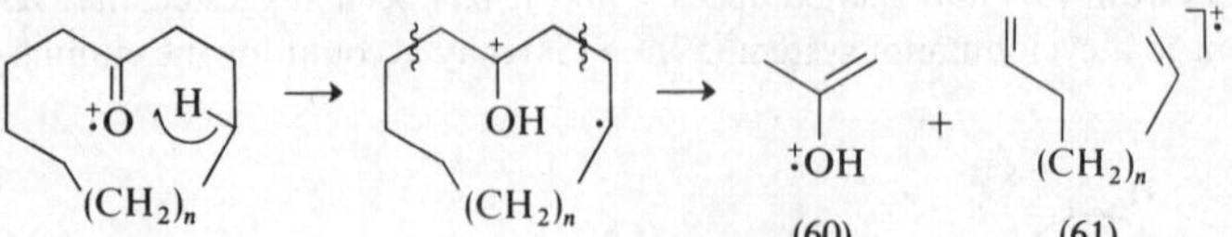

Aromatic. The losses of C_2H_2O and C_3H_4O from the hydrocinnamaldehyde molecular ion, and the formation of $C_7H_7^+$, occur with extensive hydrogen randomization.[208] Several studies of the McLafferty rearrangement in aryl systems have appeared. A comparison of the metastable transitions of the acetophenone molecular ion and of the ion produced from the McLafferty rearrangement of (63) has shown that the rearrangement ion does not exist in the keto-form, and probably occurs as the enol (64) and (65).[209] The interesting observation that (66) can eliminate successive units of C_3H_6 by two McLafferty rearrangements has been interpreted as showing that charge can be transferred through sigma bonds,[210a] in agreement with the earlier work of Mandelbaum and Biemann.[210b] Several aryl alkyl ketones which do not suffer the Norrish Type II photochemical elimination undergo the McLafferty rearrangement.[211] Hydrogen-migration reactions produce several peaks of high intensity in the low-energy spectra of phenyl-substituted $\alpha\beta$-unsaturated ketones.[212] For example, the $M - 58$ ion is the base peak of (67) and (68) at 12 eV, and the ions mainly occur by the rearrangement processes illustrated in (67) and (68).[212] The variation of fragmentation with chain length has been studied for ketones of the type (69).[213] The substituent R has an effect upon the formation of the benzoyl cation from (70) even when $n = 3$.[214] The spectra of aryl acetylenic

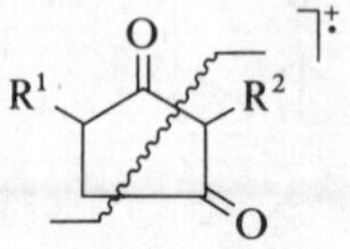

[207] E. Cant and M. Vanderwalle, *Org. Mass Spectrometry*, 1971, **5**, 1197.
[208] A. Venema, N. M. M. Nibbering, and Th. J. de Boer, *Org. Mass Spectrometry*, 1970, **3**, 583.
[209] J. H. Beynon, R. M. Caprioli, and T. W. Shannon, *Org. Mass Spectrometry*, 1971, **5**, 967.
[210] (a) A. Tatematsu, S. Naga, H. Sakurai, T. Goto, and H. Nakata, *Bull. Chem. Soc. Japan*, 1971, **44**, 3450; (b) A. Mandelbaum and K. Biemann, *J. Amer. Chem. Soc.*, 1968, **90**, 2975.
[211] M. M. Bursey, D. G. Whitten, M. T. McCall, W. E. Punch, M. K. Hoffman, and S. A. Benezra, *Org. Mass Spectrometry*, 1970, **4**, 157.
[212] R. J. Liedke, A. F. Gerrard, J. Diekman, and C. Djerassi, *J. Org. Chem.*, 1972, **37**, 776.
[213] C. Fenselau, A. A. Baum, and D. O. Cowan, *Org. Mass Spectrometry*, 1970, **4**, 229.
[214] B. M. Zolotarev, L. A. Yanovskaya, B. Umirzakov, O. S. Chizhov, and V. F. Kucherov, *Org. Mass Spectrometry*, 1971, **5**, 1043.

ketones are dominated by $M - CO$ ions,[193] and further work has appeared on benzophenones[215—217] and tetralin-1,2-diones.[218]

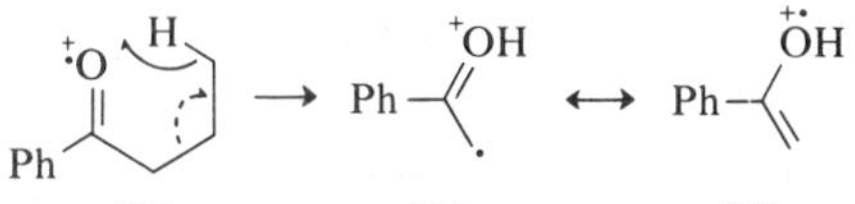

(63) (64) (65)

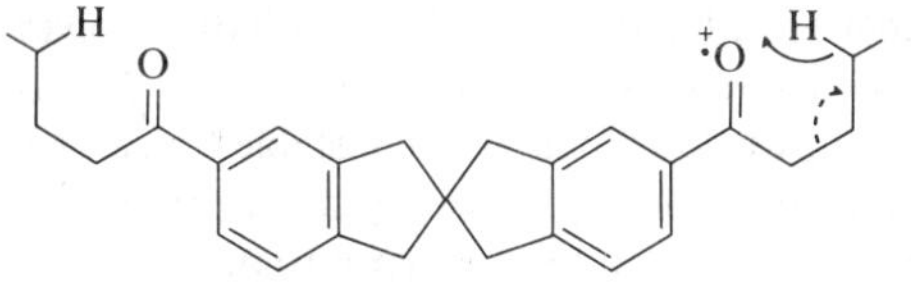

(66) 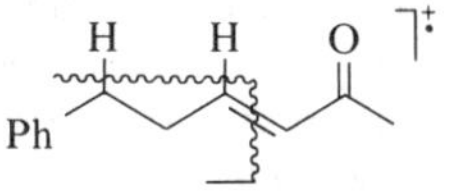 (67)

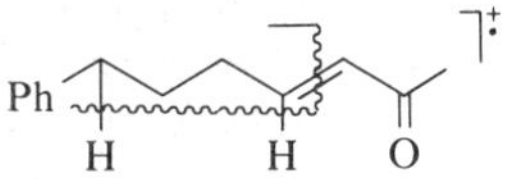

(68) 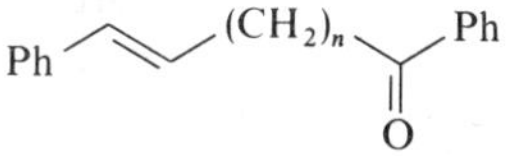 (69)

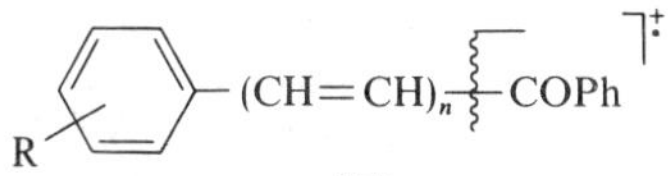

(70)

The fragmentation of (71) proceeds through an $M - C_{14}H_{12}$ species which has the properties of the stilbenol benzoate radical ion (72).[219] A re-examination by fluorine-labelling studies of the formation of the ion represented as the tetrahedrane species (74) has shown that scrambling can occur in both the C_4Ph_4 ion and in the precursor molecular ion.[220] For example, the phenyl groups of (73) and (75) have lost their positional identity prior to the formation of (74).[220] Several papers[221,222] on 1,2- and 1,4-naphthaquinones are useful additions to previous work in this field (Vol. 1, p. 110).

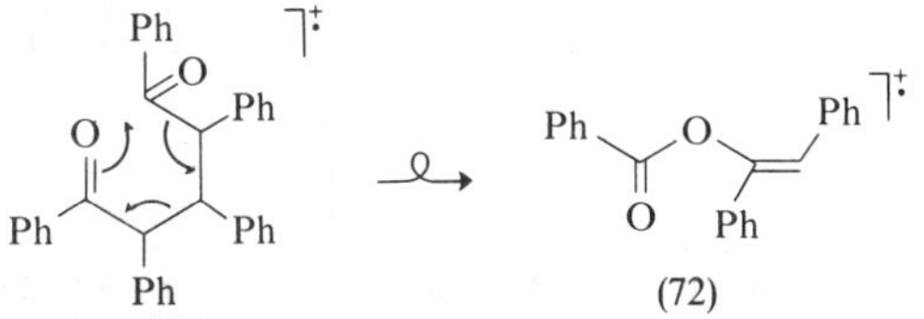

(71) (72)

[215] M. M. Bursey and C. E. Twine, *J. Org. Chem.*, 1970, **35**, 2012.
[216] N. Einolf and B. Munson, *Org. Mass Spectrometry*, 1971, **5**, 397.
[217] W. O. George, D. V. Hassid, and J. Phillips, *Org. Mass Spectrometry*, 1971, **5**, 605.
[218] R. F. C. Brown and M. Butcher, *Austral. J. Chem.*, 1970, **23**, 1907.
[219] J. H. Bowie, *Austral. J. Chem.*, 1972, **25**, 903.
[220] M. K. Hoffmann, T. A. Elwood, P. F. Rogerson, J. M. Tesarek, M. M. Bursey, and D. Rosenthal, *Org. Mass Spectrometry*, 1970, **3**, 891.
[221] T. A. Elwood, K. H. Dudley, J. M. Tesarek, P. F. Rogerson, and M. M. Bursey, *Org. Mass Spectrometry*, 1970, **3**, 841.
[222] R. W. A. Oliver and R. M. Rashman, *J. Chem. Soc. (B)*, 1971, 341.

(73) → (74) ← (75)

Acids.—The McLafferty rearrangement of butyric acid (Vol. 1, p. 110) involves a stepwise rearrangement, which occurs at a fast rate relative to $\alpha\beta$-C—C rotation, and transfer of hydrogen to the γ-carbon atom.[223] Metastable transitions show evidence of hydrogen exchange preceding the rearrangement process.[223] A mixture of carboxylic acids has been analysed by mass spectrometry[224] and the spectra of dicarboxylic acids have been described.[225] The spectra of *cis*- and *trans*-cyclohexane-1,2-dicarboxylic acids are readily distinguished by the presence of a large $M - CO_2$ ion for the *trans*-isomer,[226] whereas the spectra of the corresponding *cis*- and *trans*-cyclohex-4-ene-1,2-dicarboxylic acids are very similar.[227]

The singly and doubly charged molecular ions of benzoic acid eliminate carbon monoxide.[228] The loss of CO from the singly charged molecular ion of benzoic acid is complex, with 80% originating from the carboxy-group and 20% of the carbon coming from the phenyl ring at 15 eV.[228] No o-H–carboxyl-H exchange of the type observed for benzoic acid[229,230] occurs for thiobenzoic acid,[228] phthalic acid,[228] or salicyclic acid.[231] Reversible hydrogen exchange between the methyl and carboxy-groups precedes the loss of water from the o-toluic acid molecular ion.[232] The spectra of 2-biphenylcarboxylic acids,[233] β-phenylpropionic acids[234,235] and *trans*-pentafluorocinnamic acid[236] have been described. It has been suggested that the characteristic loss of C_4H_5O from the molecular ions of β-aroyl-α-methylpropionic acids may be rationalized in terms of a process (76) → (77).[237]

[223] J. S. Smith and F. W. McLafferty, *Org. Mass Spectrometry*, 1971, **5**, 483.
[224] V. L. Talrose, V. E. Skurat, I. G. Gorodetsky, and N. B. Zolotoi, *Zhur. analit. Khim.*, 1971, **26**, 2205.
[225] J. L. Holmes and T. St. Jean, *Org. Mass Spectrometry*, 1970, **3**, 1505.
[226] F. Benoit and J. L. Holmes, *Org. Mass Spectrometry*, 1972, **6**, 541.
[227] F. Benoit and J. L. Holmes, *Org. Mass Spectrometry*, 1972, **6**, 549.
[228] J. L. Holmes and F. Benoit, *Org. Mass Spectrometry*, 1970, **4**, 97.
[229] J. H. Beynon, B. E. Job, and A. E. Williams, *Z. Naturforsch.*, 1965, **20a**, 883.
[230] S. Meyerson and J. L. Corbin, *J. Amer. Chem. Soc.*, 1965, **87**, 3045.
[231] S. A. Benezra and M. M. Bursey, *Org. Mass Spectrometry*, 1972, **6**, 463.
[232] M. J. Lacey, C. G. MacDonald, and J. S. Shannon, *Org. Mass Spectrometry*, 1971, **5**, 1391.
[233] A. T. Balaban and A. M. Duffield, *Rev. Roumaine Chim.*, 1971, **16**, 1095.
[234] B. M. Zolotarev, B. I. Kadentsev, B. F. Kucherov, O. S. Chizhov, Kh. Shakhidayatov, and L. A. Yanovskaya, *Izvest. Akad. Nauk S.S.S.R., Ser. khim.*, 1970, 1552.
[235] J. A. Ballantine and R. F. Curtis, *Org. Mass Spectrometry*, 1970, **3**, 1215.
[236] H. Heaney and A. P. Price, *Chem. Comm.*, 1971, 894.
[237] H. M. A. Buurmans, B. van de Graaf, and A. P. G. Kieboom, *Org. Mass Spectrometry*, 1971, **5**, 1081.

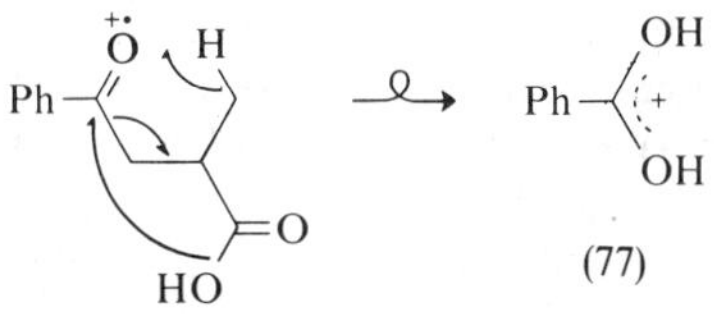

Esters, Lactones, and Anhydrides.—Hydrogen scrambling within the ethyl group and between the ethyl and acetyl substituents occurs prior to the elimination of water from the ethyl acetate molecular ion.[238] Metastable ion–molecule reactions have been observed for n-butyl acetate.[239] Long-chain esters,[240,241] esters of α-hydroxy- and α-methoxy-acids,[242] and methyl acyloxyacetates[243] have been investigated. A comparison of the products of the losses of keten from the molecular ions of enol acetates and the corresponding ketones suggests that the elimination from enol acetates proceeds through a four-membered transition state (78) to afford enolic products.[244] The differences between the spectra of dimethyl fumarate and dimethyl maleate have been re-evaluated.[245] Differences between the decompositions of stereoisomers are observed for esters of 3,4-diethylmuconic acids[246] and fused systems incorporating the 1,2-dimethoxycarbonylcyclohex-3-ene moiety.[247] For example, the molecular ions of (79) and (80) eliminate methoxyl radicals, whereas that of (81) loses methanol.[246]

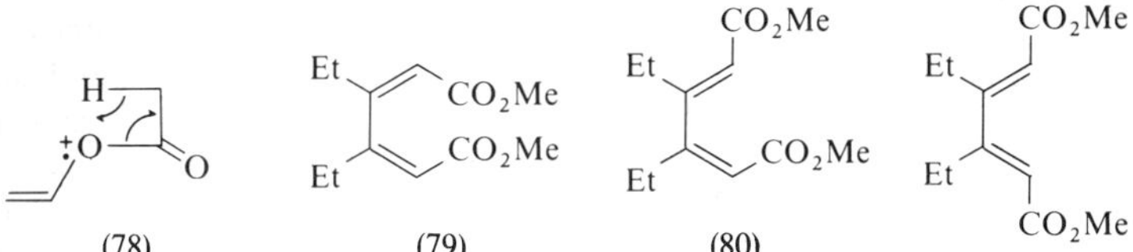

The IKE spectra of $Ph-C^{18}O-OC_2D_5$ and $PhCO_2C_2D_5$ show that the ion formed by loss of C_2D_4 from the molecular ion has the transferred deuterium attached to the carbonyl oxygen.[248] Transfer of this deuterium to the other oxygen can only take place by an exchange reaction involving the *ortho*-hydrogen atoms[249] and requires at least two rotations of the side-chain, (82) → (83).

[238] A. N. H. Yeo, *Chem. Comm.*, 1970, 1154.
[239] J. H. Beynon, R. M. Caprioli, W. E. Baitinger, and J. W. Amy, *Org. Mass Spectrometry*, 1970, **3**, 817.
[240] K. K. Sun, H. W. Hayes, and R. T. Holman, *Org. Mass Spectrometry*, 1970, **3**, 1035.
[241] D. G. Chasin and E. G. Perkins, *Chem. Phys. Lipids*, 1971, **6**, 8.
[242] J. D. S. Goulden and D. J. Manning, *Org. Mass Spectrometry*, 1970, **3**, 1467.
[243] E. K. Euranto, *Suomen. Kemi. (B)*, 1970, **43**, 324.
[244] H. Nakata and A. Tatematsu, *Org. Mass Spectrometry*, 1970, **4**, 211.
[245] S. Meyerson, P. J. Ihrig, and T. L. Hunter, *J. Org. Chem.*, 1970, **36**, 995.
[246] E. Gil-Av, J. H. Leftin, A. Mandelbaum, and S. Winstein, *Org. Mass Spectrometry*, 1970, **4**, 475.
[247] J. Deutsch and A. Mandelbaum, *J. Amer. Chem. Soc.*, 1970, **92**, 4288.
[248] R. H. Shapiro, K. B. Tomer, R. M. Caprioli, and J. H. Beynon, *Org. Mass Spectrometry*, 1970, **3**, 1333.
[249] R. H. Shapiro, K. Tomer, J. H. Beynon, and R. M. Caprioli, *Org. Mass Spectrometry*, 1970, **3**, 1593.

Various aspects of the spectra of phenyl acetates[250,251] and halogenated phenyl acetates[252,253] have been considered. The molecular ion of methyl o-$(O$-$[^2H_3]$-acetyl)-salicylate eliminates CD_2CO followed by MeOD.[254] This indicates that the initial elimination of CD_2CO occurred through a four-membered transition state (84) to yield (85) and then (86). If the elimination of deuteriated keten had occurred through a six-membered transition state (87) to produce (85) and (89), both MeOH and MeOD would have been eliminated from the

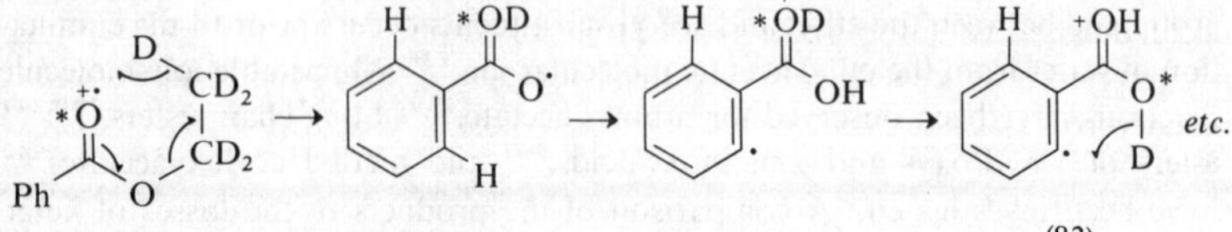

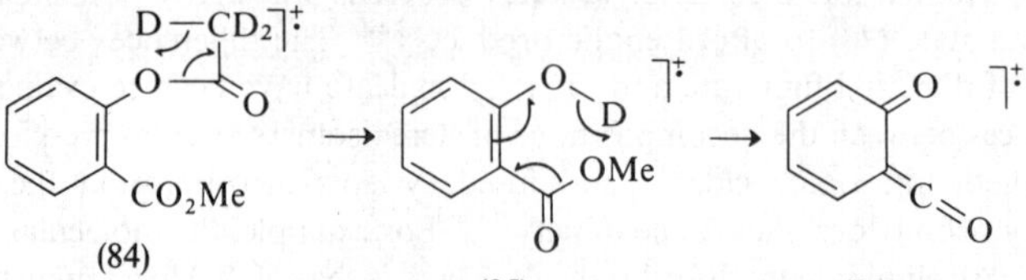

$M - CD_2CO$ ion.[254] The spectra of methyl 3-phenylpropionate,[235] phenyl-valerates,[255] methyl phenylpentanoates,[256] methyl phenylnonanoates,[256] diphenylnorbornadiene esters,[257] and substituent effects in the spectra of a series of aryl esters[258] and acetates[259] have been reported.

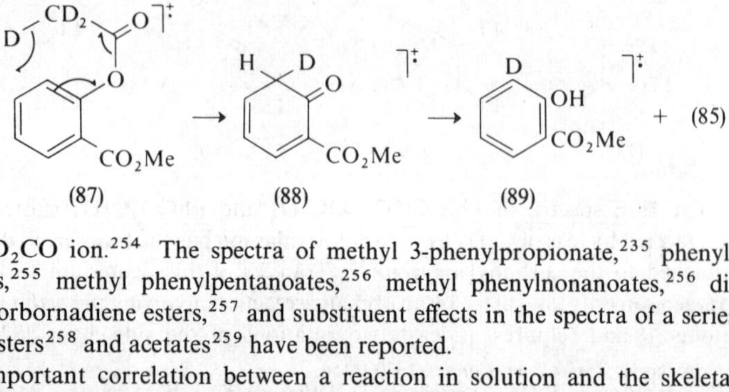

An important correlation between a reaction in solution and the skeletal rearrangement (90) → (91) has been demonstrated.[260] The reaction between keten and the α-methoxybenzylcarbonium ion (91) yields methyl cinnamate,

[250] A. A. Gamble, J. R. Gilbert, and J. G. Tillett, *Org. Mass Spectrometry*, 1971, **5**, 1093.
[251] H. Nakata and A. Tatematsu, *Org. Mass Spectrometry*, 1971, **5**, 1343.
[252] M. J. Saxby, *Org. Mass Spectrometry*, 1970, **4**, 133.
[253] S. A. Benezra and M. M. Bursey, *J. Chem. Soc.* (*B*), 1971, 1515.
[254] H. Nakata and A. Tatematsu, *Org. Mass Spectrometry*, 1971, **5**, 1343.
[255] G. G. Smith and S. W. Cowley, *Chem. Comm.*, 1971, 1066.
[256] M. F. Ansell and G. F. Whitfield, *Org. Mass Spectrometry*, 1970, **3**, 1099.
[257] H. Prinzbach and M. Thyes, *Chem. Ber.*, 1971, 2489.
[258] V. J. Feil and J. M. Sugihara, *Org. Mass Spectrometry*, 1972, **6**, 265.
[260] B. Davis and D. H. Williams, *J. Org. Chem.*, 1970, **35**, 2033.
[259] H. E. Audier, G. Bouchoux, and M. Fétizon, *Bull. Soc. chim. France*, 1971, 858.

Reactions of Specific Functional Groups

which is analogous to the reverse of the rearrangement (90) → (91).[260] Methyl benzylidene malonates undergo the characteristic reaction $M - \text{MeO}^{\bullet} - \text{C}_3\text{O}_2$, (92) → (93).[261] The spectra of pivalolactone,[262] and $\alpha\beta$-disubstituted butenolides[263] have been discussed, and (94) undergoes the characteristic elimination of H_2O involving the carbonyl group and the *peri*-methoxy-group[264] (Vol. 1, p. 110). The spectrum of (95) contains an $M - \text{CO}$ ion, whereas the corresponding ion in the spectrum of (96) is of low abundance.[265] The spectra of *o*-phenylene carbonates have been described.[266] Triple hydrogen migrations occur during the decompositions of esters of trimellitic anhydride (97; R ⩾ n-C_6H_{13}) to produce (98), with the three hydrogens originating in a non-specific manner from the side-chain.[267]

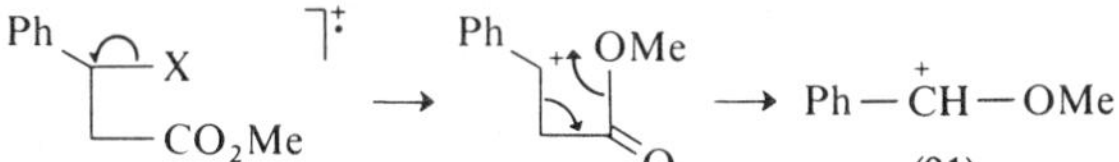

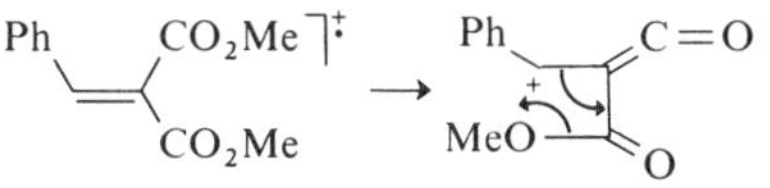

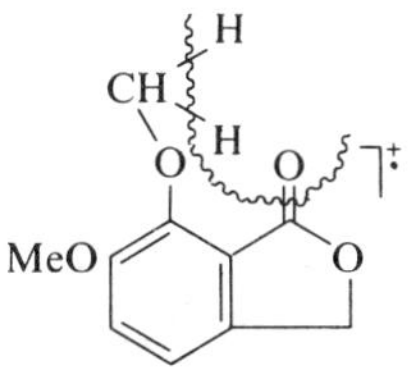

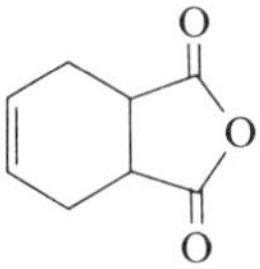

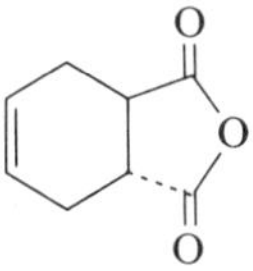

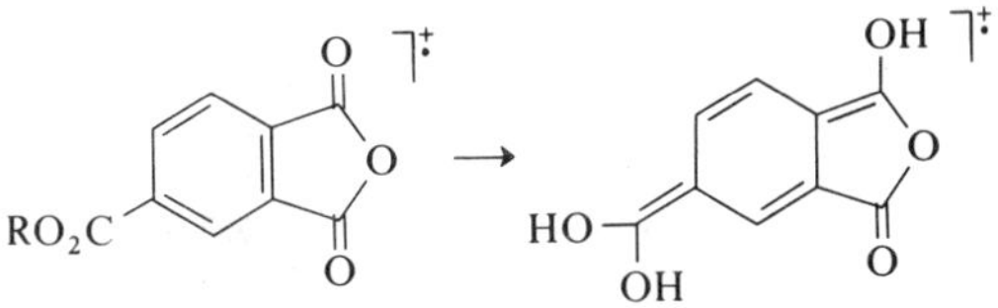

Ethers and Peroxides.—The two ions (99) and (100), produced from aliphatic ethers, do not fragment through a common intermediate.[268] The base peak in the spectra of cyclopropenyl ethers is produced as indicated in (101; R ⩾ Me).[269] Cyclization occurs after the α-cleavage process of (102; X = alkyl or OR) to form (103), and a prerequisite for this reaction is that the ether and carbonyl groups must be separated by three methylene groups.[270] The spectra

[261] Q. N. Porter and C. C. R. Ramsay, *Austral. J. Chem.*, 1971, **24**, 823.
[262] R. H. Wiley, *J. Macromol. Sci. Chem.*, 1970, **A4**, 1797.
[263] D. N. Reinhoudt and B. van de Graaf, *Rec. Trav. chim.*, 1970, **89**, 509.
[264] B. M. King, D. A. Evans, and K. Biemann, *Org. Mass Spectrometry*, 1970, **3**, 1049.
[265] A. Karpati and A. Mandelbaum, *Org. Mass Spectrometry*, 1971, **5**, 1345.
[266] D. C. DeJongh and D. A. Brent, *J. Org. Chem.*, 1970, **35**, 4204.
[267] J. Cable and C. Djerassi, *J. Amer. Chem. Soc.*, 1971, **93**, 3905.
[268] C. W. Tsang and A. G. Harrison, *Org. Mass Spectrometry*, 1970, **3**, 647.
[269] G. Salmona, J.-P. Galy, and E.-J. Vincent, *Compt. rend.*, 1971, **273**, C, 685.
[270] M. Sheehan, R. J. Spangler, M. Ikeda, and C. Djerassi, *J. Org. Chem.*, 1971, **36**, 1796.

of the methyl ethers of cyclopentane-, cyclohexane-, and cycloheptane-diols all contain a pronounced peak due to (105), produced by a methoxyl migration of the type (104) → (105).[271,272] The ion $C_2H_5O^+$ is produced from the methyl allyl ether molecular ion as shown in (106), together with some prior scrambling of hydrogen within the allyl unit.[273] A series of hydrogen-migration reactions occurs from the molecular ions of unsaturated ethers.[274] As an example, the loss of C_3H_6 from (107) involves transfer of an α-, β-, or γ-hydrogen atom in the ratio 13 : 30 : 53.[274]

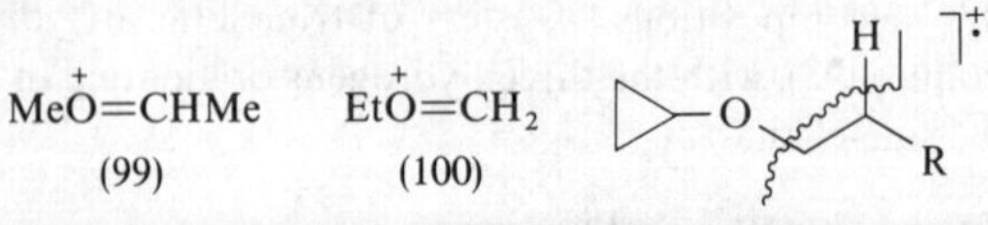

$MeO^+=CHMe$ (99)

$EtO^+=CH_2$ (100)

(101)

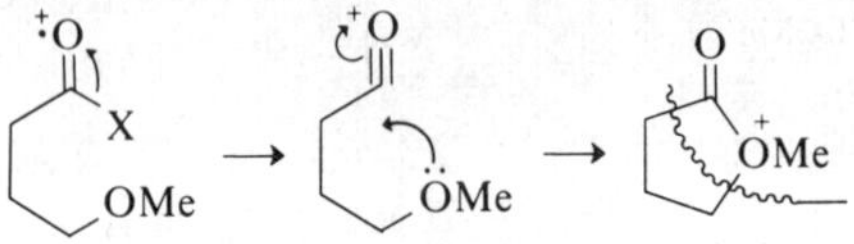

(102) → (103)

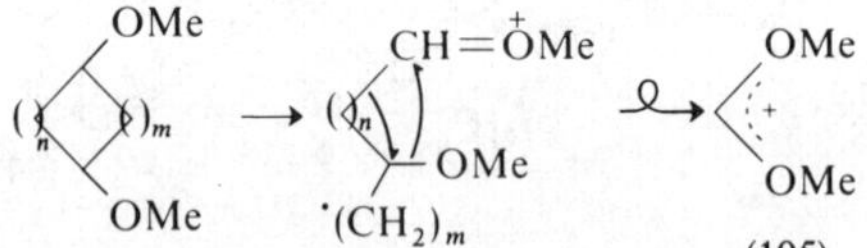

(104) → (105)

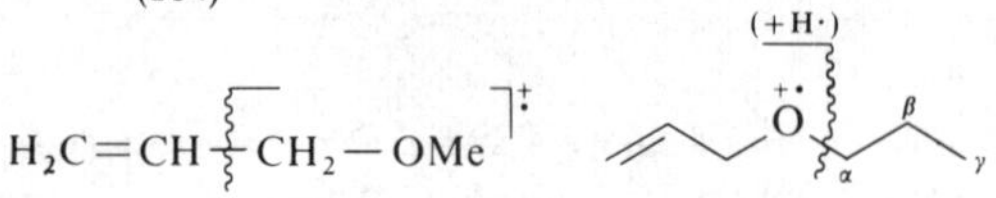

$H_2C=CH\text{---}CH_2\text{---}OMe$]⁺˙ (106)

(107)

Equilibration of the methoxyl and o-hydrogens occurs prior to the elimination of formaldehyde from the anisole molecular ion.[232] The competing eliminations of Me· and CH_2O from a series of substituted anisoles have been studied.[275,276] It has been established[277] that the $M - Me·$ ions from substituted anisoles and the $M - NO·$ ions from substituted nitrobenzenes decompose through common intermediates. The formation of benzyl and/or tropylium ions from methyl benzyl ethers and methyl tropyl ethers has been considered in terms of the quasi-equilibrium theory of mass spectra.[278] The formation of the ion

[271] J. Winkler and H.-F. Grützmacher, *Org. Mass Spectrometry*, 1970, **3**, 1117.
[272] J. Winkler and H.-F. Grützmacher, *Org. Mass Spectrometry*, 1970, **3**, 1139.
[273] D. Hasselmann and W. Kirmse, *Chem. Ber.*, 1972, **105**, 859.
[274] J.-P. Morizur and C. Djerassi, *Org. Mass Spectrometry*, 1971, **5**, 895.
[275] P. Brown, *Org. Mass Spectrometry*, 1970, **3**, 1175.
[276] P. Brown, *Org. Mass Spectrometry*, 1970, **4**, 419.
[277] B. Davis and D. H. Williams, *Chem. Comm.*, 1970, 412.
[278] M. K. Hoffman and M. M. Bursey, *Chem. Comm.*, 1971, 824.

Reactions of Specific Functional Groups   113

$C_6H_6O^{+\cdot}$ from the n-butyl phenyl ether molecular ion involves a non-specific hydrogen-atom transfer from the side-chain at high energy,[279] but metastable transitions show the transferred hydrogen to originate largely from the 3-position at lower energies.[280] This may be interpreted in terms of different transition states for the transfer processes, with the transfer from the 3-position [*e.g.* (108) → (109)] having the lowest activation energy. The formation of $C_7H_7O^+$ ions from (110) involves prior scrambling of benzylic and hydroxyl hydrogen atoms.[281] Further reports of the behaviour of aliphatic ketals,[282] phenyl-1,3-dioxans[283] and aryl alkyl diethers[284] upon electron impact have been published. The basic breakdown of the diethyl peroxide molecular ion is $M - C_2H_4 - HO\cdot$,[285] in marked contrast to the fragmentation $M - C_2H_4 - C_2H_4$ observed for the analogous disulphide.[286]

<pre>
        O   H   ⎤⁺·
       ╱ ╲   → Ph—ÖH   │ ╲
  Ph⁺· ╲ ╱ H   (109)   Ph   O—(CH₂)ₙ
                                        ╲   ╱
      (108)   (110) H
</pre>

Amines, Amides, Imides, and Related Systems.—Detailed studies of trimethyl-amine[287] and trifluoromethyldimethylamine[288] have been published. The decompositions of a variety of aliphatic amines and amides produce a series of ions $C_3H_8N^+$, and evidence from labelling and metastable studies is consistent with the existence of at least three different ion structures.[289] Dialkylamino-ethers[290] and $\alpha\omega$-diaminoalkanes[291,292] have been described. The loss of methanol from the $M - R\cdot$ ions of (111; $n \geqslant 3$) occurs by several mechanisms, and involves the hydrogens indicated in (112).[293]

The spectra of benzylamines,[294] benzylidenebenzylamines,[295] and phenyl-alkylamines[296] have been recorded. The loss of NH_3 from the molecular ion of 3-phenylpropylamine proceeds after partial equilibration of the amino and

[279] P. D. Woodgate and C. Djerassi, *Org. Mass Spectrometry*, 1970, **3**, 1093.
[280] A. N. H. Yeo and C. Djerassi, *J. Amer. Chem. Soc.*, 1972, **94**, 482.
[281] M. Sheehan, R. J. Spangler, and C. Djerassi, *J. Org. Chem.*, 1971, **36**, 3526.
[282] J. R. Dias and C. Djerassi, *Org. Mass Spectrometry*, 1972, **6**, 385.
[283] R. Böhm, N. Bild, and M. Hesse, *Helv. Chim. Acta*, 1972, **55**, 630.
[284] P. Vouros and K. Biemann, *Org. Mass Spectrometry*, 1970, **3**, 1317.
[285] R. T. M. Fraser, N. C. Paul, and L. E. Phillips, *J. Chem. Soc. (B)*, 1970, 1278.
[286] J. H. Bowie, S.-O. Lawesson, J. Ø. Madsen, C. Nolde, G. Schroll, and D. H. Williams, *J. Chem. Soc. (B)*, 1966, 946.
[287] G. Hvistendahl and K. Undheim, *Org. Mass Spectrometry*, 1970, **3**, 821.
[288] B. W. Tattershall, *J. Chem. Soc. (A)*, 1970, 3263.
[289] N. A. Uccella, I. Howe, and D. H. Williams, *J. Chem. Soc. (B)*, 1971, 1933.
[290] Ts. Ye. Aghejanian and R. T. Grigorian, *Armyan. khim. Zhur.*, 1971, **24**, 113.
[291] A. Guggisberg, H. J. Veith, and M. Hesse, *Tetrahedron Letters*, 1970, 3639.
[292] H. J. Veith, A. Guggisberg, and M. Hesse, *Helv. Chim. Acta*, 1971, **54**, 653.
[293] A. Caspar, G. Teller, and R. E. Wolff, *Org. Mass Spectrometry*, 1970, **3**, 1351.
[294] M. W. Couch and C. M. Williams, *Org. Mass Spectrometry*, 1972, **6**, 21.
[295] Yu. S. Nekrasov, V. A. Puchkov, and N. S. Wulfson, *Zhur. obshchei. Khim.*, 1970, **40**, 1506.
[296] D. A. Lightner, F. W. Sunderman, L. Hurtado, and E. Thommen, *Org. Mass Spectrometry*, 1970, **3**, 1325.

$$R\underset{\overset{+}{N}H_2}{\overset{}{\diagup\diagdown(\ )_n\diagdown}}CO_2Me \quad \longrightarrow \quad \overset{H}{\underset{H\overset{+}{N}H}{\diagup\diagdown(\ )_n\diagdown}}\overset{O}{\underset{}{\diagdown OMe}}$$

(111)   (112)

benzylic hydrogens.[296] Several articles have been published concerning the effect of charge localization on the fragmentation of aminocarbonyl compounds.[297,298] It has been suggested[297] that certain fragmentations of the molecular ions of α-dimethylaminoacetophenone and γ-dimethylaminobutyrophenone may be rationalized in terms of decomposition through the charged carbonyl group. An investigation of the fragmentations of (113; X = O or CH_2, $n = 0$ or 1) as a function of the distance between the amine and ketone (ester) groups, has led Djerassi[298] to suggest that initial ionization may occur at any site, with the positive charge then localizing at favoured sites. It has also been proposed that uncoupling of the carbonyl electrons is unimportant in the

(113)   (114)

McLafferty rearrangement of (114), in contrast to the Norrish Type II photochemical elimination.[299] The effect of substituents upon the cleavages of diarylethylenediamines has been evaluated,[300] the spectra of di-N-phenylaminetricyclo[4,2,0,0]octanes have been measured,[301] and doubly charged ions have been noted in the spectra of diphenylamines and their trimethylsilyl derivatives[302] and of phenylenediamines.[303]

The spectra of pyrrolidine amides are more useful for the characterization of long-chain carboxylic acids than those of the corresponding esters.[304] The behaviour of N-substituted β-ketoamides upon electron impact has been reported.[305] Scrambling between the amino and o-hydrogens does not precede the loss of $NH_2^{\cdot}$ from the benzamide or thiobenzamide molecular ions (cf. benzoic acid[229,230]).[306] Further studies of substituted acetanilides[307—309] and

[297] P. J. Wagner, Org. Mass Spectrometry, 1970, 3, 1307.
[298] J. Cable, G. W. Adelstein, J. Gore, and C. Djerassi, Org. Mass Spectrometry, 1970, 3, 439, and references cited therein.
[299] J. G. Calvert and J. N. Pitts, 'Photochemistry', Wiley, New York, 1966, pp. 382—385.
[300] H. Giezendanner, M. Hesse, and H. Schmid, Org. Mass Spectrometry, 1970, 4, 405.
[301] C. Lageot, Bull. Soc. chim. France, 1971, 3723.
[302] J. L. Beck, W. J. A. Van den Heuvel, and J. L. Smith, Org. Mass Spectrometry, 1970, 4, 237.
[303] R. M. Caprioli, J. H. Beynon, and T. Ast, Org. Mass Spectrometry, 1971, 5, 417.
[304] W. Vetter, W. Walther, and M. Vecchi, Helv. Chim. Acta, 1971, 54, 1599.
[305] J. Reisch and D. H. Niemeyer, Tetrahedron, 1971, 27, 4637.
[306] J. F. Holmes and F. Benoit, Org. Mass Spectrometry, 1971, 5, 525.
[307] A. A. Gamble, J. R. Gilbert, and J. G. Tillett, Org. Mass Spectrometry, 1970, 3, 1223.
[308] A. A. Gamble, J. R. Gilbert, and J. G. Tillett, J. Chem. Soc. (B), 1970, 1231.
[309] N. Uccella, I. Howe, and D. H. Williams, Org. Mass Spectrometry, 1972, 6, 229.

Reactions of Specific Functional Groups   115

toluanilides[310] have been reported. Consideration of the steric effect upon the $M - CH_2CO$ elimination[307] and of the isotope effect for the process $(M - CH_2CO) - HCN$[309] strongly supports the elimination of keten from (115) occurring through a four-membered transition state to furnish (116). Rearrangement processes are observed for NN-diphenylacetamides,[311] benzylidene hippuric esters[312] and benzylidene malonamide derivatives.[313] For example, (117) undergoes an amino-migration to produce ultimately (118).[312] Perfluorocyclobutenylethyleneimides,[314] maleimides and isomaleimides,[315] succinimides and hydantoins,[316] N-substituted cyclohexene-1,2-dicarboximides,[317] and various carbamates[318–322] and monoacylureas[323] have been investigated.

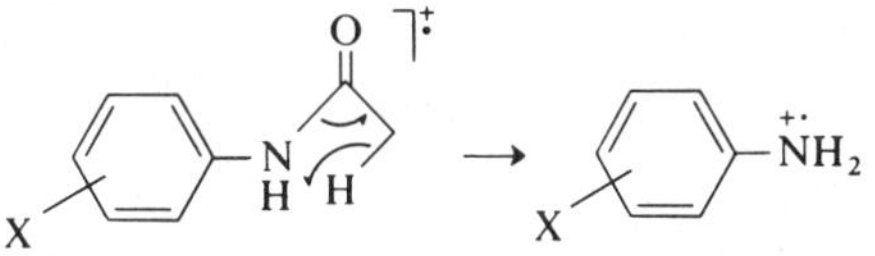

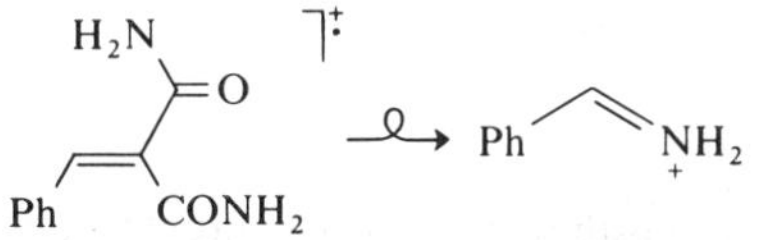

The C=N—, >N—N<, —N=N—, —CN, and —NC Groups.—The molecular ion of phenyl isocyanate eliminates CO specifically from the isocyanate moiety.[324] The spectra of acetophenone azine[325] and benzophenone azine[326] have been reported. The phenyl and methyl groups of acetophenone azine retain their structural identity upon electron impact.[325] The hydrogen atoms lost from the

[310] S. Kozuka, H. Takahashi, and S. Oae, *Bull. Chem. Soc. Japan*, 1971, **44**, 1965.
[311] H.-W. Fehlhaber and P. Welzel, *Org. Mass Spectrometry*, 1970, **4**, 545.
[312] M. McCamish and J. D. White, *Org. Mass Spectrometry*, 1971, **5**, 625.
[313] Q. N. Porter and C. C. R. Ramsay, *Tetrahedron*, 1970, **26**, 5327.
[314] Z. E. Samoilova and R. G. Kostyanovsky, *Izvest. Acad. Nauk S.S.S.R., Ser. khim.*, 1970, 1030.
[315] W. J. Feast, J. Put, F. C. de Schryver, and F. C. Compernolle, *Org. Mass Spectrometry*, 1970, **3**, 507.
[316] R. A. Locock and R. T. Coutts, *Org. Mass Spectrometry*, 1970, **3**, 735.
[317] E. D. Mitchell and G. R. Waller, *Org. Mass Spectrometry*, 1970, **3**, 519.
[318] J. A. Durden, H. W. Stollings, J. E. Casida, and M. Slade, *J. Agric. Food Chem.*, 1970, **18**, 459.
[319] W. E. Pereira, B. Halpern, M. D. Solomon, and A. M. Duffield, *Org. Mass Spectrometry*, 1971, **5**, 157.
[320] G. G. Still, *Org. Mass Spectrometry*, 1971, **5**, 977.
[321] J. A. Durden and W. J. Bartley, *J. Agric. Food Chem.*, 1971, **19**, 441.
[322] S. M. Shildcrout and C. C. Gebelein, *Org. Mass Spectrometry*, 1972, **6**, 485.
[323] G. Rücher, G. Bolm, and K. Kahrs, *Arch. Pharm.*, 1970, **303**, 601.
[324] A. S. Siegel, *Org. Mass Spectrometry*, 1970, **3**, 1471.
[325] S. E. Scheppele, R. D. Grigsby, D. W. Whitaker, S. D. Hinds, K. F. Kenneberg, and R. K. Mitchum, *Org. Mass Spectrometry*, 1970, **3**, 571.
[326] S. E. Scheppele, R. D. Grigsby, K. F. Kinneberg, E. D. Mitchell, and C. A. Mannan, *Org. Mass Spectrometry*, 1970, **3**, 557.

acetophenone azine molecular ion[325] and the NN-dimethyl-N'-phenylformamidine molecular ion (119)[327] originate from the phenyl rings. The major decompositions noted in the spectra of aliphatic oxime ethers are the α-cleavage processes (120), γ-cleavage (121), and the hydrogen rearrangement process (122).[328] Benzaldoxime ethers undergo rearrangement reactions [e.g. (123) → (124)] and no evidence was obtained for oxime ether/C-nitroso tautomerism of the molecular ions.[329,330] Further reports on hydrazones have appeared.[331—334]

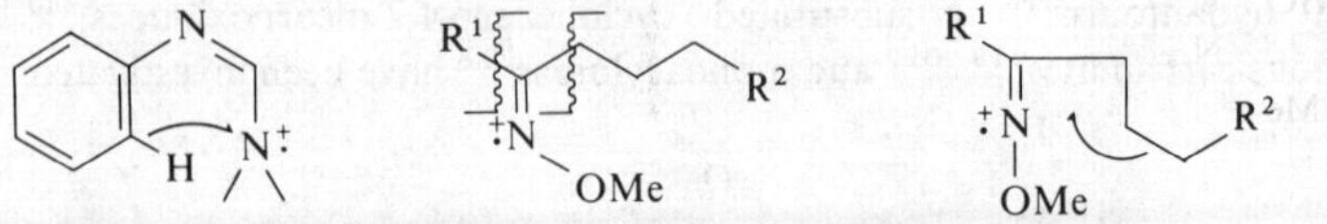

(119)   (120)   (121)

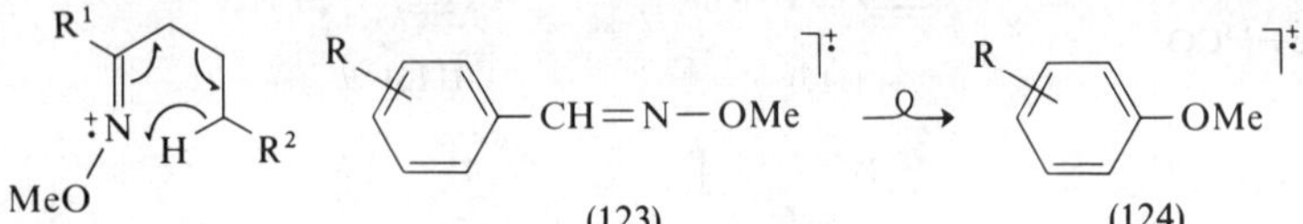

(122)   (123)   (124)

The spectra of monosubstituted hydrazine salts,[335] cis-2,3-diazabicycloalkane derivatives,[336] phenylhydrazine,[337] N-phenyl-N'-(p-acetoxyphenyl)-N'-trifluoroacetylhydrazine[338] and 2,2-dibenzoyl-1,1-dimethylhydrazine[339] have been reported. The elimination of C_2H_4N from (125) involves a double hydrogen migration from one methyl group.[339] The spectra of diazonium salts of o-aminophenols have been obtained,[340] but some decomposition is observed in the inlet system. The ion (126) eliminates N_2 followed by CO, and this has been rationalized in terms of an electron-impact-induced Wolff rearrangement, (127) → (128) + (129).[341] The spectrum of (130) shows almost equal elimina-

[327] H.-F. Grützmacher and H. Kuschel, Org. Mass Spectrometry, 1970, 3, 605.
[328] B. S. Middleditch and B. A. Knights, Org. Mass Spectrometry, 1971, 6, 179.
[329] R. G. Cooks and A. G. Varvoglis, Org. Mass Spectrometry, 1971, 5, 687.
[330] R. G. Cooks, D. W. Setser, K. R. Jennings, and S. Jones, Internat. J. Mass Spectrometry Ion Phys., 1971, 7, 493.
[331] J. Seibl, Org. Mass Spectrometry, 1970, 3, 417.
[332] D. G. I. Kingston, H. P. Tannenbaum, G. B. Baker, J. R. Dimmock, and W. G. Taylor, J. Chem. Soc. (C), 1970, 2574.
[333] R. J. Liedtke, A. M. Duffield, and C. Djerassi, Org. Mass Spectrometry, 1970, 3, 1089.
[334] J. Cable, S. A. Kagal, and J. K. MacLeod, Org. Mass Spectrometry, 1972, 6, 301.
[335] J. A. Blair and R. J. Gardner, Org. Mass Spectrometry, 1970, 4, 291.
[336] J. P. Snyder, M. L. Heyman, V. T. Bandurco, and D. N. Harpp, Tetrahedron Letters, 1971, 4693.
[337] R. G. Gillis, Org. Mass Spectrometry, 1971, 5, 1107.
[338] H. Achenbach, K. Klemom, and E. Langenscheid, Chem. Ber., 1971, 104, 1025.
[339] Q. N. Porter and A. E. Seif, Org. Mass Spectrometry, 1970, 4, 361.
[340] K. Undheim, O. Thornstad, and G. Hvistendahl, Org. Mass Spectrometry, 1971, 5, 73.
[341] K.-P. Zeller, H. Meier, and E. Müller, Annalen., 1971, 749, 178.

Reactions of Specific Functional Groups

tion of CO and ^{13}CO, demonstrating in this case that methyl and phenyl have similar migration abilities.[341] Unusual rearrangement reactions are observed in the spectra of azoacylals, (131) → (132).[342] The spectra of phenyl azide[343] and o-toluazide[344] labelled with ^{13}C show that some rearrangement of the carbon skeleton accompanies or follows the initial losses of nitrogen.

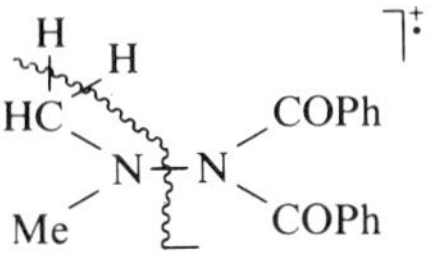

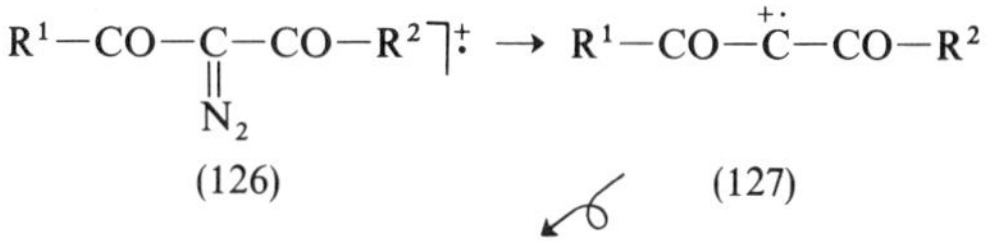

$$[\text{structures (125), (126), (127)}]$$

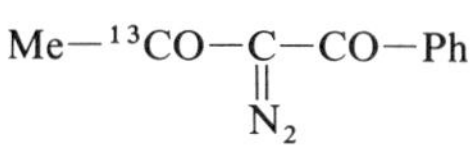

$$R^1-CO-\underset{N_2}{\overset{\|}{C}}-CO-R^2\;]^{\ddagger} \;\rightarrow\; R^1-CO-\overset{+\cdot}{C}-CO-R^2$$

$$Me-^{13}CO-\underset{N_2}{\overset{\|}{C}}-CO-Ph \qquad [\underset{R^2}{\overset{R^1CO}{>}}C=O]^{\ddagger} + [\underset{R^1}{\overset{R^2CO}{>}}C=O]^{\ddagger}$$

(130) 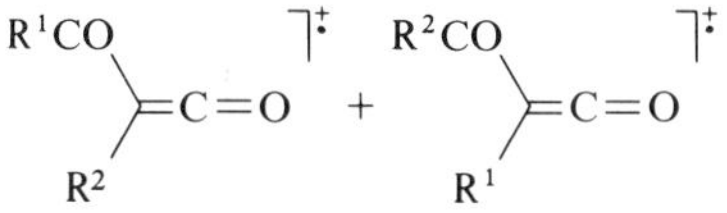   (128)   (129)

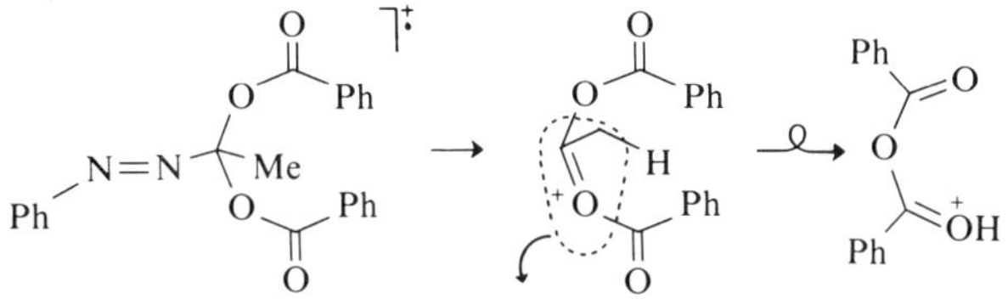

$$[\text{structures (131), (132)}]$$

The spectra of isocyanic acid,[345] alkyl cyanides,[346] alkyl isocyanides,[346] β-alkoxypropronitriles,[347] and dicyanomethylene derivatives of long-chain aliphatic acids[348] have been discussed. The loss of HCN from the benzyl cyanide molecular ion does not involve carbon scrambling between the phenyl residue and the side-chain, but all the hydrogen atoms are completely scrambled.[349] The $M - HCN$ ion produced in the ion source contains none of the cyano carbon, but metastable transitions indicate that the loss of HCN from low-energy molecular ions involves both the nitrile and benzylic carbon atoms in the ratio 78 : 22.[349] This may be rationalized[349] in terms of rearrangement of the molecular ion to a 1-phenylazirine ion radical which may then eliminate

[342] S. W. Breuer, W. A. F. Gladstone, and P. A. Hirst, *Org. Mass Spectrometry*, 1971, **5**, 757.
[343] P. D. Woodgate and C. Djerassi, *Tetrahedron Letters*, 1970, 1875.
[344] P. A. Abramovitch, E. P. Kyba, and E. F. V. Scrivin, *J. Org. Chem.*, 1971, **36**, 3796.
[345] D. J. Bogan and C. W. Hand, *J. Phys. Chem.*, 1971, **75**, 1532.
[346] W. Heerma and J. J. de Ridder, *Org. Mass Spectrometry*, 1970, **3**, 1439.
[347] R. Gross and J. Kelm, *Z. Naturforsch.*, 1972, **27b**, 327.
[348] R. E. Wolff, J. Throck Watson, and B. J. Sweetman, *Tetrahedron Letters*, 1970, 2719.
[349] T. A. Molenaar-Langeveld, N. M. M. Nibbering, and Th. J. de Boer, *Org. Mass Spectrometry*, 1971, **5**, 725.

either side-chain carbon atom. Substituted phenylacetonitriles[350,351] cyano-cyclopropanes[352,353] and 4,4-dicyanobiphenyls[354] have been studied.

The N—O Group.—The spectra of N-nitrosamines are characterized by $M - HO\cdot$ and $M - NO\cdot$ ions;[355] 1-chloro-2-nitrosocycloalkanes[356] and diaryl nitrones[357] have been studied; anilinium oxides undergo pyrolysis prior to ionization;[358] the spectrum of 1-nitroadamantane has been reported.[359] The spectrum of 3-phenylpropyl nitrite differs considerably from that of 3-phenylnitropropane.[360] The loss of water from the molecular ion of 3-phenylnitropropane occurs as shown in (133).[360] The basic fragmentation of (134) is $M - NO\cdot - C_2H_4$, with the elimination of C_2H_4 mainly involving the α and β positions.[360]

The spectrum of nitro[1-^{13}C]benzene (135) shows the processes $M - NO\cdot - {}^{13}CO$, showing that (136) undergoes no C—C or C—O bond cleavage prior to

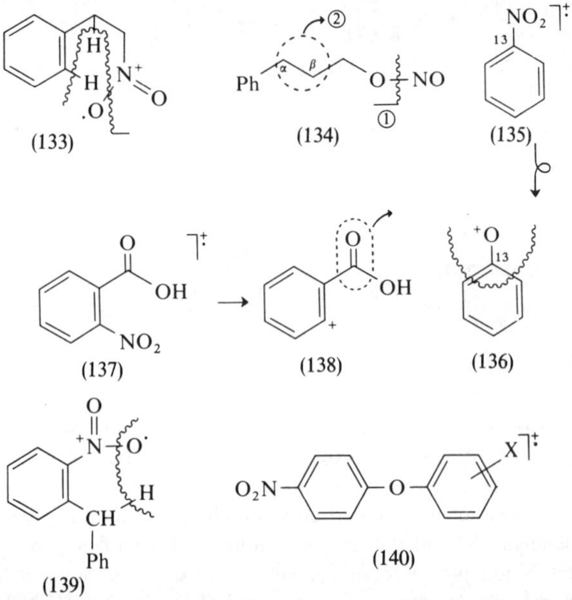

[350] M. Jimerez, M. C. Romero, and E. Cortes, *Bol. Inst. Quim. Univ. nac. auton. México*, 1970, **22**, 103.
[351] A. Buchs, *Helv. Chim. Acta*, 1970, **53**, 2026.
[352] C. Lageot, *Org. Mass Spectrometry*, 1971, **5**, 845.
[353] C. Lageot, *Org. Mass Spectrometry*, 1971, **5**, 105.
[354] M. De Bertorello, H. E. Bertorello, and N. Garcia-Martinez, *Anales. Asoc. quim. argentina*, 1970, **58**, 291.
[355] J. W. ApSimon and J. D. Cooney, *Canad. J. Chem.*, 1971, **49**, 1367.
[356] J. C. Tou and K. Y. Chang, *Org. Mass Spectrometry*, 1970, **3**, 1055.
[357] Yu. S. Nekrasov, V. A. Puchkov, and N. S. Wulfson, *Izvest. Akad. Nauk S.S.S.R., Ser. khim.*, 1970, 2483.
[358] K. Undheim and G. Hvistendahl, *Org. Mass Spectrometry*, 1970, **3**, 1423.
[359] A. I. Feinstein, E. K. Fields, P. J. Ihrig, and S. Meyerson, *J. Org. Chem.*, 1971, **36**, 996.
[360] N. M. M. Nibbering and Th. J. de Boer, *Org. Mass Spectrometry*, 1970, **3**, 487.

Reactions of Specific Functional Groups 119

fragmentation.[361] The $M - NO_2^{\cdot}$ ion $(C_6H_5^+)$ loses C_2H_2 in a completely random manner.[361,cf.85] The ions $C_7H_6NO_2^+$ produced from m- and p-alkylnitrobenzenes probably have benzylic structures,[362] and it is certain that they cannot both be nitrotropylium ions. The activation energies for the losses of $NO_2^{\cdot}$ and $NO\cdot$ from variously substituted nitrobenzenes have been measured.[363] The spectra of o-nitroanisole[364] and other o-substituted nitrobenzenes[365] usefully complement earlier work[366] in this field. For example, the fragmentation of (137) is $M - NO_2^{\cdot} - CO - CO$, with the loss of the first CO originating as indicated in (138).[365] The $M - HO\cdot$ ion from (139) is produced by a proximity effect,[367] and the effect of the substituent X on the abundances of $M - NO\cdot$ ions from a series of molecular ions (140) is interpreted to indicate that charge can migrate across phenyl rings through oxygen.[368] The spectra of all the dinitronaphthalenes have been listed.[369]

Heterocyclic Systems (excluding Sulphur Compounds).—*Three- and Four-membered Rings.* The loss of a hydrogen atom occurs from both positions of the aziridine molecular ion, and the loss of a methyl radical originates mainly as indicated in (141).[370] Both $Me\cdot$ and $CD_2H\cdot$ are eliminated from (142), the base peak in the spectrum of (143) is produced by a β-cleavage reaction, and the molecular ion of 1-phenylaziridine undergoes a rearrangement reaction to produce an ion corresponding to $C_7H_7^+$.[370] Azetidine methanols,[371] adamantyl-

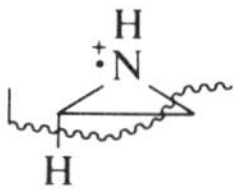

(141) 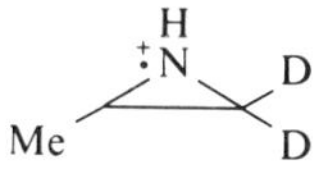   (142) 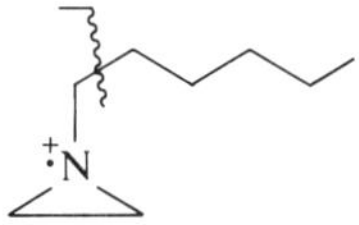   (143)

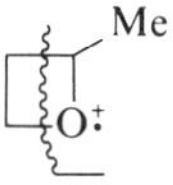

(144) 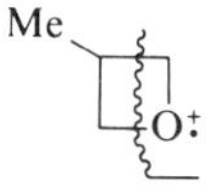   (145) 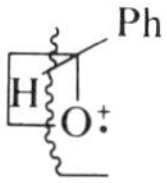   (146)

[361] F. Benoit and J. L. Holmes, *Chem. Comm.*, 1970, 1031.
[362] R. Westwood, D. H. Williams, and A. N. H. Yeo, *Org. Mass Spectrometry*, 1970, **3**, 1485.
[363] P. Brown, *Org. Mass Spectrometry*, 1970, **4**, 533.
[364] O. A. Mamer, R. J. Kominar, and F. P. Lossing, *Org. Mass Spectrometry*, 1970, **3**, 1411.
[365] F. Benoit and J. L. Holmes, *Org. Mass Spectrometry*, 1970, 3, 993.
[366] J. Harley-Mason, T. P. Toube, and D. H. Williams, *J. Chem. Soc. (B)*, 1966, 396.
[367] G. E. Robinson, C. B. Thomas, and J. M. Vernon, *J. Chem. Soc. (B)*, 1971, 1273.
[368] K. Yamada, T. Konakahara, and I. Iida, *Bull. Chem. Soc. Japan*, 1971, **44**, 3060.
[369] E. F. Brittain, C. H. J. Wells, H. M. Paisley, and D. J. Stickley, *J. Chem. Soc. (B)*, 1970, 1714.
[370] Q. N. Porter and R. J. Spear, *Org. Mass Spectrometry*, 1970, 3, 1259.
[371] J. A. Deyrup and C. L. Moyer, *J. Org. Chem.*, 1970, **35**, 3424.

bisazirldones,[372] oxadiaziridines,[373] diaziridines,[374] and 2-phenyl-1-azetine[375] have been examined. The spectra of oxetans have been reported.[376,377] The major fragmentations in the spectra of 2-methyl-, 3-methyl-, and 2-phenyl-oxetan are indicated in (144)—(146).[376]

Five-membered Rings. Furfuryl derivatives,[378] 4-oxo-4,5,6,7-tetrahydrobenzofuran,[379] aurone epoxides,[380] and N-acetylpyrrolidines[381] have been examined. A specific hydrogen-transfer reaction, occurring through an eight-membered transition state, is observed for (147).[382] Pyrrolidone hydroperoxides,[383] pyrrolines,[384] and 3-t-butylpyrroles[385] have been studied. The molecular ion of indolizine (149) eliminates HCN followed by C_2H_2,[386] whereas that of (150) loses a hydrogen atom to give a ring-expanded species.[386] The spectra of indoles,[387,388] methylindoles[387–389] and hydroxyindole-3-carboxylic acids[390] have been reported. The molecular ions of hydroxyskatoles (151) eliminate

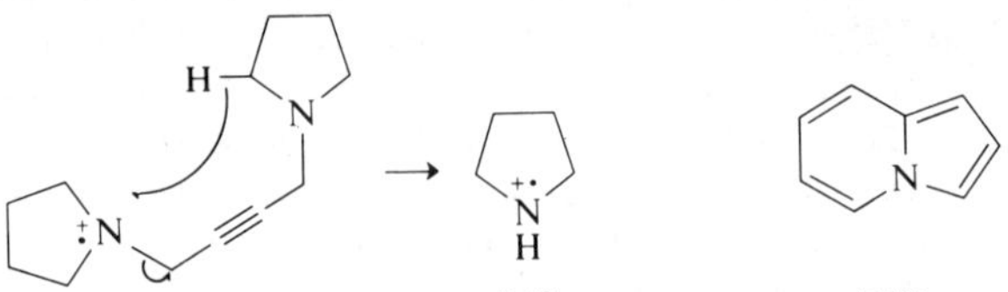

(147)   (148)   (149)

[372] I. Lengyel, D. B. Uliss, and R. V. Mark, *J. Org. Chem.*, 1970, **35**, 4077.
[373] F. D. Greene and S. S. Hecht, *J. Org. Chem.*, 1970, **35**, 2482.
[374] I. Lengyel, F. D. Greene, and J. F. Pazos, *Org. Mass Spectrometry*, 1970, **3**, 623.
[375] R. G. Kostanovsky, I. M. Gella, and Kh. Khajizov, *Izvest. Akad. Nauk S.S.S.R., Ser. khim.*, 1971, 893.
[376] P. O. I. Virtanen, A. Karyalainen, and H. Ruotsalainen, *Suomen Kemi.* (*B*), 1970, **43**, 219.
[377] K. Pihlaja, K. Polviander, R. Keskinen, and J. Jalonen, *Acta Chem. Scand.*, 1971, **25**, 765.
[378] E. D. Loughran, E. M. Wewerka, and G. J. Hammons, *J. Heterocyclic Chem.*, 1972, **9**, 57.
[379] E. Cortés and M. Salmón, *Org. Mass Spectrometry*, 1971, **6**, 85.
[380] R. A. Brady, W. I. O'Sullivan, and A. M. Duffield, *Org. Mass Spectrometry*, 1972, **6**, 199.
[381] W. J. Richter, J. M. Tesarek, and A. L. Burlingame, *Org. Mass Spectrometry*, 1971, **5**, 531.
[382] C. Bogentoft and B. Karlén, *Acta Chem. Scand.*, 1971, **25**, 754.
[383] D. W. Bristol and D. C. Dittmer, *J. Org. Chem.*, 1970, **35**, 2487.
[384] N. A. Klyner, R. A. Khmel'nitskii, S. B. Nikitina, and I. I. Grandberg, *Zhur. org. Khim.*, 1971, **7**, 2038.
[385] V. I. Vysotskii, R. A. Khmel'nitskii, and I. I. Grandberg, *Izvest. Timiryazev Sel'skokhoz. Akad.*, 1971, 227.
[386] G. Jones and J. Stanyer, *Org. Mass Spectrometry*, 1970, **3**, 1489.
[387] V. I. Vysotskii and R. A. Khmel'nitskii, *Izvest. Timiryazev Sel'skokhoz. Akad.*, 1970, 204.
[388] V. I. Vysotskii, R. A. Khmel'nitskii, I. I. Grandberg, and V. A. Budylin, *Izvest. Timiryazev Sel'skokhoz Akad.*, 1970, 221.
[389] S. Safe, W. D. Jamieson, and O. Hutzinger, *Org. Mass Spectrometry*, 1972, **6**, 33.
[390] R. Marchelli, W. D. Jamieson, S. H. Safe, O. Hutzinger, and R. A. Heacock, *Canad. J. Chem.*, 1971, **49**, 1296.

Reactions of Specific Functional Groups 121

carbon monoxide prior to loss of hydrogen cyanide.[390] 1-Oxoisoindoles have been examined;[391] isatin fragments by the processes $M - CO - HCN - CO$, and labelling studies[392] show that the first CO to be eliminated is that illustrated in (152); other aspects of substituted isatins have been discussed.[393,394] The substituted tetrapyrido[3,4-b]indole molecular ion (153) undergoes both the retro-process [see (153)], and the rearrangement to $M - RNH\cdot$.[395] The behaviour of tetracarbazole,[396] pyrazolo[3,4-c]carbazole,[397] pyromellitimides,[398] and porphrins[399] upon electron impact has been reported.

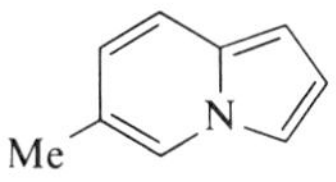

(150) 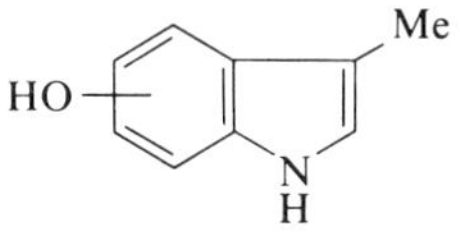 (151) 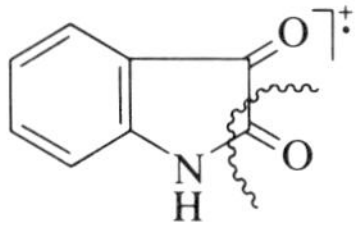 (152)

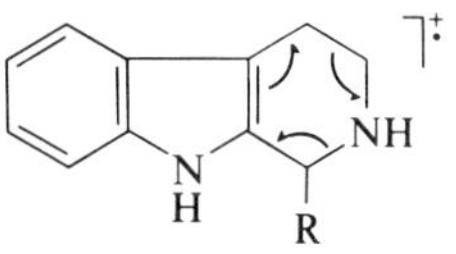

(153) 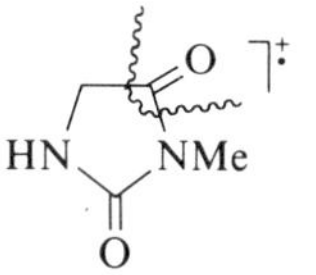 (154) 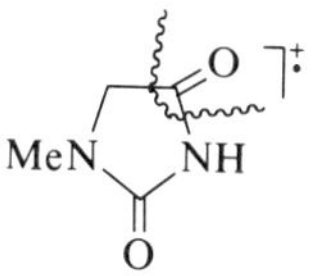 (155)

The effect of substituents upon the decompositions of 2-aryl-1,3-dioxolans has been measured.[400] The molecular ions of the hydantoins (154) and (155) eliminate two molecules of carbon monoxide, with the initial eliminations occurring as depicted in (154) and (155) with specificities of 80 and 90% respectively.[401] The major fragmentations of 1-arylhydantoins are illustrated in (156).[402] Benzimidazolidones[403] and phenanthro[9,10-d]imidazoles[404] have been studied.

[391] W. L. F. Armarego and S. C. Sharma, *J. Chem. Soc.* (C), 1970, 1600.
[392] M. Butcher, *Org. Mass Spectrometry*, 1971, 5, 759.
[393] K. Yamada, T. Konakahara, and H. Iida, *Kogyo Kagaku Zasshi*, 1970, 73, 980.
[394] J. A. Ballantine, R. G. Fenwick, and F. D. Popp, *Org. Mass Spectrometry*, 1971, 5, 1003.
[395] R. T. Coutts, R. A. Locock, and G. W. A. Slywka, *Org. Mass Spectrometry*, 1970, 3, 879.
[396] V. I. Vysotskii, R. A. Khmel'nitskii, I. I. Grandberg, and G. V. Fridlandsky, *Izvest. Timiryazev Sel'skokhoz Akad.*, 1971, 206.
[397] H.-J. Teuber and L. Vogel, *Chem. Ber.*, 1970, 103, 3302.
[398] J. L. Cotter and R. A. Dine-Hart, *Org. Mass Spectrometry*, 1970, 4, 315.
[399] A. D. Adler, J. H. Green, and M. Mautner, *Org. Mass Spectrometry*, 1970, 3, 955.
[400] J. W. Horodniak, J. Wright, and N. Indictor, *Org. Mass Spectrometry*, 1971, 5, 1287.
[401] R. A. Corral, O. O. Orazi, A. M. Duffield, and C. Djerassi, *Org. Mass Spectrometry*, 1971, 5, 551.
[402] S. Veibel, *Acta Chem. Scand.*, 1971, 25, 777.
[403] T. Kametani, S. Hirata, S. Shibuya, and M. Shio, *Org. Mass Spectrometry*, 1970, 4, 395.
[404] J. L. Cooper, S. R. Lipsky, and W. J. McMurray, *Org. Mass Spectrometry*, 1970, 3, 1355.

The spectra of alkyl-2-pyrazolines have been listed.[405] The initial loss of HCN from the pyrazole molecular ion (157) involves the 1- and 2-positions, with very little hydrogen scrambling accompanying this loss.[406] The loss of the hydrogen atom α to the nitrogen atom of (157) proceeds much faster than hydrogen scrambling.[407] Substituted pyrazoles have been examined,[408] and an unusual loss of C_2H_4 is observed from (158).[409] Both C_2H_4 and C_3H_6 are lost from (159);[409] the molecular ion of pyrazolo[1,5-a]pyridine (160) does not lose nitrogen, but eliminates successive units of HCN.[410] Benz[cd]indazole dioxides have been investigated.[411] Several studies of isoxazoline derivatives have been reported.[412,413] The ion (161) suffers elimination of PhNDO.[413] Further work on isoxazoles[414–418] constitutes a useful addition to previous work in this

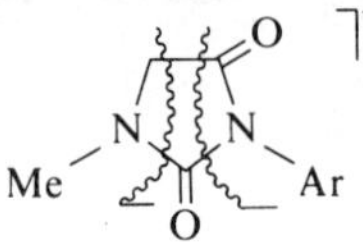

(156) 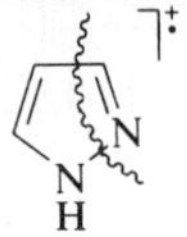   (157) 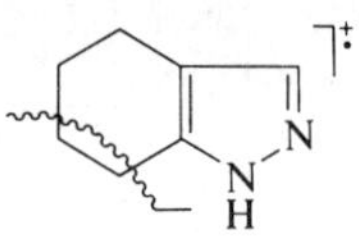   (158)

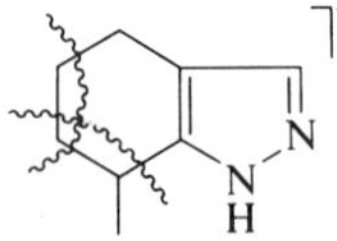

(159) 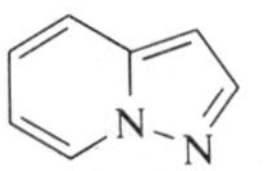   (160) 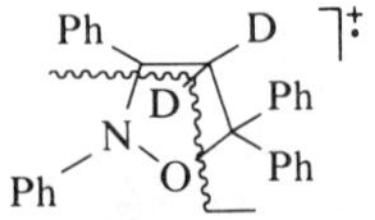   (161)

field (Vol. 1, p. 124). An interesting example of a skeletal-rarrangement reaction is afforded by the spectrum of (162), where the base peak is produced by the ion $C_7H_7N^+$ [see (162)].[416] Similar rearrangements are observed for thioalkyl-isoxazoles.[416] The spectra of alkyloxazolines have been described[419] and the

[405] K. B. Sloan and N. Rabjohn, *J. Heterocyclic Chem.*, 1970, 7, 1273.
[406] J. van Thuijl, K. J. Klebe, and J. J. van Houte, *J. Heterocyclic Chem.*, 1971, 8, 311.
[407] J. van Thuijl, K. J. Klebe, and J. J. van Houte, *Org. Mass Spectrometry*, 1971, 5, 1101.
[408] J. van Thuijl, K. J. Klebe, and J. J. van Houte, *Org. Mass Spectrometry*, 1970, 3, 1549.
[409] H.-E. Audier, J. Bottin, M. Fetizon, and J.-C. Tabet, *Bull. Soc. chim. France*, 1971, 2911.
[410] K. T. Potts and U. P. Singh, *Org. Mass Spectrometry*, 1970, 3, 433.
[411] R. W. Alder, G. A. Niazi, and M. C. Whiting, *J. Chem. Soc.* (C), 1970, 1693.
[412] A. M. Duffield and O. Buchardt, *Org. Mass Spectrometry*, 1970, 3, 1043.
[413] F. Caruso, G. Cum, and N. Uccella, *Tetrahedron Letters*, 1971, 3711.
[414] C. F. Beam, M. C. D. Dyer, R. A. Schwarz, and C. R. Hauser, *J. Org. Chem.*, 1970, 35, 1806.
[415] R. Jacquier, C. Petrus, F. Petrus, and M. Valentin, *Bull. Soc. chim. France*, 1970, 2665.
[416] T. Nishiwaki, *Org. Mass Spectrometry*, 1971, 5, 123.
[417] R. A. Khmel'nitskii, K. K. Zhigulev, S. D. Sokolov, and L. P. Tsurkanova, *Zhur. org. Khim.*, 1970, 6, 2162.
[418] D. C. Nonhebel, *Org. Mass Spectrometry*, 1970, 3, 1519.
[419] S. Osman, C. J. Dooley, T. A. Foglia, and L. M. Gregory, *Org. Mass Spectrometry*, 1970, 4, 139.

Reactions of Specific Functional Groups

basic fragmentations of 4-phenyloxazolidones are indicated in (163) and (164).[420] The spectra of 2-aryl-4-arylideneoxazol-5-ones,[421] pyranobenzoxazole carboxylic acids[422] and pyrano[1,4]benzoxazinones[423] have been reported. The molecular ion of benzoxazole undergoes competitive losses of CO and HCN, whereas that of 2-aminobenzoxazole loses only carbon monoxide.[424]

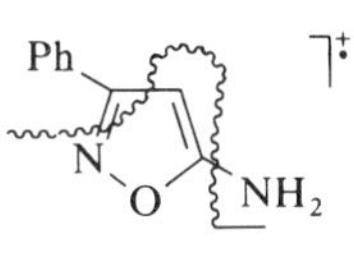

(162) 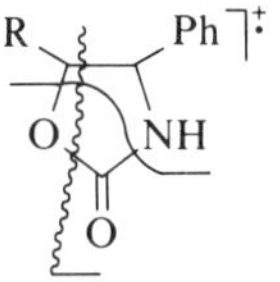 (163) 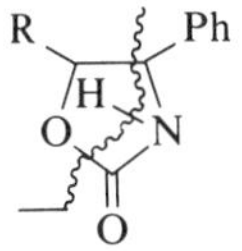 (164)

The 4,5-diphenyl-1,2,3-triazole molecular ion (165) undergoes skeletal rearrangement to form the ion $C_{13}H_9^+$ [425] [a rearrangement analogous to that described for 4,5-diphenylimidazole and 3,4-diphenylpyrazole (Vol. 1, p. 123)], and also fragments by the scheme $M - N_2 - HCN - H\cdot$. It has been suggested[425] that the $M - N_2$ ion may correspond to a 2,3-diphenylazirine ion radical (166).[cf.426] The benzotriazole molecular ion eliminates N_2 in preference to HCN,[427] and reactions have been observed between 1-chlorobenzotriazole and water in the ion source of a mass spectrometer[428] [cf. nitrobenzene[429] and 1,2-naphthaquinones (Vol. 1, pp. 109, 110)]. The base peak in the spectrum of (167) is produced by the rearrangements $M - N_2 - N_2 - H\cdot$, with the hydrogen atom mainly originating from the phenyl group.[430] The spectra of a

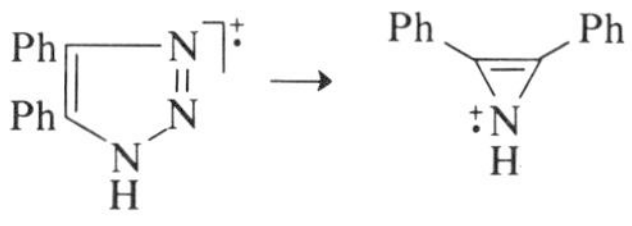

(165) (166) 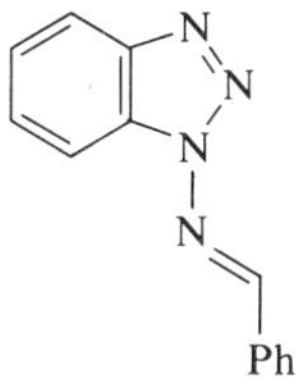 (167)

[420] R. A. Auerbach, D. L. von Minden, and C. A. Kingsbury, *Org. Mass Spectrometry*, 1970, **4**, 41.
[421] J. A. Ballantine and R. G. Fenwick, *Org. Mass Spectrometry*, 1971, **5**, 615.
[422] G. Barker and G. P. Ellis, *J. Chem. Soc. (C)*, 1971, 1482.
[423] G. Barker, G. P. Ellis, and D. A. Wilson, *J. Chem. Soc. (C)*, 1971, 2079.
[424] H. Ogura, S. Sugimoto, and T. Itoh, *Org. Mass Spectrometry*, 1970, **3**, 1341.
[425] F. Compernolle and M. Dekeirel, *Org. Mass Spectrometry*, 1971, **5**, 427.
[426] B. K. Simons, B. Nussey, and J. H. Bowie, *Org. Mass Spectrometry*, 1970, **3**, 925.
[427] M. Ohashi, K. Tsuyimoto, A. Yoshino, and T. Yonezawa, *Org. Mass Spectrometry*, 1970, **4**, 203.
[428] C. B. Thomas, *Org. Mass Spectrometry*, 1970, **3**, 1523.
[429] J. H. Beynon, J. A. Hopkinson, and G. R. Lester, *Internat. J. Mass Spectrometry Ion Phys.*, 1969, **2**, 291.
[430] U. Rapp, H. A. Staab, and C. Wünsche, *Tetrahedron*, 1971, **27**, 2679.

large number of 1,2,4-triazole derivatives have been reported, including alkyl- and aryl-1,2,4-triazoles,[431] 1,2,4-triazole-3-carbaldehydes,[432] 1-aryl-Δ^2-1,2,4-triazolin-5-ones[433,434] and 1,2,4-triazole-4-N-oxides.[435] The 1,2,4-triazole[431] and 1-aryl-1,2,4-triazolin-5-one[433] systems fragment as shown in (168) and (169) respectively.

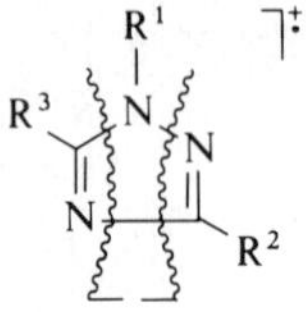

(168) 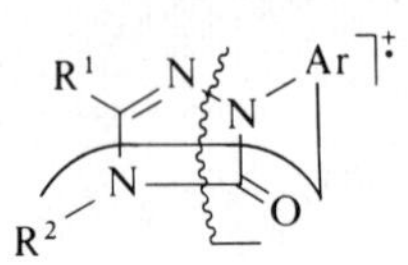   (169)

5-Triazolo[3,4-f]-1,2,4-triazines have been examined,[436] the molecular ion of sym-triazolo[4,3-a]pyridine (170) eliminates both N_2 and HCN,[437] in contrast to simple 1,2,4-triazoles which generally only lose HCN,[431] whereas both (171) and (172) decompose by loss of RCN.[437] It is of interest to compare the decomposition processes of the sym-triazolo[4,3-a]pyrazine (173),[438,439] sym-triazolo[1,5-a]-pyrazine (174),[438,439] and sym-triazolo[4,3-b]pyridazine (175)[440] molecular ions; $viz.$ (173), $M - N_2 - $ HCN; (174), $M - $ HCN $- $ HCN; and (175), $M - $ CHN$_3 - $ HCN. 2,4-Disubstituted Δ^2-1,3,4-oxadiazolin-5-ones[433,441] have been studied, further aspects of the fragmentation of sydnones and sydnonimines have been

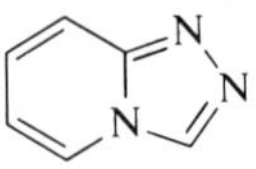

(170) 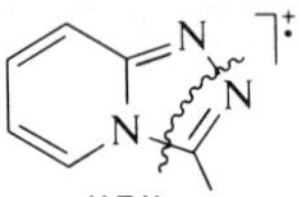   (171) R   (172)

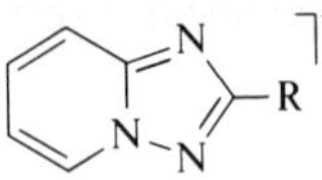

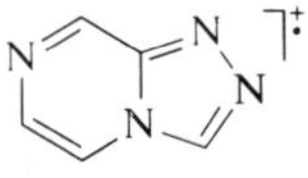

(173) 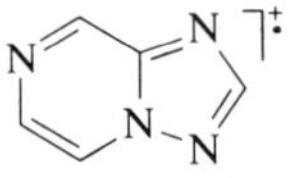   (174) 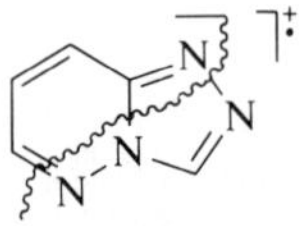   (175)

[431] K. T. Potts, R. Armbruster, and E. Houghton, *J. Heterocyclic Chem.*, 1971, **8**, 773.
[432] E. J. Browne, *Austral. J. Chem.*, 1971, **24**, 393.
[433] T. Kametani, S. Hirata, S. Shibuya, and M. Shio, *Org. Mass Spectrometry*, 1971, **5**, 117.
[434] T. Kametani, S. Shibuya, and M. Shio, *J. Heterocyclic Chem.*, 1971, **8**, 541.
[435] H. G. O. Becker, D. Beyer, and H.-J. Timpe, *J. prakt. Chem.*, 1970, **312**, 869.
[436] H. G. O. Becker, D. Beyer, G. Israel, R. Muller, W. Riediger, and H.-J. Timpe, *J. prakt. Chem.*, 1970, **312**, 669.
[437] K. T. Potts, E. Brugel, and U. P. Singh, *Org. Mass Spectrometry*, 1971, **5**, 1.
[438] K. T. Potts and C. R. Surapaneni, *J. Heterocyclic Chem.*, 1970, **7**, 1019.
[439] K. T. Potts and E. Brugel, *Org. Mass Spectrometry*, 1971, **5**, 663.
[440] V. Pirc, B. Stanovnik, M. Tišler, J. Marsel, and W. W. Paudler, *J. Heterocyclic Chem.*, 1970, **7**, 639.
[441] T. Kametani, S. Shibuya, and M. Shio, *J. Heterocyclic Chem.*, 1971, **8**, 889.

Reactions of Specific Functional Groups 125

discussed,[442] and the basic cleavages of N-[5-(1,2,3,4-oxatriazolio)]amides are shown in (176) and (177);[443] 5-aminotetrazole fragments by the processes $M - N_2 - HCN$,[444] and the tetrazolo[1,5-b]pyridazine molecular ion (178) suffers the decompositions $M - N_4 - HCN^{440}$ [$cf.$ (175)].

(176)   (177)   (178)

Six- and Seven-membered Rings. The fragmentations of Δ^3-dihydropyrans,[445,446] tetrahydropyran-3-one,[447] pyrilium salts,[448] chromones,[449] and benzochromones[450] have been recorded. Labelling studies (^{2}H, ^{13}C) of chromans and flavans have shown the complexity of their spectra.[451] For example, the loss of Me· from (179) involves the carbon atoms at positions 2, 3, and 4 in the ratio 50 : 25 : 25, and the losses of C_7H_7 from (180) utilize the carbons of the 2- and 3-positions (83 : 17) and from (181) the 2-, 3-, and 4-positions (45 : 30 : 25).[451] N-Acetylpiperidines,[452,453] N-substituted nortropinones,[454] methoxycarbonylquinuclidines and benzoquinuclidines,[455] β-quinuclidones and β-benzoquinuclidones,[456] and 2-aryl-1,2-dihydropyridines[457] have been examined.

All the carbon atoms in the pyridine molecular ion lose positional identity prior to elimination of HCN.[458] The $M - 1$ ion in the spectrum of γ-picoline may correspond to the azatropylium ion, and is produced by loss of an α, β, or methyl hydrogen in the ratio 40 : 32 : 28, with the same trend being observed for the loss

[442] T. Shima, A. Ouchida, and Y. Asahi, *Mass Spectrometry (Japan)*, 1970, **17**, 661.
[443] G. Christophersen and S. Treppendahl, *Acta Chem. Scand.*, 1971, **25**, 625.
[444] L. E. Brady, *J. Heterocyclic Chem.*, 1970, **7**, 1223.
[445] V. N. Bochkarev, B. V. Unkovsky, V. B. Mochalin, Z. I. Smolina, and A. N. Wulfson, *Izvest. Akad. Nauk S.S.S.R., Ser khim.*, 1970, 1184.
[446] V. N. Bochkarev, B. V. Unkovsky, V. B. Mochalin, Z. I. Smolina, and A. N. Wulfson, *Izvest. Akad. Nauk S.S.S.R., Ser khim.*, 1970, 1442.
[447] J. Gore and F. Guigues, *Bull. Soc. chim. France*, 1970, 3521.
[448] A. M. Duffield, C. Djerassi, and A. T. Balaban, *Org. Mass Spectrometry*, 1971, **5**, 87.
[449] G. Barker and G. P. Ellis, *Org. Mass Spectrometry*, 1971, **5**, 857.
[450] A. Tatematsu, S. Fukushimo, T. Noro, Y. Saiki, A. Ueno, Y. Akabori, and H. Nakata, *Mass Spectrometry (Japan)*, 1970, **18**, 1300.
[451] J. R. Trudell, S. D. Sample-Woodgate, and C. Djerassi, *Org. Mass Spectrometry*, 1970, **3**, 753.
[452] W. J. Richter, J. M. Bursey, and A. L. Burlingame, *Org. Mass Spectrometry*, 1971, **5**, 1295.
[453] W. L. Richter, J. G. Liehr, and A. L. Burlingame, *Org. Mass Spectrometry*, 1972, **6**, 443.
[454] Y. Kashman and S. Cherkez, *Tetrahedron*, 1972, **28**, 155.
[455] R. G. Kostyanovsky, E. E. Mikhlina, E. I. Levkoeva, and L. N. Yakhontov, *Org. Mass Spectrometry*, 1970, **3**, 1023.
[456] A. I. Ermakov, Yu. N. Sheinker, E. E. Mikhlina, L. I. Mastafanova, V. Ya. Vorobjova, A. D. Yanina, L. N. Yakhontov, and R. G. Kostyanovsky, *Org. Mass Spectrometry*, 1971, **5**, 1029.
[457] R. E. Lyle and E. White, *Tetrahedron Letters*, 1970, 1871.
[458] R. Dickinson and D. H. Williams, *J. C. S. Perkin II*, in press.

of HCN.[459] The spectra of other picolines[460] and benzyl- and tolyl-pyridines[461] have been reported; the loss of carbon dioxide from the molecular ions of pyridine carboxylic acids is influenced by the ring nitrogen;[462,463] the loss of

(179)   (180)   (181)

a hydroxyl radical from the γ-pyridine carboxylic acid molecular ion involves a benzoic acid-type exchange[229,230] between the carboxyl hydrogen and the β-hydrogens;[464] the spectra of phenylpyridyl ketones[465,466] and diketones[466] contain pronounced $M - CO$ and $M - 2CO$ ions, respectively. Studies of polycyanopicolines,[467] nitraminopyridines,[468] 4-hydroxy-6-methyl-2-pyridones,[469] alkylpyridine N-oxides,[470] pyridinium oxides,[470,471] iminopyridinium- and isoquinolinium-betaines,[473-475] a dimer of 3,4-bipyridyl,[476] and bipyridyls[477] have been reported.

The major ions observed in the spectra of 3- and 4-hydroxyquinolizidines are shown in (182) and (183).[478] The basic cleavages of decahydroquinoline produce the ions $M - C_4H_8$, $M - C_3H_7$, and $M - C_2H_5$ respectively [see (184) and (185)].[479] The spectra of 1,3-bridged 1,2,3,4-tetrahydroisoquinolines,[480] 8-

[459] N. Neeter, N. M. M. Nibbering, and Th. J. de Boer, *Org. Mass Spectrometry*, 1970, **3**, 597.
[460] V. E. Sahini, C. Podina, and V. Constantin, *Rev. Roumaine Chim.*, 1970, **15**, 495.
[461] P. B. Terent'ev, R. A. Khmel'nitskii, and A. N. Kost, *Zhur. org. Khim.*, 1971, **7**, 1878.
[462] R. J. Moser and E. V. Brown, *Org. Mass Spectrometry*, 1970, **4**, 555.
[463] E. V. Brown and R. J. Moser, *J. Heterocyclic Chem.*, 1971, **8**, 189.
[464] R. Neeter and N. M. M. Nibbering, *Org. Mass Spectrometry*, 1971, **5**, 735.
[465] E. V. Brown and M. B. Shambhu, *Org. Mass Spectrometry*, 1972, **6**, 479.
[466] C. S. Barnes, R. J. Goldrack, E. J. Halbert, J. G. Wilson, R. J. Lyall, and S. Middleton, *Tetrahedron Letters*, 1972, 705.
[467] H. Achenbach, K. Wallenfals, and P. Neumann, *Chem. Ber.*, 1972, **105**, 1646.
[468] J. G. Wilson, C. S. Barnes, and R. J. Goldsack, *Org. Mass Spectrometry*, 1970, **4**, 365.
[469] N. S. Wulfson, V. G. Zaikin, Z. S. Ziyavidinova, V. M. Burikov, and S. K. Mukerjee, *Org. Mass Spectrometry*, 1971, **5**, 743.
[470] D. A. Lightner, R. Nicoletti, G. B. Quistad, and E. Irwin, *Org. Mass Spectrometry*, 1970, **4**, 571.
[471] K. Undheim and G. Hvistendahl, *Org. Mass Spectrometry*, 1971, **5**, 325.
[472] T. Grønneberg and K. Undheim, *Acta Chem. Scand.*, 1971, **25**, 2807.
[473] M. Ikeda, N. Tsujimoto, and Y. Tamura, *Org. Mass Spectrometry*, 1971, **5**, 61.
[474] M. Ikeda, N. Tsujimoto, and Y. Tamura, *Org. Mass Spectrometry*, 1971, **5**, 389.
[475] M. Ikeda, N. Tsujimoto, and Y. Tamura, *Org. Mass Spectrometry*, 1971, **5**, 935.
[476] J. M. Kramer and R. S. Berry, *J. Amer. Chem. Soc.*, 1971, **93**, 1303.
[477] R. A. Khmel'nitskii, N. A. Klynev, and P. V. Terent'ev, *Zhur. org. Khim.*, 1971, **7**, 395.
[478] M. Hussain, J. S. Robertson, and T. R. Watson, *Org. Mass Spectrometry*, 1970, **4**, 109.
[479] C. K. Yu, D. Oldfield, and D. B. MacLean, *Org. Mass Spectrometry*, 1970, **4**, 147.
[480] S. Shiotani and K. Mitsuhashi, *Chem. and Pharm. Bull. (Japan)*, 1972, **20**, 1.

Reactions of Specific Functional Groups

hydroxyquinolines,[481] aminoquinolines,[482] Reissert compounds,[483] bisquinolines,[484] and bisisoquinolines[485] have been discussed.

(182) (183) (184) (185)

Thermodynamic data have been measured for 1,3-dioxans,[486] and unsaturated dimethyl-2,2-dioxans have been examined.[487] The isomers pyridazine, pyrimidine, and pyrazine are more readily differentiated by their IKE spectra than by the normal mass spectra,[488] and metastable peak shapes have been measured for certain decompositions of pyrimidine and pyrazine.[489] The spectra of pyridazine N-oxides contain the usual $M - O$ and $M - HO\cdot$ ions,[490] and a further study of phthalazones has appeared.[491] The fragmentations of many pyrimidine and fused pyrimidine systems have been reported, including 2-methylpyrimidines,[492,493] pyrimidones [*e.g.* (186)],[492,494] complex pyrimidines,[495,496] 5H-indeno[1,2-d]pyrimidines,[497] oxoquinazolines,[498] and N-allenic and N-propargylic 4-oxoquinazolines.[499,500] The base peak in the spectrum of diphenylpiperazide (187) is produced by a retro-Diels–Alder cleavage, and a rearrangement ion is produced by the process $M - C_2H_4N$.[501] The

---

[481] H. Budzikiewicz and E. Plöger, *Org. Mass Spectrometry*, 1970, **3**, 709.
[482] E. V. Brown, A. C. Plasz, and S. R. Mitchell, *J. Heterocyclic Chem.*, 1970, **7**, 661.
[483] F. D. Popp, K. T. Potts, and R. Armbruster, *Org. Mass Spectrometry*, 1970, **3**, 1075.
[484] R. A. Khmel'nitskii, N. A. Klynev, and K. K. Zhigulev, *Izvest. Timiryazev Sel'skokhoz Akad.*, 1970, 200.
[485] E. Ziegler, H. Mittelbach, and W. Steiger, *Monatsh.*, 1970, **101**, 1059.
[486] K. Pihlaja and J. Jalonen, *Org. Mass Spectrometry*, 1971, **5**, 1363.
[487] J. Kossanyi, J. Chuche, and A. M. Duffield, *Org. Mass Spectrometry*, 1971, **5**, 1409.
[488] J. H. Beynon, R. M. Caprioli, and T. Ast, *Org. Mass Spectrometry*, 1972, **6**, 273.
[489] J. H. Beynon, R. M. Caprioli, and T. Ast, *Org. Mass Spectrometry*, 1971, **5**, 229.
[490] H. Ogura, S. Sugimoto, H. Igeta, and T. Tsuchija, *J. Heterocyclic Chem.*, 1971, **8**, 391.
[491] A. M. Zyakun, K. V. Grablyauskas, A. N. Kost, V. I. Zaretskii, and N. S. Wulfson, *Izvest. Akad. Nauk S.S.S.R., Ser khim.*, 1970, 2208.
[492] K. Undheim and G. Hvistendahl, *Acta Chem. Scand.*, 1971, **25**, 3227.
[493] R. A. Khmel'nitskii, E. A. Kumna, A. B. Belikov, and U. P. Shvachkin, *Izvest. Timiryazev Sel'skokhoz Akad.*, 1971, 214.
[494] T. Kato, H. Yamanaka, H. Ichikawa, T. Chiba, H. Abe, and S. Sasaki, *Org. Mass Spectrometry*, 1970, **4**, 181.
[495] J. J. Dolhun and J. L. Wiebers, *Org. Mass Spectrometry*, 1970, **3**, 669.
[496] J. Clark, Z. Munawar, and A. W. Timms, *J. C. S. Perkin II*, 1972, 233.
[497] V. F. Sedova and V. F. Mamayev, *Khim. geterotsikl. Soedinenii*, 1970, 691.
[498] C. Bogentoft and B. Danielssen, *Acta Pharm. Suecica*, 1970, **7**, 257.
[499] C. Bogentoft, L. Kronberg, and B. Danielsson, *Acta Chem. Scand.*, 1970, **24**, 2244.
[500] C. Bogentoft, *Org. Mass Spectrometry*, 1970, **3**, 1527.
[501] F. Yamada, Y. Fujimoto, and T. Nishiyama, *Bull. Chem. Soc. Japan*, 1972, **45**, 280.

spectrum of dibenzyl-3,6-diketo-2,5-piperazine has been recorded.[502] The 2-methoxy-3-methylpyrazine molecular ion eliminates water, a reaction not observed for other isomers.[503] This ion must be produced by a complex mechanism as the spectrum of (188) shows losses of both D_2O and HDO. The loss of CDO· from (188) has been rationalized in terms of the formation of (189).[503] The spectra of pyrazine N-oxides contain peaks due to $M - O$ ions.[504] Alkyl-quinoxalines have been discussed;[505] the two fragmentations shown in (190) give ions of similar abundance.[505] 3-Hydroxypteridin-4-one (191) decomposes

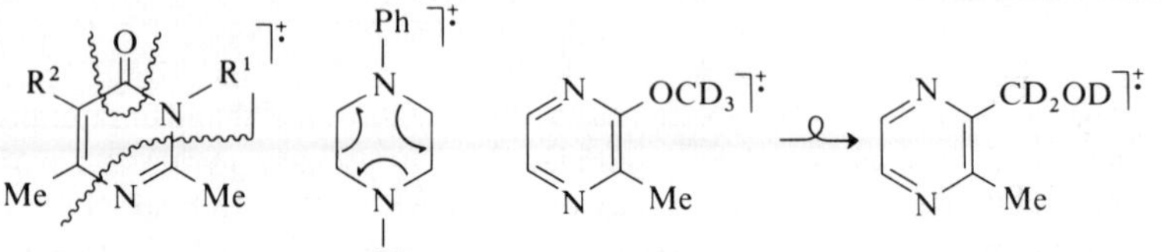

by the scheme $M - NO· - CO - HCN - HCN$.[506] Details of the spectra of 2,4,7-trihydroxypteridines and their TMS derivatives,[507] pteridin-4(3H)-ones,[508] 4-trifluoromethylpteridines,[509] and pyrimidopyrimidines[510] have been reported.

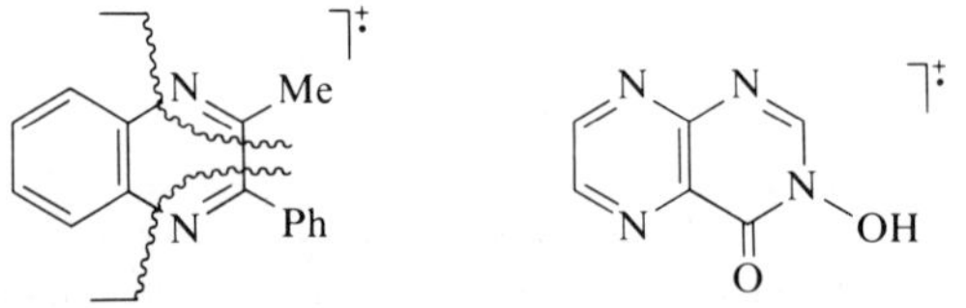

Pronounced loss of N_2 occurs from diaryl-1,2,3-triazine and benzotriazine molecular ions.[511] The base peak in the spectrum of (192) corresponds to the diphenylacetylene ion radical.[512] The spectra of variously substituted 1,3,5-

[502] K. Jankowski and L. Varfalvy, *Canad. J. Chem.*, 1971, **49**, 1583.
[503] M. G. Kolor and D. J. Rizzo, *Org. Mass Spectrometry*, 1971, **5**, 959.
[504] F. Uchimaru, S. Okada, A. Kosasayama, and T. Konno, *J. Heterocyclic Chem.*, 1971, **8**, 99.
[505] A. Karjalainen and H. Krieger, *Suomen Kemi. (B)*, 1970, **43**, 273.
[506] J. Clark and C. Smith, *Org. Mass Spectrometry*, 1971, **5**, 447.
[507] P. Haug and T. Urshibara, *Org. Mass Spectrometry*, 1970, **3**, 1365.
[508] J. Clark, R. Maynard, and C. Smith, *Org. Mass Spectrometry*, 1971, **5**, 993.
[509] J. Clark and F. S. Yates, *Org. Mass Spectrometry*, 1971, **5**, 1419.
[510] K. Yamada, K. Hayashida, and H. Iida, *Kogyo Kagaku Zasshi*, 1971, **74**, 952.
[511] R. A. W. Johnstone, D. W. Payling, P. N. Preston, H. N. E. Stevens, and M. F. G. Stevens, *J. Chem. Soc. (C)*, 1970, 1238.
[512] K. Wakabayashi, M. Tsunoda, and Y. Suzuki, *J. Synthetic Org. Chem. Japan*, 1970, **28**, 252.

Reactions of Specific Functional Groups   129

triazines have been discussed.[512–514] The molecular ion of 7-methylamino-5-trifluoromethylpyrimido[5,4-e]-asym-triazine (193) eliminates N_2 in preference to HCN,[515] and the spectra of hexahydro-1,2,4,5-tetrazines[516,517] and 2H-naphth-[1,8-bc]oxepin derivatives[518] have been listed. The base peak in the spectrum of (194) is produced by loss of $C_5H_7O\cdot$;[519] norbenzomorphans fragment mainly by elimination of Me· and C_2H_4.[520] The spectra of 1,4-benzodiazepin-2-ones,[521] 3-hydroxy-2,3,4,5-tetrahydro-1,5-benzoxazepine,[522] 3-benzylidine-3,4-dihydro-4-methyl-1H-1,4-benzodiazepin-2,5-diones,[523] and dibenzodiazepines[524,525] have been examined.

(192)   (193)   (194)

Sulphur Compounds.—The heats of formation of ions, and bond energies, have been measured for alkyl and perfluoroalkyl sulphides,[526] and the spectra of various aromatic sulphur compounds have been described.[527] The spectrum of thiacyclobutane has been calculated using the quasi-equilibrium theory.[528] 1,3-Oxathiolan,[529] α-ketohemithioketals,[530] 1,3-dithiolan,[529] bithiophthalides,[531] and thiochromans[532] have been examined. The spectrum of 1,3-oxathian (195) is dominated by sulphur-containing fragments, i.e. $C_3H_5S^+$, $C_3H_6S^{+\cdot}$ and

[513] P. N. Preston, W. Steedman, M. H. Palmer, S. M. MacKenzie, and M. F. G. Stevens, Org. Mass Spectrometry, 1970, 3, 863.
[514] J. Karliner and R. Seltzer, J. Heterocyclic Chem., 1971, 8, 629.
[515] J. Clark, Org. Mass Spectrometry, 1972, 6, 467.
[516] S. Hammerum and J. Moller, Org. Mass Spectrometry, 1971, 5, 1209.
[517] W. Sucrow, H. Bethke, and G. Chondromatidis, Tetrahedron Letters, 1971, 1481.
[518] N. P. Buu-Hoi, G. Saint-Ruf, and J.-C. Perche, J. Chem. Soc. (C), 1970, 1327.
[519] J. Mitera and V. Kubelka, Org. Mass Spectrometry, 1971, 5, 651.
[520] M. Mokotoff and A. E. Jacobson, J. Heterocyclic Chem., 1970, 7, 773.
[521] W. Sadée, J. Medicin. Chem., 1970, 13, 475.
[522] C. J. Coulson, K. R. W. Wooldridge, J. Memel, and B. J. Millard, J. Chem. Soc. (C), 1971, 1164.
[523] M. McCamish and J. D. White, Org. Mass Spectrometry, 1970, 4, 241.
[524] A. C. Casey, J. H. Green, A. Lee, and M. Mautner, J. Heterocyclic Chem., 1970, 7, 879.
[525] F. D. Popp, R. J. Dubois, K. T. Potts, and E. Brugel, Org. Mass Spectrometry, 1970, 3, 1169.
[526] W. R. Cullen, D. C. Frost, and M. T. Pun, Inorg. Chem., 1970, 9, 1976.
[527] E. S. Brodskii, R. A. Khmel'nitskii, A. A. Polyakova, and I. A. Mikhailov, Zhur. org. Khim., 1970, 5, 969.
[528] J. R. Gilbert and A. J. Stace, Org. Mass Spectrometry, 1971, 5, 1119.
[529] G. Conde-Caprace and J. E. Collin, Org. Mass Spectrometry, 1972, 6, 415.
[530] C. Fenselau, H. A. Chandler, M. Pyles, D. L. Sander, and C. H. Robinson, Org. Mass Spectrometry, 1971, 5, 697.
[531] C. W. Koch and J. H. Markgraf, J. Heterocyclic Chem., 1971, 8, 225.
[532] A. G. Harrison, M. T. Thomas, and I. W. J. Still, Org. Mass Spectrometry, 1970, 3, 899.

$C_3H_7S^+$.[533] The spectra of compounds of the type (196; X = O or S, Y = S, Se, or Te)[534] and (197; X = Se or Te)[535] have been reported. The major decompositions of 1,3-dithian are $M - CH_2S$ and $M - C_3H_6S$ (198) and $M - CH_3S\cdot$ (199; A : B ≈ 1 : 1).[536] The molecular ions of 1,3-dithians also eliminate $S_2H\cdot$ by a rearrangement reaction (Vol. 1, p. 130), and the respective ions $C_4H_7^+$, $C_5H_9^+$ and $C_6H_{11}^+$, produced from (198), 2-methyl-1,3-dithian, and 2,2-dimethyl-1,3-dithian, are completely scrambled with respect to hydrogen either before or accompanying further decomposition.[536] The spectra of 1,4-oxathian,[537] 1,3,6-dioxathiocan[537] and certain fused sulphides[538] have been studied. The base peak in the spectrum of the bicyclic thioketone (200) corresponds to a

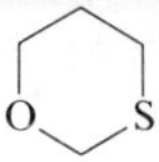

(195) 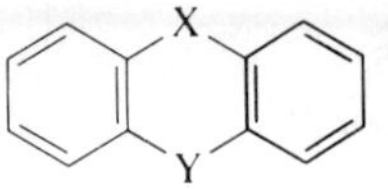   (196) 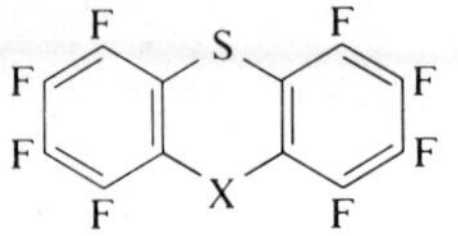   (197)

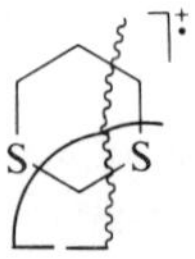

(198) 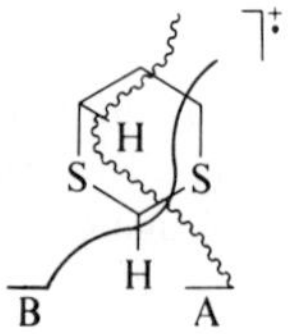   (199) 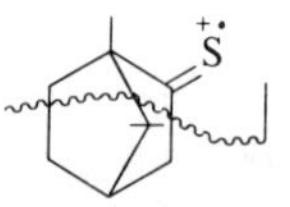   (200)

$C_4H_5S^+$ species.[539] The spectra of dithiocarboxylic acid esters have been listed.[540] Dimethylthionocarbamates undergo the molecular ion rearrangement (201) → (202) which is analogous to the thermal rearrangement of this system.[541] S-(Alkoxythiocarbonyl)thiohydroxylamines eliminate COS by the rearrangement (203) → (204),[542] and thioureas fragment as indicated in (205).[543]

[533] K. Pihlaja and P. Pasanen, *Org. Mass Spectrometry*, 1971, **5**, 763.
[534] I. C. Calder, R. B. Johns, and J. M. Desmarchelier, *Org. Mass Spectrometry*, 1970, **4**, 121.
[535] S. C. Cohen, A. G. Massey, C. F. Lanthier, and J. M. Miller, *Org. Mass Spectrometry*, 1972, **6**, 373.
[536] J. H. Bowie and P. Y. White, *Org. Mass Spectrometry*, 1972, **6**, 317.
[537] G. Condé-Caprace and J. E. Collin, *Org. Mass Spectrometry*, 1972, **6**, 341.
[538] G. H. Wahl, *Org. Mass Spectrometry*, 1970, **3**, 1349.
[539] M. M. Campbell, G. M. Anthony, and C. J. W. Brooks, *Org. Mass Spectrometry*, 1971, **5**, 297.
[540] N. H. Leon, *Org. Mass Spectrometry*, 1972, **6**, 407.
[541] J. C. Tou and R. M. Rodia, *Org. Mass Spectrometry*, 1972, **6**, 493.
[542] A. Holm and G. M. Jensen, *Acta Chem. Scand.*, 1972, **26**, 205.
[543] M. A. Baldwin, A. M. Kirkien, A. G. Loudon, and A. Maccoll, *Org. Mass Spectrometry*, 1970, **4**, 81.

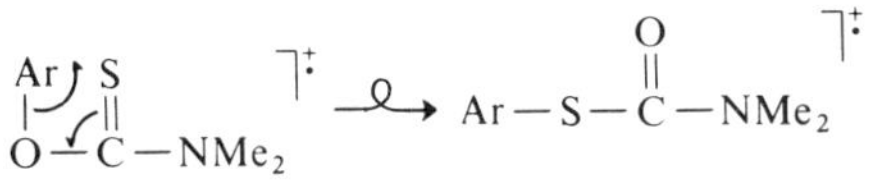

Reactions of Specific Functional Groups 131

(201) (202)

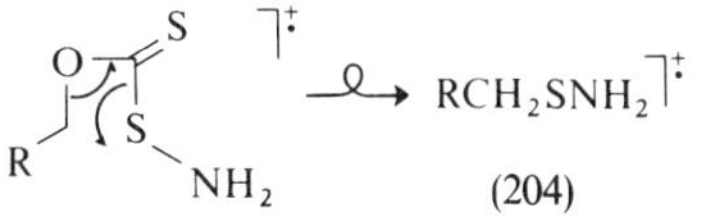

(203) (204) 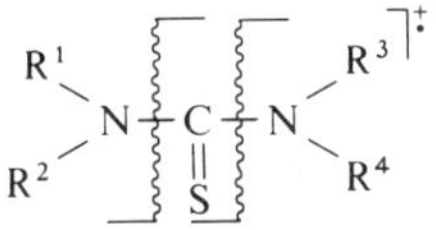 (205)

Further work on sulphones[544–547] and sulphoxides[532,544,546] has been published, and of particular interest is the proximity effect which occurs for o-hydroxyphenyl alkyl sulphones (206) → (207).[546] The molecular ions of aryl sulphonic acids eliminate SO_2 in common with other sulphonyl derivatives,[547–549] and sulphonyl azides fragment by elimination of N_3.[550] Methylene scrambling is observed for 2-(1-azulyl)ethyl tosylates.[551] The molecular ions of o-alkyl-N-arylsulphonylcarbonates eliminate SO_2 and also cleave as indicated in (208);[552] N-arylsulphonyliminopyridinium betaines also lose SO_2 from their molecular ions.[553] A correlation has been observed between the behaviour of sulphinylaniline upon electron impact and thermolysis,[554] and the spectrum of (209) shows the elimination of CO from the molecular ion, indicating that the oxygen migration reaction does not occur to the 1-position of the phenyl ring.[555] Cyclic sulphites have been studied.[556]

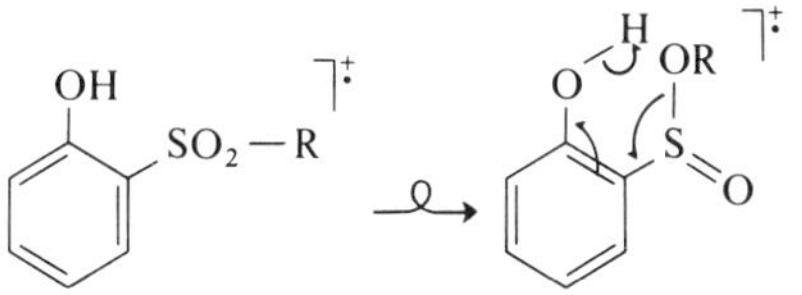

(206) (207) 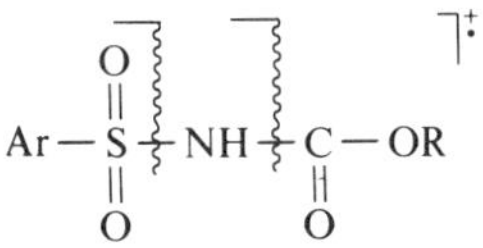 (208)

[544] R. Smakman and Th. J. de Boer, *Org. Mass Spectrometry*, 1970, **3**, 1561.
[545] H. W. Gibson and D. A. McKenzie, *J. Org. Chem.*, 1970, **35**, 2994.
[546] A. O. Pedersen, G. Schroll, S.-O. Lawesson, W. A. Laurie, and R. I. Reed, *Tetrahedron*, 1970, **26**, 4449.
[547] R. J. Soothill and L. R. Williams, *Org. Mass Spectrometry*, 1972, **6**, 141.
[548] R. H. Wiley, *Org. Mass Spectrometry*, 1970, **4**, 55.
[549] A. Heywood, A. Mathias, and A. E. Williams, *Analyt. Chem.*, 1970, **42**, 1272.
[550] M. M. Campbell and A. D. Dunn, *Org. Mass Spectrometry*, 1972, **6**, 599.
[551] R. G. Cooks, N. L. Wolfe, J. R. Curtis, H. E. Petty, and R. N. McDonald, *J. Org. Chem.*, 1970, **35**, 4048; R. G. Cooks, R. N. McDonald, J. R. Curtis, and H. E. Petty, *Org. Mass Spectrometry*, 1971, **5**, 785.
[552] W. H. Daly and C. W. Heurtevant, *Org. Mass Spectrometry*, 1970, **4**, 165.
[553] M. Ikeda, S. Kato, Y. Sumida, and Y. Tamura, *Org. Mass Spectrometry*, 1971, **5**, 1383.
[554] C. Wentrup, *Tetrahedron*, 1971, **27**, 1027.
[555] A. S. Siegel, *Org. Mass Spectrometry*, 1970, **3**, 875.
[556] A. A. Gamble and J. G. Tillet, *Tetrahedron Letters*, 1970, 3625.

The scrambling processes of thiophen (210) have been carefully studied using ^{2}H- and ^{13}C-labelling.[557,558] The formation of the ions $M - HCS\cdot$, $M - C_3H_3^+$, $M - C_2H_2$, and $M - Me\cdot$ proceeds with partial scrambling of the carbon skeleton,[557,558] and in general the hydrogen scrambling is more complete than carbon scrambling.[557] It has been suggested that the reversible formation of (211) may account for the carbon scrambling.[557,558] The problem of phenyl migration *versus* valence isomerization [*e.g.*(212)] has been considered for 2-phenyl-,[559,560] 3-phenyl-,[559] and 2,5-diphenyl-thiophen.[560] The spectra of ^{13}C-labelled phenyl-thiophens show that the fragmentations cannot be simply rationalized in terms of one mechanism, but that both phenyl migration and valence

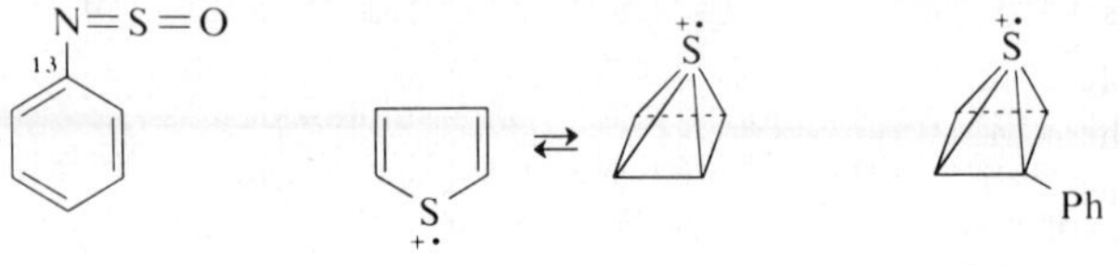

(209) (210) (211) (212)

isomerism occur to an extent dependent upon the particular fragmentation.[560] Carbon scrambling does not occur between the thiophen and phenyl rings.[560] The spectra of thioethers of furan, thiophen, and selenophen,[561] of S-alkylthiophenium salts,[562] thienothiophens,[563] and thieno[3,2-c]pyridazine derivatives[564] have been listed.

The molecular ions of alkyl 4,4-diphenyl-1,3-oxathiolan-5-ones eliminate carbon dioxide,[565] alkylisothiazoles fragment as indicated in (213),[566] and labelling studies confirm that the loss of HCN from thiazole occurs as shown in (214).[567] The spectra of substituted thiazoles,[568,569] benzothiazolium salts,[570]

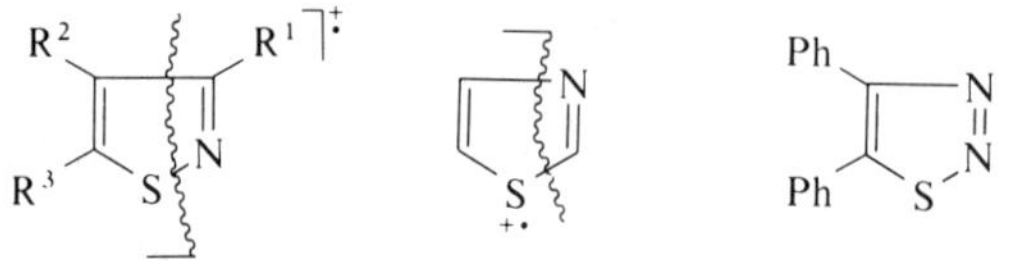

(213) (214) (215)

[557] F. de Jong, H. J. M. Sinnige, and M. J. Janssen, *Org. Mass Spectrometry*, 1970, **3**, 1539.
[558] A. S. Siegel, *Tetrahedron Letters*, 1970, 4113.
[559] M. J. Janssen and F. de Jong, *Z. Chem.*, 1970, **10**, 216.
[560] W. D. Weringa, H. J. M. Sinnige, and M. J. Janssen, *Org. Mass Spectrometry*, 1971, **5**, 1399.
[561] O. S. Chizhov, B. M. Zolotarev, A. N. Sukiasian, V. P. Litvinov, and Ya. L. Gol'dfarb, *Org. Mass Spectrometry*, 1970, **3**, 1379.
[562] R. M. Acheson and D. R. Harrison, *J. Chem. Soc. (C)*, 1970, 1764.
[563] A. Bugge, *Acta Chem. Scand.*, 1971, **25**, 1504.
[564] A. J. Poole and F. L. Rose, *J. Chem. Soc. (C)*, 1971, 1285.
[565] J. Møller and C. Th. Pedersen, *Acta Chem. Scand.*, 1970, **24**, 2489.
[566] J. C. Poite, A. Perichaut, and J. Roggero, *Compt. rend.*, 1970, **270**, C, 1677.
[567] I. N. Bojesen, J. H. Høg, J. T. Nielsen, I. B. Petersen, and K. Schaumburg, *Acta Chem. Scand.*, 1971, **25**, 2739.
[568] M. J. Rix and B. R. Webster, *Org. Mass Spectrometry*, 1971, **5**, 311.
[569] Y. Nakagawa, S. Matsumoto, and A. Takamizawa, *Mass Spectrometry (Japan)*, 1970, **18**, 1044.
[570] P. R. Briggs, T. W. Shannon, and P. Vouros, *Org. Mass Spectrometry*, 1971, **5**, 545.

Reactions of Specific Functional Groups   133

thiazolo[2,3-c]-S-triazole[571] and thiazolo[3,2-b]-S-triazole[571] have been reported. The molecular ions of 2-oxo-3-aryl-1,2,3-oxathiazolidines lose SO_2,[572] whereas those of 1,2,3-thiadiazole derivatives decompose by loss of N_2;[573–575] the fluorenyl cation ($C_{13}H_9^+$) is produced by the processes $M - N_2 - CS - H\cdot$ from (215).[575] The 1,2,4-triazoline-5-thione system has been studied by ^{2}H- and ^{15}N-labelling;[576] the nitrogens in the ring retain their positional identity.[576] The basic breakdown of 1-methyl-1,2,4-triazoline-5-thione follows the pattern $M - HCN - HCN - H\cdot$, with the initial loss of HCN originating as shown in (216) and with the second loss involving the 1- and 4-nitrogens in the ratio 3 : 1. A second process, $M - SH\cdot - HCN$, is a proximity effect, and is illustrated in (217).[576] The spectra of 5-amino-1,2,4-thiadiazoles,[577] 5-alkylthio-1,3,4-thiadiazolyl-2-amines,[578] thiazolo[3,2-a]pyridinium 3- and 8-oxides,[579,580] 2,3-dihydrothiazolo[3,2-a]pyrimidones,[581] 3H-1,2,4-thiadiazolo[4,3-a]pyridines,[582] thiopurines,[583,584] phenylthiabenzenes,[585] thionaphthenes,[586] phenothiazines[587–589] and benzopyrano[4,3-b]indoles[590] have been described. The molecular ion of 5-phenyl-1,2,3,4-thiadiazole (218) loses N_2, followed by CS.[575]

(216)   (217)

(218)

[571] K. T. Potts and S. Husain, *J. Org. Chem.*, 1971, **36**, 10.
[572] F. Yamada, T. Nishiyama, Y. Fujimoto, and M. Kimigara, *Bull. Chem. Soc. Japan*, 1971, **44**, 1152.
[573] B. J. Millard and D. L. Pain, *J. Chem. Soc.* (C), 1970, 2042.
[574] K.-P. Zeller, H. Meie, and E. Müller, *Org. Mass Spectrometry*, 1971, **5**, 373.
[575] K.-P. Zeller, H. Meier, and E. Müller, *Tetrahedron*, 1972, **28**, 1353.
[576] A. J. Blackman and J. H. Bowie, *Austral. J. Chem.*, 1972, **25**, 335.
[577] A. H. Miller and R. J. Pancirov, *J. Heterocyclic Chem.*, 1971, **8**, 163.
[578] C. S. Barnes, R. J. Goldsack, and R. P. Rao, *Org. Mass. Spectrometry*, 1971, **5**, 317.
[579] K. Undheim, P. E. Fjeldstad, and P. O. Tveita, *Acta Chem. Scand.*, 1971, **25**, 2943.
[580] K. R. Reistad and K. Undheim, *Acta Chem. Scand.*, 1971, **25**, 2954.
[581] E. Falch and T. Natvig, *Acta Chem. Scand.*, 1970, **24**, 1423.
[582] K. T. Potts and R. Armbruster, *J. Org. Chem.*, 1970, **35**, 1965.
[583] J. Deutsch, Z. Neiman, and F. Bergmann, *Org. Mass Spectrometry*, 1970, **3**, 1219.
[584] J. Deutsch, Z. Nieman, and F. Bergmann, *Org. Mass Spectrometry*, 1971, **5**, 279.
[585] C. C. Price, J. Follweiler, H. Pirelahi, and M. Siskin, *J. Org. Chem.*, 1971, **36**, 791.
[586] V. S. Fal'ko, V. I. Khvostenko, V. E. Udre, and M. G. Voronkov, *Khim. geterotsikl. Soedinenii*, 1971, 326.
[587] R. L. Mital, S. K. Jain, M. Azzaro, A. Cambon, and J.-P. Rosset, *Bull. Soc. chim. France*, 1970, 2195.
[588] D. Simov and I. G. Taulov, *Org. Mass Spectrometry*, 1971, **5**, 1133.
[589] W. Riepe and M. Zander, *Z. Naturforsch.*, 1972, **27a**, 170.
[590] N. P. Buu-Hoi, S. Croisy, P. Jacquignon, A. Martani, and A. Ricci, *J. Heterocyclic Chem.*, 1970, **7**, 931.

The P—O Group*.—The spectrum of (219) is notable for the presence of a rearrangement ion $M - PhPO_2$;[591] the phenyl-t-butylphosphinic acid ion-radical (220) eliminates C_4H_8 to produce (221);[592] the ion (222; R = H or Me) eliminates RO· possibly after the rearrangement (222) → (223);[593] rearrangement of the phenoxaphosphinic acid molecular ion (224) produces an $M - CO$ ion.[594] Benzenephosphonic acid (225) suffers pyrolysis at high temperatures,

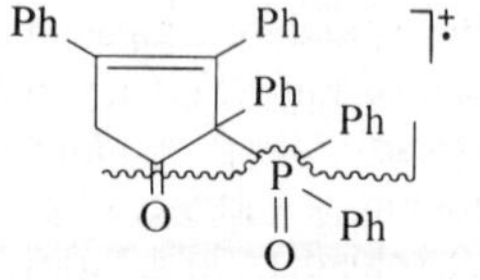

(219) 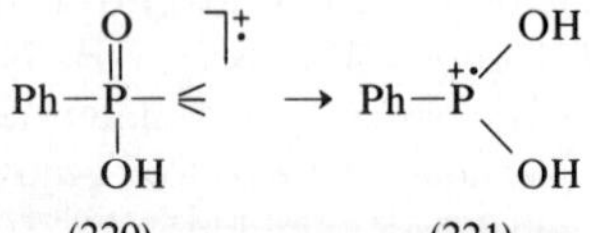 (220) (221)

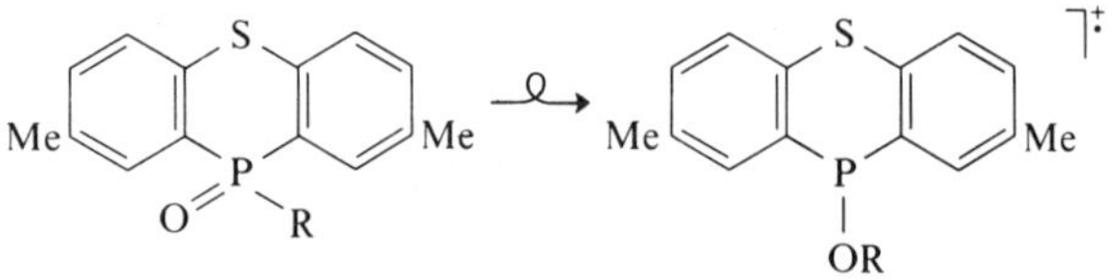

(222) (223)

but when the spectrum is measured at low temperature, the molecular ion is the base peak, and $M - H_2O$, $C_6H_5^+$ and $C_6H_5O^+$ ions are observed.[595] The spectra of diphosphonates[596] and phosphonates from norbornane derivatives[597] have been described. The fragmentations of N-phenylphosphoramidate esters

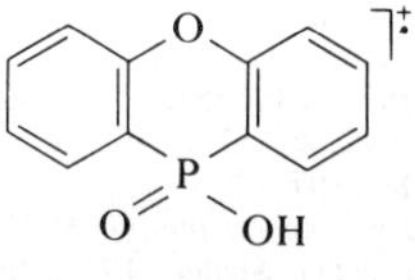

(224) 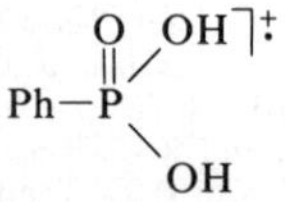 (225) 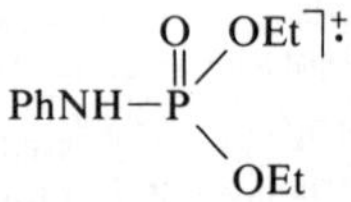 (226)

[591] J. A. Miller, G. M. Stevenson, and B. C. Williams, *J. Chem. Soc.* (C), 1971, 2714.
[592] R. Brooks and C. A. Bunton, *J. Org. Chem.*, 1970, **35**, 2642.
[593] I. Granoth, A. Kalir, Z. Pelah, and E. D. Bergmann, *Org. Mass Spectrometry*, 1970, **3**, 1359.
[594] I. Granoth and J. B. Jevy, *J. Chem. Soc.* (B), 1971, 2391.
[595] R. H. Wiley, *Org. Mass Spectrometry*, 1971, **5**, 675.
[596] D. J. Whelan and J. C. Johannessen, *Austral. J. Chem.*, 1971, **24**, 887.
[597] H. J. Callot and C. Benezra, *Org. Mass Spectrometry*, 1971, **5**, 343.

* Some other aspects of the mass spectra of phosphorus compounds are covered in Chapter 5.

Reactions of Specific Functional Groups 135

are similar to those of phosphate esters.[598] For example, (226) decomposes by the route $M - C_2H_4 - C_2H_4 - H_2O$, but the aryl ester (227) undergoes cleavage α to P=O [see (227)], and suffers the rearrangement (227) → (228) (compare triphenyl phosphate[599]).

$$\text{PhNH}-\overset{\overset{O}{\|}}{\underset{(227)}{P}}\!\!\begin{Bmatrix}OPh\\OPh\end{Bmatrix}^{\ddagger} \xrightarrow{-Q} \underset{(228)}{Ph-\overset{+\cdot}{O}-Ph}$$

3 Hydrogen Scrambling and Skeletal Rearrangement—A Summary

The distinction between scrambling and rearrangement processes was discussed in Volume 1 (Chapter 2 and pp. 134—136), as were the relevance and importance of such processes.

The occurrence of hydrogen scrambling in the molecular ions of aliphatic systems, as opposed to specific hydrogen migration, is still relatively uncommon, although further instances have been reported, including cyclohexane-1,2-diol,[168] undecan-2-one,[181] ethyl acetate,[238] and various ethers.[274]

The definitive studies of hydrogen and carbon scrambling in aromatic and heterocyclic systems has led to some insight into the mechanisms of such processes. It is now known that the benzene molecular ion undergoes both independent carbon and hydrogen scrambling,[84,85] that all the carbon atoms in the $M - 1$ ion from toluene are identical and presumably scramble through benzyl–tropylium ion interconversions,[89] that similar processes precede the loss of Me· from the diphenylmethane molecular ion with all C—H bonds being broken and reformed in a statistical manner,[107] that the carbon atoms of pyridine[458] and thiophen[557,558] scramble, and that both phenyl migration and randomization of the thiophen carbons occur for the phenylthiophens.[560] Other instances of scrambling include styrene,[113] phenylacetylene,[119] substituted diphenyls,[149] hydrocinnamaldehyde,[208] tetraphenylpentadienone,[220] and tetraphenyl-1,4-benzoquinone and related systems.[220] Chlorine scrambling has been proposed for chlorodiphenyls[151,152] and chloronaphthalenes.[150] Further discoveries have been made of cations which undergo either partial or complete carbon and/or hydrogen scrambling prior to further decomposition. These include $C_4H_7^+$,[130,536] $C_4H_9^+$,[129] $C_5H_9^+$,[130,536] $C_6H_5^+$,[85,361] $C_6H_9^+$,[130] $C_6H_{11}^+$,[536] $C_8H_9^+$,[113] $C_9H_9^+$,[139] $C_{10}H_{11}^+$,[139,600] $C_{11}H_{13}^+$,[139] and $C_{12}H_{15}^+$.[139]

Rearrangement reactions which involve migrations of groups other than hydrogen (skeletal rearrangements) continue to be discovered. These rearrangements, described throughout Section 2, generally fall into categories described

---

[598] W. J. McMurray, S. R. Lipsky, C. Zioudrou, and G. L. Schmir, *Org. Mass Spectrometry*, 1970, **3**, 1031.
[599] A. Quayle, in 'Advances in Mass Spectrometry' ed. J. D. Waldron, Pergamon Press, London, 1959, p. 365.
[600] J. H. Bowie and P. Y. White, *Austral. J. Chem.*, 1972, **25**, 439.

either in Volume 1 (p. 135) or in review articles.[601,602] Some highlights include the rearrangements of arylacetylenic ketones,[193] 1,2,3,4,5-pentaphenylpentane-1,5-dione,[219] cyclo-octatetraene derivatives,[139] β-aryl-α-methylpropionic acids,[237] benzylidenemalonates,[261] polymethoxy-compounds,[271,272] azodicarbonyl compounds (Wolff rearrangement),[341] flavans,[451] and aminoisoxazoles.[416] Sulphur and phosphorus compounds are noted for skeletal-rearrangement reactions, and examples of these are to be found in the spectra of 1,3-dithians,[536,600] thionocarbamates,[541] thiohydroxylamines,[542] various sulphonyl compounds,[547—553] phosphine oxides,[593] phosphinic acids,[594] phosphonic acids,[595] and phosphoramidate esters.[598] Organometallic compounds (see Chapter 5) also undergo extensive rearrangement reactions. Of particular interest are the rearrangements of phenylboronates.[603,604] For example, (229) eliminates $C_2H_2D_2BO_2$ to produce the ion $C_7H_7^+$, the base peak of the spectrum.[604] Similar rearrangements do not occur for the corresponding 2-phenyl-1,3-dioxan or -1,3-dithian.

Trimethylsilyl (TMS) derivatives are widely used for the mass spectral analysis of non-volatile compounds. The molecular ions of such compounds undergo

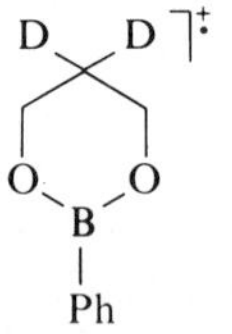

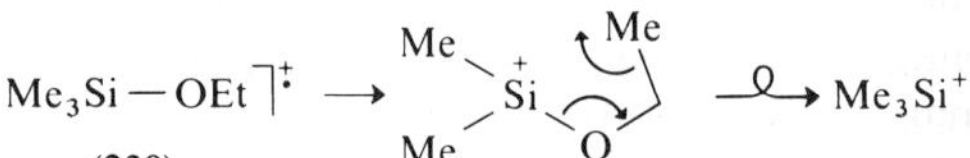

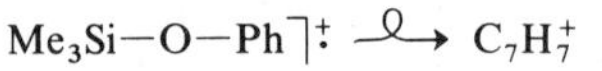

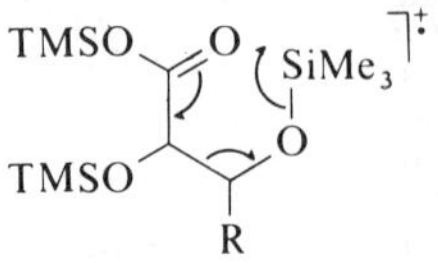

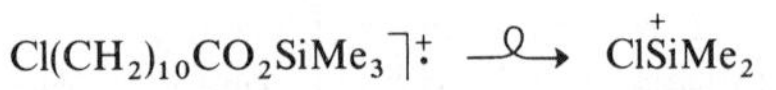

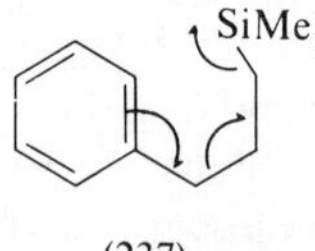

[601] R. G. Cooks, *Org. Mass Spectrometry*, 1969, **2**, 481.
[602] J. H. Bowie, B. K. Simons, and S.-O. Lawesson, *Rev. Pure Appl. Chem. (Australia)*, 1969, **19**, 61.
[603] I. R. McKinley and H. Weigel, *Chem. Comm.*, 1970, 1022.
[604] B. J. Bose and M. D. Peters, *Canad. J. Chem.*, 1971, **49**, 1766.

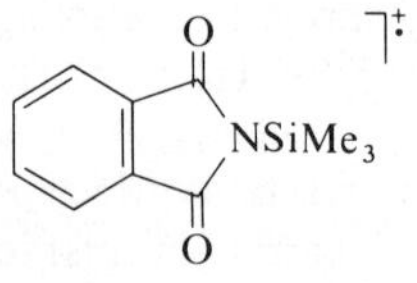

Reactions of Specific Functional Groups   137

a variety of rearrangement reactions, and a selection of such reactions is included here. No attempt is made to give a general coverage of TMS derivatives, but silicon derivatives are considered in more detail in Chapter 5. Skeletal-rearrangement ions are observed in the spectra of TMS derivatives of alcohols [*e.g.* (230) → (231)][605] and phenols [*e.g.* (232) → (233)].[606] A McLafferty-type elimination occurs for many silylated hydroxycarbonyl compounds [*e.g.* (234)],[607,608] and TMS derivatives of halogeno-acids and halogeno-alcohols suffer unusual rearrangements [*e.g.* (235) → (236)].[609] Many rearrangements occur for TMS cyclohexanol derivatives;[610,611] migration of the TMS group is observed for (237)[612] and silylated hydroxyalkylphenones;[613] (238) eliminates a methyl radical followed by carbon dioxide.[614] Other rearrangement ions are observed in the spectra of cyclocarbosiloxanes,[615] silylhydrazines,[616] and silylpyrimidines and purines.[617]

4 Negative-ion Mass Spectrometry

The problems inherent in the production and measurement of negative ions were outlined in Volume 1 (pp. 136, 137). Two books[14,15] on negative-ion mass spectrometry have been published since 1970. Further reports of the negative-ion spectra of various fluorine-containing molecules have appeared;[618—622] the trifluoromethylcarboxylate anion is the major ion in the spectrum of perfluoroacetic anhydride.[620] Molecular anions are generally absent from the spectra of aliphatic compounds, but fragment ions are observed for keten,[623] long-chain aldehydes and esters,[624] 3-methoxy-2-hydroperoxypropane,[625] dimethyl-

[605] G. Dube, *Z. Chem.*, 1970, **10**, 301.
[606] R. H. Cragg, K. J. A. Hargreaves, J. F. J. Todd, and R. B. Turner, *Chem. Comm.*, 1972, 336.
[607] G. Petersson, *Org. Mass Spectrometry*, 1972, **6**, 577.
[608] G. Petersson, *Tetrahedron*, 1970, **26**, 3413.
[609] D. C. De Jongh, D. A. Brent, and R. Y. van Fossen, *J. Org. Chem.*, 1971, **36**, 1469.
[610] R. T. Gray, J. Diekman, G. L. Larson, W. K. Musker, and C. Djerassi, *Org. Mass Spectrometry*, 1970, **3**, 973.
[611] P. D. Woodgate, R. T. Gray, and C. Djerassi, *Org. Mass Spectrometry*, 1970, **4**, 257.
[612] W. P. Weber, A. K. Willard, and H. G. Boettger, *J. Org. Chem.*, 1971, **36**, 1620.
[613] G. G. Smith and C. Djerassi, *Org. Mass Spectrometry*, 1971, **5**, 505.
[614] A. F. Janzen and E. A. Kramer, *Canad. J. Chem.*, 1971, **49**, 1011.
[615] V. Yu. Orlov, N. S. Nametkin, L. E. Gusel'nikov, and T. H. Islamov, *Org. Mass Spectrometry*, 1970, **4**, 195.
[616] K. G. Das, P. S. Kulkarni, V. Kalyanaraman, and M. V. George, *J. Org. Chem.*, 1970, **35**, 2140.
[617] E. White, V. P. M. Krueger, and J. A. McCloskey, *J. Org. Chem.*, 1972, **37**, 430.
[618] G. D. Cady, A. L. Crittenden, and F. B. Dudley, *U.S. Govt. Res. Dev. Rep.*, 1970, **70**, 67.
[619] G. D. Cady, A. L. Crittenden, and F. B. Dudley, *Nasa Star*, 1970, **8**, 3488.
[620] F. B. Dudley, G. H. Cady, and A. L. Crittenden, *Org. Mass Spectrometry*, 1971, **5**, 953.
[621] C. Lifshitz and R. Grajower, *Internat. J. Mass Spectrometry Ion Phys.*, 1970, **4**, 92.
[622] K. A. G. Macneil and J. C. J. Thynne, *Internat. J. Mass Spectrometry Ion Phys.*, 1970, **3**, 455.
[623] J. E. Collin and R. Locht, *Internat. J. Mass Spectrometry Ion Phys.*, 1970, **3**, 465.
[624] P. C. Rankin, *Lipids*, 1970, **5**, 825.
[625] K. Jager, A. Henglein, and M. Deumont, *Z. Naturforsch.*, 1970, **25a**, 202.

sulphoxide,[626] and ethyl nitrate.[627] The ions CN^- and C_2H^- are observed in the spectra of alkyl cyanides.[628] Trinitromethanes of the type (239; X = halogen) give $M^{\bar{\cdot}} - NO_2\cdot$ anions,[629] whereas diazocyclopentadiene-2-carboxylic acid (240) produces $M^{\bar{\cdot}} - N_2$ and $M^{\bar{\cdot}} - (N_2 + CO_2)$ radical-anions.[630]

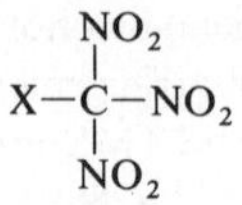

$$X-\underset{\underset{NO_2}{|}}{\overset{\overset{NO_2}{|}}{C}}-NO_2$$

(239)

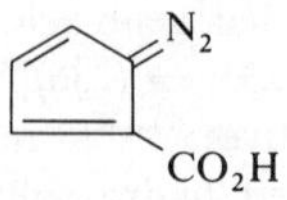

(240)

Negative ions are produced from triptycene,[631] hexahydrotryptycene[631] and molecular polymers,[632] and $M - 1$ anions have been noted in the high-pressure spectra of cyclopentadiene,[633] pyrrole,[633] indole,[633] pyridine,[634] and pyridazine.[634] The negative-ion spectra of riboflavin derivatives have been reported.[635]

Bowie and his colleagues have initiated a programme[636] to study the fragmentations of functional groups in the negative mode. In order to study the fragmentations of X in the system R—X, the unit R was initially chosen such that it would accept an electron to produce a molecular anion $RX^{\bar{\cdot}}$ at 70 eV, produce a large negative-ion current at low pressures, and not fragment through R but allow the fragmentation of X. Many organic compounds accept an electron in solution, and using this analogy the moiety chosen was 9,10-anthraquinone, which forms an intense molecular anion (241) but does not fragment. The

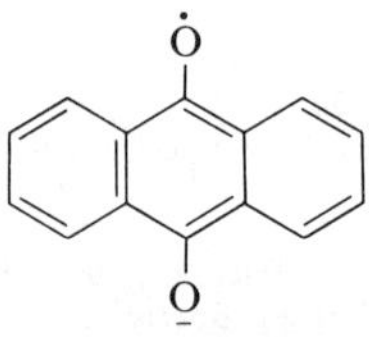

(241)

[626] J. C. Blas, M. Cottin, and B. Gitton, *J. Chim. phys.*, 1970, **67**, 1475.
[627] R. Kriemler and S. E. Buttrill, *J. Amer. Chem. Soc.*, 1970, **92**, 1123.
[628] S. Tsuda, A. Yokohata, and T. Umaba, *Bull. Chem. Soc. Japan*, 1970, **43**, 3383.
[629] J. T. Larkins, J. M. Nicholson, and F. E. Saalfeld, *Org. Mass Spectrometry*, 1971, **5**, 265.
[630] J. C. Martin and D. R. Block, *J. Amer. Chem. Soc.*, 1971, **93**, 451.
[631] J. Eloranta and H. Joela, *Suomen Kemi. (B)*, 1969, **42**, 416.
[632] H. Knof and D. Krafft, *Z. Naturforsch.*, 1970, **25a**, 849.
[633] V. I. Khvostenko, I. I. Furlei, A. N. Kost, V. A. Budylin, and L. G. Yudin, *Doklady Phys. Chem.*, 1969, **189**, 778.
[634] W. W. Pauden and S. A. Humphrey, *Org. Mass Spectrometry*, 1970, **4**, 513.
[635] R. Tümmler, K. Steinfelder, E. C. Owen, and D. W. West, *Org. Mass Spectrometry*, 1971, **5**, 40.
[636] A. C. Ho, J. H. Bowie, and A. Fry, *J. Chem. Soc. (B)*, 1971, 530.

spectra of substituted anthraquinones[636,637] are intense, give characteristic fragmentations of the substituent, contain metastable ions, and in many cases give information which is not obtainable from the corresponding positive-ion spectra. The negative-ion fragmentation of functional groups are generally (but not always) different from the decompositions in the positive mode.

The basic fragmentation of alkyl esters is $M^{\bar{\cdot}} - R\cdot - CO_2$ [see (242)].[636] The fragmentation of the ester group depends upon whether it occupies the 2-(242) or 1-position (243) of the anthraquinone system. The ester group of (243) undergoes the same fragmentations as that of (242), but in addition, suffers the proximity effect $M - RO\cdot$.[636] Alkyl ethers (244; R = alkyl) show the elimination $M^{\bar{\cdot}} - R\cdot$ irrespective of the position of the alkoxy-substituent.[637] The spectra of 1- and 2-acetoxyanthraquinones are different, with the 2-acetate losing an acetyl radical (244; R = COMe), whereas the 1-acetate eliminates keten as illustrated in (245).[636] This should be contrasted with the losses of keten from the molecular cations of both 1- and 2-acetoxyanthraquinone. Fragmentation of the 1-substituent precedes that of the 2-substituent as shown by the decompositions $M^{\bar{\cdot}} - CD_2CO - CH_2CO$ of (246).[636] This is useful for structure determination and is further illustrated by (247) and (248), which cleave the processes $M^{\bar{\cdot}} - Me\cdot - CH_2CO$ (247) and $M^{\bar{\cdot}} - CH_2CO - Me\cdot$ (248).[637]

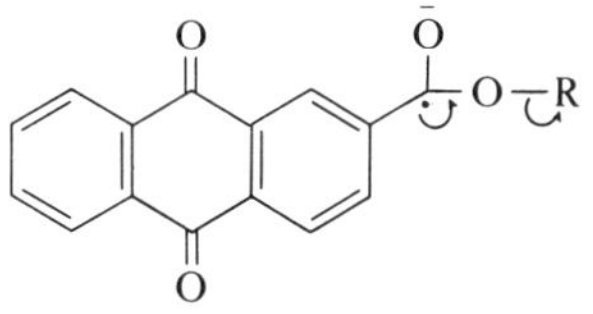

(242) 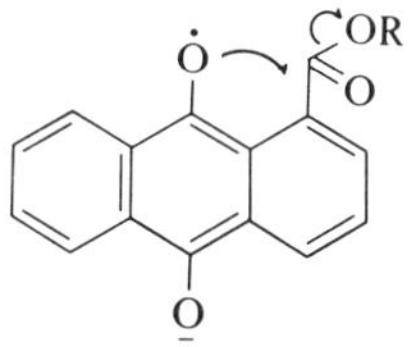 (243)

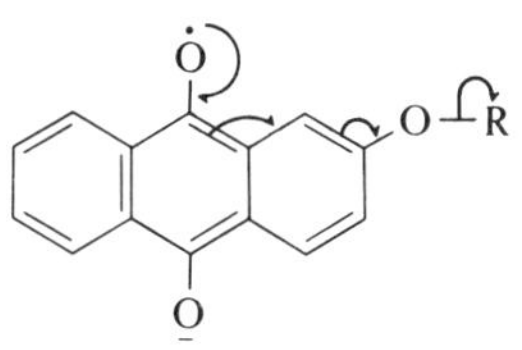

(244) 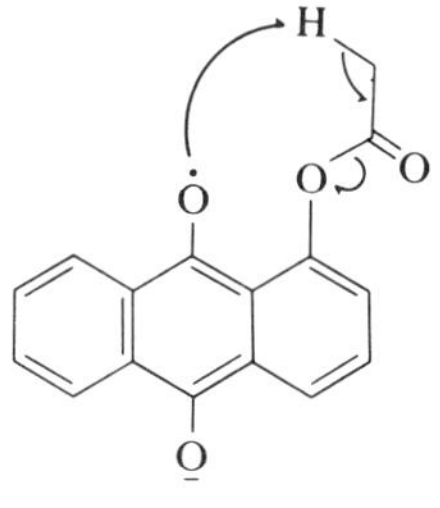 (245)

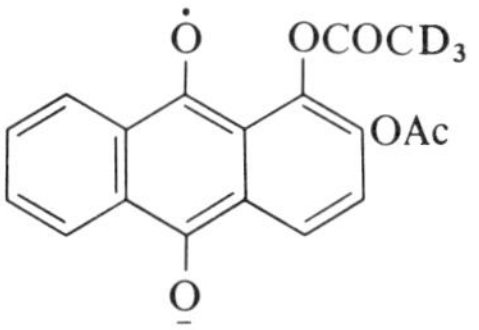

(246) 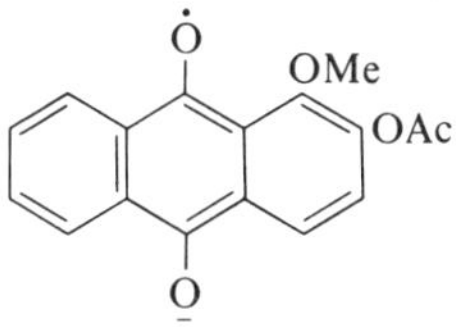 (247) 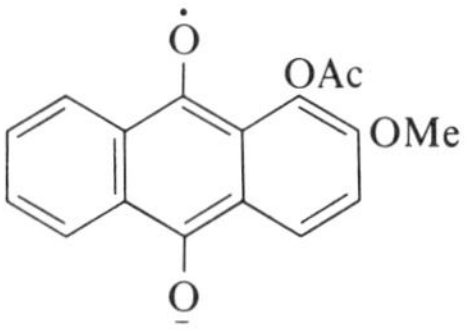 (248)

[637] J. H. Bowie and A. C. Ho, *Austral. J. Chem.*, 1971, **24**, 1093.

The nitrophenyl group can also be used as a stabilizing unit,[638] with (249) being a contributor to the molecular anion. The fragmentations of a large number of functional groups have been described.[638—645] The fragmentations depend upon the relative position of the substituent, with o-, m-, and p-isomers often being more readily distinguished by negative-ion than by positive-ion mass spectrometry. It would not be practical to describe all fragmentations, but the —CO—alkyl group does not fragment,[638] halogeno-groups give rise to X^- ions, and the nitro-group of the nitrophenyl unit undergoes a skeletal rearrangement with the elimination of NO· to give phenoxide anions.[638—645] Proximity effects occur for many o-substituted nitrobenzenes. Highlights include o-nitrobenzaldehyde, $M^{\bar{\cdot}} - NO\cdot - CO$ (250) → (251)[638] and o-nitrobenzoic acid, $M^{\bar{\cdot}} - NO\cdot - CO_2$ (252) → (253).[638] The elimination of keten from the $M^{\bar{\cdot}} - NO\cdot$ ion (254) of o-nitroacetanilide occurs by a non-concerted mechanism, with transfer of hydrogen to give (255), followed by equilibration of the methylene and amino hydrogens and then elimination of keten. The characteristic loss of HO· is noted from molecular anions of the type (256;

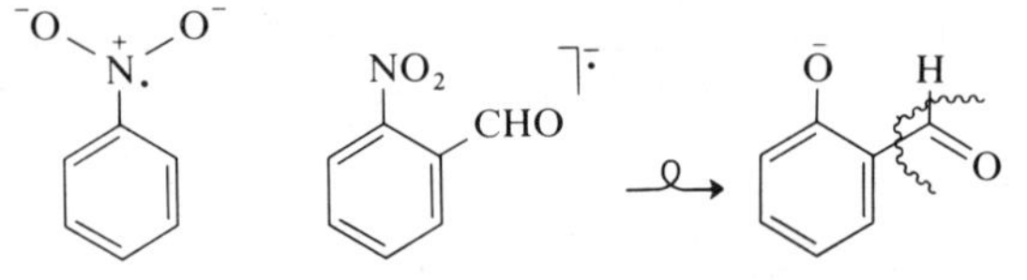

(249)   (250)   (251)

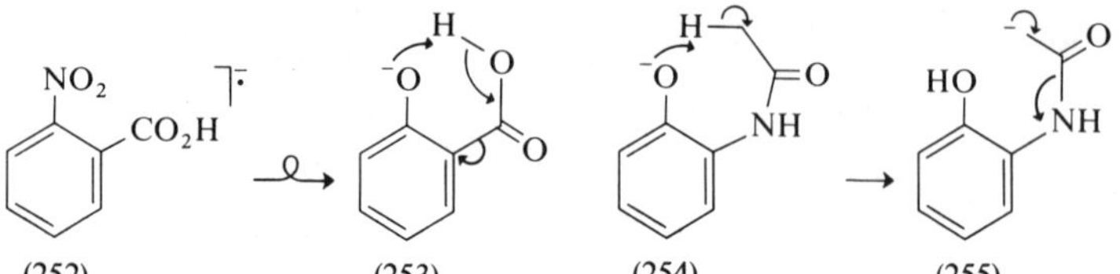

(252)   (253)   (254)   (255)

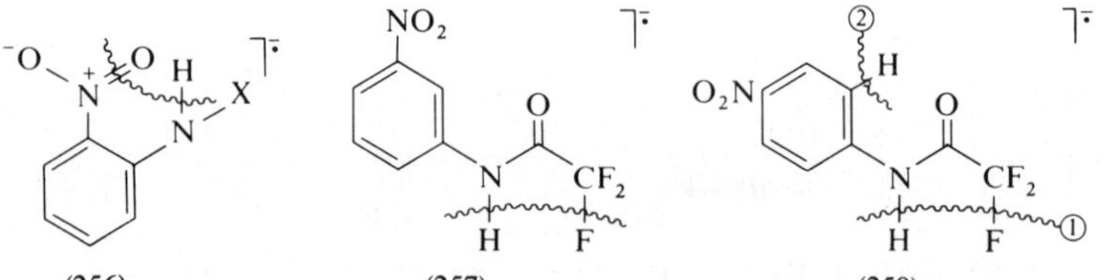

(256)   (257)   (258)

[638] J. H. Bowie, *Org. Mass Spectrometry*, 1971, **5**, 945.
[639] C. L. Brown and W. P. Weber, *J. Amer. Chem. Soc.*, 1970, **92**, 5775.
[640] J. H. Bowie, T. Blumenthal, and I. Walsh, *Org. Mass Spectrometry*, 1971, **5**, 777.
[641] J. H. Bowie, *Austral. J. Chem.*, 1971, **24**, 989.
[642] J. H. Bowie and B. Nussey, *Org. Mass Spectrometry*, 1972, **6**, 429.
[643] J. H. Bowie and A. J. Blackman, *Austral. J. Chem.*, 1972, **25**, 1335.
[644] T. Blumenthal and J. H. Bowie, *Austral. J. Chem.*, 1971, **24**, 1853.
[645] J. H. Bowie and P. Y. White, *Org. Mass Spectrometry*, 1972, **6**, 75.

Reactions of Specific Functional Groups   141

$X = COR$ or SO_2R).[640] The positive-ion mass spectra of the three isomers of nitrophenyltrifluoracetanilide are very similar, but the negative-ion spectra are completely different.[641] The *o*-isomer eliminates HO·, the *m*-isomer (257) fragments by the process $M - HF - HO·$, and the *p*-isomer (258) decomposes by loss of HF followed by H·.

The six nitrophenylbenzoates and phenylnitrobenzoates are readily differentiated by negative-ion mass spectrometry.[642] An unusual proximity effect is observed for *o*-nitrophenylbenzoate, where the base peak ($PhCO_2^-$) is produced as illustrated in (259).[642] The basic fragmentations of substituted *p*-nitrophenylbenzoates are shown in (260) and the abundances of these ions are a function of the Hammett value of the substituent X.[642] The nitrophenyl group may be used in order to examine the fragmentations of heterocyclic systems,[643] and as an example, (261) fragments by the cleavages $M^{\bar{\cdot}} - HCN - HCN$. The major decompositions noted in the spectrum of 2-(*p*-nitrophenyl)-1,3-dithian are shown in (262) and (263), and both processes are preceded by the equilibration of the five hydrogen atoms at positions 2, 4, and 6.[645] A striking rearrangement occurs from the 4-nitrophthalic anhydride molecular anion (264); this ion eliminates two molecules of carbon dioxide with the initial loss involving one of the nitro oxygen atoms.[644]

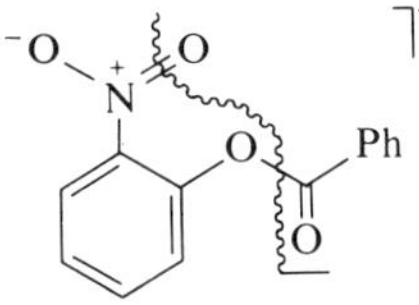

(259) 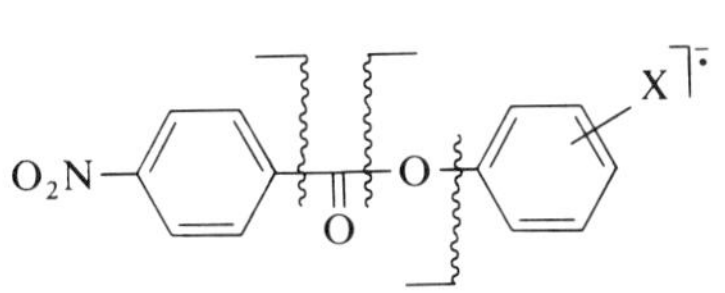   (260) 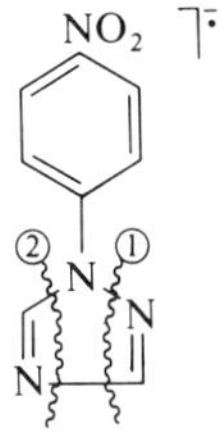   (261)

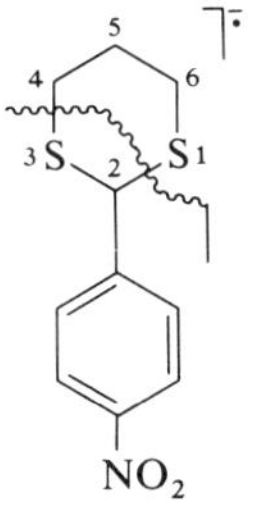

(262) 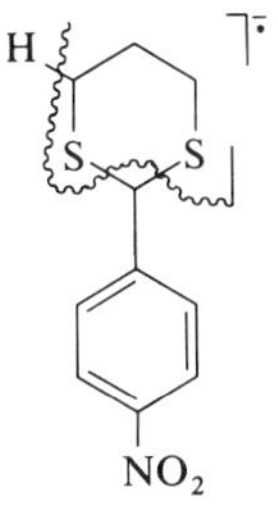   (263) 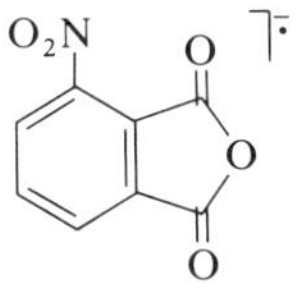   (264)

It has been suggested from ICR studies that the molecular anions of nitrobenzene derivatives are produced by the capture of secondary electrons,[646] and this is in accord with earlier results with sulphur hexafluoride[647] and hexafluoro-

[646] T. McAlister, *Chem. Comm.*, 1972, 245.
[647] G. Jacobs and A. Henglein, in 'Advances in Mass Spectrometry', ed. W. L. Mead, Institute of Petroleum, London, 1966, p. 286.

acetone.[648] The capture of low-energy electrons has previously been noted in other ICR studies.[649,650] The stage is now set for further investigations of the electron-capture process, and for continuing studies of the varied and unusual fragmentations of negatively-charged species.

This chapter was written during a period of Study Leave spent at the University Chemical Laboratory, Cambridge. Grateful acknowledgement is made to Professor Lord Todd and Dr. D. H. Williams for facilities, and to the University of Adelaide for support through the Study Leave Fund.

[648] P. W. Harland, K. A. G. Macneil, and J. C. J. Thynne, 'Studies of Negative-Ion Formation at Low Electron Energies', in 'Dynamic Mass Spectrometry', Heyden, London, and Sadtler Research Laboratories Inc., Philadelphia Pa., 1970, Vol. 1, p. 122.
[649] D. P. Ridge and J. L. Beauchamp, *J. Chem. Phys.*, 1969, **51**, 470.
[650] R. M. O'Malley and K. R. Jennings, *Internat. J. Mass Spectrometry Ion Phys.*, 1969, **2**, App. 1.

4
Natural Products

BY T. J. MEAD, H. R. MORRIS, J. H. BOWIE, and I. HOWE

1 Introduction

The past two years has seen a period of consolidation in the application of mass spectrometry to the structure determination of natural products. Few really new developments have been reported during the reviewing period, and all major advances have occurred in areas specifically defined in Volume 1. The number of papers describing aspects of the mass spectra of natural products is quite enormous, and because of the limited period available for the preparation of this chapter (see the Foreword), it has been necessary to apply selectivity both to the areas chosen for review, and within each particular section. References were scanned using the Mass Spectrometry Bulletin (July 1970—June 1972) but it must be stressed that references to certain journals from this Bulletin may be up to a year out of date. The more important areas of advancement discussed include (i) sequencing of various derivatives of peptides, carbohydrates, and related systems, and (ii) the application of other methods of ionization, including field ionization, field desorption, and chemical ionization.

The material contained in the majority of sections records and illustrates the more interesting applications of mass spectrometry in that area. The amino-acid and peptide section has been dealt with in more detail, and in a more critical manner, because of the obvious importance and applicability of research in this field. It is anticipated that topics not dealt with in this chapter will be covered in future reports. The widespread application of gas chromatography–mass spectrometry is discussed in Chapter 7.

2 Lipids

This section reports the recent progress in mass spectrometry as applied to naturally occurring and synthetic lipids. Since complex lipid mixtures are amenable to analysis by g.c.–m.s., reference to their mass spectra may also be found in Chapter 7.

Mass spectrometry has continued to be used to locate the position of the second functional group in long-chain functionalized fatty acids. For example, three methyl esters of long-chain fatty acids from the sphingolipids of *Tetrahymena* were identified as having an α-hydroxy-substituent by intense peaks in the mass spectrum at m/e 90 and 103, attributable to the ions (1) and (2) respectively.[1] Facile loss of 59 mass units (CO_2Me) also suggests that the hydroxy-function is α to the ester grouping. Some long-chain β-hydroxy-acids have been identified from the mass spectra of their trimethylsilyl (TMS) ether methyl esters[2] or simply as their methyl esters.[3] Identification of 3-hydroxy-octadecanoic acid by mass spectrometry from an ornithine-containing lipid,[4] and of poly-substituted methyl and hydroxy fatty acid components of a sulpholipid[5] has been reported.

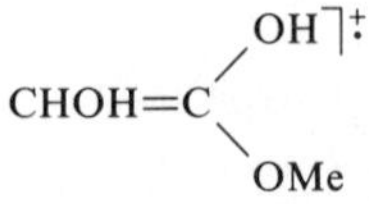

$$CHOH=C{\overset{OH}{\underset{OMe}{\Big/}}}\Big]^{\ddagger} \qquad \overset{+}{C}H_2CHOHCO_2Me$$

(1)　　　　　　　(2)

A convenient novel technique for the structural analysis of the hydroxy fatty acid polymer cutin, the major component of plant cuticle, has been described.[6,7] Reduction of cutin with $LiAlH_4$ yielded long-chain alkanols, diols, and triols whose TMS ethers were conveniently analysed by mass spectrometry. For example, the TMS ether (3) of a reduction product triol was identified readily from its molecular ion and the intense α-cleavage ions at m/e 275 and 319.[7] The carboxylic acid function in the original cutin was identified by reduction with $LiAlD_4$. D_2 was incorporated as shown [see (3)], thus helping to identify the original fatty acid as 10,16-dihydroxypalmitic acid (4).

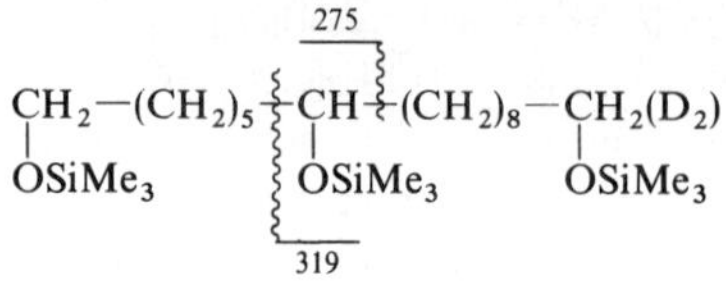

$$\underset{OSiMe_3}{CH_2}-(CH_2)_5\overset{275}{\Big\{}\underset{OSiMe_3}{CH}\Big\}(CH_2)_8-\underset{OSiMe_3}{CH_2(D_2)}$$
　　　　　　　　319

(3)

$$\underset{OH}{CH_2}-(CH_2)_5-\underset{OH}{CH}-(CH_2)_8CO_2H$$

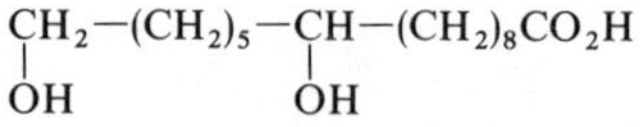

(4)

The mass spectra of the 1,4-epoxides (5) formed from methyl linoleate and related esters by reaction with toluene-p-sulphonic acid have been reported.[8]

---

[1] K. A. Ferguson, R. L. Conner, F. B. Mallory, and C. W. Mallory, *Biochim. Biophys. Acta*, 1972, **270**, 111.
[2] R. D. Batt, R. Hodges, and J. G. Robertson, *Biochim. Biophys. Acta*, 1971, **239**, 368.
[3] D. E. Koeltzow and H. E. Conrad, *Biochemistry*, 1971, **10**, 214.
[4] H. W. Knoche and J. M. Shively, *J. Biol. Chem.*, 1972, **247**, 170.
[5] M. B. Goren, O. Brokl, B. C. Das, and E. Lederer, *Biochemistry*, 1971, **10**, 72.
[6] T. J. Walton and P. E. Kolattukudy, *Biochemistry*, 1972, **11**, 1885.
[7] P. E. Kolattukudy and T. J. Walton, *Biochemistry*, 1972, **11**, 1897.
[8] G. G. Abbot, F. D. Gunstone, and S. D. Hoyes, *Chem. Phys. Lipids*, 1970, **4**, 351.

Cleavage of the C—C bond next to the ether function yields intense ions. In the low eV spectra of the complete series of methyl epoxyoctadecanoates (31 isomers) the position of the epoxy-function along the chain was determinable,[9] but there is very little difference between the spectra of a pair of *cis–trans* isomers.

$$\text{Me(CH}_2)_n\overset{\cdot}{\xi}\underset{O}{\diagdown\diagup}\overset{\cdot}{\xi}(CH_2)_m CO_2 Me$$

(5)

A series of branched-chain C_{14}, C_{16}, and C_{18} fatty acids has been identified by mass spectrometry.[10] Fatty acid methyl and ethyl esters have been found as natural products in *Tetrahymena pyriformis* and characterized by mass spectrometry.[11] Wax esters from insect cuticular lipids have been shown to incorporate long-chain secondary alcohols, with the hydroxy-group near the middle of the chain.[12] The free alcohols were identified from their mass spectra after hydrolysis of the esters. Molecular ions were absent but $M^{+\cdot} - H_2O$ was a constant feature. Mass spectrometry and deuterium labelling has been used for the identification of synthetic wax esters.[13] The intensities of certain characteristic ions were selected for the quantitative analysis of wax mixtures. Simple[14] and complex[15] esters have been identified in the wax of the honeybee.

Free fatty alcohols extracted from the lipid fraction of *E. coli* have been analysed as their acetate esters (6) by mass spectrometry.[16] Two series of alcohols [primary and secondary, see (6)] were identified. In particular, cleavage between C-1 and C-2 of the alkyl moiety to yield m/e 73 or 87 identified the alcohol as belonging to the primary or secondary series respectively. The positions of the hydroxy- and keto-groups in long-chain hydroxyketones have been identified

$$\text{R}-\text{CH}_2\overset{73 \text{ or } 87}{\underset{\underset{X}{|}}{\overline{\xi}}}\text{CH}-\text{O}\overset{O}{\overset{\|}{\text{C}}}-\text{Me}$$

(6) X = H or Me

from their mass spectra.[17] The negative-ion mass spectra of a number of long-chain aldehydes and alcohols have been reported.[18] The main advantage of

[9] F. D. Gunstone and F. R. Jacobsberg, *Chem. Phys. Lipids*, 1972, **9**, 26.
[10] T. Kaneda, *Biochemistry*, 1971, **10**, 340.
[11] I. Ming Chu, M. A. Wheeler, and C. E. Holmlund, *Biochim. Biophys. Acta*, 1972, **270**, 18.
[12] G. J. Blomquist, C. L. Soliday, B. A. Byers, J. W. Brakke, and L. L. Jackson, *Lipids*, 1972, **7**, 356.
[13] A. J. Aasen, H. H. Hofstetter, B. T. R. Iyengar, and R. T. Holman, *Lipids*, 1971, **6**, 502.
[14] K. Stransky and M. Streibl, *Coll. Czech. Chem. Comm.*, 1971, **36**, 2267.
[15] K. Stransky, M. Streibl, and V. Kubelka, *Coll. Czech. Chem. Comm.*, 1971, **36**, 2281.
[16] W. F. Naccarato, R. A. Gelman, J. C. Kawalek, and J. R. Gilbertson, *Lipids*, 1972, **7**, 275.
[17] H. H. O. Schmid and P. C. Bandi, *J. Lipid Res.*, 1971, **12**, 198.
[18] P. C. Rankin, *Lipids*, 1970, **5**, 825.

this spectrum over the positive-ion spectrum in these cases lies in the identification of the molecular ion. Both aldehydes and alcohols yield a prominent $M^{\overline{\cdot}} - H\cdot$ ion. Mass spectrometry has been used to determine the position of branching in naturally occurring long-chain methyl-branched alkanes from the eggs of the tobacco hornworm.[19] Cleavage occurs preferentially adjacent to the secondary carbon atom.

The mass spectra of the C_{30}-isoprenoid hydrocarbons squalene, dihydrosqualene, and tetrahydrosqualene all exhibit abundant molecular ion peaks.[20] One notable difference between the spectra is that loss of the terminal C_5H_9 group from the molecular ion of squalene (bis-allylic cleavage) is replaced in tetrahydrosqualene (7) by loss of C_5H_{11}, thus suggesting that the terminal isoprenoid units are saturated in the latter. The identification of a polyisoprenol (C_{55}) by mass spectrometry has been reported.[21]

The mass spectra of triglycerides (8) have been extensively investigated both for different acyl groupings and for deuterium-labelled compounds.[22,23] Addition of deuterium to unsaturated acyl groups helped to locate the position of the double-bond by mass spectrometry.

Glycerophospholipids have been characterized by mass spectrometry as their TMS derivatives [see (9), where X is a naturally occurring residue such as serine,

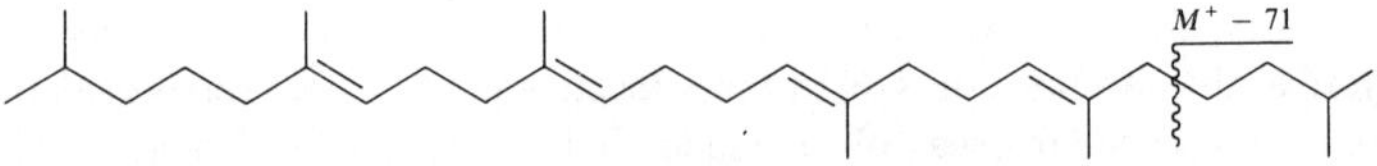

(7)

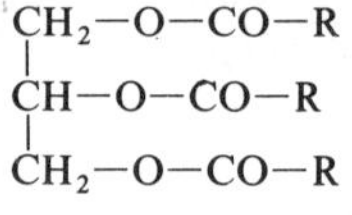

$CH_2-O-CO-R$
$CH-O-CO-R$
$CH_2-O-CO-R$

(8)

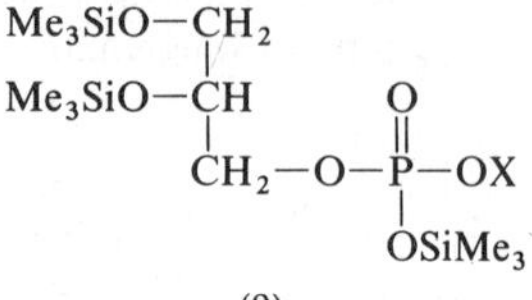

$Me_3SiO-CH_2$
$Me_3SiO-CH$
$CH_2-O-\overset{O}{\underset{OSiMe_3}{\overset{\|}{P}}}-OX$

(9)

inositol, or ethanolamine, attached at its hydroxy-group].[24] Molecular ions are found only in trace amounts, but characteristic fragment ions involving the TMS group serve to confirm the molecular weight. Hydrogen and TMS rearrangements are a feature of the spectra. The structurally informative direct cleavage reaction with charge-retention on the X-group is particularly important for the ethanolamine derivatives, becoming the most important process at low electron energies. Some phosphatidylcholines [(10), where R is a long-chain

[19] D. R. Nelson, D. R. Sukkestad, and R. G. Zaylskie, *J. Lipid Res.*, 1972, **13**, 413.
[20] J. K. G. Kramer, S. C. Kushwaha, and M. Kates, *Biochim. Biophys. Acta*, 1972, **270**, 103.
[21] F. A. Troy, F. E. Frerman, and E. C. Heath, *J. Biol. Chem.*, 1971, **246**, 118.
[22] W. M. Lauer, A. J. Aasen, G. Graff, and R. T. Holman, *Lipids*, 1970, **5**, 861.
[23] A. J. Aasen, W. M. Lauer, and R. T. Holman, *Lipids*, 1970, **5**, 869.
[24] J. H. Duncan, W. J. Lennarz, and C. C. Fenselau, *Biochemistry*, 1971, **10**, 927.

Natural Products

hydrocarbon grouping] have been characterized by their mass spectra.[25] Molecular ion peaks are either very weak or absent, although ions associated with the hydrocarbon chains, glycerol esters, and phosphorylcholine have been identified. The reproducibility of the spectra appears to refute the possibility of pyrolysis on the probe. Furthermore, observation of metastable transitions in the first drift region has shown that ions which might be supposed to arise by pyrolysis are formed by an electron-impact process.[26]

$$\begin{array}{l} CH_2OCOR^1 \\ | \\ CHOCOR^2 \\ | \\ CH_2OP(OH)OCH_2CH_2\overset{+}{N}Me_3 \\ | \\ O^- \end{array}$$

(10)

The mass spectra of tocopherols [see α-tocopherol (11)] have been reported.[27] Being aromatic, the compounds exhibit abundant molecular ions and yield very simple breakdown patterns. The principal fragmentation involves formation of an ion having a benzylic (or tropylium) ion structure, e.g. (12) in the spectrum of (11).

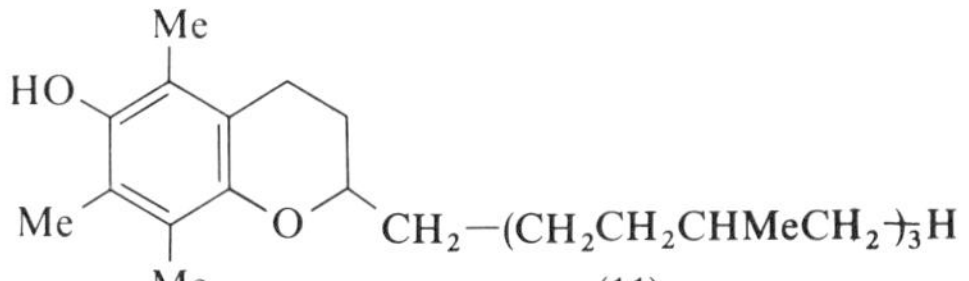

(11)

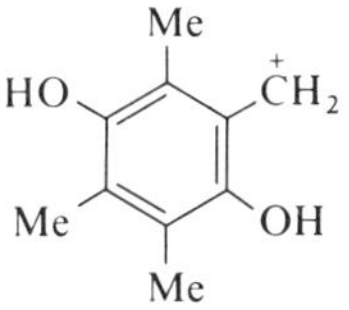

(12)

neo-Inositol has been isolated from brain and identified, as its TMS ether and as its acetate ester, by mass spectrometry.[28] An inositol phosphate has been identified by the mass spectrum of its TMS ether ester (13). A series of phosphoinositides has been analysed by mass spectrometry by first deacylating the extracted lipids, followed by full trimethylsilylation.[29] The TMS deacyl cardiolipin (14) from brain was identified by this method.[30]

$$(Me_3SiO)_5C_6H_6O\overset{O}{\underset{\|}{P}}(OSiMe_3)_2$$

(13)

[25] R. A. Klein, *J. Lipid Res.*, 1971, **12**, 123.
[26] R. A. Klein, *J. Lipid Res.*, 1971, **12**, 628.
[27] S. E. Scheppele. R. K. Mitchum, C. J. Rudolph, K. F. Kinneberg, and G. V. Odell, *Lipids*, 1972, **7**, 297.
[28] W. R. Sherman, S. L. Goodwin, and K. D. Gunnell, *Biochemistry*, 1971, **10**, 3491.
[29] T. J. Cicero and W. R. Sherman, *Biochem. Biophys. Res. Comm.*, 1971, **42**, 428.
[30] T. J. Cicero and W. R. Sherman, *Biochem. Biophys. Res. Comm.*, 1971, **43**, 451.

$$\underset{\underset{\text{OTMS}}{|}}{\text{CH}_2}\text{―}\underset{\underset{\text{OTMS}}{|}}{\text{CH}}\text{―CH}_2\text{O}\text{―}\overset{\overset{\text{O}}{\|}}{\underset{\underset{\text{OTMS}}{|}}{\text{P}}}\text{―O―CH}_2\text{―}\underset{\underset{\text{OTMS}}{|}}{\text{CH}}\text{―CH}_2\text{―O―}\overset{\overset{\text{O}}{\|}}{\underset{\underset{\text{OTMS}}{|}}{\text{P}}}\text{―O―CH}_2\text{―}\underset{\underset{\text{OTMS}}{|}}{\text{CH}}\text{―}\underset{\underset{\text{OTMS}}{|}}{\text{CH}_2}$$

(14)

Microgram quantities of complex glycosphingolipids have been fully trimethylsilylated and analysed by mass spectrometry. Information regarding all three constituents (*i.e.* sphingosine, fatty acid, and sugar moieties) of these derivatized glycolipids (15) is forthcoming from the mass spectra.[31] The number of monosaccharides in the oligosaccharide moiety was determinable from the relative intensities of several characteristic peaks. For example, the relative intensity of m/e 103 (16) which is derived from C-6 of the sugar was proportional to the number of monosaccharides present in the lipid. It was also possible to identify 1 → 3 glycosidic linkages, but distinction between 1 → 2 and 1 → 4 linkages was not possible. Cleavage of the long-chain base between the carbon atoms bearing the amide and TMS groupings yields the ion (17) which gives information on the structure of the base. Characteristic ions were also observed which helped to identify the acid moiety (normal or α-hydroxy fatty acid).

$$\text{R―}\underset{\underset{\text{OTMS}}{|}}{\text{CH}}\text{―}\underset{\underset{\underset{\underset{\underset{\text{R}^1}{|}}{\text{CO}}}{|}}{\underset{\text{NH}}{|}}}{\text{CH}}\text{―CH}_2\text{O―oligosaccharide TMS}$$

(15)

$$\text{CH}_2\!\!=\!\!\overset{+}{\text{O}}\text{SiMe}_3 \qquad \text{RCH}\!\!=\!\!\overset{+}{\text{O}}\text{SiMe}_3$$
(16)   (17)

The use of mass spectrometry in the structural microanalysis of glycosphingolipids is further illustrated by the identification of a xylose-containing cerebroside [see (18) for the fully trimethylsilylated derivative[32]]. The strong m/e 339 peak indicates that the long-chain base is the C_{20} homologue of sphingosine, although the position of unsaturation was not determinable from the mass spectrum. The fatty acid is identified as 2-hydroxytetracosanoic acid. The presence of a pentose residue was indicated not only from the TMS derivative (18) but also from the fully acetylated derivative. At present it is not possible to interpret the stereochemistry of glycolipids from mass spectra. However, it was determinable from the mass spectra of acetyl derivatives of oxidized and reduced cerebroside that the pentose residue is in the pyranose form. Xylose was identified in the above example by g.l.c. techniques. A new glycosphingolipid from human adenocarcinomas has been identified with the aid of mass spectrometry.[33]

[31] G. Dawson and C. C. Sweeley, *J. Lipid Res.*, 1971, **12**, 56.
[32] K.-A. Karlsson, B. E. Samuelsson, and G. O. Steen, *J. Lipid Res.*, 1972, **13**, 169.
[33] H.-J. Yang and S.-I. Hakomori, *J. Biol. Chem.*, 1971, **246**, 1192.

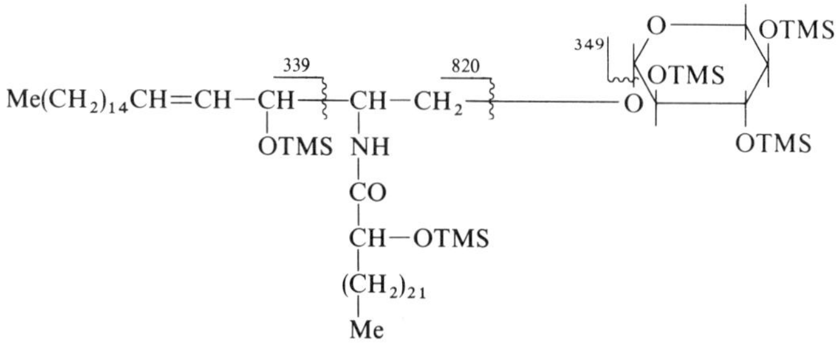

Me(CH$_2$)$_{14}$CH=CH—CH$\overset{339}{-\}-}$CH—CH$_2\overset{820}{-\}-}$——————$\overset{349}{\underset{O}{|}}\underset{}{\overset{O}{\diagup}}\underset{OTMS}{\overset{}{\diagdown}}$OTMS

OTMS NH   OTMS
          |
          CO
          |
          CH—OTMS
          |
          (CH$_2$)$_{21}$
          |
          Me

(18)

The *in vitro* conversion of a deuterium-labelled ceramide (along with an equal quantity of the corresponding protium compound) into cerebrosides has been studied and the products analysed by mass spectrometry.[34] The galactosyl and glucosyl ceramides formed were analysed as their TMS derivatives both as intact molecules and after degradation to ceramides. The position of the deuterium atoms in the incubation products (as determined by mass spectrometry) confirmed that the ceramide was transformed intact into products and not hydrolysed to acid and long-chain base prior to reaction.

The position of the double-bond in unsaturated long-chain lipids has been located by hydroxylation with OsO$_4$, followed by formation of the di-TMS ether.[35] This procedure creates a site for facile cleavage in the mass spectrometer.[36] The long-chain basic constituent (19) of sphingomyelin from the honey-bee was identified from the mass spectrum of the derivative (20).

Me(CH$_2$)$_8$CH=CHCH(OH)CH(NH$_2$)CH$_2$OH

(19)

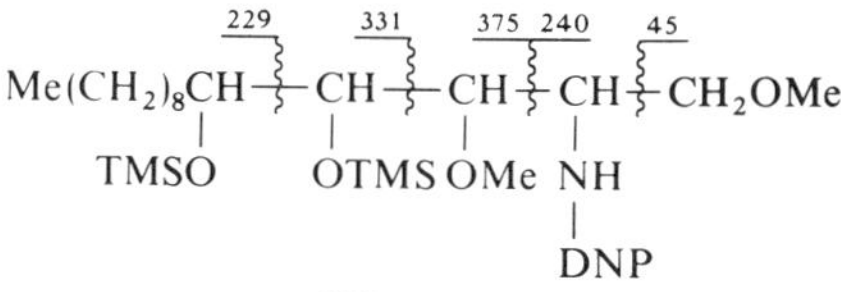

Me(CH$_2$)$_8$CH$\overset{229}{-\}-}$CH$\overset{331}{-\}-}$CH$\overset{375}{-\}-}$CH$\overset{240}{-\}-}$$\overset{45}{}CH_2$OMe
                    | | | |
                   TMSO   OTMS OMe NH
                                                              |
                                                              DNP

(20)

Isolations of steroid glycoside mixtures in the past have been hydrolysed and the steroid and sugar moieties analysed separately. Intact steryl glucosides have now been identified directly from biological sources by preparing TMS ethers of the glycolipid.[37]

Mass spectral evidence has helped to identify 1,25-dihydroxycholecalciferol (21), a new kidney hormone controlling calcium metabolism.[38] The molecular

[34] S. Hammarström and B. Samuelsson, *J. Biol. Chem.*, 1972, **247**, 1001.
[35] S.-G. Karlander, K.-A. Karlsson, H. Leffler, A. Lilja, B. E. Samuelsson, and G. O. Steen, *Biochim. Biophys. Acta*, 1972, **270**, 117.
[36] P. Capella and C. M. Zorzut, *Analyt. Chem.*, 1968, **40**, 1458.
[37] R. A. Laine and A. D. Elbein, *Biochemistry*, 1971, **10**, 2547.
[38] D. E. M. Lawson, D. R. Fraser, E. Kodicek, H. R. Morris, and D. H. Williams, *Nature*, 1971, **230**, 228.

ions of the unknown and its TMS ether established that two additional oxygen atoms had been inserted into the cholecalciferol molecule in the form of two hydroxy-groups. A hydroxy-group was located at C-25, from the intense peaks due to cleavage of the C-24—C-25 bond in (21) (to give m/e 59) and its TMS ether. Other mass spectral evidence located the third hydroxy-group in ring A and radiochemical evidence indicated that it is carried on C-1.

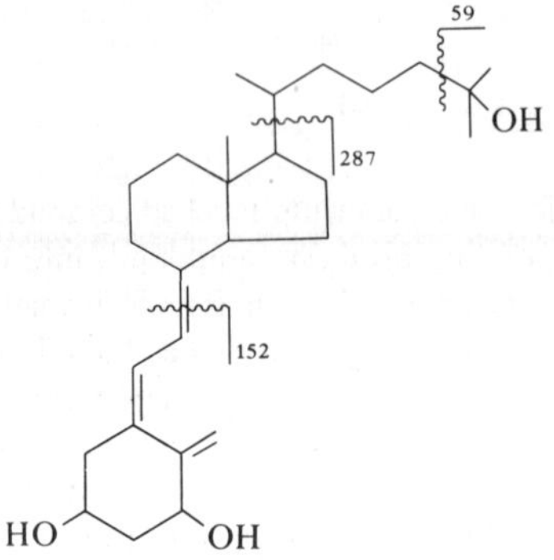

(21)

A novel derivative (22) of prostanoic acid has been isolated during the biosynthetic conversion of arachidonic acid into prostaglandins, and its structure has been elucidated with the aid of mass spectrometry of several derivatives and oxidation products.[39] The mass spectrum of the methyl ester showed a molecular ion at m/e 366 consistent with the molecular formula $C_{21}H_{34}O_5$. The molecular ion shifted to m/e 510 in the TMS ether methyl ester, indicating the presence of two hydroxy-groups. In addition, characteristic cleavage adjacent to the acyclic OTMS function gave information about the position of one of the hydroxy-groups. The mass spectrum of the hydrogenation product revealed the presence of two double bonds. Detailed analysis of the mass spectra of these three derivatives enabled structure (22) to be proposed, and the acetyl methyl ester (23) of the oxidative ozonolysis product helped to confirm the assignment. Other polyunsaturated fatty acids involved in prostaglandin biosynthesis have been identified with the aid of mass spectrometry.[39–41]

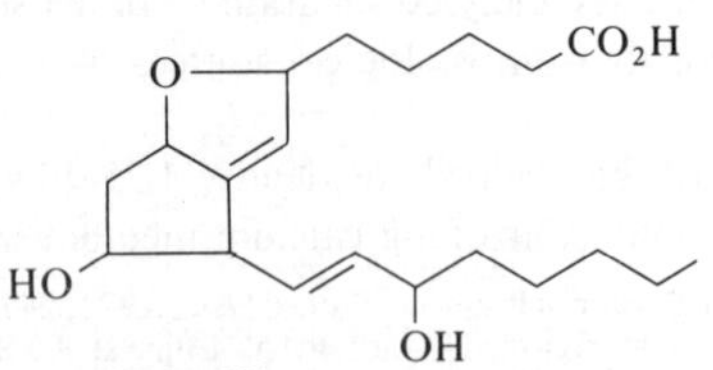

(22)

[39] C. Pace-Asciak and L. S. Wolfe, *Biochemistry*, 1971, **10**, 3657.
[40] C. Pace-Asciak, *Biochemistry*, 1971, **10**, 3664.
[41] C. Pace-Asciak and L. S. Wolfe, *Biochim. Biophys. Acta*, 1970, **218**, 539.

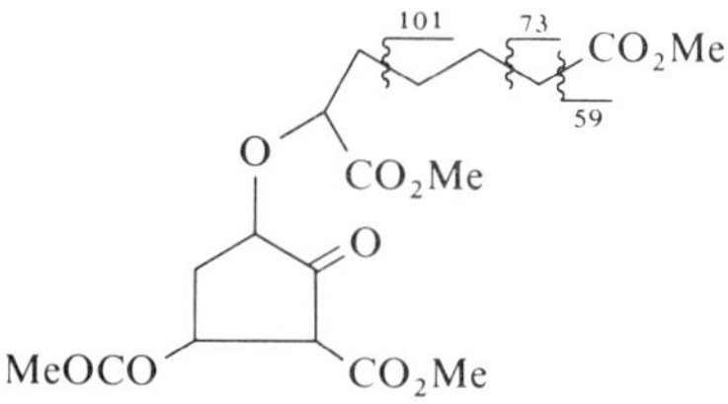

3 Amino-acids and Peptides

General.—The 2,4-dinitrophenyl derivatives of twenty amino-acids have been prepared and their mass spectra determined.[42] The spectra allowed characterization of each amino-acid, the most difficult one being arginine which decomposed. A combination of paper chromatographic separation and mass spectrometric identification is being used for the analysis of deuteriated amino-acids formed by abiotic synthesis from gas mixtures of D_2, ND_3, and CO. The spectra were determined on a time-of-flight mass spectrometer and the three isomeric leucines could be differentiated by their side-chain fragmentations.

A further study has been conducted on the fragmentations and mass spectral behaviour of 1-dimethylaminonaphthalene-5-sulphonyl (DNS) amino-acids[43] (see ref. 44 for earlier work). It was shown that 1 nmol of DNS amino-acid was sufficient to provide a good spectrum. The derivatives were examined between 180 and 280 °C using the direct inlet system, and all but bis-DNS cystine and bis-DNS histidine gave molecular ions whose intensities were strongly dependent upon source temperature. Fragmentations included the formation of dimethylaminonaphthalene and dimethyl sulphonyl cation, dansyl amide, and the usual β-fragmentation of amines. The side-chain is lost only from the aromatic amino-acid derivatives. Elimination of CO_2H by homolytic dissociation occurs to a small extent. Assessing the applicability to amino-acid analysis, it was felt that the method would be a powerful tool for identification, when coupled with t.l.c.

Experiments with C_3 and C_4 aliphatic amino-acids[45] have provided diagnostic criteria for protein and, particularly, non-protein amino-acids where normal amino-acid analysis is of little use. The compounds were studied as N-TFA n-butyl ester derivatives. The complexity of the spectra increased on going from α- through β- to γ-amino-acids. The β- and γ-amino-acids gave intense $M - 73$ signals (loss of butoxy-group), the resulting ions probably being stabilized by electron donation from nitrogen (24).

---

[42] M. H. Studier, L. P. Moore, R. Hayatsu, and S. Matsuoka, *Biochem. Biophys. Res. Comm.*, 1970, **40**, 894.
[43] N. Seiler, H. H. Schneider, and K.-D. Sonnenberg, *Analyt. Biochem.*, 1971, **44**, 451.
[44] G. Marino and V. Buonocore, *Biochem. J.*, 1968, **110**, 603.
[45] J. D. Lawless and M. S. Chadha, *Analyt. Biochem.*, 1971, **44**, 473.

$$R^3-\overset{H}{\underset{|}{C}}-C\equiv\overset{+}{O} \quad \leftrightarrow \quad R^3-\overset{H}{\underset{|}{C}}-\overset{O}{\overset{\parallel}{C}}$$
$$R^2-\underset{|}{\underset{H}{C}}-\underset{|}{\underset{R^1}{N}}-COCF_3 \qquad R^2-\underset{|}{\underset{H}{C}}-\underset{|}{\underset{R^1}{N^+}}-COCF_3$$

R^1, R^2, R^3 = H, Me
(24)

The chemical ionization spectra of 21 of the 23 naturally occurring amino-acids have been measured,[46] and all give quasi-molecular ions and simple fragmentation patterns which can be rationalized as arising from decomposition of the molecule protonated at several sites. Methane was used as the reactant gas, and it was noticed that rearrangement processes normally prevalent in EI spectra were generally absent from the CI counterparts. This could well be an advantage when studying unknown compounds. Major ions in the CI spectra were the quasi-molecular ion QM^+, $QM^+ - H_2O$, $QM^+ - NH_3$, and $QM^+ - COOH_2$ in contrast with the normal 'amine' fragments derived from EI spectra, $R-CH=NH_2^+$.

A combined g.c.–m.s. study of specifically labelled deuterium amino-acids[47] examined as their TMS derivatives has shown that the g.c. retention times were measurably shorter than their protium analogues, as has been observed for other classes of compound.[48–51]

The mass spectral properties of NN-bis(trifluoroacetyl)-L-cystine dimethyl ester and its lanthionyl analogue have been reported in detail.[52] The formation of a unique oxazoline ion, not observed in the acetyl or trifluoroacetyl spectra of other amino-acids, arises from cleavage α to the disulphide in the cystine derivative (25). The spectrum of the lanthionyl analogue is radically different. Here the major fragmentation is the common cleavage, β to the sulphide (26). Of most interest in this spectrum was the loss of elements of trifluoroacetamide to give m/e 315, 40% of the base peak (27).

$$\begin{array}{c}
O \quad\quad CO_2Me \\
\parallel \quad\quad\quad | \\
CF_3C-NH-CHCH_2 \\
| \\
S \\
| \\
S \\
| \\
CF_3C-NH-CHCH_2 \\
\parallel \quad\quad\quad | \\
O \quad\quad CO_2Me \\
(25)
\end{array}
\quad \rightarrow \quad
\begin{array}{c}
O \quad\quad CH_2 \\
\parallel \quad\;\; / \\
CF_3CNH-CH \\
\quad\quad\quad \backslash \\
\quad\quad\quad CO_2Me \\
m/e \; 198, \text{ base peak}
\end{array}
\quad \text{or} \quad
\begin{array}{c}
O \\
\parallel \\
CF_3 \diagdown N^+ \diagup CO_2Me \\
| \\
H
\end{array}$$

---

[46] G. W. A. Milne, T. Axenrod, and H. M. Fales, *J. Amer. Chem. Soc.*, 1970, **92**, 5170.
[47] W. J. A. Vandenheuvel, J. L. Smith, I. Putter, and J. S. Cohen, *J. Chromatog.*, 1970, **50**, 405.
[48] J. A. McCloskey, A. M. Lawson, and F. A. J. M. Leemans, *Chem. Comm.*, 1967, 285.
[49] R. Bentley, N. C. Saha, and C. C. Sweeley, *Analyt. Chem.*, 1965, **37**, 1118.
[50] C. C. Sweeley, W. H. Elliott, I. Fries, and R. Ryhage, *Analyt. Chem.*, 1966, **38**, 1549.
[51] G. R. Waller, S. D. Sastry, and K. Kinneberg, *J. Chromatog. Sci.*, 1969, **7**, 577.
[52] D. N. Harpp and J. G. Gleason, *J. Org. Chem.*, 1971, **36**, 73.

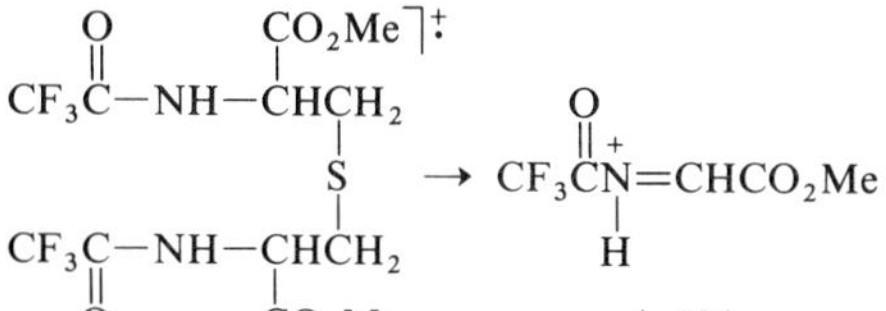

Natural Products

[Structure (26): CF₃C(O)—NH—CHCH₂—S / CF₃C(O)—NH—CHCH₂ with CO₂Me groups]⁺·

→ CF₃C(=N⁺H)=CHCO₂Me

m/e 184

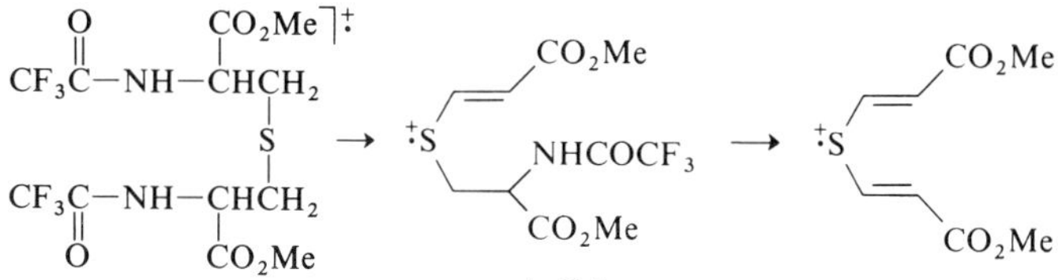

(26)

[Structure (26) → fragment *m/e* 315 (27) → fragment *m/e* 202]

The differences in spectra can be rationalized by consideration of the processes depicted (28a—c), together with the fact that sulphides are better electron donors than disulphides.

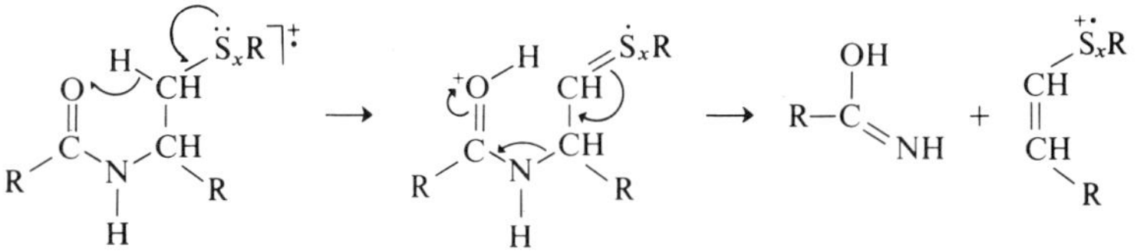

(28a)

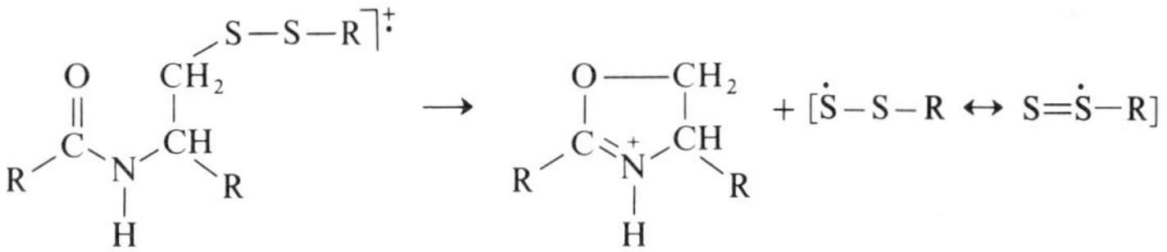

(28b) [with + [Ṡ—S—R ↔ S=Ṡ—R]]

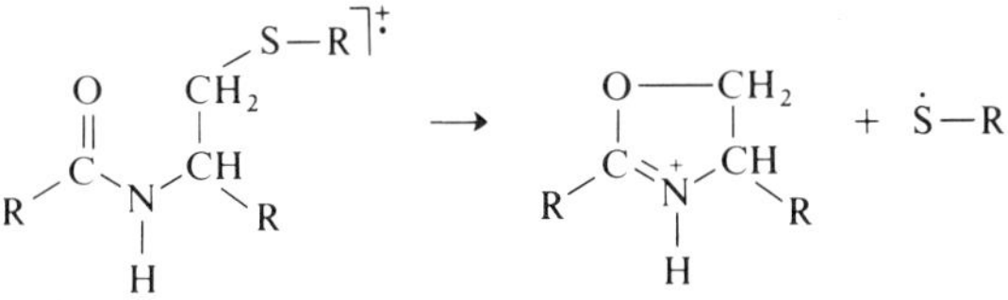

(28c) [with + Ṡ—R]

Mass spectrometry has played a part in the structural elucidation or confirmation of several new amino-acids and related compounds. Amongst these have been Indospicine,[53] a hepatonic amino-acid from *Indigofera Spicata* identified as L-2-amino-6-amidinohexanoic acid (29), and Carzinophilin II,[54] from an antitumour antibiotic. The latter amino-acid has been identified on the basis of mass spectrometry and n.m.r. as *dl-erythro*-4-amino-2,3-dihydroxy-3-methyl butanoic acid (30).

$$H_2N-\overset{\overset{NH}{\|}}{C}-(CH_2)_4CH(NH_2)CO_2H$$
(29)

$$H_2NCH_2\underset{\underset{OH}{|}}{C}(Me)-\underset{\underset{OH}{|}}{C}HCO_2H$$
(30)

The mass spectra of some amino-acids have been studied using a special application of field ionization: field-desorption mass spectrometry.[55] This ionization technique is particularly useful for compounds of low volatility, or which decompose extensively under electron impact. A molecular ion M^{+} or quasi-molecular ion $M + 1^{+}$ was present for each of the α-amino-acids studied, and both fragmentation and thermal decomposition were considerably reduced in comparison to the respective EI or CI spectra.

The structure of several synthetic substituted quinones containing α-amino-acid ester substituents has been established by mass spectrometry and n.m.r.[56] These compounds are believed to contribute to the stability of soil organic nitrogen. The amino-acid *erythro*-β-hydroxy-L-asparagine has been isolated from unripe seeds of *Astragalus sinius L*, and characterized as the diethyl ester by the well-documented amine and ester fragments.[57]

A dipeptide which probably plays an important role in cell biochemistry has been isolated as the S-carboxymethyl derivative from wheatgerm, and its structure identified by mass spectrometry and amino-acid analysis.[58] The trifluoroacetyl methyl ester gave a molecular ion at m/e 360, and the structure was deduced as cysteinylglycine, though the normal peptide bond cleavage did not occur. The base peak in the spectrum was at m/e 90 due to $\cdot S-CH_2-C(O)-O^+$.

Mass spectrometry has been used to study the racemization mechanism of two *N*-benzoxycarbonylamino-acid active esters in non-polar solvents, in the presence of triethylamine.[59] The proposed mechanism of racemization by α-hydrogen abstraction was investigated in a deuterium exchange experiment;

[53] M. P. Hegarty and A. W. Pound, *Austral. J. Biol. Sci.*, 1970, **23**, 831.
[54] M. Onda, Y. Konda, S. Omura, and T. Hata, *Chem. and Pharm. Bull. (Japan)*, 1971, **19**, 2013.
[55] H. U. Winkler and H. D. Beckey, *Org. Mass Spectrometry*, 1972, **6**, 655.
[56] P. A. Cranwell and R. D. Haworth, *Tetrahedron*, 1971, **27**, 1831.
[57] H. Inatomi, F. Inukai, and T. Murakami, *Chem. and Pharm. Bull. (Japan)*, 1971, **19**, 216.
[58] R. Tkachuk, *Canad. J. Biochem.*, 1970, **48**, 1029.
[59] J. Kovacs, H. Cortegiano, R. E. Cover, and G. L. Mayers, *J. Amer. Chem. Soc.*, 1971, **93**, 1541.

the deuteriated active ester content of the isolated active ester was determined mass spectrometrically. The results of a kinetic study supported the hypothesis that racemization of N-benzoxycarbonyl-S-benzyl cysteine occurs *via* isoracemization, a process in which racemization occurs without exchange of the proton.

A differentiation between *meso* and racemic diaminopimelic acid (DAP) has been made on the basis of differing fragmentation patterns following electron impact.[60] The mass spectrum could not be interpreted by the usual 'amine type' fragmentation mechanism, but thermal cyclization appeared to have taken place in the ion source, with losses of either one or two molecules of water. Consideration of the structures of *meso* and racemic shows that the racemate is in a better conformation to form a second internal amide linkage than is the *meso* form, and can then lose predominantly a second molecule of water to give m/e 154 (31). The authors are investigating to what extent mass spectrometry can be used for the similar but somewhat more subtle problem of distinguishing dipeptide diastereomers.

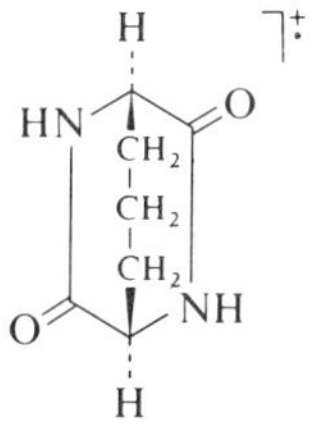

m/e 154

(31)

Peptides isolated from micro-organisms frequently contain residues of D-amino-acids. Several previous studies have shown that these are derived from L-precursors. Two mechanisms may be envisaged (a) derivatization at the carboxy-group to form an activated compound, followed by racemization or inversion to form the D-amino-acid which is then used in the synthesis, (b) L-amino-acid incorporated in the synthesis followed by inversion to the D-form. Mass spectrometry has played an important role[61] in a study which has provided evidence for the latter theory. The subject of the study was angolide (32), a cyclotetra-depsipeptide isolated from the fungus *Pithomyces sacchari* which was grown on a medium containing L-valine. The various products were separated and examined

cyclo-L-Hiv-L-Ile-L-Hiv-D-aIle-

(32)   Hiv = α-hydroxyisovaleric acid

[60] H. Falter, M. Madaiah, and R. A. Day, *Tetrahedron Letters*, 1970, 4463.
[61] R. O. Okotore and D. W. Russell, *Experientia*, 1971, **27**, 380.

by mass spectrometry (see refs. 62 and 63 for previous m.s. studies on angolide). A possible mechanism for the biosynthesis of the peptide could be deduced from the results.

Mass spectrometry has been used to confirm the structure of the insecticidal cyclodepsipeptides Destruxins A and B[64] produced by *metarrhizium anisopliae* (see ref. 65 for a description of the mass spectrometry of cyclodepsipeptides). Ring-opening occurs at more than one site, making the spectra difficult to interpret. The lactone linkage was therefore cleaved to afford the open-chain structure which was then acetylated and methylated. The spectra could be rationalized in terms of the structures already established.

In a similar study, the structures of Staphylomycin S components (33) have been confirmed by mass spectrometry.[66] The mass spectrometry of Staphylomycin S has been previously discussed,[67,68] but the present authors have disputed

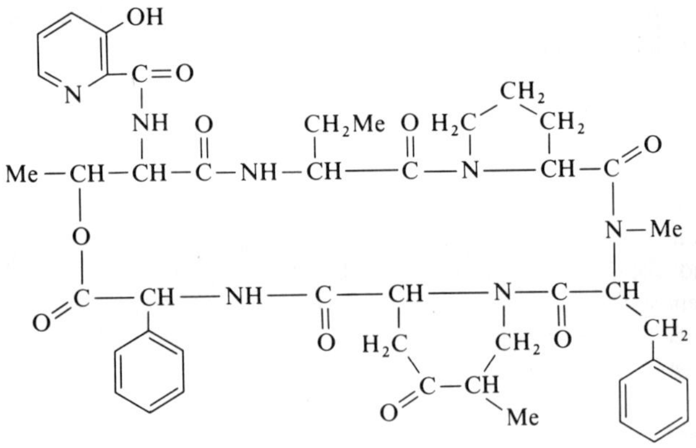

Staphylomycin S
(33)

the fragmentation mechanisms postulated in the original study. These were loss of CO_2 and H from the lactone group with formation of a linear peptide ion from which amino-acid residues were eliminated in a consecutive manner. This CO_2 type fragmentation predominates in cyclodepsipeptides with several

[62] N. S. Wulfson, V. A. Puchkov, V. N. Bochkarev, B. V. Rozynov, A. M. Zyakoon, M. M. Shemyakin, Yu. A. Ovchinnikov, V. T. Ivanov, A. A. Kiryushkin, E. I. Vinogradova, M. Yu. Feigina, and N. A. Aldanova, *Tetrahedron Letters*, 1964, 951.
[63] C. G. Macdonald and J. S. Shannon, *Tetrahedron Letters*, 1964, 3113.
[64] A. Suzuki, N. Takahashi, and S. Tamura, *Org. Mass Spectrometry*, 1970, **4**, 175.
[65] D. Russell, *Quart. Rev.*, 1966, **20**, 559.
[66] F. Compernolle, H. Vanderhaeghe, and G. Janssen, *Org. Mass Spectrometry*, 1972, **6**, 151.
[67] A. A. Kiryushkin, V. N. Burikov, and B. V. Rosinov, *Tetrahedron Letters*, 1967, 2675.
[68] B. V. Rozynov, V. N. Burikov, I. A. Bogdanova, and A. A. Kiryushkin, *Zhur. obshchei Khim.*, 1969, **39**, 891.

hydroxyamino-acids. However, in the spectrum obtained by the authors the corresponding losses could not be detected, but low intensity ions were found 1 mass unit higher. They found another fragmentation predominating, namely elimination of an azomethine molecule, $R-CH=N-R^1$. A particularly predominant degradation pathway was shown to start with elimination of a 1,2-dehydro-4-oxopiperidine to form (34). This was followed by loss of NHCO and a concerted loss of CO_2 and a benzyl radical involving hydrogen rearrangement. Further losses of amino-acid residues could then be characterized. As

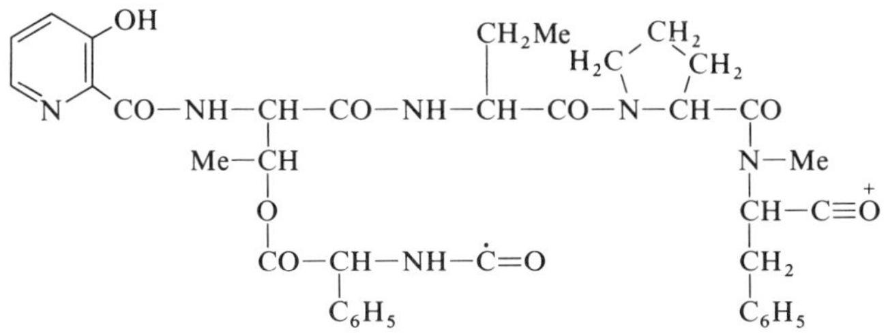

$$
\begin{array}{c}
\text{(structure 34)}
\end{array}
$$

mentioned earlier, a simpler approach to structural analysis with these compounds involves the formation of a linear molecule by opening of the lactone ring, and this was also carried out in the above study.

Azulene chromophoric substrates have proved to be particularly useful[69] for a precise identification of the chromophore-containing cleavage products which result from an enzymatic hydrolysis: they permit a rapid and unequivocal localization of enzymatic attack. If the substances are volatile, or are made so by esterification, the spectra of N-[(7-isopropyl-1-methylazulen-4-yl)acetyl] amino-acid or peptide derivatives show surprisingly intense molecular ions, and little fragmentation in the upper mass region. A precise identification of the N-terminal cleavage products is thus possible.

Peptide alkaloids are peptides with two or more amide groups of plant origin. They have so far been found in fifteen plants from seven different species, and their structures are normally determined by n.m.r., u.v., i.r., and changes following hydrogenation. More recently high-resolution mass spectrometry has proved useful,[70] and the structures of twenty peptide alkaloids are described.

Mass spectrometry can be used for amino-acid sequence determination in β-lysine containing peptides, and for the identification of α- and β-lysine residues in peptides, as well as for amide bond type determination in β-lysine peptides.[71] A distinguishing feature of β-lysine-containing peptides, at least in the spectra of N-acyl methyl esters, is the cyclic ion (35) at m/e 70 as distinct from the corresponding α-lysine fragment (36) at m/e 84. A $C_\beta-C_\gamma$ cleavage was used to

---

[69] E. Wünsch and E. Jaeger, *Z. Physiol. Chem.*, 1971, **352**, 1584.
[70] E. W. Warnhoff, *Fortschr. Chem. org. Naturstoffe*, 1970, **28**, 162.
[71] L. I. Rostovtseva and A. A. Kiryushkin, *Org. Mass Spectrometry*, 1972, **6**, 1.

diagnose ε-amino participation in peptide bond formation in C-terminal β-lysine peptides, although this could not always be deduced easily due to interfering fragmentations from the N-decanoyl protecting group.

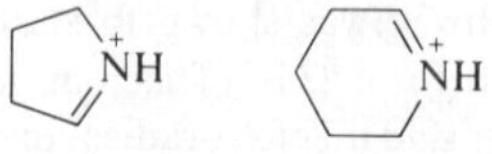

(35)   (36)

Peptides and glycopeptides isolated by mild hydrolysis of mycoside C_2 from *mycobacterium avium* have been characterized by Edman degradation and mass spectrometry.[72] In the proposed structure (37) the link between the alaninol and 3,4-di-O-methyl rhamnose is most likely α-glycosidic.

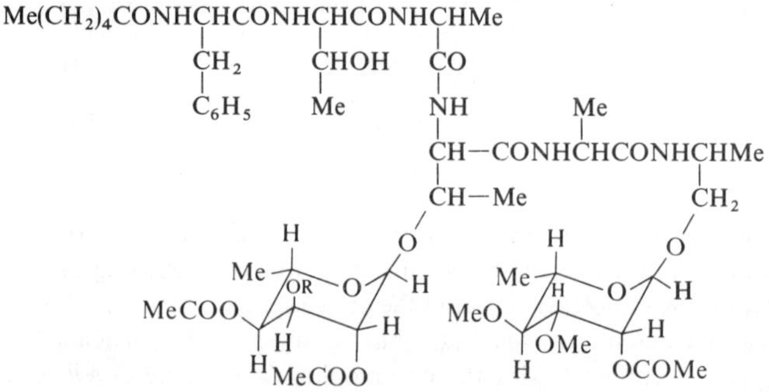

$$Me(CH_2)_4CONHCHCONHCHCONHCHMe$$

(37) R = H or Me

The mass spectrometry of acetylated O-glycosyl derivatives of serine and threonine methyl esters have been studied,[73] and the spectra show characteristic features which allow identification of the amino-acid and sugar moieties and the type of linkage between them. Some information on the sequence of sugar units and the position of the glycosidic bond in the carbohydrate moiety of disaccharide O-glycosyl derivatives of serine methyl ester is also provided by mass spectrometry. We may anticipate a total structural analysis of suitably derivatized, probably permethylated, glycopeptides by mass spectrometry in the near future.

Further progress has been made in determining the nature of the major cross-linking moieties in collagens from mineralized tissues. It has now been reported that the major reducible cross-link is dihydroxylysinonorleucine,[74] and its

---

[72] A. Voiland, M. Bruneteau, and G. Michel, *European J. Biochem.*, 1971, **21**, 285.
[73] O. S. Chizhov, V. A. Derevitskaya, B. M. Zolotarev, L. M. Likhosherstov, O. S. Novikova, and N. K. Kochetkov, *Carbohydrate Res.*, 1971, **20**, 275.
[74] G. Mechanic, P. M. Gallop, and M. L. Tanzer, *Biochem. Biophys. Res. Comm.*, 1971, **45**, 644.

structure has been confirmed by mass spectrometric examination of the trifluoroacetyl methyl ester (38). It was concluded that a previous mass spectrometric study[75] was in error in interpreting the spectrum as arising from an aldol condensation product. The spectrum showed no molecular ion, but the predominant loss of one or two molecules of trifluoroacetic acid. A characteristic ion formed by loss of CO_2Me was the highest mass ion in the spectrum. Further confirmation of the proposed structure was derived from the fragmentation pattern of the isobutoxycarbonyl permethylated derivative.

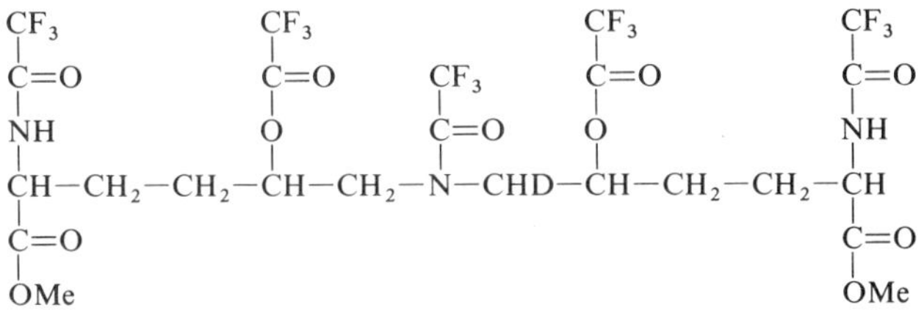

$$\begin{array}{cccc}
CF_3 & CF_3 & CF_3 & CF_3 \\
| & | & | & | \\
C=O & C=O & CF_3 & C=O & C=O \\
| & | & | & | & | \\
NH & O & C=O & O & NH \\
| & | & | & | & | \\
CH-CH_2-CH_2-CH-CH_2-N-CHD-CH-CH_2-CH_2-CH \\
| & & & & | \\
C=O & & & & C=O \\
| & & & & | \\
OMe & & & & OMe
\end{array}$$

The deuterium atom results from reduction of the Schiff-base precursor with $NaBD_4$

(38)

Peptide Sequencing.—During the past two years there has been a continued interest in protein sequencing by mass spectrometry, despite the previous lack of general application of the method to genuine structural problems. Critics of the technique have been able, justifiably, to point to the poor sensitivity of mass spectrometry (despite many claims to the contrary[76]) compared with classical sequencing techniques, and the inability to handle certain naturally occurring amino-acids effectively.

Much progress has now been made in both these areas, and those studies which have contributed or may contribute to the development of the method are discussed in this section.

Derivatization. Derivative formation, and the correct choice of derivatives, is an important part of the work since the peptide volatility and its fragmentation characteristics depend upon the derivative chosen. The most widely used and successful derivative has been the *N*-acetyl permethylated peptide, the permethylation conditions used being those described by Vilkas and Lederer,[77] and modified by Thomas.[78] The Russian group has been a notable exception to this rule, and they have continued to use *N*-decanoyl peptide methyl esters, avoiding permethylation. It has been clear for some time now that the decanoyl

---

[75] A. J. Bailey, L. J. Fowler, and C. M. Peach, *Biochem. Biophys. Res. Comm.*, 1969, **35**, 663.
[76] See, for example, A. L. Burlingame, and G. A. Johanson, *Analyt. Chem. Rev.*, 1971, **44**, 345R. 'Much less time and sample is needed for analysis with this technique rather than conventional methods'.
[77] E. Vilkas and E. Lederer, *Tetrahedron Letters*, 1968, 3089.
[78] D. W. Thomas, *Biochem. Biophys. Res. Comm.*, 1968, **33**, 483.

acyl group has no significant advantages, and some obvious disadvantages over the simple acetyl group; avoidance of the permethylation reaction reduces the size of peptide able to be studied, and contributes to the complexity of the spectra produced.

Despite the success of the permethyl derivative, until recently it has not been applicable to several of the common amino-acids, including histidine, cysteine or its derivatives, and methionine. Arginine has also remained a problem, since the pyrimidyl ornithine derivative[79] is not amenable to permethylation because of salt formation.[80]

A solution to the problem with the sulphur-containing amino-acids, namely desulphurization, has been further developed. A milder Raney Nickel treatment than previously suggested has been proposed,[81] although somewhat unrealistic quantities of peptide (5—10 mg) were used to demonstrate the method. An alternative catalyst, of approximate composition Ni_2B, has been suggested for desulphurization,[82] and this is claimed to be more efficient and applicable to smaller quantities of material than the Raney Nickel method. The products formed from cysteine and methionine are alanine and α-aminobutyric acid respectively, and the use of D_2O in catalyst preparation will remove the ambiguity in the alanine case by a shift of one mass unit if the precursor was cysteine. Quaternary salt formation on methionine has been avoided by its temporary conversion into the sulphoxide, followed by permethylation under normal conditions, and reduction back to N-methyl methionine (39).[83] Several peptide

$$-NH-CH-CO- \xrightarrow[HOAc]{H_2O_2} -NH-CH-CO- \xrightarrow[DMF]{MeI, NaH}$$
$$\quad\quad\ \ |\quad\quad\quad\quad\quad\quad\quad\quad\quad\ \ |$$
$$\quad\quad\ \ CH_2\quad\quad\quad\quad\quad\quad\quad\ \ CH_2$$
$$\quad\quad\ \ |\quad\quad\quad\quad\quad\quad\quad\quad\quad\ \ |$$
$$\quad\quad\ \ CH_2\quad\quad\quad\quad\quad\quad\quad\ \ CH_2$$
$$\quad\quad\ \ |\quad\quad\quad\quad\quad\quad\quad\quad\quad\ \ |$$
$$\quad\quad\ \ S\quad\quad\quad\quad\quad\quad\quad\quad\ \ O=S$$
$$\quad\quad\ \ |\quad\quad\quad\quad\quad\quad\quad\quad\quad\ \ |$$
$$\quad\quad\ \ Me\quad\quad\quad\quad\quad\quad\quad\quad Me$$

$$\quad\quad\quad\quad\quad Me\quad\quad\quad\quad\quad\quad\quad\quad Me$$
$$\quad\quad\quad\quad\quad\ |\quad\quad\quad\ \ i, KI\quad\quad\quad\ |$$
$$\quad\quad -N-CH-CO- \xrightarrow[ii, Na_2S_2O_3]{MeCOCl} -N-CH-CO-$$
$$\quad\quad\quad\quad\quad\quad\ |\quad\quad\quad\quad\quad\quad\quad\quad\quad |$$
$$\quad\quad\quad\quad\quad\ CH_2\quad\quad\quad\quad\quad\quad\quad CH_2$$
$$\quad\quad\quad\quad\quad\quad\ |\quad\quad\quad\quad\quad\quad\quad\quad\quad |$$
$$\quad\quad\quad\quad\quad\ CH_2\quad\quad\quad\quad\quad\quad\quad CH_2$$
$$\quad\quad\quad\quad\quad\quad\ |\quad\quad\quad\quad\quad\quad\quad\quad\quad |$$
$$\quad\quad\quad\quad\quad O=S\quad\quad\quad\quad\quad\quad\quad\quad S$$
$$\quad\quad\quad\quad\quad\quad\ |\quad\quad\quad\quad\quad\quad\quad\quad\quad |$$
$$\quad\quad\quad\quad\quad\ Me\quad\quad\quad\quad\quad\quad\quad\quad Me$$
$$\quad\quad\quad\quad\quad\ (39)$$

[79] J. Lenard and P. M. Gallop, *Analyt. Biochem.*, 1969, **29**, 203.
[80] D. W. Thomas, B. C. Das, S. D. Géro, and E. Lederer, *Biochem. Biophys. Res. Comm.*, 1968, **32**, 199.
[81] R. Toubiana, J. E. G. Barnett, E. Sach, B. C. Das, and E. Lederer, *F.E.B.S. Letters*, 1970, **8**, 207.
[82] M. A. Paz, A. Bernath, E. Henson, O. O. Blumenfeld, and P. M. Gallop, *Analyt. Biochem.*, 1970, **36**, 527.
[83] P. Roepstorff, K. Norris, S. Severinsen, and K. Brunfeldt, *F.E.B.S. Letters*, 1970, **9**, 235.

Natural Products 161

examples were given, and all gave interpretable spectra following this procedure. An approach which seemed sensible for dealing with the cysteine problem, namely deliberate sulphonium salt formation, followed by an attempted elimination to produce dehydroalanine, was not successful.[84] However, the same authors were able to produce interpretable spectra by balancing the amount of base used in the deprotonation step of the permethylation reaction, with the amount of methyl iodide added.[84]

The basic principle of 'equimolar' reaction employed here has since been used by other groups in an effort to prevent salt formation with the problematical amino-acids. The reaction has been applied at the 100 nmol level to a series of synthetic methionine-,[85] histidine-,[86] and arginine-[87] containing peptides. The methionine peptides studied included Met-Gly-Met-Met and Thr-Met. Sequence analysis was aided by a computer program[88] modified to accept low-resolution data. The peptides (50 μg) were acetylated and treated with base (measured amount prepared under N_2) for 15 min followed by a measured amount of MeI for 1 h. The mass spectra were run on 20 μg of sample. The spectra produced would have been difficult to determine as unknowns and the Thr-Met sequence was confirmed by high-resolution mass spectrometry. For the arginine peptides, conversion into N-2-(4,6-dimethyl)pyrimidyl ornithine[89] as previously suggested, was recommended. The modification is accomplished by condensation of the guanidine with acetyl acetone (40). The amino-acid and three synthetic peptides

$$-NH-C\begin{matrix}\nearrow NH \\ \searrow NH_2\end{matrix} \xrightarrow{MeCOCH_2COMe} -NH-C\begin{matrix}\nearrow N=C\diagdown Me \\ \searrow N-C \diagup Me\end{matrix}$$

(40)

were studied; Arg-Arg, Ser-Arg-His-Pro, and the nonapeptide bradykinin Arg-Pro-Pro-Gly-Phe-Ser-Pro-Phe-Arg. The spectrum of bradykinin was complex, but ions corresponding to the sequence of the first seven residues could be assigned in the spectrum. Given the fact that many of the published spectra would have been difficult to interpret without a prior knowledge of the peptide sequences, it may be somewhat premature to share the enthusiasm of the authors for this technique. One obvious difficulty is the problem of measuring and

[84] M. L. Polan, W. J. McMurray, S. R. Lipsky, and S. Lande, *Biochem. Biophys. Res. Comm.*, 1970, **38**, 1127.
[85] P. A. Leclercq and D. M. Desiderio, *Biochem. Biophys. Res. Comm.*, 1971, **45**, 308.
[86] P. A. White and D. M. Desiderio, *Analyt. Letters*, 1971, **4**, 141.
[87] P. A. Leclercq, L. C. Smith, and D. M. Desiderio, *Biochem. Biophys. Res. Comm.*, 1971, **45**, 937.
[88] K. Biemann, C. Cone, B. R. Webster, and G. P. Arsenault, *J. Amer. Chem. Soc.*, 1966, **88**, 5598.
[89] H. Vetter-Diechtl, W. Vetter, W. Richter, and K. Biemann, *Experientia*, 1968, **24**, 340.

handling extremely small amounts of reagent, particularly the volatile methyl iodide.

Another example of the use of the 'equimolar' principle has been on the tetrapeptide Trp-Met-Asp-Phe.[90] In this study sodium hydride–DMF–methyl iodide was used to permethylate the N-acetyl peptide methyl ester. Quaternization of the methionine was prevented, and the sequence could be established. 0.06 mmol (36 mg) of peptide was derivatized, considerably more than is normally available to the biochemist.

A different approach to the derivatization of the problematical amino-acids has been made,[91] following a study of the rate of permethylation using the Hakamori procedure. An excess of methyl iodide and the recommended reaction time of 1 h were used, but initially deuterium-labelled methyl iodide was added for a short time, followed by ordinary methyl iodide, making up the total time of 1 h. The results showed that with a CD_3I time of 60 s (followed by a further 59 min of CD_3I–CH_3I) the standard peptides used were virtually completely labelled only with CD_3, as determined by mass spectrometry. This means that the previously accepted 'optimum reaction time' is in fact well in excess of that needed for essentially quantitative methylation of the peptide molecule. Using the shortened reaction time at the methylation stage, it was shown that quaternization on histidine is appreciably slower than the rate of methylation at other sites on the peptide, and several protein-derived histidine-containing peptides were sequenced as 'unknowns' at the 100 nmol level.[91] The success of this approach has since been extended to methionine- and arginine-containing peptides isolated from proteins,[92] and preliminary results with carboxymethylcysteine have been encouraging.

Structural Elucidation. Mass spectrometry has continued to play a useful role in the confirmation of the structures of several peptides sequenced by classical methods. The fragmentation pattern of porcine thyrotropin releasing hormone[93] supported the structure obtained by enzymic and chemical degradation as pyroglutaminyl-histidyl-proline amide. A peptide extracted from the mycelium of a *Cephalosporium Sp2*, and whose structure had been determined classically as δ-(α-aminoadipyl)cysteinyl-valine has been examined in the mass spectrometer as the N-ethoxycarbonyl methyl ester.[94] The fragmentation pattern confirmed the structure shown in (41). A peak at m/e 230 was indicative of the δ-linkage. (An aldimine type fragment would have been expected for the normal α-linkage.)

---

[90] G. Marino, L. Valente, R. A. W. Johnstone, F. Mohammedi-Tabrizi, and G. C. Sodini, *J.C.S. Chem. Comm.*, 1972, 357.
[91] H. R. Morris, *F.E.B.S. Letters*, 1972, **22**, 257.
[92] (a) H. R. Morris, paper presented to the 20th Annual Conference on Mass Spectrometry and Allied Topics, Dallas, June 1972, in press; (b) H. R. Morris, R. J. Dickinson, and D. H. Williams, paper submitted to *Biochem. Biophys. Res. Comm.*
[93] R. M. G. Nair, J. F. Barrett, C. Y. Bowers, and A. V. Schally, *Biochemistry*, 1970, **9**, 1103.
[94] R. Bronwen Loder and E. P. Abraham, *Biochem. J.*, 1971, **123**, 471.

An unusual strategy has been used in the structural elucidation of a biologically important peptide extracted from rat brain, scotophobin.[95] The peptide, fifteen

$$\underset{(41)}{\text{EtO}-\overset{\|}{\underset{\text{O}}{\text{C}}}\text{\textonehalf NH\textonehalf CHCH}_2\text{CH}_2\text{CH}_2\overset{230}{\overset{\|}{\underset{\text{O}}{\text{C}}}}\text{\textonehalf NHCH\textonehalf }\overset{\|}{\underset{\text{O}}{\text{C}}}\text{\textonehalf NHCH\textonehalf CO}_2\text{Me}}$$

(schematic structure 41 with side chains: CO₂Me; CH₂–S–C(=O)–OEt; CH(Me)(Me))

residues long, was split in two and each section was pyrolysed in the mass spectrometer ion source and the resulting fragments characterized by high-resolution mass spectrometry. From the complex data obtained, together with a partial sequence derived chemically, several possible sequences were computed. These sequences were then synthesized, and a biologically active one was found, and postulated to have the same structure as scotophobin. Few details of the high-resolution spectra of the fragments were published, but these must have been very complex, giving rise to many ambiguities. The strategy used here cannot be recommended as a general procedure since with the choice of a more suitable derivative, a unique sequence may have been obtained using low-resolution mass spectrometry.

There has been further progress, along several lines, towards the ultimate goal of determining a protein sequence by mass spectrometry. Using low-resolution mass spectrometry and manual interpretation, a number of protein-derived peptides have been sequenced as complete 'unknowns', that is without a knowledge of their amino-acid compositions or lengths.[96] Working down to the 100 nmol level on N-acetyl permethylated derivatives, 42 of the 46 residues present in the peptides were unambiguously placed in sequence, showing that, contrary to some predictions,[97] low-resolution mass spectrometry and manual interpretation without an amino-acid analysis is a feasible way to sequence peptides isolated from proteins. The remaining four residues were not incorrect assignments, but were missed owing to the absence of molecular ions from some spectra, in turn due to the small quantities available.

In an entirely different strategy, the chemistry of the classical Edman degradation has been combined with a mass spectrometric determination of the resulting thiohydantoins.[98] A modification of the procedure makes it applicable to

[95] D. M. Desiderio, G. Ungar, and P. A. White, *Chem. Comm.*, 1971, 432.
[96] H. R. Morris, D. H. Williams, and R. P. Ambler, *Biochem. J.*, 1971, **125**, 189.
[97] Amongst others see, for example, 'Prior to mass spectrometry the amino-acid analysis must be quantitatively determined, . . .', M. M. Shemyakin, Y. A. Ovchinnikov, E. I. Vinogradova, A. A. Kiryushkin, M. Y. Feigina, N. A. Aldonova, Y. B. Alakhov, V. M. Lipkin, A. I. Miroshnikov, B. V. Rosinov, and S. A. Kazaryan, *F.E.B.S. Letters*, 1970, **7**, 8.
[98] T. Fairwell, W. T. Barnes, F. F. Richards, and R. E. Lovins, *Biochemistry*, 1970, **9**, 2260.

protein or peptide mixtures, and the technique is as follows. A standard mixture of ^{15}N-enriched amino-acids or their hydantoins is added to the sample prior to the Edman degradation. The N-terminal amino-acid(s) released is identified quantitatively by monitoring changes in the ^{15}N : ^{14}N ratios of several relevant fragment ions in the spectrum produced. In this way ten or more amino-acids from the N-terminus may be placed in sequence. Limitations are apparent when very minor components are present, or when two components are present in the same amounts. In the latter case ambiguities arise at each step of the degradation, leading to many possible sequences.

Another approach to this same strategy has been the use of *p*-bromophenyl-isothiocyanate for the initial attack on the N-terminal free amino-group in the Edman degradation.[99] It is claimed that the resulting bromophenylthiohydantoins (BrPTH) are more easily differentiated from background interference, by the recognizable 1 : 1 bromine isotope ratios. Characteristic fragment ions were given for each of the common amino-acids, except cysteine. Most BrPTH's gave strong M^+ signals or $M^+ - H_2O$ (serine and threonine), with the exceptions of arginine, lysine, and tryptophan. The isomeric leucines could be differentiated by their side-chain fragmentations. The procedure was evaluated by the degradation of 3 μM of an octapeptide isolated from Cyclononalinopeptide A, and 4 μM of glucagon. In the latter case, the 16 N-terminal amino-acids were placed in sequence. A limitation of the method appears to be the large quantity of material used, relative to that required for classical studies, although the authors claimed 'high sensitivity'. Further work along these lines has concerned the mass spectrometric analysis of thiazolinones,[100] the precursors of the thiohydantoins which are formed by an acid-catalysed rearrangement. The study showed the spectra to be virtually identical with the hydantoins, indicating a thermal rearrangement in the mass spectrometer ion source, prior to fragmentation. It should now be possible, therefore, to take the thiazolinones directly from an automated sequenator, and analyse them without prior conversion into the thiohydantoins. A comparative study has also been made between methyl and phenyl thiohydantoins at low and high ionizing voltages.[101] Eighteen MTH's and 13 PTH's were examined, and it was found that ionizing voltages between 11 and 15 eV provided the best compromise between minimum fragmentation and low sensitivity. The mass spectrometer used was designed to operate at constant total emission from the filament during variations of electron voltage. An interface for directly coupling an automated protein sequenator to a low-resolution mass spectrometer has been developed,[102] and we can look forward to further progress in this field.

A further development of interest in the attack on protein sequencing has been in the concept of 'mixture analysis', the first example of which was given in some

---

[99] F. Weygand and R. Obermeier, *European J. Biochem.*, 1971, **20**, 72.
[100] T. Fairwell and R. E. Lovins, *Biochem. Biophys. Res. Comm.*, 1971, **43**, 1280.
[101] T. Sun and R. E. Lovins, *Analyt. Biochem.*, 1972, **45**, 176.
[102] R. E. Lovins, J. Craig, and T. Fairwell, paper presented to the 20th Annual Conference on Mass Spectrometry and Allied Topics, Dallas, June 1972, in press.

work on the protein silk fibroin several years ago.[103] Results have now appeared of studies on mixtures of synthetic peptides[104] and on mixtures derived by enzymic digestion of proteins.[96,105] The components of a mixture are sequenced by attempting to promote at least a partial separation by using a temperature gradient in the ion source.[96,103,104] Peptide sequence ions belonging to each individual component of the mixture are then picked out by visual comparisons of consecutive scans. Again it has been demonstrated that low-resolution mass spectrometry and manual interpretation is a perfectly feasible approach to the sequencing even of 'unknown' peptide mixtures.[96] The importance of peptide mixture analysis lies in the fact that the rate-determining step in classical sequencing is the laborious isolation and purification procedures needed to produce pure single peptides, and not the actual sequencing stage. The capacity for mass spectrometry to handle peptide mixtures can therefore revolutionize the whole sequencing process.

A strategy has been suggested, involving mixture analysis, for the complete sequence analysis of a protein by mass spectrometry;[96] preliminary results have been given[92a,106] of this procedure applied to the enzyme Ribitol Dehydrogenase (mol.wt. 30 000), in which over 200 of the 310 amino-acid residues have been placed in sequence, and following the recent advances in derivatization,[91] several 'overlaps' in the individual peptide sequences have been assigned. This study involved the use of enzymic and chemical degradations followed by partial separations and further enzymic degradation of high molecular weight fragments, leading eventually to mass spectrometric examination of peptide mixtures. The work was undertaken without a knowledge of the classically determined structure, which is only now nearing completion.

One of the problems associated with mixture analysis is the possibility of ambiguities arising, especially since the amino-acid compositions of the components are not known. These ambiguities may arise in a complex mixture if a partial separation cannot be achieved. An example of this would be a mass difference of 198 on a spectrum between two possible sequence ions. This could be due to lysine or Leu,Gly or Ala,Val, the sequence ion between the last two being either weak or perhaps corresponding to some other sequence in the mixture. There are a number of other possible ambiguities but these arise surprisingly infrequently, and can normally be resolved either by a manual mass measurement, or by the use of isotopic derivatives.[96,107] For example, the ambiguity mentioned above may be partially resolved by the use of deuterioacetic anhydride for acetylation, which will cause a mass shift to 201 for lysine whereas the other possibilities remain unchanged.

[103] A. J. Geddes, G. N. Graham, H. R. Morris, F. Lucas, M. Barber, and W. A. Wolstenholme, *Biochem. J.*, 1969, **114**, 695.
[104] P. Roepstorff, R. K. Spear, and K. Brunfeldt, *F.E.B.S. Letters*, 1971, **15**, 237.
[105] H. R. Morris and D. H. Williams, *J.C.S. Chem. Comm.*, 1972, 114.
[106] H. R. Morris, D. H. Williams, G. M. Midwinter, and B. S. Hartley, *Biochem. J.*, submitted for publication.
[107] P. Roepstorff and K. Brunfeldt, *F.E.B.S. Letters*, 1972, **21**, 320.

An ingenious use of permethylation has been suggested for the determination of the N-terminal sequence of a protein by mass spectrometry.[108] This involves initial N-acetylation of the protein (perhaps not so easy in practice as on paper) followed by digestion with a suitable protease, after which the whole digest is permethylated using the Hakamori procedure.[77] The argument is that the N-terminal peptide will be extracted in the chloroform phase in the normal manner whereas all the other peptides, having possessed free amino-groups, will have formed quaternary salts and be extracted into the aqueous layer. Of course, depending upon the actual sequence of the protein, one may extract anything from a single amino-acid to a large peptide, and the method has a strong competitor in the automated sequenator.

A new development in protein sequence analysis has arisen from the discovery of a dipeptidase, Cathepsin C. This may be used to break proteins down into dipeptides which can then be characterized by a variety of methods, including mass spectrometry.[109a] The protein is then taken through one step of the Edman degradation, followed by a further treatment with the enzyme to produce a new set of overlapping peptides, from which the total sequence may be deduced. Problems can arise from non-cleavage at certain positions in the peptide chain (e.g. bonds involving proline) which effectively blocks the enzyme, and from ambiguities which may derive from the large numbers of dipeptide sequences. The use of the enzyme has been tested on the peptide Val-Gly-Gly-Val-Glu-Ser-Leu-Gly-Gly-Thr-Gly-Ala-Leu-Arg by gas chromatographic separation of dipeptide derivatives, followed by mass spectrometric identification.[109b] Although the study was not completely successful, a partial sequence could be derived: Val-Gly-Gly-(Ala; Thr-Gly; Leu-Gly-Gly; Val-Glu-Ser)-Leu-Arg. The authors stated that 0.01—0.03 μmol of starting material would give perfectly interpretable spectra, but this remains to be demonstrated since 0.5 μmol was used for each digest (× 2) of the above peptide. Should the current problems be overcome, the enzyme provides a very attractive alternative route to protein sequence analysis.

The reader is referred to a review on the application of computers in chemical analysis which includes a section on computer-aided sequence assignment of the high-resolution spectra of peptides.[110] More recently, a new concept for the sequence analysis of peptides and other oligomers by computer-aided examination of high-resolution spectra called 'submolecular group analysis' has been described.[111] The normal approach previously described consists of locating the N-terminal sequence ion, or loss of CO from it, (or alternatively the molecular ion) and progressively testing for the next amino-acid in the chain by adding the accurate masses of all the possibilities. Where more than one sequence has resulted, a technique was devised to sum the intensities of all the

---

[108] W. R. Gray and U. E. del Valle, *Biochemistry*, 1970, **9**, 2134.
[109] (a) R. J. Rowlands and H. Lindley, *Biochem. J.*, 1972, **126**, 685; (b) Yu. A. Ovchinnikov, and A. A. Kiryushkin, *F.E.B.S. Letters*, 1972, **21**, 320.
[110] B. Sheldrick, *Quart. Rev.*, 1970, **24**, 454.
[111] A. Kunderd, R. B. Spencer, and W. L. Budde, *Analyt. Chem.*, 1971, **43**, 1086.

sequence ions that contribute to a particular sequence. The 'total intensity maximum' is preferred as the correct sequence. The submolecular group approach consists of selecting relevant submolecular group accurate masses, *i.e.* all amino-acids, blocking groups, possible fragments, CO to account for aldimine type ions, and H to take account of a hydrogen rearrangement. Combinations of the submolecular groups are then compared with the measured masses, and in this way structures are assigned to fragment ions. A peptide C_4H_9OCO-Ala-Pro-Ser(Benzyl)-Gly-OEt was used to demonstrate the method,[111] which is claimed to use more of the information present in the spectrum than is taken into account by other programs. Computers and high-resolution mass spectrometry may perhaps be of help in more complex structural problems (*e.g.* mixture analysis[112]) although the cost of instrumentation would appear to be prohibitive to the biochemical laboratory, and the use of high-resolution mass spectrometry reduces the much-needed sensitivity of the method.

New Methods of Ionization.—The development of new ionization techniques (Chapter 1) has prompted several groups to investigate the behaviour of peptides following different ionization processes. Chemical ionization in particular is showing promise with regard to the 'intensity decay' problem towards the high-mass end of peptide spectra.[113] The use of methane reagent gas leads to the formation of a quasimolecular ion '$M + 1$' which fragments in one of two characteristic modes, both involving peptide bond cleavage, (a) formation of an acylium ion $R-C\equiv O^+$ plus neutral amine, (b) proton transfer from the α-carbon to the nitrogen giving a charged ammonium fragment (the reagent proton is believed to be held on the nitrogen of the peptide bond) $Me-\overset{+}{N}H_2-CH(R)CO-$ and a neutral keten derivative. A comparison of the EI and CI spectra of several synthetic *N*-acetyl permethylated peptides (examples of penta-alanine and Thr-Gly were given), showed that the molecular ions, and the high-mass region in general, were more intense in the CI spectra, as would be expected from the 'milder' ionization process giving rise to an even-electron species. However, the examples chosen to express this point (penta-alanine for instance) were something less than fair in representing a typical electron-impact spectrum. Nevertheless, the method obviously shows promise in improving the signal-to-noise ratio at high masses.

Another paper dealing with chemical ionization of peptides[114] must be commended for its clear, easily understandable approach and the results again confirm the general improvement of high-mass intensities. In contrast to the previous paper[113] only (non-permethylated) *N*-acyl peptide esters were examined, and once again several of the statements made should not be taken too literally

---

[112] F. W. McLafferty, R. Venkataraghavan, and P. Irving, *Biochem. Biophys. Res. Comm.*, 1970, **39**, 274.
[113] W. R. Gray, L. H. Wojcik, and J. H. Futrell, *Biochem. Biophys. Res. Comm.*, 1970, **41**, 1111.
[114] A. A. Kiryushkin, H. M. Fales, T. Axenrod, E. J. Gilbert, and G. W. A. Milne, *Org. Mass Spectrometry*, 1971, **5**, 19.

by the beginner in this field of research. For example, referring to electron-impact spectra—'If an attempt is made to determine the amino-acid sequences of typical N-acyl tetrapeptide methyl esters, the correct sequence will be found, but only among 20 or so other possibles, because of coincident masses and side-chain losses. Chemical ionization eliminates these'. No doubt CI does eliminate many of the non-peptide bond fragmentations, but the fact remains that during the last decade a very large number of N-acyl peptide esters have been sequenced quite unambiguously by electron impact, and one would not have thought it 'typical' to find 20 possible sequences! More to the point, however, is the fact that permethylated derivatives, in addition to other advantages, also normally provide easily interpretable spectra, devoid of many of the fragmentations of N-acyl peptide esters. It would therefore appear most sensible to combine the advantages of permethylation with those of chemical ionization.[113]

The mass spectrum of N-acetyl trialanine methyl ester has been studied by proton and deuteron transfer from hydrated species in single ion–molecule collisions, using a tandem mass spectrometer.[115] Fragmentation was sufficient to provide the necessary information to sequence the molecule. Since this single collision technique operates at very low pressures compared to conventional CI, it may afford an advantage in the study of relatively non-volatile materials. The study further showed that secondary collisions, a feature of CI spectroscopy, are not required to stabilize quasi-molecular ions when these ions are generated with very little excess energy. Single-ion impact using low velocity $CH_3NH_3^+$ ions has also been used in a study to determine the effects of ionization chamber surfaces on the volatility of thyrotropin releasing hormone, again in a tandem mass spectrometer.[116] Activation energies of desorption from Teflon surfaces were $30\,\text{kcal mol}^{-1}$ lower than values obtained with powder samples or those dispersed on glass from dilute solutions. The spectra obtained are similar to CI spectra, except that with single-ion impact only one quasi-parent ion is obtained.

A comparative study of photo-ionization and electron-impact spectra of peptide derivatives[117] has shown that the total number of PI peaks is less than those obtained by EI, especially in the low-mass region ($<m/e$ 250). Moreover, the relative abundance of the heavier fragments is increased. Amino-acid type fragmentation is found to be the same, thus facilitating sequence analysis. The aldimine type sequence ions increase with PI, especially for aliphatic amino-acids. Photo-ionization was performed with a monochromatic photon beam producing $10.2\,\text{eV}$, as compared to $70\,\text{eV}$ for EI. Ten synthetic N-decanoyl and N-propionyl peptide methyl esters were studied, and in all but one case (Gly-Trp-Ala-Ala) molecular ions were in greater abundance in the PI spectra. In one peptide the Gly—X bond did not fragment under PI, producing an ambiguity of sequence in the first two positions.

[115] R. J. Beuhler, L. J. Greene, and L. Friedman, *J. Amer. Chem. Soc.*, 1971, **93**, 4307.
[116] R. J. Beuhler, E. Flanigan, L. J. Greene, and L. Friedman, *Biochem. Biophys. Res. Comm.*, 1972, **46**, 1082.
[117] V. M. Orlov, Ya. M. Varshavsky, and A. A. Kiryushkin, *Org. Mass Spectrometry*, 1972, **6**, 9.

An interesting development over the past two years has been field-desorption mass spectrometry, and an initial study has been made of this ionization technique on a few small peptides.[118] The results have been most encouraging in that strong molecular ions have been obtained on underivatized peptides, some containing arginine, which would not of course have volatilized under normal conditions. The sensitivity of the method seems to be good (10^{-8} g) and the technique may well prove to be an excellent complement to electron-impact studies of peptides. In a recent report,[119] thermal fragmentations were produced during a field-desorption study of some simple peptide derivatives, and as a consequence some sequence information was also present in the spectra.

4 Carbohydrates

Advances in the carbohydrate field during the past two years have been largely concerned with the application and development of various principles outlined in the previous report.[120] The major advance concerns further successes with the sequence determination of oligosaccharides (see below). Certain aspects of the mass spectra of carbohydrates have been reviewed.[121,122]

The majority of carbohydrates studied by mass spectrometry are first converted into more volatile derivatives, including peracetyl, permethyl, trimethylsilyl (TMS), and acetal derivatives. The spectra of acetals of D-allose and D-psicose have been listed.[123] The mass spectra of permethylated 3,6-anhydrogalactose derivatives confirm the presence of the 3,6-anhydro-ring, and allow the elucidation of both the nature and position of the substituents at C-1, C-2, and C-4 of the anhydro-sugar derivative in both mono- and di-saccharides.[124] The spectra of the TMS derivatives of D-fructose, L-sorbose, methyl-D-fructopyranose, and methyl-L-sorbofuranose have been described, and characteristic peaks have been listed for 2-ketohexoses, aldohexoses, pyranose rings, and furanose rings.[125] As an example, the base peaks in the spectra of α-methyl-1,3,4,6-tetra-D-trimethylsilyl-D-fructofuranose and methyl-1,3,4,5-tetra-D-trimethylsilyl-L-sorbopyranose are indicated in (42) and (43) respectively. Ions produced by rearrangement processes are common in these spectra[125] (cf. Vol. 1, pp. 135, 136).

---

[118] H. U. Winkler and H. D. Beckey, *Biochem. Biophys. Res. Comm.*, 1972, **46**, 391.
[119] H. U. Winkler and H. D. Beckey, paper presented to the 20th Annual Conference on Mass Spectrometry and Allied Topics, Dallas, June 1972, in press.
[120] R. G. Cooks and G. S. Johnson, in 'Mass Spectrometry', ed. D. H. Williams (Specialist Periodical Reports), The Chemical Society, London, 1970, Vol. 1, p. 164.
[121] 'Carbohydrate Chemistry', ed. J. S. Brimacombe (Specialist Periodical Reports), The Chemical Society, London, 1971, Vol. 4, p. 158.
[122] S. Hanessian, 'Mass Spectrometry in the Determination of Structure of certain Natural Products Containing Sugars', in 'Methods of Biochemical Analysis', ed. D. Glick, Interscience, 1971, Vol. 19, p. 105; see also A. L. Burlingame and G. A. Johanson, *Analyt. Chem.*, 1972, **44**, 337R.
[123] M. Haga, M. Takano, and S. Tejima, *Carbohydrate Res.*, 1970, **14**, 237.
[124] O. S. Chizhov, B. M. Zolotarev, A. I. Usov, M. A. Rechter, and N. K. Kochetkov, *Carbohydrate Res.*, 1971, **16**, 29.
[125] S. Karady and S. H. Pines, *Tetrahedron*, 1970, **26**, 4527.

The behaviour of pentodialose derivatives,[126] and N'-acyl derivatives of 2-acetamino-3,4,6-tri-O-acetyl-2-deoxy-β-D-glucopyranosylamine[127] upon electron impact has been reported, and it has been noted[128] that the mass spectra of methyl-4-azido-4-deoxypentosides often contain molecular ions and characteristic fragment ions. The spectra of acetyl, benzyl, isopropylidene, and methyl derivatives of 4-thio-α-D-xyloses have been compared with the O-analogues.[129] For example, the spectrum of methyl 2,3-di-O-acetyl-4-S-acetyl-4-thio-α-D-xylopyranoside exhibits fragmentations similar to those of the methyl 2,3,4-tri-O-acetylpentopyranosides, but, in addition, undergoes the decompositions (44) → (45). The spectrum of D-mannopyranodihydrobenzothiazine has been recorded,[130] the fragmentations of the molecular ions of acetylated deoxyfluoro-D-glucitols have been rationalized in terms of the position of the fluorine substituent

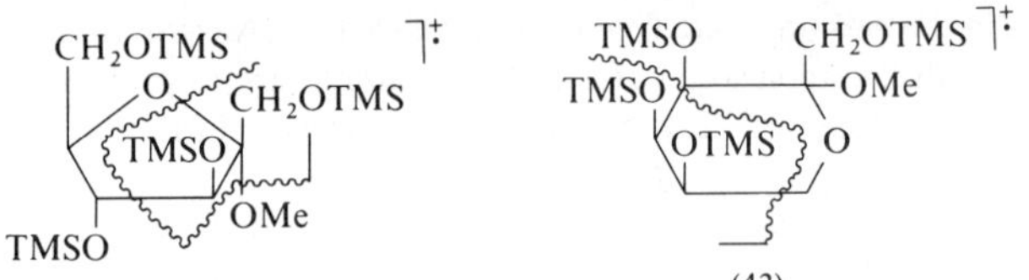

on the carbon chain,[131] and the effect of various halogen substituents on the fragmentations of 6-deoxy-6-halogeno-α-D-glucopyranose tetra-acetates has been reported.[132]

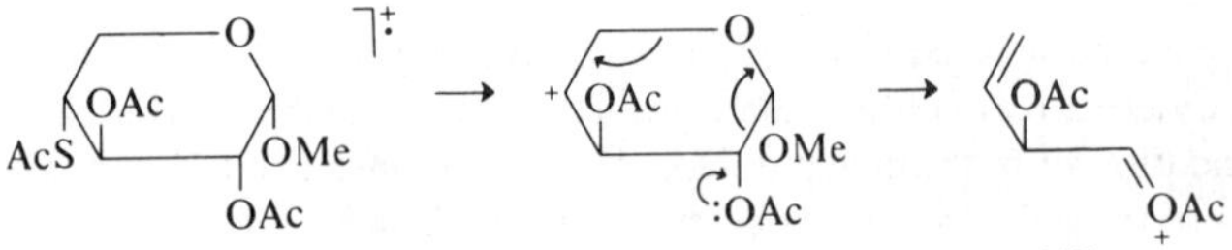

A detailed study of aldonic and deoxyaldonic acids has shown that structural isomers can be distinguished by recourse to the mass spectra of their TMS ethers, but that the spectra of diastereoisomers are similar.[133] Deuterium labelling has aided the interpretation of the spectra of partially methylated alditol

[126] N. K. Kochetkov, O. S. Chizhov, and A. F. Sviridov, *Carbohydrate Res.*, 1970, **14**, 277.
[127] T. Komori, Y. Ida, Y. Inatsu, M. Kiyozumi, K. Kato, and T. Kawasaki, *Annalen*, 1970, **741**, 33.
[128] V. Kováčik, C. Peciar, Š. Bauer, and H. F. Grützmacher, *Carbohydrate Res.*, 1971, **19**, 169.
[129] V. Kováčik, P. Kováč, and R. L. Whistler, *Carbohydrate Res.*, 1971, **16**, 353.
[130] M. Sekija and S. Ishiguro, *Tetrahedron Letters*, 1971, 431.
[131] J. Adamson, A. D. Barford, E. M. Bessell, A. B. Foster, M. Jarman, and J. H. Westwood, *Org. Mass Spectrometry*, 1971, **5**, 865.
[132] O. S. Chizhov, A. B. Foster, M. Jarman, and J. H. Westwood, *Carbohydrate Res.*, 1972, **22**, 37.
[133] G. Petersonn, *Tetrahedron*, 1970, **26**, 3413.

Natural Products

acetates.[134,135] The molecular ion of 3-deoxy-2,4,6-tri-*O*-methyl-D-*arabino*-hexitol acetate cleaves as shown in (46), and the ion at m/e 175 decomposes by loss of methanol through a four-centre state (47).[135]

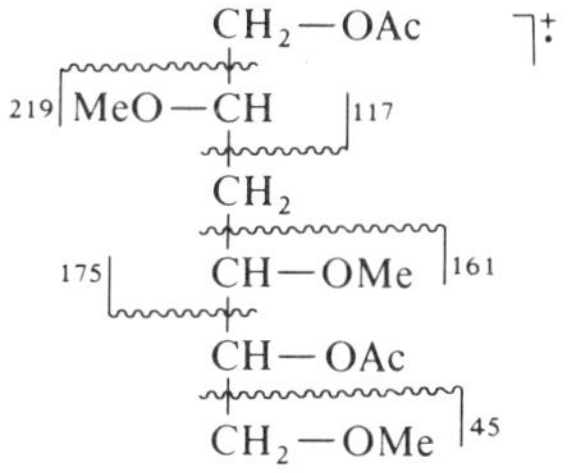

$$\begin{array}{c}
\text{CH}_2-\text{OAc} \\
219\,|\,\text{MeO}-\text{CH}\quad|_{117} \\
\text{CH}_2 \\
175\,|\quad\text{CH}-\text{OMe}\quad|_{161} \\
\text{CH}-\text{OAc} \\
\text{CH}_2-\text{OMe}\quad|_{45}
\end{array}\right]^{\ddagger}$$

(46)

$$\begin{array}{c}
\text{CH}_2-\text{OAc} \\
\text{MeO}-\text{CH} \\
\text{H}-\text{CH} \\
+\text{CH}-\text{OMe}
\end{array}$$

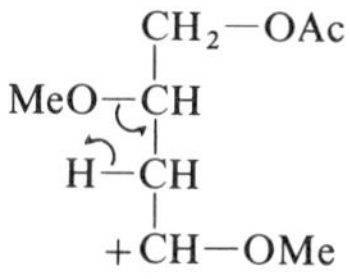

(47)

TMS derivatives of α-lactose,[136] β-lactose,[136] β-cellobiose,[136] 2-acetamido-2-deoxyaldohexosylaldohexoses,[137] and fructose-containing oligosaccharides[138] have been studied. There has been a continuing search for methods which allow the sequence determination of polysaccharides. The spectra of TMS aldosyl oligosaccharides have been studied[139] and the decompositions of the molecular ions of three aldopentose disaccharides allowed the determination of the monosaccharide sequence, as well as the position of the glycosidic link.[139] The spectra of both TMS and peracetate derivatives of sucrose, melibiose, 1-kestose, planteose, raffinose, nystose, and stachyose demonstrate that it is possible to recognize the presence or absence of a ketohexofuranose residue, its terminal or non-terminal nature, and the presence of such residues attached to adjacent moieties through methylene bridges.[140] A most elegant illustration of the use of mass spectrometry for the sequence determination of sugars occurs for aminocyclitols.[141,142] As an example, some of the cleavages occurring from the molecular ion of the

---

[134] H. Björdal, B. Lindberg, Å. Pilotti, and S. Svensson, *Carbohydrate Res.*, 1970, **15**, 339.
[135] H. B. Borén, P. J. Garegg, B. Lindberg, and S. Svensson, *Acta Chem. Scand.*, 1971, **25**, 3299.
[136] J. Vink, J. J. de Ridder, J. P. Kamerling, and J. F. G. Vliegenthart, *Biochem. Biophys. Res. Comm.*, 1971, **42**, 1050.
[137] J. P. Kamerling, J. F. G. Vliegenthart, J. Vink, and J. J. de Ridder, *Tetrahedron*, 1971, **27**, 4749.
[138] J. P. Kamerling, J. F. G. Vliegenthart, J. Vink, and J. J. de Ridder, *Tetrahedron Letters*, 1971, 2367.
[139] J. P. Kamerling, J. F. G. Vliegenthart, J. Vink, and J. J. de Ridder, *Tetrahedron*, 1971, **27**, 4275.
[140] W. Binkley, R. C. Dougherty, D. Horton, and J. D. Wander, *Carbohydrate Res.*, 1971, **17**, 127.
[141] P. W. K. Woo, *Tetrahedron Letters*, 1971, 2621.
[142] P. J. L. Daniels, M. Kugelman, A. K. Mallams, P. W. Tkach, H. F. Vernay, J. Weinstein, and A. Yehaskel, *Chem. Comm.*, 1971, 1629.

TMS derivative of butirosin A are shown in (48).[141] The sugar sequence of cardenolide glycosides can also be determined by mass spectrometry.[143,144]

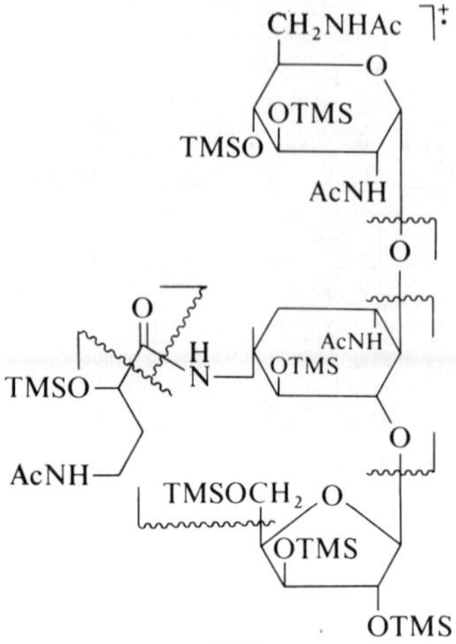

(48)

Several schools have been able to enhance the abundances of molecular ions and to determine the monosaccharide sequence of oligosaccharides by the incorporation of a heterocyclic unit into the terminal sugar. Most promising results have been obtained for peracetylated di-pentasaccharides containing a 1-phenylflavazole unit (49),[145,146] an N-arylglycosylamine substituent,[147] or an N-phenylisotriazole moiety (50).[148,149] A simplified picture of the various cleavage processes is depicted in (51).[148]

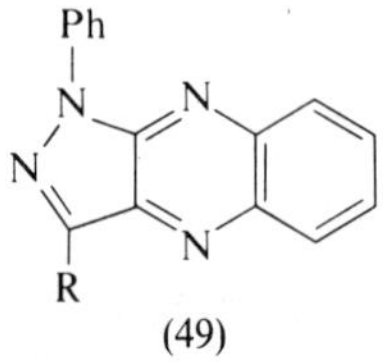

(49) 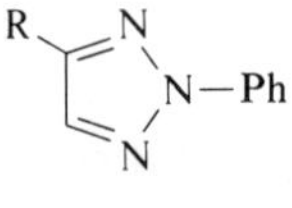   (50)

[143] P. Brown, F. R. Brüschweiler, G. R. Pettit, and T. Reichstein, *Org. Mass Spectrometry*, 1971, **5**, 573.
[144] B. Blessington, Y. Nakagawa, and D. Satoh, *Org. Mass Spectrometry*, 1970, **4**, 215.
[145] G. S. Johnson, W. S. Ruliffson, and R. G. Cooks, *Carbohydrate Res.*, 1971, **18**, 233.
[146] G. S. Johnson, W. S. Ruliffson, and R. G. Cooks, *Carbohydrate Res.*, 1971, **18**, 243.
[147] N. K. Kochetkov, O. S. Chizhov, N. N. Malysheva, and A. I. Shiyonok, *Org. Mass Spectrometry*, 1971, **5**, 481.
[148] O. S. Chizhov, N. K. Kochetkov, N. N. Malysheva, A. I. Shiyonok, and V. L. Chashchin, *Org. Mass Spectrometry*, 1971, **5**, 1145.
[149] O. S. Chizhov, N. K. Kochetkov, N. N. Malysheva, A. I. Shiyonok, and V. L. Chashchin, *Org. Mass Spectrometry*, 1971, **5**, 1157.

$$G_1 \}O\{ G_2\}O\{ G_3\}O\{ G_4\}O\{ GM\ ]^+$$

(51)

The structure elucidation of a new sphingoglycolipid isolated from human adenocarcinomas was aided by the mass spectra of degradation products.[33] TMS derivatives of neutral, mono-, and di-sialoglycosphingolipids give spectra which provide information concerning the numbers of sialic acid, N-acetylated hexosamine, and monosaccharide residues in the lipid and glycosyl moieties.[31] The behaviour of N-acyl derivatives of 2-acetamido-3,4,6-tri-O-acetyl-2-deoxy-β-D-glucopyranosylamines upon electron impact has been investigated as a guide to the behaviour of the terminal amino-groups in glycopeptides.[150] The spectra of O-glycosyl derivatives of L-serine and L-threonine allow identification of both sugar and amino-acid units, and of the type of linkage between them.[73]

Mass spectrometry has aided the structure determination of a large number of synthetic and natural glycosides, and a selection of these follows. Helminthosporoside (52), a host-specific toxin from the fungus *Helminthosporium sacchari* contains a cyclopropane ring, and eliminates two molecules of water from its molecular ion.[151] The mass spectra of 3,5-dihydroxy-5-ethoxy-2-syringoyl-1-methyl-4-O-β-D-glucopyranosyl cyclopentane (from *Galipae officinales*)[152] and of sisocin, an unsaturated aminoglycoside from *Micromonospora inyoensis*,[153] have been recorded. Two closely related glycosides, parasorboside (53; R = H)[154] from *Sorbus aucuparia* L., and mevaloside (from *Mespilus germanica* L.)[155] give spectra in which the base peak is produced by cleavage of the glycosidic link [*e.g.* (53) for R = Ac]. Mass spectrometry has been used for the elucidation of structure of a complex terpene,[156] several cardenolide glycosides,[157,158] a saponoside,[159] the livestock poison 3-nitro-1-propyl-β-D-glycopyranoside (from *Astragalus miser*),[160] and the medicinal glucoside Kutkin from *Picrorhiza kurro* Royle ex Benthe.[161] The mass spectra of peracetates of bacterial carotenoid glycosides (*e.g.* rhodopin β-D-glucoside) contain molecular ions, and fragment

[150] T. Komori, Y. Ida, Y. Inatsu, M. Kiyozumi, K. Kato, and T. Kawasaki, *Annalen*, 1970, **741**, 33.
[151] G. W. Steiner and G. A. Strobel, *J. Biol. Chem.*, 1971, **246**, 4350.
[152] C. H. Brieskorn and V. Beck, *Phytochemistry*, 1971, **10**, 3205.
[153] D. J. Cooper, R. S. Jaret, and H. Reimann, *Chem. Comm.*, 1971, 285.
[154] R. Tschesche, H.-J. Hoppe, G. Snatzke, G. Wulff, and H.-W. Fehlhaber, *Chem. Ber.*, 1971, **104**, 1420.
[155] R. Tschesche, K. Struckmeyer, and G. Wulff, *Chem. Ber.*, 1971, **104**, 3567.
[156] T. Sévenet, C. Thal, and P. Potier, *Tetrahedron*, 1971, **27**, 663.
[157] J. Polonia, H. Jäger, J. von Euw, and T. Reichstein, *Helv. Chim. Acta*, 1970, **53**, 1253.
[158] E. V. Rao, D. V. Rao, S. K. Pavanaram, J. von Euw, and T. Reichstein, *Helv. Chim. Acta*, 1971, **54**, 1960.
[159] M. Vanhaelen, *Phytochemistry*, 1972, **11**, 1111.
[160] F. R. Stermitz, W. T. Lowry, F. A. Norris, F. A. Buckeridge, and M. C. Williams, *Phytochemistry*, 1972, **11**, 1117.
[161] K. Basu, B. Dasgupta, and S. Ghosal, *J. Org. Chem.*, 1970, **35**, 3159.

ions which arise mainly by cleavage of the carotenoid unit.[162] Mannosyl-1-phosphoryldecaprenol has been produced by an enzymic synthesis, and the structure was confirmed by the mass spectra of degradation products.[163]

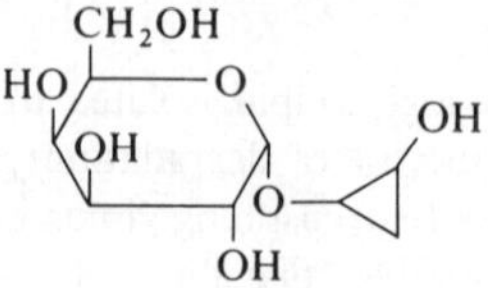

(52) 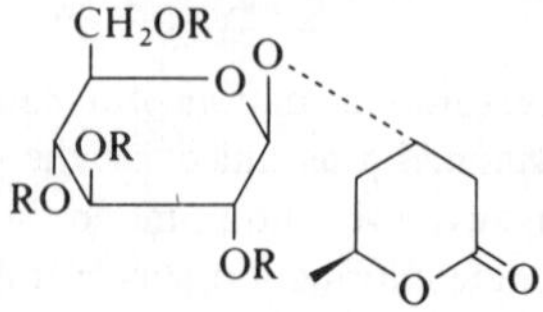   (53)

Mass spectrometry has confirmed that the major bilirubin conjugates in human bile occur as acyl glycosides of aldobiouronic acid, pseudoaldobiouronic acid, and hexuronosylhexuronic acid.[164] The mass spectra of adenine nucleosides,[165] 4′,6′-di-O-acetyl-2′,3′-dideoxy-D-erythro-hex-2-enopyranosyl nucleosides,[166] and hydroxymethyl-3-β-D-erythro-furanosyl-9-adenine[167] have been recorded, and differences in ion intensities in the spectra of the α- and β-anomers of various trifluoracetylated analogues of adenosine have been ascribed to different steric relationships.[168] Pyrimidine cyclonucleosides and their TMS derivatives have been studied, including 2,2′-anhydro-1-(β-D-arabinofuranosyl)-uracil which suffers the major decompositions indicated in (54).[169]

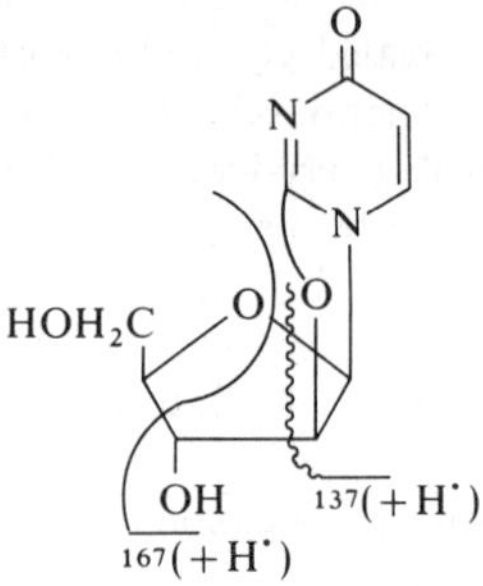

(54)

FI mass spectrometry continues to show promise for carbohydrate studies. The FI spectra of aryl β-D-glucopyranosides show both molecular and $M + 1$ ions.[170] Beckey[171] has noted that monosaccharides exhibit an intense $M + 1$

[162] K. Schmidt, G. W. Francis, and S. Liaaen-Jensen, *Acta Chem. Scand.*, 1971, **25**, 2476.
[163] T. Takayama and D. S. Goldman, *J. Biol. Chem.*, 1970, **245**, 6251.
[164] C. C. Kuenzle, *Biochem. J.*, 1970, **119**, 411.
[165] S. J. Shaw, D. M. Desiderio, K. Tsuboyama, and J. A. McCloskey, *J. Amer. Chem. Soc.*, 1970, **92**, 2510.
[166] R. J. Ferrier and M. M. Ponpipom, *J. Chem. Soc. (C)*, 1971, 553.
[167] J. M. J. Tronchet and J. Tronchet, *Helv. Chim. Acta*, 1971, **54**, 1466.
[168] W. A. König, K. Zech, R. Ulmann, and W. Voelter, *Chem. Ber.*, 1972, **105**, 262.
[169] S. Tsuboyama and J. A. McCloskey, *J. Org. Chem.*, 1972, **37**, 166.
[170] G. O. Phillips, W. G. Filby, and W. L. Mead, *Carbohydrate Res.*, 1971, **18**, 165.
[171] H. Krone and H. D. Beckey, *Org. Mass Spectrometry*, 1971, **5**, 983.

peak, but that such compounds are generally thermally unstable. Permethylation is the method of choice for obtaining FI spectra of oligosaccharides. The FI spectra of permethylated 1 → 4 and 1 → 6 linked disaccharides show characteristic differences. In general, the field-desorption method (Chapter 1) is more suitable for thermally unstable molecules than the normal FI method.[171] A continuation of initial work[172] has shown that field-ion mass spectrometry is generally more informative than EI mass spectrometry for the monosaccharide sequencing of cardiac glycosides.[143] The FI spectra contain enhanced molecular ions and definitive fragment ions produced by cleavage of glycosidic bonds. The primary fragmentation of K-strophantosid is illustrated in (55).[143] The CI mass spectrum of 1,2,5-O-ethylidene-α-D-galactofuranose contains a pronounced $M + 1$ peak, with various fragments proceeding from this ion.[173]

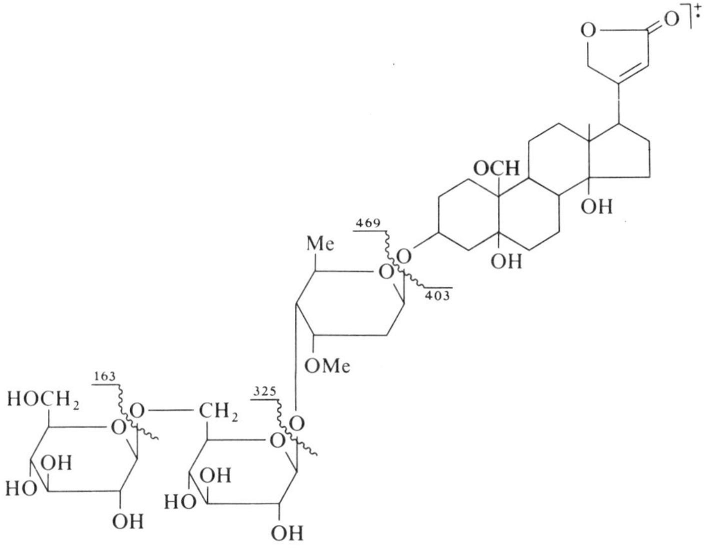

(55)

5 Antibiotics

The structural elucidation of antibiotics has been greatly aided by mass spectrometry, and a recent review covers the fragmentation patterns of many classes of antibiotics.[174] In many cases, the complex polycyclic nature of the antibiotic has meant that useful mass spectral data could only be obtained after chemical

[172] P. Brown, F. R. Brüschweiler, G. R. Pettit, and T. Reichstein, *J. Amer. Chem. Soc.*, 1970, **92**, 4470.
[173] M. Bertolini and C. P. J. Glaudemans, *Carbohydrate Res.*, 1971, **18**, 131.
[174] K. L. Rinehart and G. VanLear, in 'Biochemical Applications of Mass Spectrometry', ed. G. R. Waller, Wiley, New York, 1972, Chap. 17.

degradation. This has been especially the case for peptide and macrolide antibiotics.

Mass spectrometric sequencing of peptides is covered in another section of this chapter, and only brief mention of peptide antibiotics will be made here. The molecular structures of the cyclodepsipeptides, monamycin D_1 and H_1, have been determined by partial hydrolysis of the antibiotics to di- and tripeptides, and by mass spectral studies of the corresponding linear hexadepsipeptides.[175] The mass spectra of underivatized Staphylomycin S and related compounds have been studied, and found to be structurally informative.[66] Mass spectrometry has also been important for the structural elucidation of cycloheptamycin,[176] a novel peptide antibiotic containing 5-hydroxytryptophan and β-hydroxynorvaline, amino-acids not previously detected in natural products, and the polypeptide antibiotic, griselmycin.[177]

There have been several recent reports of the usefulness of mass spectrometry in the field of macrolide antibiotics. Since the EI spectra of these compounds are very complex, most structural elucidation has involved the use of mass spectrometry on chemical degradation products. In this way the structures of Amphotericin B,[178] Nystatin A_1,[179] Tetrin A^{180} and B,[181] Streptovaricins A—G,[182] and the part structure of oligomycin B_1,[183] have been determined.

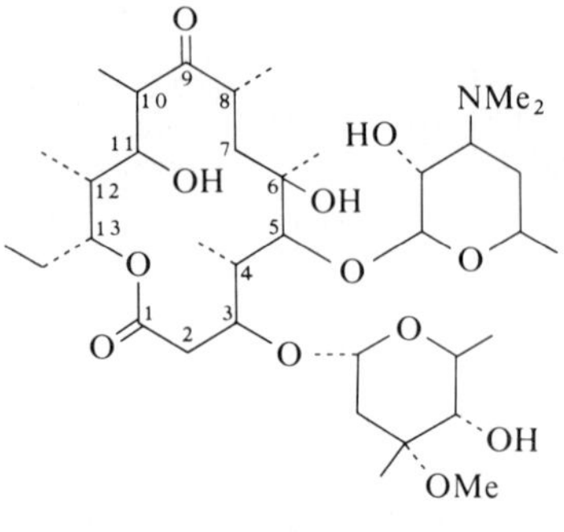

(56)

[175] C. H. Hassall, R. B. Morton, Y. Ogihara, and D. A. S. Phillips, *J. Chem. Soc. (C)*, 1971, 526.
[176] W. O. Godtfredsen, S. Vergedal, and D. W. Thomas, *Tetrahedron*, 1970, **26**, 4931.
[177] B. Terlain and J.-P. Thomas, *Bull. Soc. chim. France*, 1971, 2357.
[178] E. Borowski, J. Zieliński, T. Zimiński, L. Falkowski, P. Kolodziejczyk, J. Golik, E. Jereczek, and H. Adlercreutz, *Tetrahedron Letters*, 1970, 3909.
[179] E. Borowski, J. Zieliński, L. Falkowski, T. Zimiński, J. Golik, P. Kolodziejczyk, E. Jereczek, M. Gdulenwicz, Y. Shenin, and T. Kotienko, *Tetrahedron Letters*, 1971, 685.
[180] R. C. Pandey, V. F. German, Y. Nishikawa, and K. L. Rinehart, *J. Amer. Chem. Soc.*, 1971, **93**, 3738.
[181] K. L. Rinehart, W. P. Tucker, and R. C. Pandey, *J. Amer. Chem. Soc.*, 1971, **93**, 3747.
[182] K. L. Rinehart, M. L. Maheshwari, F. J. Antosz, H. H. Mathur, K. Sasaki, and R. J. Schacht, *J. Amer. Chem. Soc.*, 1971, **93**, 6273.
[183] W. F. Prouty, R. M. Thompson, H. K. Schnoes, and F. M. Strong, *Biochem. Biophys. Res. Comm.*, 1971, **44**, 619.

Natural Products

The isobutane CI spectra of macrolide antibiotics are much simpler than the corresponding EI spectra, and contain much structurally useful information.[184] Thus in the iso-C_4H_{10} CI spectrum of erythromycin B (56),[184] there are an intense $M^+ + 1$ ion (m/e 718), ions corresponding to loss of cladinose from $M^+ + 1$ with (m/e 542) or without (m/e 560) its glycosidic oxygen, and an ion corresponding to loss of desosamine from $M^+ + 1$ without its glycosidic oxygen, and with charge retention by the desosamine residue (m/e 158). In addition, two diagnostically important reactions of $M^+ + 1$ are observed. The C-6 hydroxy-group is involved in water loss from $M^+ + 1$ [rationalized as (57) → (58)],

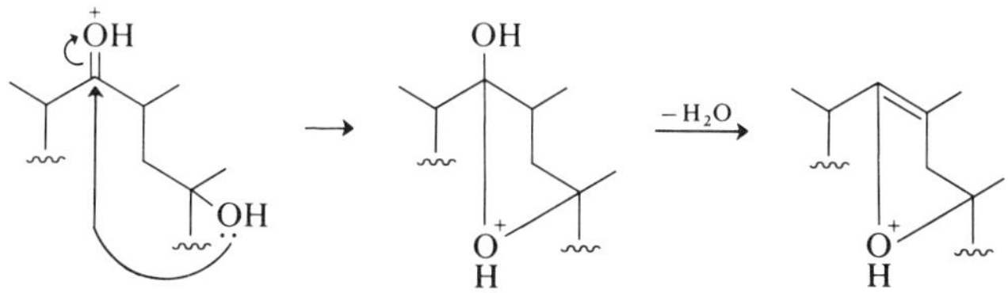

to give an intense peak at m/e 700. In the CI mass spectra of macrolide antibiotics that do not contain a C-6 hydroxy-group, there is no $M^+ + 1 - H_2O$ peak of comparable intensity. The second diagnostically important reaction is rationalized as protonation of the C-9 ketone, ring-opening at the lactone linkage, and cleavage of the C-10—C-11 bond *via* a retro-aldol type fragmentation [(59) → (60) + (61)]. Other macrolide antibiotics give the expected shifts

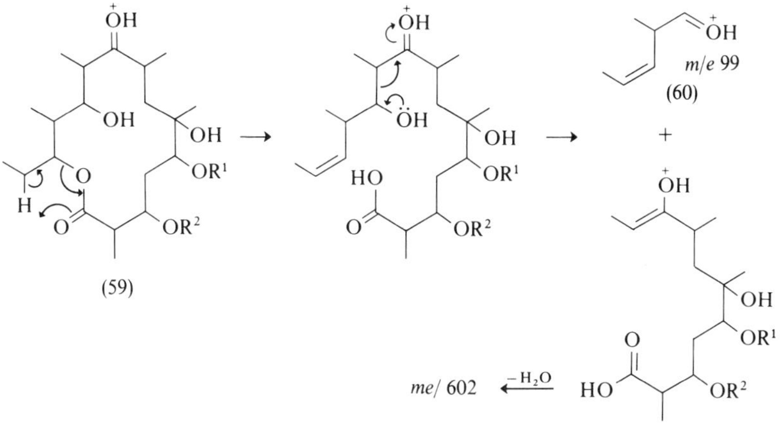

---

[184] L. A. Mitscher, H. D. H. Showalter, and R. L. Foltz, *J.C.S. Chem. Comm.*, 1972, 796.

in m/e value of the products of this retro-aldol fragmentation, e.g. oleandomycin, which has a C-13 methyl group, gives rise to ions m/e 85 and 604,[184] and lankamycin, which has an acetyl group instead of a hydroxy-group at C-11, gives no ions due to the retro-aldol type cleavage.[184] CI mass spectrometry would, therefore, seem to be of some potential importance in the structure elucidation of macrolide antibiotics. Another approach adopted to obtain more information from macrolide antibiotic mass spectra is suitable only for those compounds containing an amino-sugar moiety. Formation of the N-oxide (in situ) suppresses fragmentation of the amino-sugar, which normally dominates the EI spectrum of an amino-sugar macrolide antibiotic, and allows the non-sugar portion of the molecule to make a more substantial contribution to the spectrum.[185]

In the field of aminocyclitol antibiotics, high-resolution mass spectrometry has been used to determine the molecular formulae of underivatized aminocyclitols, and the composition of the sugar and cyclitol units.[142] Determination of molecular formulae is of particular use, since this is often difficult by conventional microanalytical techniques. The fragmentation pattern of aminocyclitol antibiotics (except those containing a 2-deoxy-sugar) may also demonstrate that the sugar units are glycosidically linked to the aglycone in antibiotics of unknown structure.[142] Aminocyclitol antibiotics whose structures have been determined with the aid of mass spectrometry include sisomicin, a novel unsaturated amino-glycoside antibiotic,[153] and the components of gentamicin C.[186,187] The latter were established as 4,6-disubstituted-2-deoxystreptamine derivatives by the use of mass spectrometry on the per-N-acyl-per-NO-methylgentamicins.

Many other antibiotics, or antibiotic components, have been studied by mass spectrometry. The structure elucidation of three new antibiotics of the cephalosporin C type, with the aid of high-resolution mass spectrometry, has been reported.[188] LL-Z1220 (62), an antibiotic and the first naturally occurring

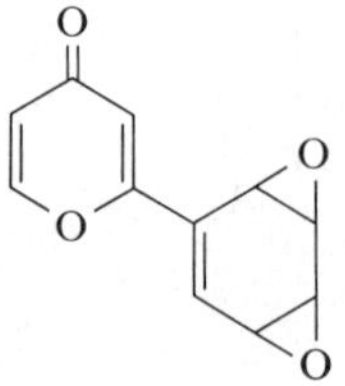

(62)

[185] L. A. Mitscher, R. L. Foltz, and M. I. Levenberg, Org. Mass Spectrometry, 1971, **5**, 1229.
[186] D. J. Cooper, M. D. Yudis, H. M. Marigliano, and T. Traubel, J. Chem. Soc. (C), 1971, 2876.
[187] D. J. Copper, P. J. L. Daniels, M. D. Yudis, H. M. Marigliano, R. D. Guthrie, and S. T. K. Bukhari, J. Chem. Soc. (C), 1971, 3126.
[188] R. Nagarajan, L. D. Boeck, M. Gorman, R. L. Hamill, C. E. Higgens, M. M. Hoehn, W. M. Stark, and J. G. Whitney, J. Amer. Chem. Soc., 1971, **93**, 2308.

compound found to contain a cyclohexene–diepoxide ring, has been characterized.[189] The structure of deoxynigericin has been determined by mass spectrometry.[190] Field ionization was used in the structural determination of 3-amino-2,3,6-trideoxy-3-C-methyl-*lyxo*-pyranose, a novel amino-sugar liberated from vancomycin by mild acid hydrolysis,[191] to confirm the molecular ion, inferred from EI and acetylation–trideuterioacetylation studies.

6 Isoprenoids

Terpenes have been mentioned in a recent review[192] and two reviews on field-ion mass spectrometry contain reference to isoprenoid problems.[193,194] Isoprene has been identified by mass spectrometry as a forest-type emission into the atmosphere.[195] The mass spectra of bicyclic monoterpenes of the bornane type have been discussed.[196]

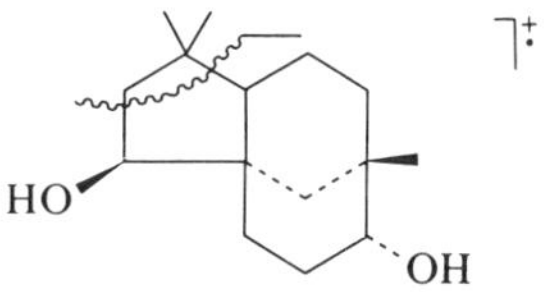

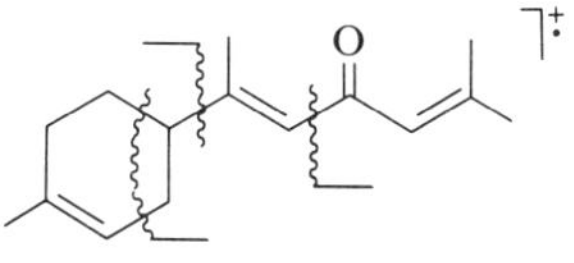

The spectra of a variety of sesquiterpenes have been reported, including α- and β-farnesene,[197] cyclonerodiol,[198] bazzanene,[199] prutenin,[200] and cartleyoside.[201] The mass spectrum of clovane diol (63) shows the decompositions

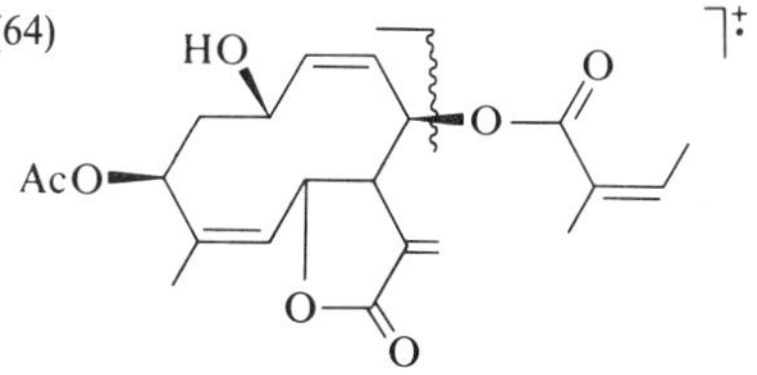

[189] D. B. Borders, P. Shu, and J. E. Lancaster, *J. Amer. Chem. Soc.*, 1972, **94**, 2540.
[190] G. Horváth and J. Gyimesi, *Biochem. Biophys. Res. Comm.*, 1971, **44**, 639.
[191] W. D. Weringa, D. H. Williams, J. Feeney, J. P. Brown, and R. W. King, *J.C.S. Perkin I*, 1972, 443.
[192] A. Quayle, 'Industrial Applications of Mass Spectrometry', in 'Recent Topics in Mass Spectrometry', ed. R. I. Reed, Gordon and Breach, London, New York, 1971, p. 267.
[193] H. D. Beckey, 'Recent Studies in Field Ion Mass Spectrometry', in 'Recent Developments in Mass Spectrometry' (Proceedings of the International Conference at Kyoto, Japan 1969), ed. K. Ogata and T. Hayakawa, Univ. Park Press, Baltimore, London, and Tokyo, 1970, 1154.
[194] H. D. Beckey, 'Field Ion Mass Spectrometry', Pergamon Press, Oxford, New York, Toronto, Sydney, 1971.
[195] R. A. Rasmussen, *Environ. Sci. Technol.*, 1970, **4**, 669.
[196] A. Daniel and A. A. Pavia, *Org. Mass Spectrometry*, 1971, **5**, 1237.
[197] E. F. L. J. Anet, *Austral. J. Chem.*, 1970, **23**, 2101.
[198] B. E. Cross, R. E. Markwell, and J. C. Stewart, *Tetrahedron*, 1971, **27**, 1663.
[199] A. Matsuo, *Tetrahedron*, 1971, **27**, 2757.
[200] F. Bohlmann and C. Zdero, *Chem. Ber.*, 1971, **104**, 1611.
[201] T. Sévenet, C. Thal, and P. Potier, *Tetrahedron*, 1971, **27**, 663.

$M - C_4H_8 - H_2O$.[202] Mass spectrometry has aided the structure determination of atlantone which fragments as shown in (64),[203] and of several antileukemic sesquiterpenes[204,205] including eupacunin, which cleaves as in (65) followed by loss of CH_3CO_2H and H_2O.[204]

The mass spectra of furanoterpenes are particularly suited to structure determination, as evidenced by the toxic furanoid ketones from *Myoporium deserti* [*e.g.* isomyodesmone (66)],[206] furosongin-1 (67),[207] linderalactone (68),[208] and odoratin.[209]

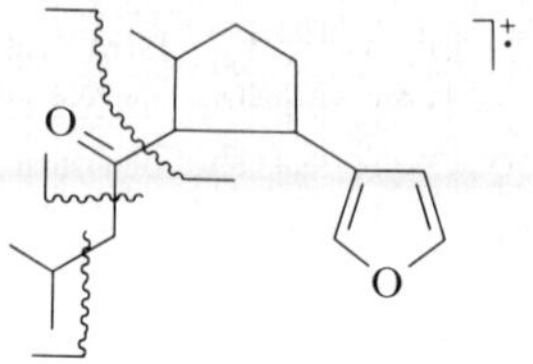

(66)

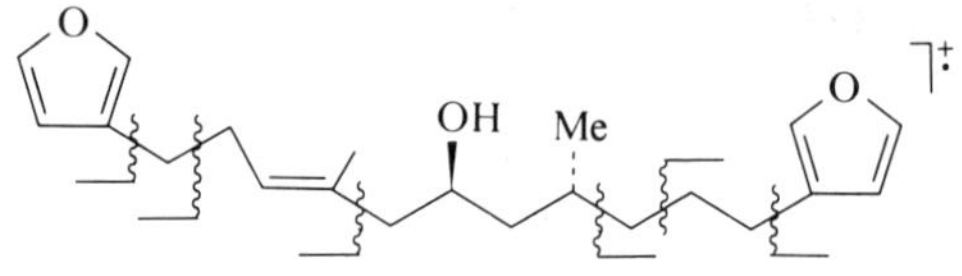

(67)

An interesting example of the use of mass spectrometry for distinguishing between stereoisomers has been reported for the diterpene lactones 12-norambrienolide (69) and 8-epi-12-norambrienolide (70).[210] The base peaks in the spectra of (69) and (70) are produced respectively by an $M - CO_2$ species and an ion $C_{10}H_{16}$]⁺·, with their counterparts in the other spectrum being of small abundance. Mass spectrometry has been successfully employed for the structure

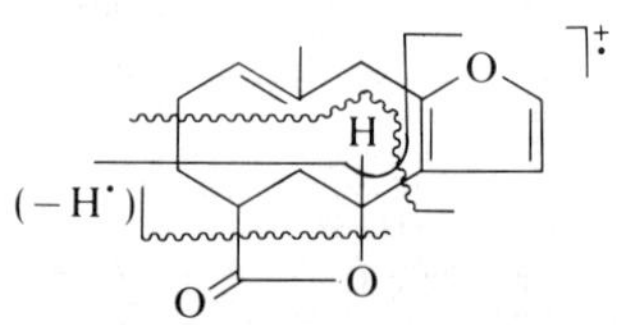

(68) 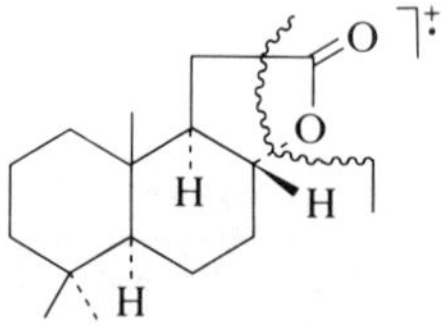   (69) 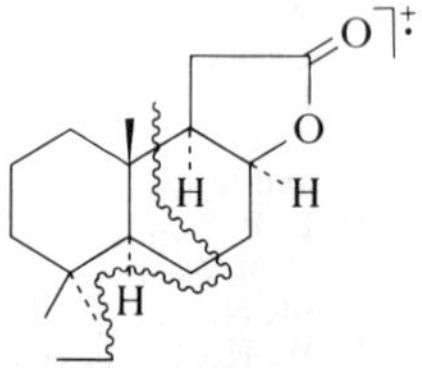   (70)

[202] A. S. Gupta and S. Dev, *Tetrahedron*, 1971, **27**, 635.
[203] B. S. Pande, S. Krishnappa, S. C. Bisarya, and S. Dev, *Tetrahedron*, 1971, **27**, 841.
[204] S. M. Kupchan, M. Maruyama, R. J. Hemingway, J. C. Hemingway, S. Shibuya, T. Fujita, P. D. Cradwick, A. D. U. Hardy, and G. A. Sim, *J. Amer. Chem. Soc.*, 1971, **93**, 4914.
[205] S. M. Kupchan, V. H. Davies, T. Fujita, M. R. Cox, and R. F. Bryan, *J. Amer. Chem. Soc.*, 1971, **93**, 4916.
[206] I. D. Blackburne, R. J. Park, and M. D. Sutherland, *Austral. J. Chem.*, 1971, **24**, 995.
[207] G. Cimino, S. DeStefano, L. Minale, and E. Fattorusso, *Tetrahedron*, 1971, 27, 4673.
[208] K. G. Das, M. S. B. Nayar, and B. S. Joshi, *Org. Mass Spectrometry*, 1971, **5**, 187.
[209] W. R. Chan, D. R. Taylor, and R. T. Aplin, *Tetrahedron*, 1972, **28**, 431.
[210] R. Hodges, R. C. Cambie, and K. N. Joblin, *Org. Mass Spectrometry*, 1970, **3**, 1473.

elucidation of gutierolide, a chlorodiFerpenoid lactone,[211] diterpenes of the kauran-16-ene family,[212] kauran-16-ols,[213] kauran-19-oic acids,[214] the first isolated diterpene sulphoxide,[215] the diterpene ketol macarangonol,[216] and pododacric acid.[217] A detailed investigation of the complex fragmentations of aromatic diterpenes of the podocarpa-8,11,13-trien-7-one family has been aided by deuterium labelling.[218]

The tetracyclic triterpenes litsomentol[219] and cyclomahogenol,[220] have been investigated by mass spectrometry. The major cleavages of the molecular ion of nimbidinin are represented in (71).[221] Carvings from Zapote wood have resisted fungus and termite attack for more than 1200 years. Mass spectral investigation of the 'Basle plate' from Tikal (751 AD) has shown that this resistance is due to the presence of hydroxy- and dihydroxy-oleanolic acid.[222] The mass spectrum of the new pentacyclic triterpene lup-20(29)-ene-3α,23-diol has been reported.[223]

The molecular ions of the unsaturated pentacyclic triterpenes pentanolic acid,[224] 3-oxo-6β-hydroxyolean-12-en-28-oic acid,[225] and spinosic acid A[226]

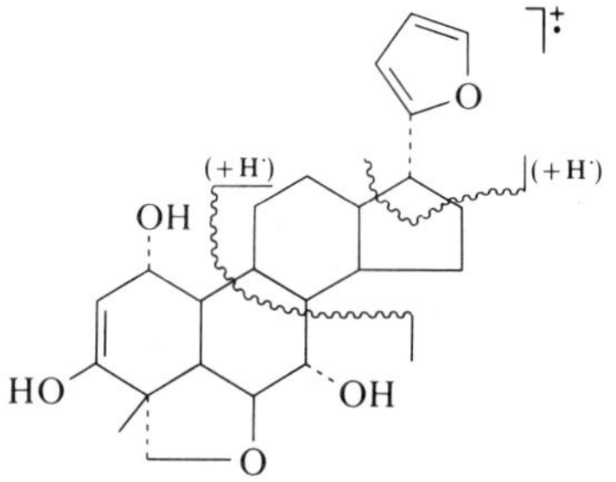

(71)

[211] W. B. T. Cruse, M. N. G. James, A. A. Al-Shamma, J. K. Beal, and R. W. Doskotch, *Chem. Comm.*, 1971, 1278.
[212] B. Rodriguez-Gonzalez, *Anales Quím.*, 1971, **67**, 59, 73, 85.
[213] A. I. Kalinovsky, E. P. Serebryakov, B. M. Zolotarev, A. V. Simolin, V. F. Kucherov, and O. S. Chizhov, *Org. Mass Spectrometry*, 1970, **3**, 1393.
[214] E. P. Serebryakov, A. V. Simolin, V. F. Kucherov, and B. V. Rosinov, *Tetrahedron*, 1970, **26**, 5215.
[215] M. N. Galbraith, D. H. S. Horn, and J. M. Sasse, *Chem. Comm.*, 1971, 1362.
[216] W. H. Hui, K. K. Ng, N. Fukamiya, M. Koreeda, and K. Nakanishi, *Phytochemistry*, 1971, **10**, 1617.
[217] R. C. Cambie and K. P. Mathai, *Austral. J. Chem.*, 1971, **24**, 1251.
[218] C. R. Enzell and I. Wahlberg, *Acta Chem. Scand.*, 1970, **24**, 2498.
[219] T. R. Govindachari, N. Viswanathan, and P. A. Mohamed, *Tetrahedron*, 1971, **27**, 4991.
[220] D. P. Chakraborty, S. P. Basak, B. C. Das, and R. Beugelmans, *Phytochemistry*, 1971, **10**, 1367.
[221] C. R. Mitra, H. S. Garg, and G. N. Pandey, *Phytochemistry*, 1971, **10**, 857.
[222] W. Sandermann and H. Funke, *Naturwissenhaften*, 1970, **57**, 408.
[223] W. H. Hui and W. K. Lee, *J. Chem. Soc.* (C), 1971, 1004.
[224] A. K. Barua, P. Chakrabarti, S. P. Dutta, D. K. Mukherjee, and B. C. Das, *Tetrahedron*, 1971, **27**, 1141.
[225] I. Wahlberg and C. R. Enzell, *Acta Chem. Scand.*, 1971, **25**, 70, 352.
[226] R. T. Aplin, W. H. Hui, C. T. Ho, and C. W. Yee, *J. Chem. Soc.* (C), 1971, 1067.

undergo the characteristic retro-Diels–Alder cleavage of the C ring, whereas those of triterpenes of the serrantenediol family suffer a similar cleavage of the D ring.[227] Mass spectrometry has aided the identification of squalene and norsolanese ($C_{44}H_{72}$) from Greek tobacco,[228] and has shown that squalene homologues can be formed enzymically from farnesyl pyrophosphate homologues.[229] The characteristic skeletal rearrangement ions $M - 92$, $M - 106$, and $M - 158$ in carotenoid spectra have been investigated using deuterium labelling.[230] Other rearrangement processes of isoprenoids have been investigated.[231] The mass spectra of a variety of carotenoids have been described including carotenoids of the Rhodopinal series,[232] carotenoids from the starfish *Asterias rubens*,[233] and a new $C_{40}H_{56}$ carotenoid.[234] Carotenes of the siphonaxanthin type [*e.g.* (72)]

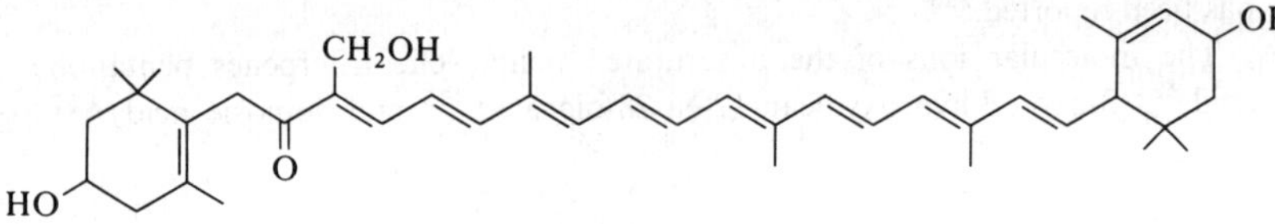

(72)

undergo a skeletal rearrangement in their molecular ions with the formation of an $M - CO_2$ species, which is characteristic of the arrangement of carbonyl and hydroxymethyl substituents.[235]

7 Steroids

The major fragmentation routes of steroids are now well defined, and a recent review covers the numerous types of fragmentation pattern found in different steroids.[236] However, there is still much effort directed towards the formation of steroidal derivatives suitable for the location of functional groups and the determination of stereochemistry, and for g.c.–m.s. (A full review of the applications of g.c.–m.s. in the steroid field appears in Chapter 7.)

---

[227] Y. Inubushi, T. Hibino, T. Harayama, T. Hasegawa, and R. Somanathan, *J. Chem. Soc. (C)*, 1971, 3109.
[228] C. R. Enzell, B. Kimland, and L.-E. Gunnarsson, *Tetrahedron Letters*, 1971, 1983.
[229] K. Ogura, T. Koyama, and S. Seto, *J. Amer. Chem. Soc.*, 1972, **94**, 308.
[230] H. Kløsen and S. Liaaen-Jensen, *Acta Chem. Scand.*, 1971, **25**, 85.
[231] U. T. Bhalerao and H. Rapoport, *J. Amer. Chem. Soc.*, 1971, **93**, 105.
[232] G. W. Francis and S. Liaaen-Jensen, *Acta Chem. Scand.*, 1970, **24**, 2705.
[233] G. W. Francis, R. R. Upadhyay, and S. Liaaen-Jensen, *Acta Chem. Scand.*, 1970, **24**, 3050.
[234] N. Arpin, J.-L. Fiasson, M. P. Bouchez-Dangye-Caye, G. W. Francis, and S. Liaaen-Jensen, *Phytochemistry*, 1971, **10**, 1595.
[235] H. H. Strain, W. A. Svec, K. Aitzetmüller, B. T. Cope, A. L. Harkness, and J. J. Katz, *Org. Mass Spectrometry*, 1971, **5**, 565.
[236] H. Budzikiewicz, in 'Biochemical Applications of Mass Spectrometry', ed. G. R. Waller, Wiley, New York, 1972, Chap. 10.

Natural Products
183

A novel derivative suggested for 3-keto-steroids is the corresponding pyrazole (73).[237] These compounds fragment by a retro-Diels–Alder mechanism in ring A, permitting the determination of the substitution of ring A.[237] However,

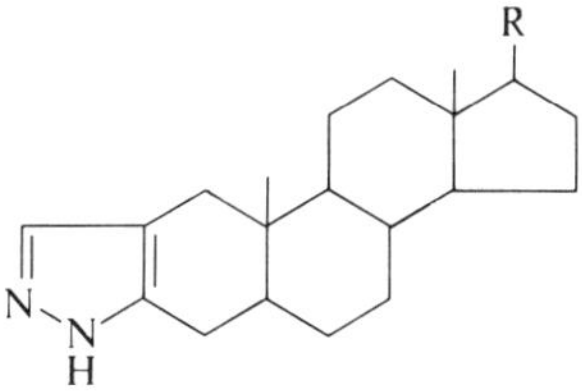

(73)

this may also be accomplished by the more readily synthesized ethylene ketals. Retention of charge by the fragment containing C-3 has been studied for a wide range of 3-substituents.[238,239] The 3-ethylene ketal and 3-amino (secondary or tertiary) substituents gave the greatest degree of charge retention and so would be most suitable for the determination of functionalization of ring A. High-resolution studies have been carried out on a variety of ring A substituted steroids,[240,241] and the 2-spiro-2'-(1',3'-dithian) derivative has been shown to have a fragmentation pattern diagnostic for 19-nor or 10-methyl steroids.[241] Indolosteranes, whether with further substitution of the steroid nucleus or not, give little diagnostically useful information.[242] The pyrrolidide group has also been studied as a potentially useful derivative.[243] It induces fragmentation at the more distant β- and γ-bonds, in contrast to the α-fragmentations associated with the more common derivatives. The influence of a hydroxy-group on the fragmentation patterns of steroidal amines, amides, and imines has been suggested as a possible means of locating the hydroxy-group on the steroid nucleus.[244]

The determination of stereochemistry of substituents on the steroid nucleus has been approached by the study of rearrangement ions [*e.g.* (74), m/e 147] in the mass spectra of the trimethylsilyl ethers of steroidal phosphates [*e.g.* (75)],[245] and steroidal diols and triols.[246] The trimethylsilyl group migrations involved were found to be specific processes dependent on the relative positions of the interacting groups. Thus, formation of rearrangement ions [*e.g.* (74)] reflects

---

[237] H.-É. Audier, J. Bottin, M. Fetizon, and J.-C. Gramain, *Bull. Soc. chim France*, 1971, 4027.
[238] P. Longevialle, M. Tassel, N. Zylber, and F. Khuong-Huu, *Compt. rend.*, 1971, **272**, C, 2073.
[239] C. R. Narayanan and A. K. Lala, *Org. Mass Spectrometry*, 1972, **6**, 119.
[240] J. M. Midgley, B. J. Millard, and W. B. Whalley, *J. Chem. Soc. (C)*, 1971, 13.
[241] J. M. Midgley, B. J. Millard, and W. B. Whalley, *J. Chem. Soc. (C)*, 1971, 19.
[242] D. J. Harvey, W. A. Laurie, and R. I. Reed, *Org. Mass Spectrometry*, 1971, **5**, 1183, 1189.
[243] W. Vetter, W. Walther, and M. Vecchi, *Helv. Chim. Acta*, 1971, **54**, 1599.
[244] P. Longevialle, J. Einhorn, J. P. Alazard, L. Diatta, P. Milliet, C. Monneret, Q. Khuong-Huu, and X. Lusinchi, *Org. Mass Spectrometry*, 1971, **5**, 171.
[245] D. J. Harvey, M. G. Horning, and P. Vouros, *Tetrahedron*, 1971, **27**, 4231.
[246] S. Sloan, D. J. Harvey, and P. Vouros, *Org. Mass Spectrometry*, 1971, **5**, 789.

the stereochemistry of the substituents, and its occurrence could be structurally informative. The elimination of trimethylsilanol in the mass spectra of a number of trimethylsilyl derivatives of steroidal di- and tri-ols has recently been investigated.[247] This appears to be a specific process, dependent on the structural

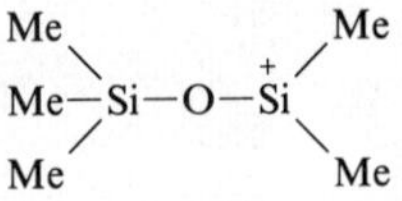

environment of the trimethylsiloxy-groups. Differences in the stereochemistry of the steroid nucleus appear to be reflected in different energy requirements for bond cleavage.[248] Thus, ionization and appearance potential measurements on the molecular ion and m/e 224 in the spectra of D-homoequilenine (76; R = α-H) and 14β-iso-D-homoequilenine (76; R = β-H) suggest that less energy

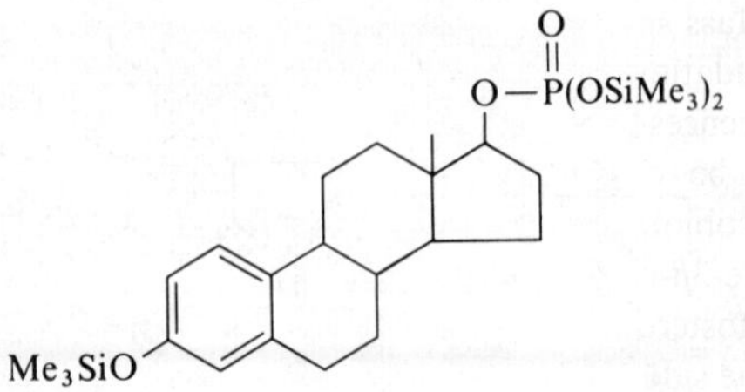

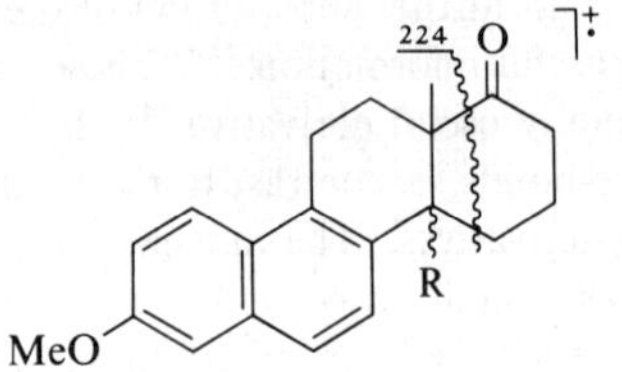

is needed for cleavage of the same bonds in the strained *cis* fused ring than in the less strained *trans* fused isomer. Stereochemical effects on the fragmentation of other steroids have also been observed.[249] Spiteller *et al.* have carried out much work on the location of functional groups, and stereochemical effects in trihydroxyandrostanes,[250] dihydroxyandrostan-17-ones,[251,252] hydroxyandro-

---

[247] P. Vouros and D. J. Harvey, *J.C.S. Chem. Comm.*, 1972, 765.
[248] V. I. Zaretskii, V. L. Sadovskaya, N. S. Wulfson, V. F. Sizoy, and V. G. Merimson, *Org. Mass Spectrometry*, 1971, **5**, 1179.
[249] H.-W. Fehlhaber, D. Lenoir, and P. Welzel, in 'Advances in Mass Spectrometry', ed. A. Quayle, Institute of Petroleum, London, 1971, p. 689.
[250] H. Obermann, M. Spiteller-Friedmann, and G. Spiteller, *Tetrahedron*, 1971, **27**, 1093.
[251] H. Obermann, M. Spiteller-Friedmann, and G. Spiteller, *Tetrahedron*, 1971, **27**, 1101.
[252] H. Obermann, M. Spiteller-Friedmann, and G. Spiteller, *Tetrahedron*, 1971, **27**, 1747.

Natural Products 185

stanediones,[253] and hydroxypregnanediones.[254] Djerassi *et al.* have investigated the role of the relief of ring strain in determining the site of charge localization in fragmentation of ring D of steroidal ketones.[255] This is a diagnostically important fragmentation and they conclude that the relative stabilities of product ions and neutrals expelled exert the greater effect.

Mass spectrometry has continued to be of great importance in the structural elucidation of steroids isolated from natural sources. There are too many references for individual mention here, but a few of the more interesting examples will be cited. The structures of androst-4-en-18-ol-3,17-dione, an unusual microbiological 18-hydroxylation product,[256] (24S)-24-methyl-5α-cholest-7,16-diene-3β-ol, the first natural sterol with a Δ^{16}-double bond,[257] 23-demethylgorgosterol, the second marine sterol isolated containing a cyclopropane ring in the side-chain,[258] 5α-pregn-9(11)-ene-3β,6α-diol-20-one and 5α-cholesta-9(11), 20(22)-diene-3β,6α-diol-23-one, two marine sterols with the unusual $\Delta^{9(11)}$-double bond,[259] and bufotalin, a bufadienolide,[260] have all been confirmed with the aid of mass spectrometry. In addition, the structure elucidation of a number of important vitamin D metabolites, 21,25-dihydroxycholecalciferol,[261] 25,26-dihydroxycholecalciferol,[262] and 1,25-dihydroxycholecalciferol,[38,263] has been achieved with considerable help from mass spectrometry. The mass spectra of several classes of naturally occurring steroids have been studied in detail. In a survey of the high-resolution mass spectra of several bufadienolides and derivatives, *e.g.* bufalin (77), Brown *et al.* used the 17β-(2-pyrone) functionality as an internal label, and concluded that the prognosis for structure elucidation of bufadienolides by mass spectrometry was good.[264]

The FI spectra of cardenolides have been studied, and the molecular weights and sequences of up to three underivatized monosaccharide residues determined.[172] Important and structurally diagnostic fragmentations of cardenolides have been discovered, especially using FI.[143] Unambiguous location of hydroxy-groups upon any of the digitoxose residues of the cardenolide, gitoxin, by a combination of acetylation, deuterioacetylation, and mass spectrometry has been reported.[144] The mass spectra of steroidal sapogenins have been studied

---

[253] H. Obermann, M. Spiteller-Friedmann, and G. Spiteller, *Tetrahedron*, 1971, **27**, 1737.
[254] M. Ende and G. Spiteller, *Monatsh.*, 1971, **102**, 929.
[255] S. Popov, G. Eadon, and C. Djerassi, *J. Org. Chem.*, 1972, **37**, 155.
[256] B. J. Auret and H. L. Holland, *Chem. Comm.*, 1971, 1157.
[257] L. I. Strigina, Y. N. Elkin, and G. B. Elyakov, *Phytochemistry*, 1971, **10**, 2361.
[258] F. J. Schmitz and T. Pattabhiraman, *J. Amer. Chem. Soc.*, 1970, **92**, 6073.
[259] Y. M. Sheikh, B. M. Tursch, and C. Djerassi, *J. Amer. Chem. Soc.*, 1972, **94**, 3278.
[260] G. R. Pettit, P. Brown, F. R. Brüschweiler, and L. E. Houghton, *Chem. Comm.*, 1970, 1566.
[261] T. Suda, H. F. DeLuca, H. K. Schnoes, G. Ponchon, Y. Tanaka, and M. F. Holick, *Biochemistry*, 1970, **9**, 2917.
[262] T. Suda, H. F. DeLuca, H. K. Schnoes, Y. Tanaka, and M. F. Holick, *Biochemistry*, 1970, **9**, 4776.
[263] M. F. Holick, H. K. Schnoes, H. F. DeLuca, T. Suda, and R. J. Cousins, *Biochemistry*, 1971, **10**, 2799.
[264] P. Brown, Y. Kamano, and G. R. Pettit, *Org. Mass Spectrometry*, 1972, **6**, 47, 613.

in detail,[265,266] and the effects of changes in stereochemistry,[265] and the influence of substitution in rings D[266] and F,[265,266] on their fragmentation patterns determined.

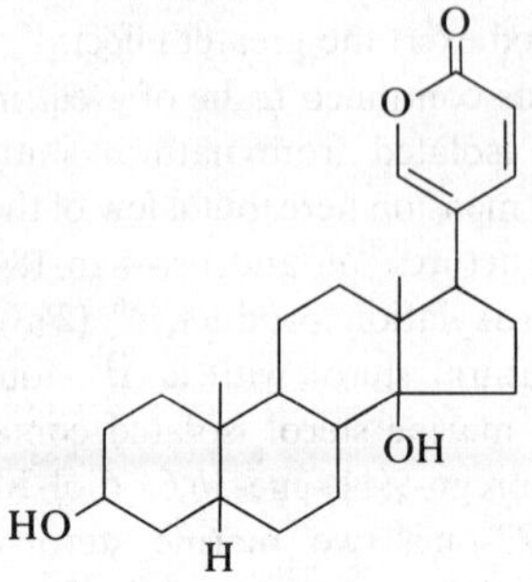

(77)

8 Alkaloids

It is now unusual for the structure of an alkaloid to be published without prior consideration of its mass spectral characteristics. The fragmentations of the major classes of alkaloids have been described,[267] and a recent compilation containing references to alkaloids is available.[268] Few new principles have appeared during the reviewing period, but a selection of the more interesting examples of the application of mass spectrometry to the alkaloid field follow.

Pyrrole and Indole Alkaloids.—The initial cleavage of ipalbidine (78)[269] occurs by a simple retro-Diels–Alder reaction, and the spectra of the related mesembrenol[270] and erythrina alkaloids[271] have been described. The structure of a complex pyrrolizidine alkaloid has been reported,[272] and a detailed discussion of the fragmentations of pyrrolizidines has appeared.[273] For example, the primary decompositions of the 7-angeloylheliotridine molecular ion are shown in (79).[274]

[265] W. H. Faul and C. Djerassi, *Org. Mass Spectrometry*, 1970, **3**, 1187.
[266] H. Budzikiewicz, K. Takeda, and K. Schreiber, *Monatsch.*, 1970, **101**, 1003.
[267] H. Budzikiewicz, C. Djerassi, and D. H. Williams, 'Structure Elucidation of Natural Products by Mass Spectrometry', Vol. 1, Alkaloids, Holden Day, San Francisco, 1964; see also 'The Alkaloids', ed. J. E. Saxton (Specialist Periodical Reports), The Chemical Society, London, 1972, Vol. 2, pp. 49, 93, 175—177, 204, 205, 213, 264, and 284.
[268] A. L. Burlingame and G. A. Johanson, *Analyt. Chem.*, 1972, **44**, 337R.
[269] A. E. Wick, P. A. Bartlett, and D. Dolphin, *Helv. Chim. Acta*, 1971, **54**, 513.
[270] P. W. Jeffs, G. Ahmann, H. F. Campbell, D. S. Farrier, G. Ganguli, and R. L. Hawks, *J. Org. Chem.*, 1970, **35**, 3512.
[271] A. Mondon and P.-R. Seidel, *Chem. Ber.*, 1971, **104**, 2937.
[272] K. B. Birnbaum, A. Klásek, P. Sedmera, G. Snatzke, L. F. Johnson, and F. Šantavý, *Tetrahedron Letters*, 1971, 3421.
[273] E. Peterson and E. Larsen, *Org. Mass Spectrometry*, 1970, **4**, 249.
[274] W. D. Jamiesen and O. Hutzinger, *Phytochemistry*, 1970, **9**, 2029.

Natural Products

It has been shown that simple 3-substituted naturally occurring indoles may be identified by mass spectrometry.[274] The basic cleavages of bis-secodihydrocyclopiazonic acid have been described[275] and the primary breakdown of the

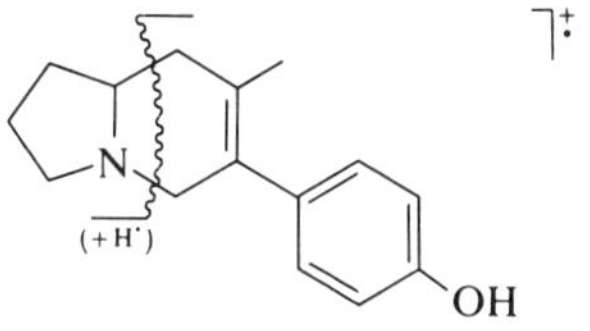

(78)

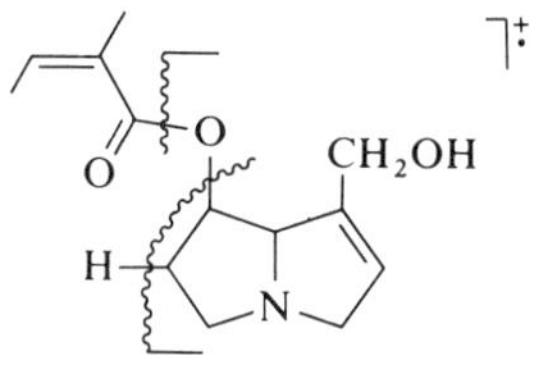

(79)

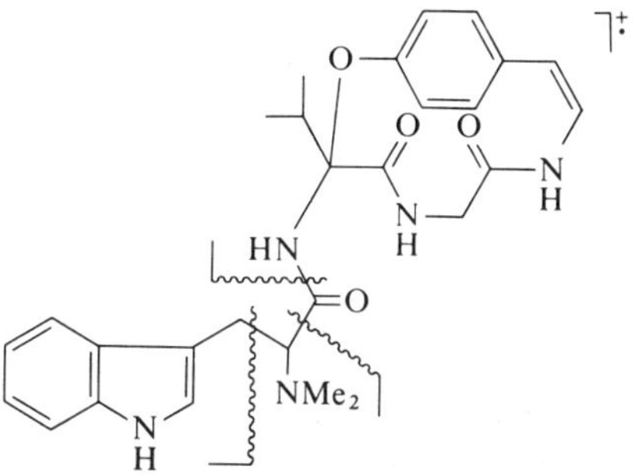

(80)

peptide alkaloid amphibin A is indicated in (80).[276] Mass spectrometry has been used extensively for the determination of the structures of a quebrachamine like alkaloid from *Amsonia tabernaemontana*,[277] the modified tetrahydro-β-carboline alkaloids talbotin,[278] usambarine,[279] alstonisidine,[280] N^a-demethylseredamin,[281] and O-3,4,5-trimethoxycinnamoylvincamajine,[282] and the aspidospermum alkaloid cathanneine.[283] The power of mass spectrometry as a structural aid is elegantly demonstrated by the behaviour of gozilin (81)[284] and spermostrychine (82) → (83)[285] upon electron impact.

[275] C. W. Holzapfel, R. D. Hutchison, and D. C. Wilkins, *Tetrahedron*, 1970, **26**, 5239.
[276] R. Tschesche, E. U. Kaussmann, and H.-W. Fehlhaber, *Tetrahedron Letters*, 1972, 865.
[277] B. Zsadon and J. Tamás, *Chem. and Ind.*, 1972, 32.
[278] M. Pinar, M. Hanaoka, M. Hesse, and H. Schmid, *Helv. Chim. Acta*, 1971, **54**, 15.
[279] M. Koch and M. Plat, *Compt. rend.*, 1971, **273**, C, 753.
[280] J. M. Cook and P. W. LeQuesne, *J. Org. Chem.*, 1971, **36**, 582.
[281] M. Hanaoka, M. Hesse, and H. Schmid, *Helv. Chim. Acta*, 1970, **53**, 1723.
[282] W. D. Crow, N. C. Hancox, S. R. Johns, and J. A. Lamberton, *Austral. J. Chem.*, 1970, **23**, 2489.
[283] G. H. Aynilian, M. Tin-Wa, N. R. Farnsworth, and M. Gorman, *Tetrahedron Letters*, 1972, 89.
[284] A. Agwada, M. B. Patel, M. Hesse, and H. Schmid, *Helv. Chim. Acta*, 1970, **53**, 1567.
[285] I. Iwataki and J. Comin, *Tetrahedron*, 1971, **27**, 2541.

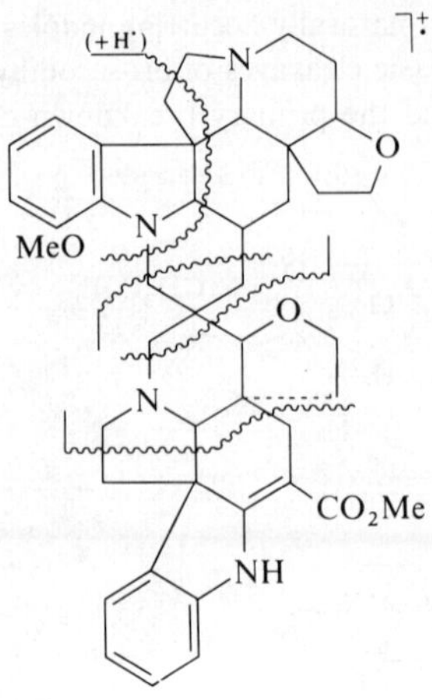

Pyridine, Quinoline, and Isoquinoline Alkaloids.—Securinol C fragments by the cleavages outlined in (84).[286] The spectra of cantleyine[287] a monoterpene alkaloid artefact, patchonerpyridine[288] a sesquiterpene alkaloid, alkaloids of the evonic acid family,[289] and evonine[290] have been reported. The characteristic decompositions of the molecular ion of skeletrum alkaloid A_4 are shown in (85).[291,292]

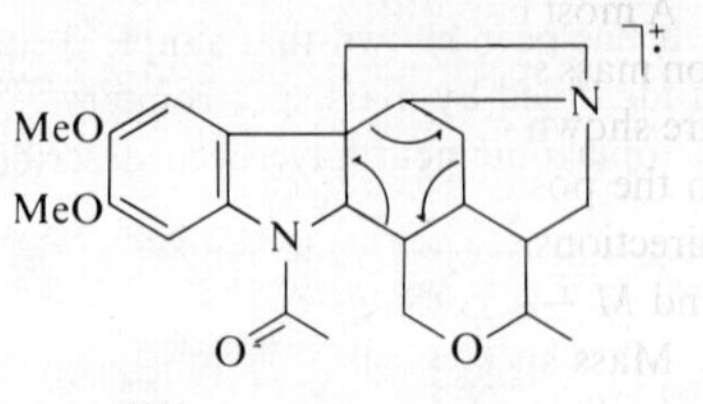

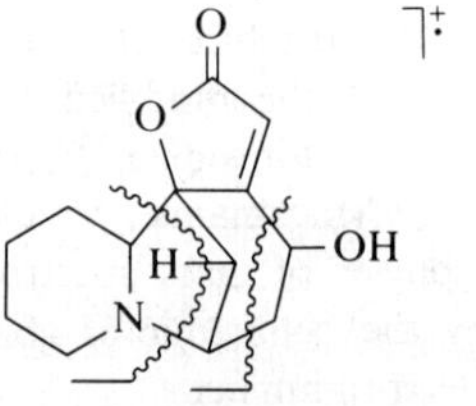

[286] Z. I. Horri, M. Yamauchi, M. Ikeda, and T. Momose, *Chem. and Pharm. Bull. (Japan)*, 1970, **18**, 2009.
[287] T. Sévenet, B. C. Das, J. Parello, and P. Potier, *Bull. Soc. chim. France*, 1970, 3120.
[288] M.-C. Cren, C. Defaye, and M. Fetizon, *Bull. Soc. chim France*, 1970, 3020.
[289] A. Klásek, F. Šantavy, A. M. Duffield, and T. Reichstein, *Helv. Chim. Acta*, 1971, **54**, 2144.
[290] M. Pailer, W. Streicher, and J. Leitich, *Monatsch.*, 1971, **102**, 1873.
[291] P. W. Jeffs, P. A. Luhan, A. T. McPhail, and N. H. Martin, *Chem. Comm.*, 1971, 1466.
[292] F. O. Snyckers, F. Strelow, and A. Wiechers, *Chem. Comm.*, 1971, 1467.

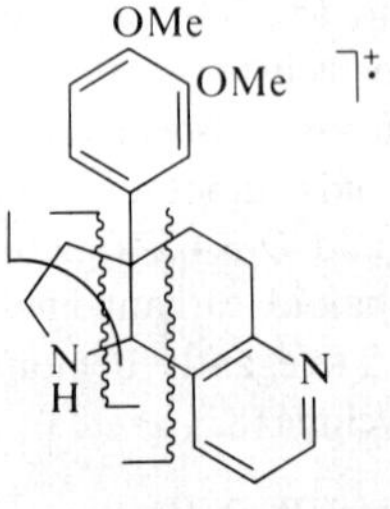

A most interesting example of the complementary use of negative- and positive-ion mass spectrometry has been reported for veralkamine.[293] The main cleavages are shown in (86) and it should be noted that the position of cleavage is the same in the positive and negative mode, but that charge retention occurs in opposite directions. The negative-ion spectrum also exhibits pronounced molecular and $M - 1$ peaks.

Mass spectrometry has aided the structure elucidations of a rare quaternary quinoline alkaloid from *Ptelia trifoliata*,[294] thalictrum alkaloids,[295] tetrahydroisoquinoline alkaloids,[296] bisbenzylisoquinoline alkaloids,[297–299] and an isoquinoline erythrina alkaloid,[300] The molecular ion of bulbocodin (87) is reported

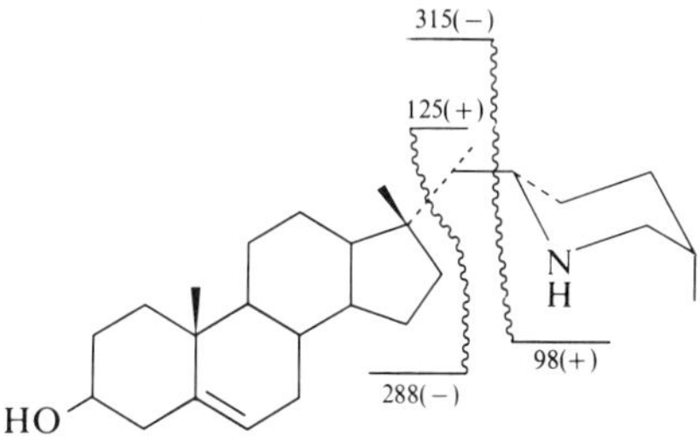

(86) 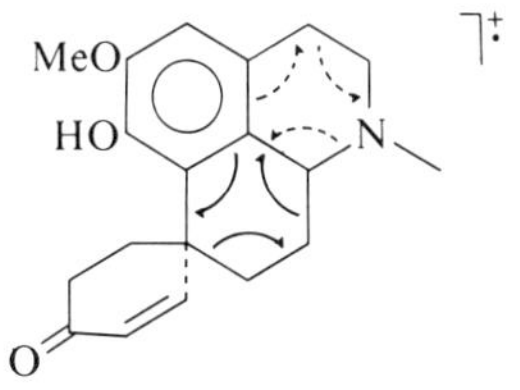　　(87)

to undergo two competing retro-Diels–Alder eliminations.[301] The mass spectra of several berberine-like alkaloids[302–304] and 2,3 : 7,8-bismethylenedioxybenz[c]phenanthridine methochloride[305] have been listed.

Miscellaneous Alkaloids.—The cleavages of the imidazole alkaloid cyphlophine are illustrated in (88),[306] and the molecular ion of vasicolin suffers the characteristic cyclization depicted in (89).[307] Dibromophakellin (90), an alkaloid isolated from a marine sponge (*Phakellia flabellata*) shows $M - NH_3$ and $M - NH_2CN$

[293] G. Adam, *Z. Chem.*, 1970, **10**, 241.
[294] L. A. Mitscher, M. S. Bathala, and J. L. Beal, *Chem. Comm.*, 1971, 1040.
[295] M. Shamma and M. A. Podczasy, *Tetrahedron*, 1971, **27**, 727.
[296] H. Koshiyama, H. Ohkuma, H. Kawaguchi, H.-Y. Hsü, and Y.-P. Chen, *Chem. and Pharm. Bull. (Japan)*, 1970, **18**, 2564.
[297] W. Sneddon, R. B. Parker, and C. Gorinsky, *Org. Mass Spectrometry*, 1970, **4**, 607.
[298] C. K. Yu, J. K. Saunders, D. B. Maclean, and R. H. F. Manske, *Canad. J. Chem.*, 1971, **49**, 3020.
[299] C. K. Yu and D. B. Maclean, *Canad. J. Chem.*, 1971, **49**, 3025.
[300] R. B. Boar and D. A. Widdowson, *J. Chem. Soc. (B)*, 1970, 1591.
[301] F. Šantavy, P. Sedmera, G. Snatzke, and T. Reichstein, *Helv. Chim. Acta*, 1971, **54**, 1084.
[302] M. Shamma and M. J. Hillman, *Tetrahedron*, 1971, **27**, 1363.
[303] C. K. Yu, D. B. Maclean, R. G. A. Rodrigo, and R. H. F. Manske, *Canad. J. Chem.*, 1970, **48**, 3673.
[304] T. Kametani, M. Ihara, and T. Honda, *J. Chem. Soc. (C)*, 1970, 2342.
[305] M. Sainsbury, S. F. Dyke, and B. J. Moon, *J. Chem. Soc. (C)*, 1970, 1797.
[306] N. K. Hart, S. R. Johns, J. A. Lamberton, J. W. Loder, and R. H. Nearn, *Austral. J. Chem.*, 1971, **24**, 857.
[307] S. Johne, D. Gröger, and M. Hesse, *Helv. Chim. Acta*, 1971, **54**, 826.

ions in its mass spectrum, and these peaks are indicative of a guanidine unit.[308] The fragmentation of a peptide alkaloid has been considered in detail,[309] and the mass spectra of several steroidal alkaloids have been recorded.[310,311,244]

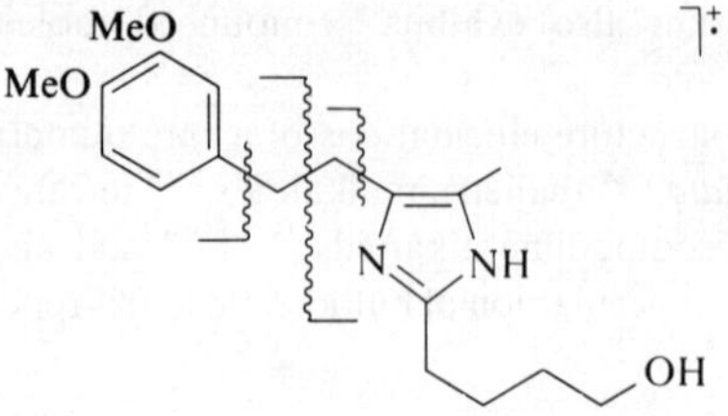

(88) 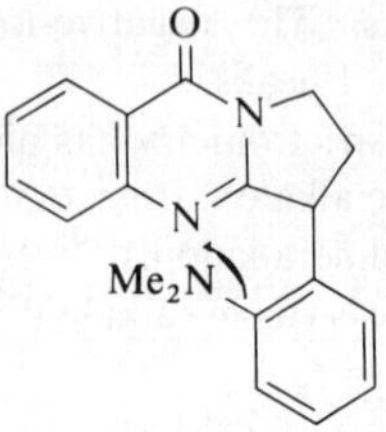   (89)

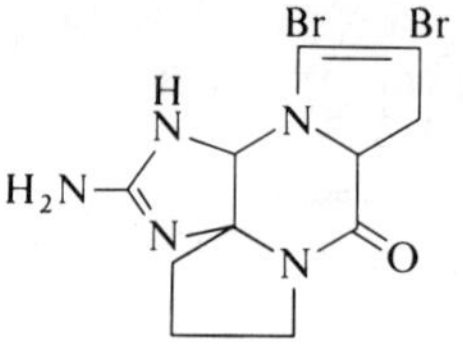

(90)

9 Flavonoids and Quinones

An unusual flavonol precursor from *Myrica gale* L. fragments as illustrated in (91).[312] The spectra of chalcones from *Flemingia chaptar* Ham have been reported.[313]

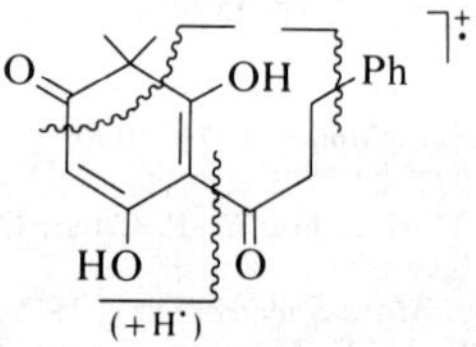

(91) 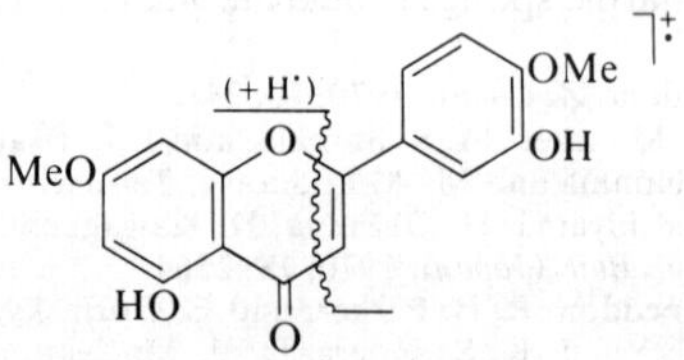   (92)

---

[308] G. M. Sharma and P. R. Burkholder, *Chem. Comm.*, 1971, 151.
[309] O. A. Mascaretti, V. M. Merkuza, G. M. Ferraro, E. A. Ruveda, C.-J. Chang, and E. Wenkert, *Phytochemistry*, 1972, **11**, 1133.
[310] C. Lukacs, P. Longevialle, and X. Lusinchi, *Tetrahedron*, 1971, **27**, 1891.
[311] P. Longevialle, A. Picot, and X. Lusinchi, *Compt. rend.*, 1970, **271**, C, 859.
[312] T. Anthonsen, I. Falkenberg, M. Laake, A. Midelfart, and T. Mortensen, *Acta Chem. Scand.*, 1971, **25**, 1929.
[313] B. Cardillo, A. Gennaro, L. Merlini, G. Nasini, and S. Servi, *Tetrahedron Letters*, 1970, 4367.

Natural Products

The retro-Diels–Alder cleavage of flavonol (or isoflavonol) systems generally occurs only when there are four (or less) substituents on the chromophore. Two examples of systems which undergo this cleavage are pelloin (92)[314] and alpinumisoflavone (93),[315] but in the latter case, the reaction occurs only after elimination of a methyl radical. A recent study[316] has confirmed (cf. ref. 317) that polymethoxyflavonols eliminate a methyl radical from the 6-methoxy-group, but it has been noted that if the flavonol contains a 3-hydroxy-6-methoxy substitution pattern the loss of methyl from the 6-position does not occur.[318] A similar elimination of methyl occurs from the molecular ions of 8-methoxyflavonols.[319,320] The substitution pattern indicated in (94; X = H, OH, or OMe) can be detected by the presence of an $M - X\cdot$ peak arising by a cyclization process.[316,321] Mass spectrometry has been used for the structure determination of bisflavones,[322,323] coumarins,[324–326] including (95),[325] and xanthones.[327,328]

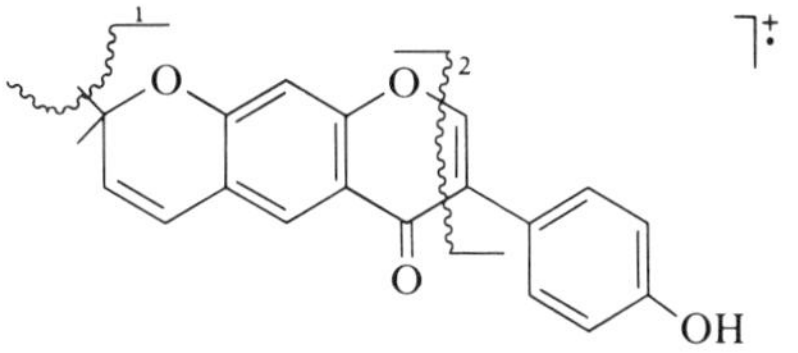

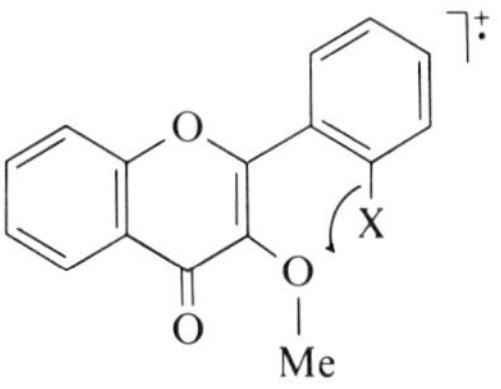

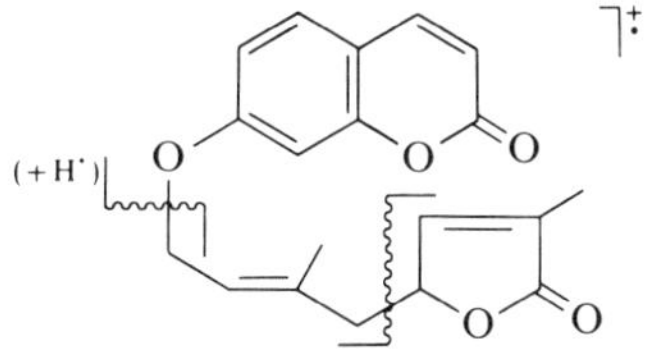

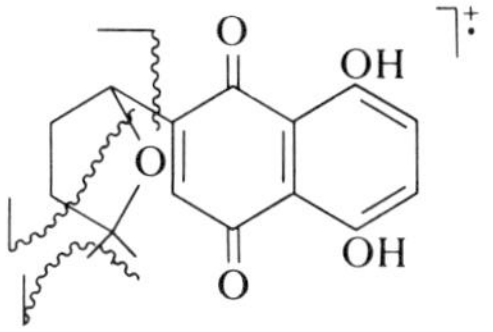

[314] J. Núñez-Alarcón, *J. Org. Chem.*, 1971, **36**, 3829.
[315] B. Jackson, P. J. Owen, and F. Scheinmann, *J. Chem. Soc.* (C), 1971, 3389.
[316] D. G. I. Kingston, *Tetrahedron*, 1971, **27**, 2691.
[317] J. H. Bowie and D. W. Cameron, *Austral. J. Chem.*, 1966, **19**, 1627.
[318] D. M. Smith, G. W. Glennie, and J. B. Harborne, *Phytochemistry*, 1971, **10**, 3115.
[319] J. G. Nielsen, P. Nørgaard, and H. Hjeds, *Acta Chem. Scand.*, 1970, **24**, 724.
[320] J. G. Nielsen and J. Møller, *Acta Chem. Scand.*, 1970, **24**, 2665.
[321] R. B. Filho and O. R. Gottlieb, *Phytochemistry*, 1971, **10**, 2433.
[322] H. D. Locksley and I. G. Murray, *J. Chem. Soc.* (C), 1971, 1332.
[323] H. S. Garg and C. R. Mitra, *Phytochemistry*, 1971, **10**, 2787.
[324] I. Carpenter, E. J. McGarry, and F. Scheinmann, *J. Chem. Soc.* (C), 1971, 3783.
[325] I. Carpenter, E. J. McGarry, and F. Scheinmann, *Tetrahedron Letters*, 1970, 3983.
[326] F. Bohlmann and E. Clausen, *Chem. Ber.*, 1970, **103**, 3619.
[327] R. K. Chaudhuri and S. Ghosal, *Phytochemistry*, 1971, **10**, 2425.
[328] T. R. Govindachari, P. S. Kalyanaraman, N. Muthukumaraswamry, and B. R. Pai, *Tetrahedron*, 1971, **27**, 3919.

Certain aspects of the mass spectra of quinones have been reviewed.[329,330] The molecular ion of an unusual naphthaquinone from *Arnebia nobilis* undergoes the primary cleavages depicted in (96).[331] The red pigment of the sea-worm *Hallaparthenopeia* shows $M - \text{Me}\cdot - \text{CO} - \text{CO}$ and $M + 2$ peaks in its mass spectrum, and is thought to correspond to the *o*-quinone (97).[332]

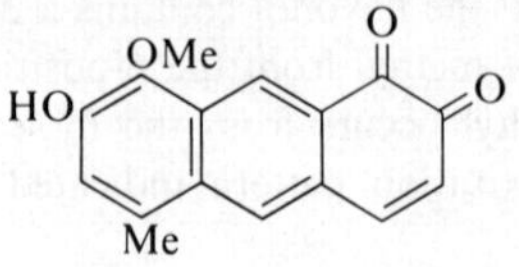

(97)

---

[329] J. H. Bowie, 'Mass Spectrometry of Carbonyl Compounds', in 'The Chemistry of the Carbonyl Group', ed. J. Zabicky, Interscience, London, 1970, Vol. 2.
[330] R. H. Thomson, 'Naturally Occurring Quinones', Academic Press, London, New York, 1971, pp. 78—89.
[331] Y. N. Shukla, J. S. Tandon, D. S. Bhakuni, and M. M. Dhar, *Phytochemistry*, 1971, **10**, 1909.
[332] G. Prota, M. D'Agostino, and G. Misuraca, *Experientia*, 1971, **27**, 15.

5
Organometallic and Co-ordination Compounds

BY M. I. BRUCE

1 Introduction

This account of the mass spectra of organometallic and co-ordination compounds generally follows the pattern established in Volume 1 of this series,[1] and continues the survey up to approximately March 1972. The majority of appropriate references in the *Mass Spectrometry Bulletin* up to and including the May 1972 issue have been consulted. The reviewer has had to compromise between a consideration of the ever-growing number of papers which report either partial or even complete mass spectral data, with little or no comment or analysis, and the considerably fewer publications which are giving serious consideration to the mechanisms of reaction occurring after ionization. The main discussion which follows is therefore confined to details of fragmentation patterns, and an Appendix indicates a majority of the many papers in which mass spectral data are listed but not further considered.

Generally, the past two years have seen few major advances in our understanding of the effect of metals on the fragmentation of attached ligands. Rather, new accounts have been limited to detailed studies of the spectra of new classes of metal derivatives. Some studies of ion–molecule reactions have also been reported recently.

Reviews.—Recent reviews include King's account of his work in this area,[2] a summary of the situation in 1969 with regard to metallocene and related derivatives,[3] and uses of mass spectrometry in the identification of transition-metal cluster compounds containing π-acid ligands.[3a] The proceedings of a conference held in Kyoto, Japan, in 1969, contain papers on studies of negative ions from metal complexes,[4] and from cobalt and manganese carbonyl derivatives;[5] a summary of the results obtained with substituted benzene-Cr(CO)$_3$ complexes;[6] and an account of the alteration of fragmentation patterns of

---
[1] M. I. Bruce, in 'Mass Spectrometry', ed. D. H. Williams (Specialist Periodical Reports), The Chemical Society, London, 1971, Vol. 1, p. 182.
[2] R. B. King, *Fortschr. Chem. Forsch.*, 1970, **14**, 92.
[3] G. A. Junk and H. J. Svec, in 'Recent Topics in Mass Spectrometry', ed. R. I. Reed, Gordon and Breach, New York, 1971, p. 85.
[3a] R. D. Johnston, *Adv. Inorg. Chem. Radiochem.*, 1970, **13**, 471.
[4] R. W. Kiser, in 'Recent Developments in Mass Spectroscopy', ed. K. Ogata and T. Hayakawa, University Park Press, Baltimore, 1970, p. 844.
[5] F. E. Saalfeld, M. V. McDowell, and A. G. MacDiarmid, ref. 4, p. 1073.
[6] J. Müller and P. Gose, ref. 4, p. 1175.

organic molecules (mainly porphin derivatives) by the formation of metal complexes.[7] A brief general review of applications of mass spectrometry to organometallic chemistry has appeared.[8]

General Considerations.—A method of discerning multiple sources of fragment ions, by deconvolution of ionization efficiency data, has been developed,[9] and its use is described in an analysis of the spectra of CrO_2F_2 and CrO_2Cl_2.

Two further studies of the replacement of metals in their complexes by other metals present in the mass spectrometer have been reported, and it is apparent that this phenomenon can cause significant abundances of other 'mystery' ions. Thus replacement of copper by iron in a series of β-diketonate complexes was confirmed by high-resolution and IP measurements.[10] No replacement of chromium by iron was found. Metal–metal chelate exchange reactions have also been observed in the ion source and recorded in the spectra of transition-metal chelates containing diethyldithiocarbamate.[11] In particular, the iron(III) complex exhibits peaks arising from fragmentation of the analogous nickel complex, the amount of these depending on the condition of the source. Several experiments pointed to the filament region as the site of exchange.

Throughout this chapter, the symbol P^+ will be used for the parent peaks (molecular ions), to avoid confusion with M^+, which is reserved to indicate a metal ion.

2 Main-group Organometallics

Group I.—The mass spectra of mixtures of LiBu^t with LiOBu^t, and of LiBu^t with LiCH$_2$SiMe$_3$, demonstrate the existence of mixed species in the vapour phase, formed by exchange of groups.[12] Alkoxide groups are lost preferentially from tetrameric $Li_4R_n(OR)_{4-n}$ species. The relative volatilities of the tetramers are important in determining the nature of the ions present. The spectra of [LiBu^t]$_4$ containing both ^{6}Li and ^{7}Li isotopes showed a slow dissociative exchange of lithium atoms. In contrast, all species Li$_4$R$_3$(OR) instantaneously attain the equilibrium distribution of lithium isotopes. These differences are ascribed to the presence or absence of hexameric molecules, which dissociate readily:

$$R_6Li_6 \rightarrow R_2Li_2 + R_4Li_4$$

in contrast with the relatively stable tetramers.

Group II.—Further extensive studies on the spectra of organomercury compounds confirm the major process as simple bond cleavage with loss of an

---
[7] H. Budzikiewicz, ref. 4, p. 1210.
[8] T. R. Spalding, in 'Spectroscopic Methods in Organometallic Chemistry', ed. W. O. George, Butterworths, London, 1970, p. 97.
[9] G. D. Flesch and H. J. Svec, *J. Chem. Phys.*, 1971, **54**, 2681.
[10] H. F. Holtzclaw, R. L. Lintvedt, H. E. Baumgarten, R. G. Parker, M. M. Bursey, and P. F. Rogerson, *J. Amer. Chem. Soc.*, 1969, **91**, 3774.
[11] J. K. Terlouw and J. J. de Ridder, *Org. Mass Spectrometry*, 1971, **5**, 1127.
[12] M. Y. Darensbourg, B. Y. Kimura, G. E. Hartwell, and T. L. Brown, *J. Amer. Chem. Soc.*, 1970, **92**, 1236.

organic group. Generally, metal-containing ions form only a small proportion of the total ion current, e.g. 30% in $HgPh_2$, compared with 90% in $SnPh_4$. A variety of structural types were studied by Bryant and Kinstle.[13] In many cases, structures involving mercurinium ions are proposed for some of the ions, e.g. (1). Hydroxyalkylmercuric chlorides exhibit loss of ROHgCl, i.e. reversal of the oxymercuration reaction (see Scheme 1). Loss of olefin also occurs in a favoured

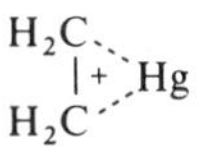

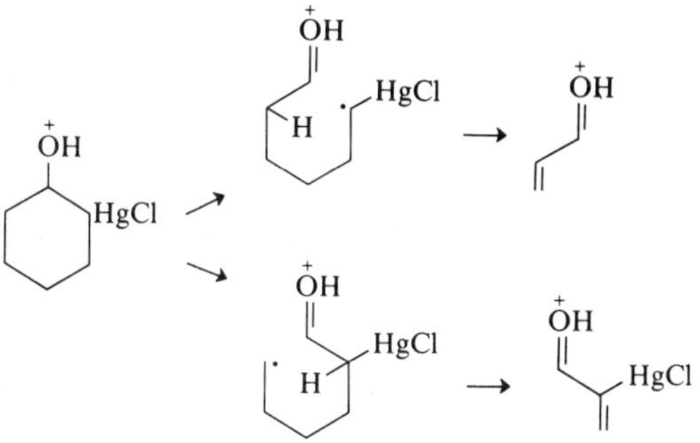

Scheme 1

process: the Hg—C bond is weak, and is readily cleaved in more sterically crowded structures.

Arylmercuric halides, ArHgX, lose ArX, Ar, or HgX, and the P^+ ions are often very strong, and may be the base peak. However, charge is retained on carbon rather than the metal. No 'ortho-effects' were observed in the spectra of ortho-substituted compounds, although loss of Me from $[MeC_6H_4HgCl]^+$ only occurs with the ortho-isomer. As expected, $[C_7H_7]^+$ is the predominant ion in the spectrum of dibenzylmercury.

In symmetrical diarylmercury compounds,[14] loss of an aryl group affords $[HgAr]^+$, which then eliminates mercury to give the aryl cation. Loss of mercury from the P^+ ion gives $[Ar_2]^+$. In the lower m/e regions, the fragmentation of this ion largely resembles that of the 'external standard' Ar_2. With these mercury compounds, fragmentation occurs by modes found in related metal-free systems, i.e. the presence of mercury does not direct the breakdown of these ions, in agreement with the postulate of charge located on carbon.

In the spectra of α-mercurated carbonyl compounds,[14] α-cleavage reactions are important, with loss of stable neutral moieties, e.g. CO, $CH_2=C=O$, Hg. With some compounds, such as $Hg[PhCH(CO_2Et)]_2$, extrusion of mercury is so ready that the spectrum is essentially that of $(PhCHCO_2Et)_2$. Introduction of mixtures of symmetrical HgR_2 compounds into the spectrometer results in the appearance of ions of the type $[HgR^1R^2]^+$, as well as of the mixed extrusion products, $[R^1R^2]^+$.

[13] W. F. Bryant and T. H. Kinstle, *J. Organometal. Chem.*, 1970, **24**, 573.
[14] S. W. Breuer, T. E. Fear, P. H. Lindsay, and F. G. Thorpe, *J. Chem. Soc.* (C), 1971, 3519.

The major positive ions in $Me_2C=C(HgCl)_2$ and $MeCH(HgCl)_2$ arise by loss of inorganic fragments, or are inorganic, e.g. Hg^+, $[Hg_2Cl]^+$.[15] Ions containing Hg—Hg bonds may be formed by a process such as

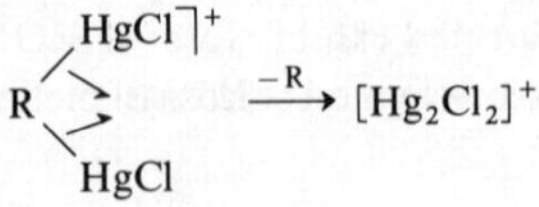

$$R \underset{HgCl}{\overset{HgCl}{\diagup}}\Bigg]^+ \xrightarrow{-R} [Hg_2Cl_2]^+$$

In both examples, ions formed by loss of methyl are found in substantial quantities, and loss of MeCl from $[Me_2C=CHgCl]^+$ gives an abundant $[MeC_2Hg]^+$ ion.

Elimination of $HgCl_2$ from trichlorovinylmercury compounds gives $[C_2Cl_2]^+$, although with $Hg(CCl=CCl_2)_2$, loss of C_2Cl_2 is favoured, giving $[HgC_2Cl_4]^+$; in these compounds the usual Hg—C bond cleavage reactions are not found.[16] The negative-ion spectra are also discussed. In the aromatic series, major ions from C_6Cl_5HgCl arise by the processes shown in Scheme 2. This contrasts

$$[C_6Cl_5HgCl]^+ \xrightarrow{-Cl} [C_6Cl_5Hg]^+$$
$$\downarrow \qquad\qquad 100\%$$
$$[C_6Cl_4]^+ \xrightarrow{-Cl} [C_6Cl_3]^+ \xrightarrow{-Cl} [C_6Cl_2]^+$$
$$95\%$$

Scheme 2

with the fluorocarbon analogues, where the base peak is formed by a multi-step reaction. The spectrum of the dichlorocarbene precursor, $PhHgCCl_3$, contains a P^+ ion, and other ions formed by loss of Cl or CCl_2, including $[HgPh]^+$. An important metal-free ion is $[CCl_3]^+$. Much simpler is the spectrum of $HgPh_2$, which shows a strong P^+ ion, $[HgC_6H_5]^+$, and $[C_{12}H_{10}]^+$.[17]

The spectrum of $Hg[C(N_2)CO_2Et]_2$ exhibits the usual ions, albeit of low intensity, associated with the fragmentation of esters.[18] Loss of the nitrogen atoms is followed by loss of Et, CO, and oxygen to give $[HgC_2]^+$ and $[HgC]^+$. Other diazoalkane derivatives, e.g. $(MeHg)_2CN_2$, give low abundance P^+ ions, which lose nitrogen; the resulting ions undergo cleavage of the Hg—C bond. In the corresponding ethyl compound, the nitrogen is retained in ions such as $[HgCN_2]^+$ and $[EtHgCN_2]^+$.

Analyses of many methylmercury compounds have been reported in connection with their occurrence in biological materials, and their detection via g.c.–m.s. techniques. In the halides, all possible ions formed by loss of X or Me groups, and by elimination of mercury, are found.[19] In other cases the ion $[HgMe_2]^+$ is found, although $[HgMe]^+$ is the base peak. Photo-ionization

---

[15] S. C. Cohen, *J. Chem. Soc. (A)*, 1971, 1571.
[16] S. C. Cohen, *J. Chem. Soc. (A)*, 1971, 632.
[17] A. K. Mal'tsev, R. G. Mikaelyan, and O. M. Nefedov, *Izvest. Akad. Nauk S.S.S.R., Ser. khim.*, 1971, 1179.
[18] J. Lorberth, *J. Organometallic Chem.*, 1971, **27**, 303.
[19] B. Johansson, R. Ryhage, and G. Westöö, *Acta Chem. Sound.*, 1970, **24**, 2349.

studies of MMe_2 (M = Zn, Cd, or Hg)[20,21] and of $MeHgX$ (X = Cl, Br, or I)[21] have also been reported. In the dialkyls, the principal ions are P^+, $[P - Me]^+$, M^+, and $[CH_3]^+$.

In $MeHgN_3$, loss of nitrogen followed by alkyl gives $[HgMe]^+$ and Hg^+ ions; $[HgN_3]^+$ is of low abundance.[22] The cyanamide $(MeHg)_2NCN$ fragments by loss of MeHgNCN; the ethyl derivative behaves similarly, but loss of ethylene and mercury also gives $[EtHgNHCN]^+$. Similar features were found with $N(HgMe)_3$ and $O(HgMe)_2$.[23]

Even in $RHgC_5H_5$ (R = Me or Et), formation of $[RHg]^+$ is the major process; $[HgC_5H_5]^+$ is of relatively low abundance in both cases.[22] Subsequent fragmentation gives R^+, $[C_5H_5]^+$, and Hg^+ ions; some coupling to form $[RC_5H_5]^+$ and derived ions is found.

Other organomercury compounds which have been studied show fragmentation patterns which are generally similar to those described above, processes involving Hg—C bond cleavage predominating, with charge retained on carbon. Photolytic and mass spectrometric processes have been compared[24] in $Hg(CH_2CH_2CN)_2$; the former process gave adiponitrile, and a fragmentation pattern of this compound is superimposed on that of the mercurial. A rapid equilibrium is achieved[25] in the spectrum of $Hg(GeMe_3)(SiMe_3)$ showing the ions $[C_6H_{18}Si_2Hg]^+$, $[C_6H_{18}GeSiHg]^+$, and $[C_6H_{18}Ge_2Hg]^+$.

Group III.—*Alkyl- and Aryl-borons, etc.* IP values for the series Et_nBCl_{3-n} (n = 1—3) compare well with similar data obtained by an empirical MO method.[26] The energy is a combination of an antibonding interaction between p-orbitals on two ligand atoms, and a bonding interaction between $p\pi$ ligand orbitals and the boron atom.

Although borenium ions, $[BX_2]^+$, are apparently stabilized by substituents bearing lone pairs, and are abundant in the spectra of such compounds as $B(OR)_3$ and $(Me_2N)_2BB(NMe_2)_2$, these ions are not formed when they would be present as five-membered rings. Thus in a series of compounds (2), no borenium ion has an abundance >1%. It is concluded that cyclic borenium ions of this type are thermodynamically unstable relative to their acyclic counterparts.[27] This agrees with thermodynamic arguments, and has been used to explain a number of features of the spectra of boron-containing compounds. Mass spectrometric evidence for the formation of $Me_2B_2O_3$ (3) in the gas-phase oxidation of BMe_3, includes the observation of P^+ and $[P - Me]^+$ ions.[28] It

[20] G. Distefano and V. H. Dibeler, *Internat. J. Mass Spectrometry Ion Phys.*, 1970, **4**, 59.
[21] J. H. D. Eland, *Internat. J. Mass Spectrometry Ion Phys.*, 1970, **4**, 37.
[22] J. Lorberth and F. Weller, *J. Organometallic Chem.*, 1971, **32**, 145.
[23] W. Thiele, F. Weller, J. Lorberth, and K. Dehnicke, *Z. anorg. Chem.*, 1971, **381**, 57.
[24] G. Ahlgren, B. Akermark, and M. Nilsson, *J. Organometallic Chem.*, 1971, **30**, 303.
[25] S. W. Bennett, H. J. Clase, C. Eaborn, and R. A. Jackson, *J. Organometallic Chem.*, 1970, **23**, 403.
[26] M. F. Lappert, M. R. Litzow, J. B. Pedley, P. N. K. Riley, T. R. Spalding, and A. Tweedale, *J. Chem. Soc. (A)*, 1970, 2320.
[27] J. C. Kotz, R. J. Vanderzanden, and R. G. Cooks, *Chem. Comm.*, 1970, 923.
[28] L. Barton and G. T. Bohn, *Chem. Comm.*, 1971, 77.

is suggested that the latter ion is a relatively stable cyclic borenium ion, with partial aromatic character conferred by the six-electron system.

The spectra of sterically hindered diarylboranes, *e.g.* $Ar_2B(OBu)$ (Ar = $C_6H_2Bu_3^t$) show loss of either aryl group and elimination of C_4H_8, leaving $[ArBOH]^+$ as the base peak in most cases.[29] Other ions are of relatively low abundance.

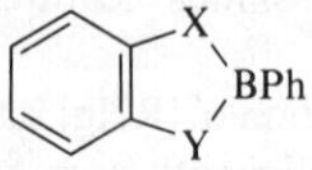

(2) *e.g.* X = Y = NH, O; X = NH, Y = O or S 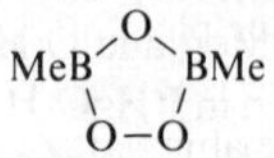   (3)

Boron Heterocycles. The 9-boranthracene cation (4) in the spectrum of (5) is observed to be accompanied by the corresponding doubly charged ion, with about one-third the intensity.[30] Loss of boron gives fluorene and fluorenyl cations. The spectra of fused-ring diboracycles, *e.g.* (6), have also been recorded.[31] Benzoborepin-3-ol shows only a low abundance P^+ ion, but loss of HBO results in aromatization to give the base peak $[C_{10}H_8]^{+}$.[32]

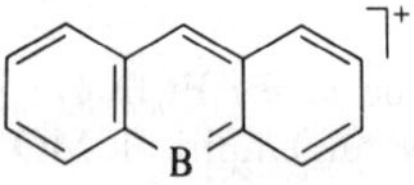

(4) 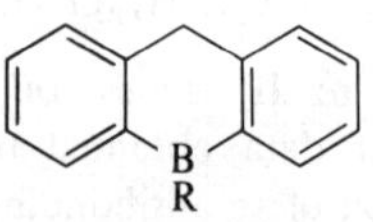   (5) R = Cl, O/2 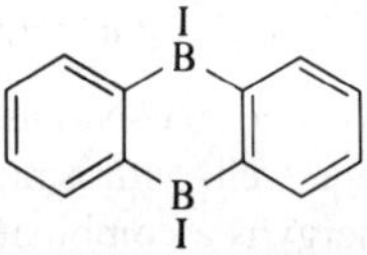   (6)

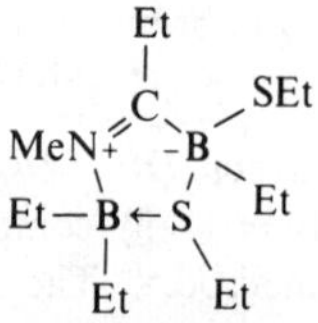

Under electron impact, fragmentation of (7) proceeds by loss of Et or SEt groups; both ions so formed then lose Et_2BSEt, HSEt, or MeN=CEt fragments.[33]

The negative-ion spectrum of 2,3-dicarbahexaborane(8) correlates with photochemical and thermal reaction products, with major breakdown routes being loss of H_2 and BH_3 from $[C_2B_4H_7]^-$, formed by electron capture and elimination of H·.[34]

Boron–Nitrogen Compounds. The spectra of cyclic and acyclic organoboranes containing B—N bonds exhibit ions corresponding to boratropylium $[C_6H_6B]^+$ and boracyclopentadienyl $[C_4H_4B]^+$ ions, related by loss of acetylene, and analogous to the $[C_7H_7]^+$ and $[C_5H_5]^+$ counterparts.[35]

The spectra of monomeric compounds $R_2C=NBR_2^1$ (R, R^1 = aryl) do not show cleavage of the B—N bond as an important process.[36] Instead, the expected

---

[29] H. Staab and B. Meissner, *Annalen*, 1971, **753**, 80.
[30] R. van Veen and F. Bickelhaupt, *J. Organometallic Chem.*, 1970, **24**, 589.
[31] W. Siebert, *Chem. Ber.*, 1970, **103**, 2308.
[32] G. Axelrod and D. Halpern, *Chem. Comm.*, 1971, 291.
[33] A. Grote, A. Haag, and G. Hesse, *Annalen*, 1972, **755**, 67.
[34] C. L. Brown, K. P. Gross, and T. P. Onak, *J.C.S. Chem. Comm.*, 1972, 68.
[35] R. H. Cragg, J. F. J. Todd, and R. B. Turner, *J.C.S. Chem. Comm.*, 1972, 206.
[36] C. Summerford and K. Wade, *J. Chem. Soc. (A)*, 1970, 2010.

loss of R or R^1 groups is found, as well as strong P^{2+} ions. The predominant feature found with 1,3,5-trialkylborazines is a group of ions formed by loss of an alkyl radical from the α-carbon of a substituent on nitrogen.[37] These ions are isoelectronic with $[C_7H_7]^+$ (cf. alkylbenzenes), and dominate the spectra obtained at low ionizing energies. Further fragmentation gives ions containing the B_2N_2 skeleton, which then form $[RN\equiv B]^+$, or lose HCN.

The spectra of alkylazaborolidines show evidence for association in the vapour phase at 55 °C; the n-butyl derivative shows a strong peak at m/e 250. In the isobutyl compound, this peak is weaker, with a corresponding increase in the abundance of fragment ions from the monomer.[38] The tri-isoindolohexa-azaboraphenalenes show loss of substituents, and of $C_8H_4N_2$ fragments; other features are similar to those found in phthalocyanine derivatives, which are structurally related to these boron compounds.[39] Neutral fragments lost in the breakdown of $PhCBr_2N[B(NMe_2)_2]_2$ include halogen, Me, Ph, and $B(NMe_2)_2$ groups.[40]

Reactions of acid chlorides with $KBCNEt_3$ give derivatives containing the CNCOB(CN) ring system; in the mass spectrometer, this ring fragments principally by extrusion of HOBCN, to give (8). Detailed breakdown schemes for several compounds of this type are given.[41]

$$\begin{bmatrix} Et & Et_2 \\ _{\bar{}}B^{-}C^{+}_{}NH \\ NC & _{}// \\ & O-CR \end{bmatrix}^+ \xrightarrow{-Et} \xrightarrow{-HOBCN} \begin{array}{c} Et_2 \\ C \\ /\backslash + \\ RC=N \end{array}$$

(8)

Boronates and other Boron–Oxygen Compounds. Under electron impact, cyclic esters of phenylboronic acids and analogous compounds give rise to hydrocarbon ions of high abundance, particularly $[C_7H_7]^+$, *via* rearrangement reactions; several recent papers have reported the formation of ions of this type. Examples include 2-phenyl-1,3,2-dioxaborolane,[42,43] and the C—O—B—OPh system (in a phenoxy-1,3,2-dioxaborinan)[44] and this feature of these spectra reflects the high B—O bond strength. Bose and Peters[45] find loss of CH_2OBO from P^+ in a one-step process, possibly as in Scheme 3. These observations have been extended to other boron heterocycles, and the formation of $[C_7H_7]^+$ and other hydrocarbon ions was rationalized in terms of relative bond energies in the

---

[37] P. Powell, P. J. Sherwood, M. Stephens, and E. F. H. Brittain, *J. Chem. Soc. (A)*, 1971, 2951.
[38] V. A. Dorokhov, O. G. Boldyreva, and B. M. Mikhailov, *Zhur. obshchei Khim.*, 1970, **40**, 1528.
[39] A. Meller and A. Ossko, *Monatsh.*, 1972, **103**, 150.
[40] G. Schmid and L. Weber, *Z. Naturforsch.*, 1971, **26b**, 994.
[41] E. Brehm, A. Haag, G. Hesse, and H. Witte, *Annalen*, 1970, **737**, 70.
[42] R. H. Cragg and J. F. J. Todd, *Chem. Comm.*, 1970, 386.
[43] I. R. McKinley and H. Weigel, *Chem. Comm.*, 1970, 1022.
[44] P. B. Brindley and R. Davis, *Chem. Comm.*, 1971, 1165.
[45] R. J. Bose and M. D. Peters, *Canad. J. Chem.*, 1971, **49**, 1766.

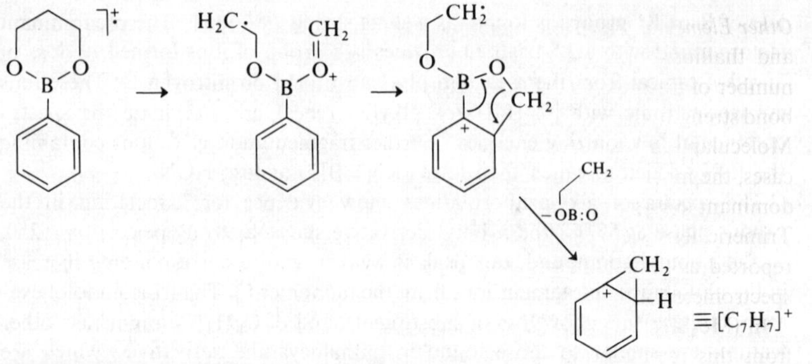

Scheme 3

heterocycles.[46,47] Whereas the one-step mechanism is confirmed for (9; X = Y = O), with the oxathioborolan fragmentation follows the pattern shown in Scheme 4.

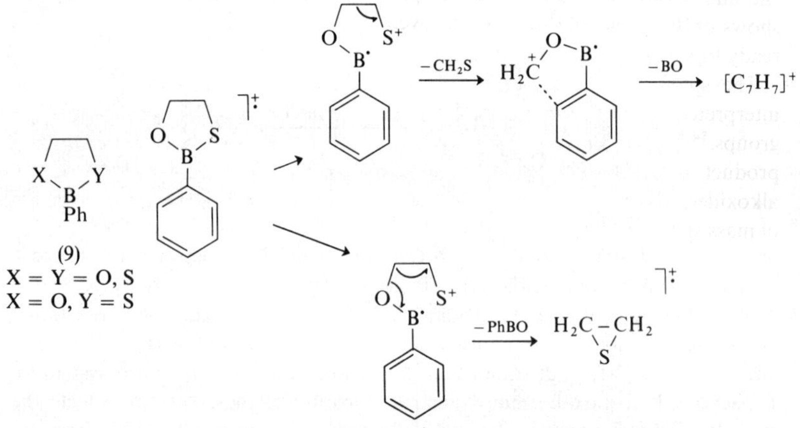

Scheme 4

The dithiaborolan gives only a small yield of $[C_7H_7]^+$, and this is consistent with a weaker B—S bond (over B—O) encouraging the formation of the ethylene sulphide ion. Peaks due to tropylium ion are also found in the spectra of acyclic benzeneboronic acid derivatives, and in both cyclic and acyclic phenylbisaminoboranes, e.g. $PhBNMe(CH_2)_2NMe$. In addition, an ion at m/e 89 is assigned the boratropylium structure.[48]

[46] R. H. Cragg, D. A. Gallagher, J. P. N. Husband, G. Lawson, and J. F. J. Todd, *Chem. Comm.*, 1970, 1562.
[47] R. H. Cragg, G. Lawson, and J. F. J. Todd, *J.C.S. Dalton*, 1972, 878.
[48] C. Cone, M. J. S. Dewar, R. Golden, F. Maseles, and P. Rona, *Chem. Comm.*, 1971, 1522.

Other Elements. The trimethyl derivatives of boron, aluminium, gallium, indium, and thallium form most ions $[MC_mH_n]^+$ consistent with their formulae, the number of fragment ions being influenced by the progressive decrease in M—C bond strengths (B → Tl); with TlMe$_3$, few ions result from C—H bond cleavage.[49] Molecular ion abundances follow the order: Al > B > Ga > In > Tl. In all cases, the most abundant ion is $[MMe_2]^+$; loss of methyl groups gives the predominant ions, with ethane being eliminated from all $[MMe_2]^+$ except $[BMe_2]^+$. Trimeric ions were found for one sample of GaMe$_3$; the trivinyl compound, reported as dimeric in benzene, shows no evidence of association in the mass spectrometer, with a very low abundance P^+ ion.

In $(Me_3Si)_2NAl_2Me_5$, loss of alkyl groups follows the elimination of AlMe$_3$ from this methyl-bridged compound; the remaining AlMe$_2$ moiety is more readily lost than SiMe$_3$, indicating the relative bond strengths: N—Si > N—Al.[50] Fragmentation patterns for the cyclic imino derivatives $[CH_2(CH_2)_xNMMe_2]_n$ (M = Al, Ga, or In) confirm $n = 3$, found by cryoscopic measurements, and involve loss of Me, MMe, and the cyclic imino-group.[51]

The spectra of the compounds MeGa$_2$Br$_5$ and Me$_2$Ga$_2$Br$_4$ show loss of Me, Br, and GaBr$_2$ fragments in a stepwise manner.[52] Diethylgallium oxinate shows only a weak P^+ peak, and other expected fragment ions, resulting from ready loss of Et, oxine, and break-up of the ligand.[53]

The spectrum of a sulphur dioxide insertion product of InMe$_3$ has been interpreted in terms of a polymeric structure with intermolecular OO'-sulphinato-groups.[54,55] No ions contain two or more indium atoms, however. Another product is Me$_3$InSO$_2$, which shows weak dinuclear ions.[56] With indium alkoxides, SO$_2$ forms a compound suggested to have structure (10) on the basis of mass spectral evidence.

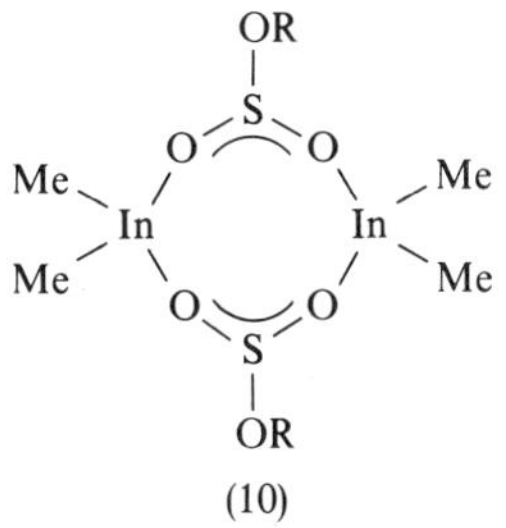

(10)

Further examples of the spectra of methylthallium compounds, including Me$_2$Tl(im) (imH = imidazole),[57] and derivatives of chelating oxygen and

[49] F. Glockling and R. G. Strafford, *J. Chem. Soc. (A)*, 1971, 1761.
[50] N. Wiberg and W. Baumeister, *J. Organometallic Chem.*, 1972, **36**, 277.
[51] A. Storr and B. S. Thomas, *J. Chem. Soc. (A)*, 1971, 3850.
[52] W. Lind and I. J. Worrall, *J. Organometallic Chem.*, 1972, **36**, 35.
[53] B. Sen and G. White, *Inorg. Nuclear Chem. Letters*, 1971, **7**, 79.
[54] A. T. T. Hsieh, *Inorg. Nuclear Chem. Letters*, 1970, **6**, 767.
[55] A. T. T. Hsieh, *J. Organometallic Chem.*, 1971, **27**, 293.
[56] J. Weidlein, *J. Organometallic Chem.*, 1970, **24**, 63.
[57] A. G. Lee, *J. Chem. Soc. (A)*, 1971, 880.

nitrogen ligands,[58] show the major features previously reported. In the imidazole derivative, loss of HCN found in the free molecule is completely suppressed by the metal; dimeric ions are present in the vapours of most complexes. In contrast, diphenylthallium derivatives[59] do not give very informative spectra, containing no P^+ ions, but showing ions such as Tl^+, $[TlPh]^+$, and $[TlPh_2]^+$.

The exchange of cyclopentadienyl groups, between indium and thallium, including deuteriated and methylated derivatives, has been studied mass spectrometrically.[60]

Group IV.—A complete analysis[61] of the spectra of MMe_4, and the series $Me_3MM'Me_3$ (M, M' = C, Si, Ge, Sn, Pb) has given IP's and AP's of $[M(or\ M')Me_3]^+$ ions. These data, combined with known or estimated gas-phase standard enthalpies $[\Delta H_f^0(g)]$ of $Me\cdot$, $Me_3C\cdot$, Me_3C^+, Me_4M, Me_6C_2, Me_6Si_2, and Me_6Sn_2, have given $\Delta H_f^0(g)$ data for all species $Me_3MM'Me_3$, Me_4M, $Me_3M\cdot$, and $[Me_3M]^+$. This paper is an important contribution to the thermochemical studies of Group IV organometallic compounds, and provides a reliable set of bond dissociation energies and thermochemical bond energy terms which are also internally consistent.

The spectra of several hexa-alkyldimetals of Group IV elements exhibit features which are common to other Me_3MX derivatives.[62] Weak P^+ peaks, transfer of groups between metals, and stability of the M—M' bond (bimetallic ions amount to 30—40% of the total ion current) are the major features present. This paper also records the spectra of the hydrides R_3SiH (R = Me, Et, or Pr^i).

Group IV tetra-allyls, $M(CH_2CH{=}CH_2)_4$, show features previously noted in the analogous alkyls for M = Ge or Sn, whereas the corresponding silicon compound shows marked differences.[63] In the germanium and tin compounds, loss of C_3H_4, C_3H_5, and C_6H_{10} fragments occurs; with silicon, elimination of a variety of groups, including C_2H_2, C_2H_4, C_3H_4, C_4H_6, C_5H_8, and C_6H_{10}, was found. Detailed breakdown patterns were deduced for the silicon compound, using an enhanced metastable peak spectrum. Loss of radicals from even-electron ions is difficult; from odd-electron ions, loss of C_3H_5 and H· is found. These differences are rationalized in terms of increasing M—C bond strength: Sn—C < Ge—C < Si—C. The spectrum of $SnMe_4$ is also discussed.

The 9-metallodihydroanthracenes of silicon, germanium, and tin show similar ions in their mass spectra; the germanium compound is detailed.[64] Low abundance P^+ ions lose HCl readily to give a supposedly aromatic ion (but see below). In the mass spectra of 5,5-diphenyl-10,11-dihydrodibenzo[bf]metallepins, most of the ion current is carried by metal-containing ions.[65] The heteroatom is

[58] A. G. Lee, J. Chem. Soc. (A), 1971, 2007.
[59] A. G. Lee, J. Organometallic Chem., 1970, **22**, 537.
[60] J. M. Lalancette and A. Lachance, Canad. J. Chem., 1971, **49**, 2996.
[61] M. F. Lappert, J. B. Pedley, J. Simpson, and T. R. Spalding, J. Organometallic Chem., 1971, **29**, 195.
[62] M. Gielen, J. Nasielski, and G. Vandendunghen, Bull. Soc. chim. belges, 1971, **80**, 175.
[63] M. Fishwick and M. G. H. Wallbridge, J. Chem. Soc. (A), 1971, 57.
[64] P. Jutzi, Chem. Ber., 1971, **104**, 1455.
[65] J. Y. Corey, M. Dueber, and M. Malaidza, J. Organometallic Chem., 1972, **36**, 49.

eliminated to form the dihydrophenanthrene cation; loss of benzene is also noted. In general, relationships are similar to those found for non-heterocyclic Group IV derivatives, the major difference arising from the low abundance of M^{III} ions. Some discussion of various mechanisms for the fragmentation of the parent ions (11) is also given.

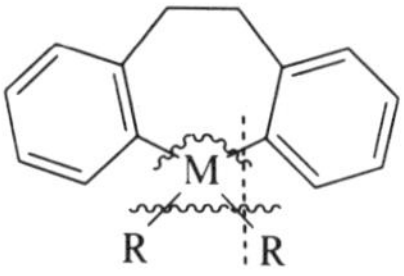

(11)

Silicon. A photo-ionization study of $SiMe_4$ shows formation of ions by loss of Me groups or olefin; threshold values and ionic heats of formation have been measured for the three most abundant ions.[66] Ionization probably occurs by loss of a Si—Me bonding electron. Ion–molecule reactions have been found with Me_2SiH_2, Me_3SiH, and $SiMe_4$ in the spectrometer, and reaction rates have been measured.[67] The most reactive ions in Me_2SiH_2 are $[Me_2Si]^+$, $[MeSiH]^+$, and $[SiH]^+$, *i.e.* those containing bivalent silicon. These react by H- or Me-transfer to produce $[Me_2SiH]^+$. Ions at higher m/e than the parent ion include $[Me_4Si_2]^+$ and $[Me_3Si_2H_2]^+$, formed by reactions such as:

$$[Me_2Si]^+ + Me_2SiH_2 \rightarrow [Me_2Si_2Me_2]^+ + H_2$$

In Me_3SiH, the reactant ions are $[Me_2Si]^+$ or $[Me_2SiH]^+$, with $[Me_3Si]^+$ being a product ion. As hydrogen is replaced by methyl groups there is an increasing tendency for hydride transfer, although this reaction is not shown by Me_4Si. With this compound, the most energetically favoured reaction is:

$$[Me_4Si]^+ + Me_4Si \rightarrow [Me_3Si]^+ + Me_3Si + C_2H_6$$

IP's for silylbenzenes have been correlated with charge transfer energies of highest occupied MO's obtained from spectra of suitable donor–acceptor complexes; CH_2SiR_3 and SiR_2SiR_3 groups cause perturbations of the benzene π-system comparable to those found in alkoxy-substituents; the SiH_3 group shows no donor properties. Evidence is given for strong electron back-donation from the benzene π-system into appropriate empty silicon orbitals.[68]

A general conclusion which emerges from a study of some 29 organosilicon compounds of the types $PhSiH_3$, $XC_6H_4CH_2SiMe_3$, and $XC_6H_4SiMe_3$ is that the cracking patterns are broadly similar to the analogous carbon compounds.[69] However, no fragmentation is observed which requires formation

[66] G. Distefano, *Inorg. Chem.*, 1970, **9**, 1919.
[67] P. Potzinger and F. W. Lampe, *J. Phys. Chem.*, 1971, **75**, 13.
[68] H. Bock and H. Alt, *J. Amer. Chem. Soc.*, 1970, **92**, 1569.
[69] M. E. Freeburger, B. M. Hughes, G. R. Buell, T. O. Tiernan, and L. Spialter, *J. Org. Chem.*, 1971, **36**, 933.

of a C=Si double bond in either the ion or the neutral fragment. Extensive rearrangements occur in the $XC_6H_4SiMe_3$ series leading to the formation of $[C_7H_7]^+$ and $[SiX]^+$ species. In particular, no resonance stabilization of the ion $[P-1]^+$ occurs in the spectrum of $PhSiH_3$, in contrast to that found in toluene. Migration of fluorine from carbon to silicon occurs in all three isomers of $FC_6H_4CH_2SiMe_3$, though this is most favoured for the *ortho*-isomer. Other substituents are similarly transferred in $XC_6H_4SiMe_3$, with formation of ions $[XSiMe_2]^+$. With oxygen substituents such as OMe, prominent ions of type (12) are found in the spectra of the *ortho*-isomers only. Ortho-activation in

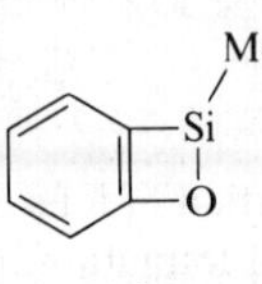

(12) 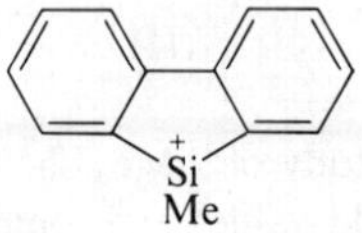   (13) 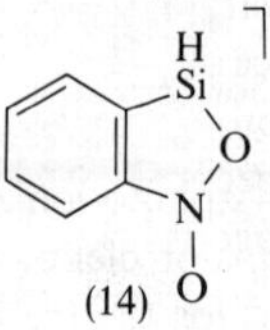   (14)

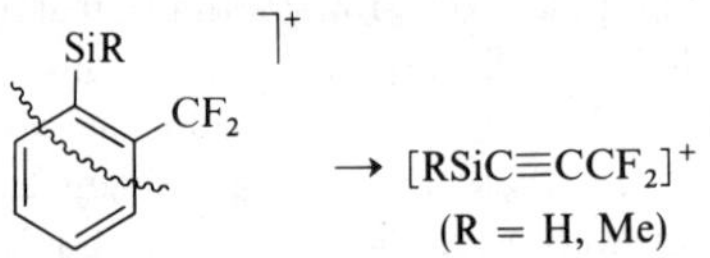

(15)   → $[RSiC\equiv CCF_2]^+$
(R = H, Me)

biphenyl derivatives results in the formation of an abundant $[C_{12}H_8SiMe]^+$ ion (13). In the nitro-compounds, Si—O bond formation again occurs to give ions such as (14). Both nitro and CF_3 derivatives undergo fragmentation of the aromatic ring, *e.g.* (15). High abundance tropylium ions are also formed by migration and ring insertion of one of the Si—Me groups.

Mass spectra and AP measurements on methyl- and phenyl-substituted disilanes have been reported.[70] The usual fragmentation processes, involving loss of a substituent, or of olefin, or of benzene (for the phenyl compounds) were found. In addition, cleavage of the Si—Si bond resulted in loss of $SiMe_2$ groups, or of $PhSiMe_n$ (n = 1 or 2). Heats of formation were determined for R·, R$^+$, R$_2$, and $RSiMe_3$ (R = Ph_nSiMe_{3-n}). For Ph_3Si^+, the stabilization energy, while appreciable, is only half that found for Ph_3C^+. For the corresponding radical, little resonance interaction between Ph and Si, and a probable pyramidal structure, are indicated by the low stabilization energy.

In the spectrum of methyl γ-trimethylsilylcrotonate,[71] the major feature is migration of MeO from the ester group to silicon in the $[P-\text{Me}]^+$ siliconium ion:

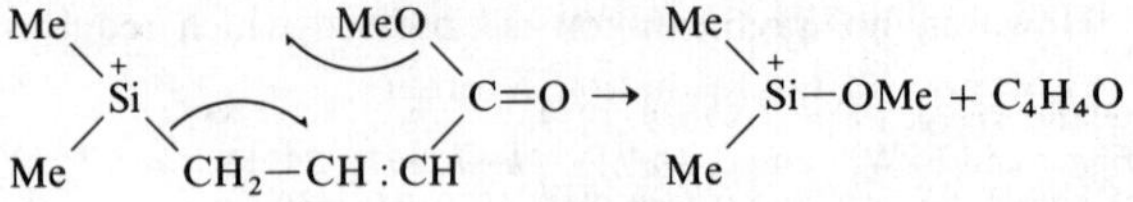

$$\text{Me}_2\overset{+}{\text{Si}}(\text{Me})\text{CH}_2\text{—CH:CH—C(=O)—OMe} \rightarrow \text{Me}_2\overset{+}{\text{Si}}(\text{Me})\text{—OMe} + C_4H_4O$$

[70] J. M. Gaidis, P. R. Briggs, and T. W. Shannon, *J. Phys. Chem.*, 1971, **75**, 974.
[71] W. P. Weber and R. A. Felix, *Tetrahedron Letters*, 1971, 1445.

Organometallic and Co-ordination Compounds

Rearrangements found[72] in the spectra of Ph(CH$_2$)$_3$SiMe$_3$ and PhCH=CHSiMe$_3$ illustrate the high migratory aptitude of the SiMe$_3$ group, transfer from a saturated carbon to an unsaturated carbon possibly occurring:

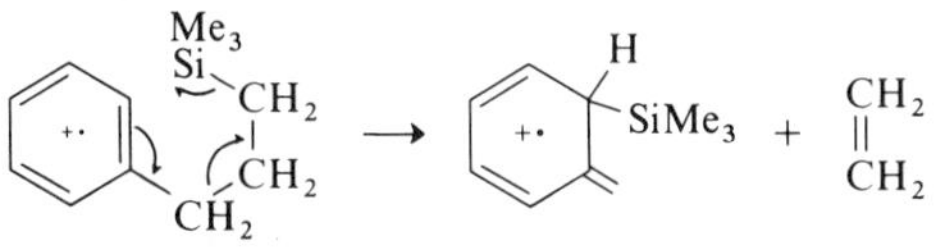

The illustrated process only occurs when the silicon group is on the γ-carbon, and hence a six-membered cyclic transition state is proposed. Also observed is the formation of [PhSiMe$_2$]$^+$ by elimination of olefin, a process additionally observed in the spectra of other Ph(CH$_2$)$_n$SiMe$_3$ ($n = 2$ or 4) compounds. In the styrene, an analogous loss of acetylene is found:

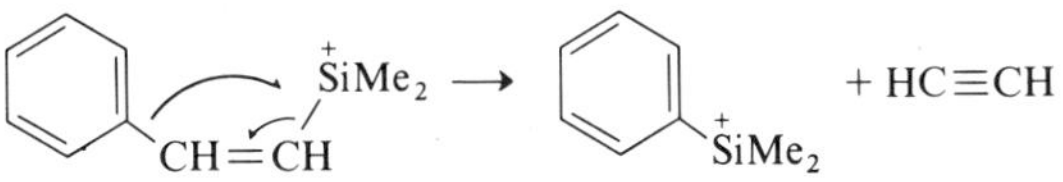

More important, however, is the loss of CH$_4$. Labelling experiments demonstrated that this involves a Si—Me group and any of the seven hydrogens of the styryl group. Although the formation of a sila-indene can be postulated, an alternative explanation may involve random interaction of the silicon centre with the phenyl ring (either open or closed) accompanied by rapid proton shifts.

The mass spectra of some products of addition between CH$_2$=CHN(CF$_3$)$_2$ and various silanes HSiR$_3$ are reported.[73] As R changed from electron-donating to electron-withdrawing, Si—Et bond fission decreased, while C—C bonds were more easily broken. Rearrangements involving migration of fluorine from carbon to silicon have a favourable energy change, but such ions are generally of only weak intensity. Similar ions are found in the spectra of perfluoroalkylene-linked silanes and siloxans; differences in relative intensities can be related to the number of CF$_2$ groups in the silanes, but not in the siloxan series.[74]

Peaks at m/e 91, 105, 121, 135, and 145 are major contributors to the spectra of compounds containing phenyl groups α and α' to a SiMe$_2$ group (16), and are found in products obtained from diphenylacetylene and a substituted

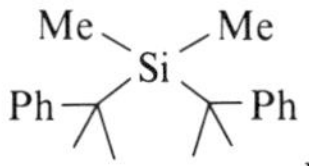

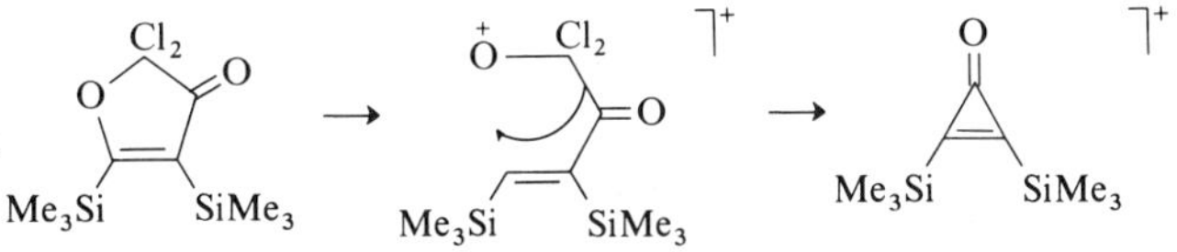

---

[72] W. P. Weber, A. K. Willard, and H. Boettger, *J. Org. Chem.*, 1971, **36**, 1620.
[73] E. S. Alexander, R. N. Haszeldine, M. J. Newlands, and A. E. Tipping, *J. Chem. Soc. (A)*, 1970, 2285.
[74] J. L. Cotter, *Org. Mass Spectrometry*, 1972, **6**, 425.

silacyclopentadiene.[75] The heterocycle (17) loses phosgene from the P^+ ion to give m/e 200, perhaps *via* ring-opening, without fragmentation of the $SiMe_3$ group.[76]

Silicon Heterocycles. The spectra of silicon- and germanium-containing four- and five-membered cyclic compounds have been interpreted and compared with products obtained by thermal decomposition.[77] Germanium derivatives contain more lower m/e fragment ions. Generally a fairly intense P^+ ion is observed and this breaks down by loss of organic radicals. Loss of olefin (C_2H_4 or C_3H_6) is found, the latter being preferred with the germanium derivatives. Several ions and neutral fragments correspond to postulated unstable intermediates in the thermal decomposition, *e.g.* $[Me_4Si_2]^+$, $[Me_3Ge_2]^+$.

The mass spectra of five- and six-membered silicon heterocycles have been studied[78] with respect to the fragmentations in which silicon plays a role; a considerable number of earlier references to related work are also noted. Most ions are formed by cleavage of C—C or C—Si single bonds, and by rearrangements of ions so produced. Silicon usually remains attached to a carbon involved in π-bonding to a second carbon atom. In these compounds, there are nine important ions, although which ones are formed in largest amount depends on the particular compound. A further important process is migration of hydrogen from carbon to silicon. In the disilacyclohexadiene, the process shown in Scheme 5 is noted.

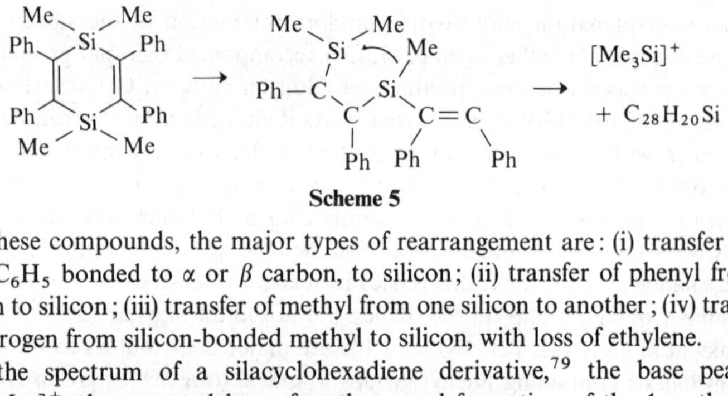

Scheme 5

In these compounds, the major types of rearrangement are: (i) transfer of H from C_6H_5 bonded to α or β carbon, to silicon; (ii) transfer of phenyl from α carbon to silicon; (iii) transfer of methyl from one silicon to another; (iv) transfer of hydrogen from silicon-bonded methyl to silicon, with loss of ethylene.

In the spectrum of a silacyclohexadiene derivative,[79] the base peak is $[PhSiMe_2]^+$, the expected loss of methane and formation of the 1-methylsilabenzene cation (18) being only one-twentieth as important.

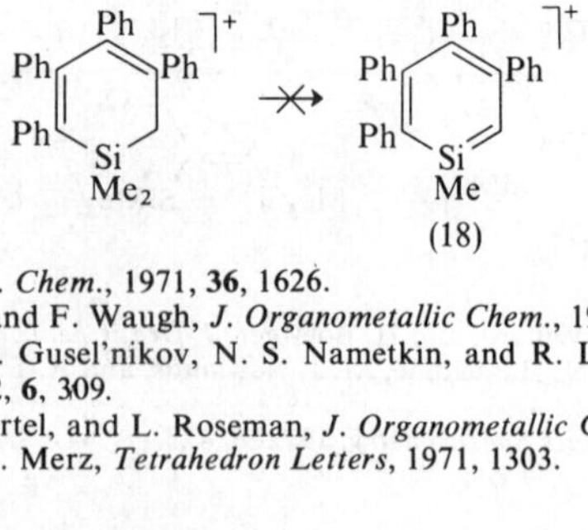

(18)

[75] R. Maruca, *J. Org. Chem.*, 1971, **36**, 1626.
[76] D. R. M. Walton and F. Waugh, *J. Organometallic Chem.*, 1972, **37**, 45.
[77] V. Y. Orlov, L. E. Gusel'nikov, N. S. Nametkin, and R. L. Ushakova, *Org. Mass Spectrometry*, 1972, **6**, 309.
[78] R. Maruca, M. Oertel, and L. Roseman, *J. Organometallic Chem.*, 1972, **35**, 253.
[79] G. Märkl and P. L. Merz, *Tetrahedron Letters*, 1971, 1303.

With 4,4-dimethylsilacyclohexadien-1-one, loss of methyl and CO are competitive, the latter process giving methylcyclopentadienyl cations.[80] The 2-silanorcarane (19) shows a P^+ ion, and an intense $[P - Me]^+$; the base peak at m/e 97 is formed by a McLafferty-type rearrangement:[81]

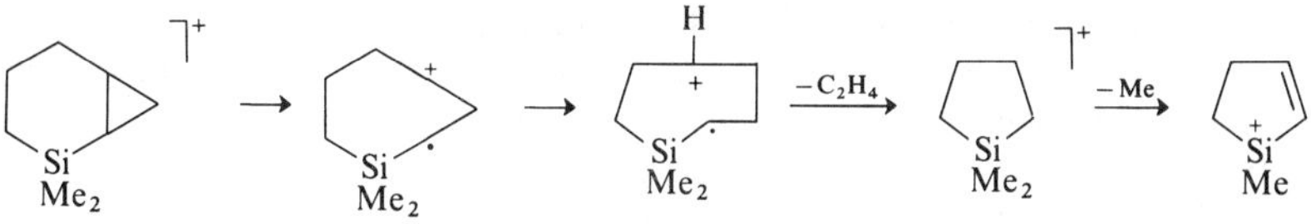

The unusual [1,1]paracyclophane (20) has a parent ion base peak, and an expectedly strong $[P - Me]^+$. A prominent fragment ion at m/e 97 is assigned structure (21) (see above also).[82]

The spectrum of (22)[83] contains a strong $[P - Cl]^+$ peak, which may be formulated as (22a). The suggested ready formation of an aromatic silicon-containing system in the spectra of derivatives of dihydrosila-anthracene[64] by loss of HX, could also be due to the stereochemistry of the system, and more recent studies on the spectra of dibenzosilepin derivatives, where loss of HX is also facile, but does not result in aromatization of the system, support this argument.[84] For some dimethyl derivatives,[85] the $[P - Me]^+$ peaks are probably best represented as siliconium ions.

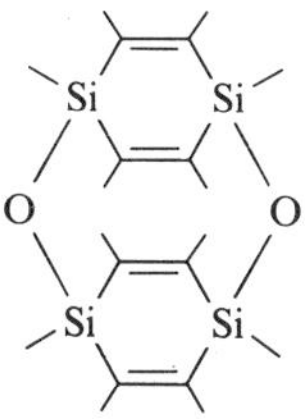

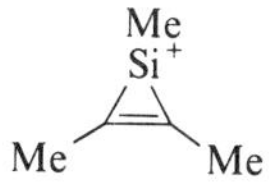

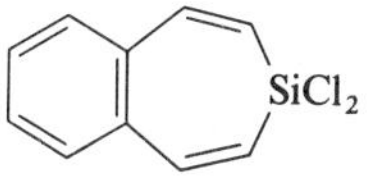

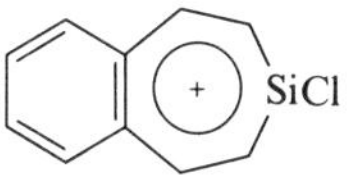

A parent ion is generally present in the spectra of tetrasila-adamantanes, with abundant ions formed by loss of bridgehead (Si) substituents;[86] the tetrachloro-derivative is notable in showing a strong $[P - Me]^+$ ion, which does not further decompose. This process is even easier than loss of Cl (66.7 vs 22.2%,

---

[80] W. P. Weber and R. Laine, *Tetrahedron Letters*, 1970, 4169.
[81] E. Rosenberg and J. J. Zuckerman, *J. Organometallic Chem.*, 1971, **33**, 321.
[82] F. Wudl, R. D. Allendoerfer, J. Dermirgian, and J. M. Robbins, *J. Amer. Chem. Soc.*, 1971, **93**, 3160.
[83] L. Birkhofer, H. Haddad, and H. Zamarlik, *J. Organometallic Chem.*, 1970, **25**, C57.
[84] J. Y. Corey, M. Dueber, and B. Bichlmeir, *J. Organometallic Chem.*, 1971, **26**, 167.
[85] F. K. Cartledge and P. D. Mollère, *J. Organometallic Chem.*, 1971, **26**, 175.
[86] C. L. Frye, J. M. Klosowski, and D. R. Weyenberg, *J. Amer. Chem. Soc.*, 1970, **92**, 6379.

respectively). Similar processes were observed in the parent tetrasila-compound ($[P - Me]^+$, 27%), and in adamantane itself ($[P - Me]^+$, 10.3%).[87]

The initial observations on the formation of tropylium ions in the spectra of cyclic phenylborolanes have been extended to silicon derivatives.[88] Relatively abundant tropylium ions (5—25%) have been found in the spectra of TMS derivatives of phenols; with $PhNHSiMe_3$, the abundance is 1.8%. The presence of electronegative substituents on the C_6 ring reduces the intensity of the corresponding tropylium ion. In all cases, the expected $[P - Me]^+$ ion is the base peak.

The spectra of branched-chain methylsiloxane polymers[89] contain abundant $[P - Me]^+$ ions, which fragment by further loss of methyl groups, and rearrangement of the Si—O—Si skeleton. The various processes are compared with those occurring on pyrolysis.

Stepwise loss of two substituents from the parent ions of (23; R = Me or Ph) occurs.[90] This is followed by elimination of SiO to give $[C_{12}H_8]^+$. The $[P - Me]^+$ ion in the 10,10-dimethylphenoxasilin also fragments by competitive loss

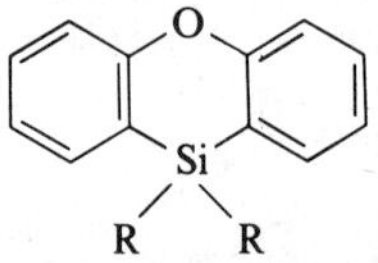

(23) 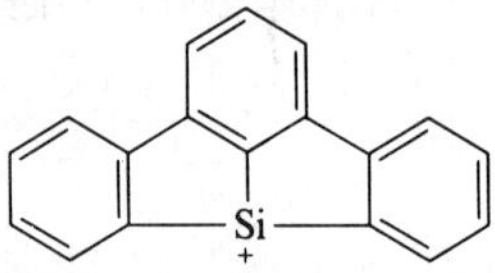   (24)

of CH_2O, SiH_2O, or SiHO groups. Rearrangements also afford a strong $[C_7H_7]^+$ peak. Other features of the spectra include loss of H_2O from $[P - Ph]^+$ to give an even-electron siliconium ion (24), and loss of dibenzofuran giving $[SiR_2]^+$ with both compounds.

The sultines (25) exhibit loss of olefin, or of an R group; subsequent fragmentation by loss of SO or SO_2 also occurs.[91]

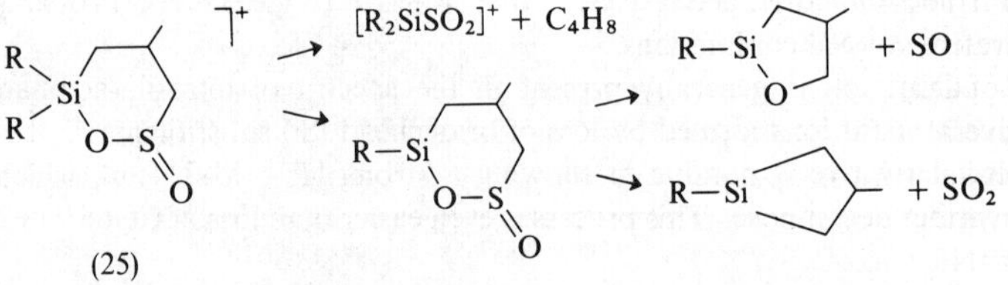

(25) $[R_2SiSO_2]^+ + C_4H_8$ + SO   + SO_2

---

[87] R. S. Gohlke and R. J. Robinson, *Org. Mass Spectrometry*, 1970, **3**, 967.
[88] R. H. Cragg, K. J. A. Hargreaves, J. F. J. Todd, and R. B. Turner, *J.C.S. Chem. Comm.*, 1972, 336.
[89] G. Garzo and K. Pehrsson, *J. Organometallic Chem.*, 1971, **30**, 187.
[90] I. Lengyel and M. J. Aaronson, *J. Chem. Soc. (B)*, 1971, 177.
[91] J. Dubac, P. Mazerolles, M. Joly, W. Kitching, C. W. Fong, and W. H. Atwell, *J. Organometallic Chem.*, 1972, **34**, 17.

Dimethylsilacycloalkanones give spectra in which the majority of ions result by interaction of the silyl centre with the carbonyl group.[92] Elimination of C_2H_4 is a common process, transannular bond formation *via* a McLafferty-type rearrangement being suggested, for example:

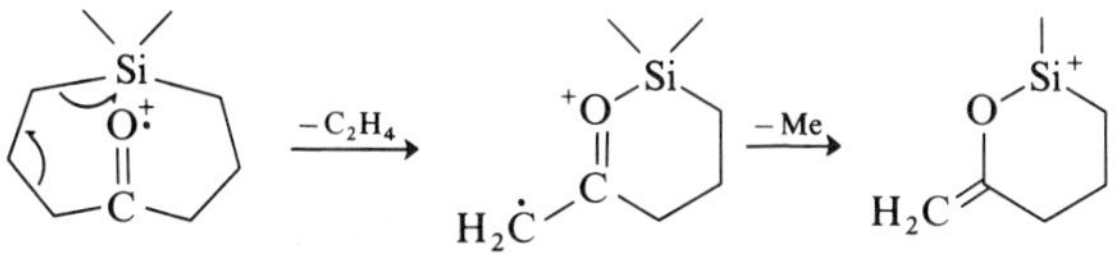

The fragmentation patterns of cyclocarbosiloxanes $Me_6Si_3O_{3-n}(CH_2)_n$ ($n = 1$ or 2) and $Me_8Si_4O_{4-n}(CH_2)_n$ ($n = 1, 2,$ or 3) are intermediate between those found for cyclosiloxanes and cyclocarbosilanes.[93] The most intense singly and doubly charged peaks correspond to loss of one or two methyl groups, respectively; break-up of the cyclic system is much reduced, compared with the cyclosiloxanes. This latter process involves elimination of $SiMe_4$, and this may also occur by separate losses of CH_4 and Me_2SiCH_2. The ready formation of cyclic ions of type (25) in the spectra of cyclosiloxanes has its counterpart in the formation of (26) in the carbosiloxanes.

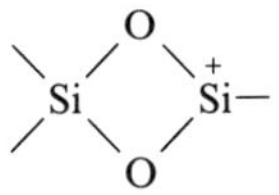

(25) 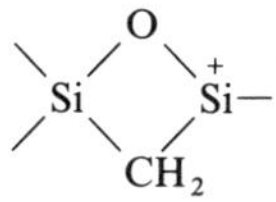   (26) 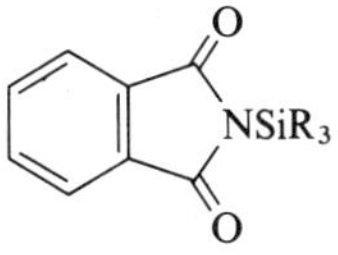   (27)

Silicon–Nitrogen Compounds. The spectrum of $(Me_3Si)_2NMe$ shows loss of Me as the major process, together with loss of Me_3SiH; other ions are related to these.[94] The parent ion of $(Me_3Si)N(Me)COCF_3$ loses Me and CF_2, or CF_3, to give ions at m/e 134 and 130, respectively. Migration of fluorine to silicon gives $[Me_2SiF]^+$ and related ions.

The spectra of a series of *N*-silylphthalimides (27; $R_3 = Me_3$, Me_2Ph, or $MePh_2$) exhibit ions arising by loss of methyl, followed by CO_2; in the $SiMe_2Ph$ derivative, loss of HCN gives the siliconium ion as shown in Scheme 6.[95] The loss of CO_2 can occur *via* an intermediate acyl isocyanate.

Several silicon–nitrogen compounds, including silylhydrazines and tetra-azadisilacyclohexanes, show significant P^+ ions; *N*-phenyl derivatives lose phenylnitrene as a characteristic process.[96] Other neutral molecules lost include RN=NR, also found in the thermal decomposition of these compounds, and

[92] W. B. Weber, R. A. Felix, A. K. Willard, and H. G. Boettger, *J. Org. Chem.*, 1971, **36**, 4060.
[93] V. Y. Orlov, N. S. Nametkin, L. E. Gusel'nikov, and T. H. Islamov, *Org. Mass Spectrometry*, 1970, **4**, 195.
[94] L. Birkofer and G. Schmidtberg, *Chem. Ber.*, 1971, **104**, 3831.
[95] A. F. Janzen and E. A. Kramer, *Canad. J. Chem.*, 1971, **49**, 1011.
[96] K. G. Das, P. S. Kulkarni, V. Kalyanaraman, and M. V. George, *J. Org. Chem.*, 1970, **35**, 2140.

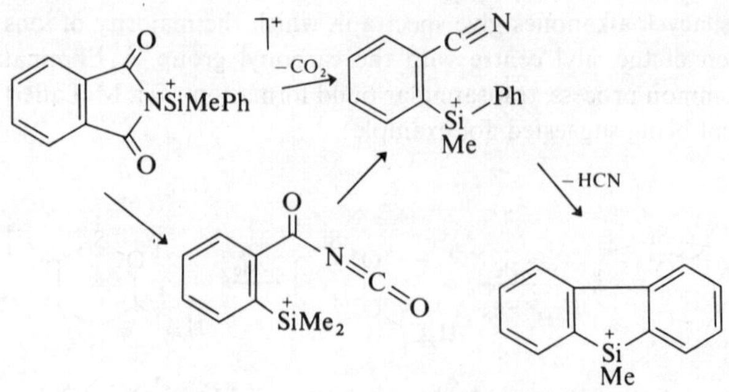

[Scheme 6]

Scheme 6

CH_4 or C_6H_6 from the hydrazines. After loss of nitrogen-containing fragments, most ions can be formulated as siliconium ions. Hydrogen transfer and skeletal rearrangements are also found, as well as the process:

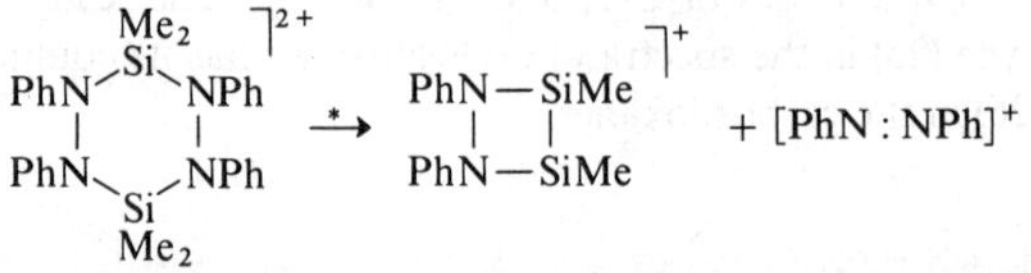

[Structure showing dimeric cyclic silicon-nitrogen compound dissociating]

$$\begin{bmatrix} \text{PhN} \overset{\text{Me}_2}{\underset{\text{Si}}{\diagup}} \text{NPh} \\ \text{PhN} \underset{\text{Si}}{\diagdown} \text{NPh} \\ \text{Me}_2 \end{bmatrix}^{2+} \rightarrow \begin{bmatrix} \text{PhN—SiMe} \\ | \quad | \\ \text{PhN—SiMe} \end{bmatrix}^{+} + [\text{PhN:NPh}]^{+}$$

Trimethylsilyl Derivatives. Detailed discussion of the use of TMS derivatives for the characterization of many natural (and unnatural) products can be found elsewhere in this volume. However, it has become increasingly apparent that a variety of functional-group interactions can occur in these derivatives, involving migration of the $SiMe_3$ moiety. Some examples are mentioned above.

In the example of benzyloxycyclohexyl TMS ethers,[97] the formation of m/e 179 is envisaged to occur *via* the intermediate (28). This paper incidentally corrects

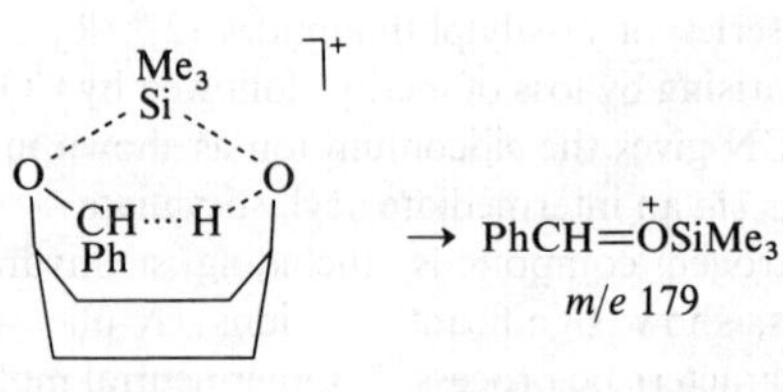

[Structure (28) showing cyclic intermediate] → $\text{PhCH}=\overset{+}{\text{O}}\text{SiMe}_3$

m/e 179

(28)

the assignment of $[\text{Me}_2\text{SiO(OMe)}]^+$ to an ion m/e 105 in the spectrum of methyl-3-trimethylsilylpropionate,[98] to that of $[\text{Me}_3\text{SiOHMe}]^+$ formed by the reaction shown in Scheme 7.

[97] P. B. Woodgate, R. T. Gray, and C. Djerassi, *Org. Mass Spectrometry*, 1970, **4**, 257.
[98] W. P. Weber, R. A. Felix, and A. K. Willard, *J. Amer. Chem. Soc.*, 1969, **91**, 6544.

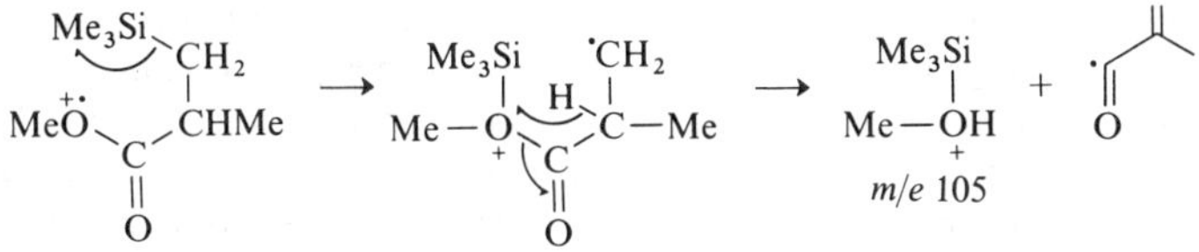

Scheme 7

Me₃Si-CH₂ / MeO-C(+•)-CHMe / ‖O → Me₃Si 'CH₂ / Me-O(+)...H...C-Me / ‖C / ‖O → Me₃Si / Me-OH(+) / m/e 105 + •CH₂C(=O)Me

Requirements for TMS group interactions and rearrangements are also discussed in a paper by Vouros and co-workers.[99] In particular, the formation of the ion at m/e 147, with structure [Me₃SiOSiMe₂]⁺, is a function of stereochemistry, steric hindrance, and separation of the TMS groups, and seems to occur independently of ring cleavage. Particularly, the occurrence of this ion is structurally informative for steroidal alcohols, and in principle can be extended to other products.

Some evidence has been provided that ions with a positively charged atom can react under some conditions with neutral species containing an electronegative site.[100] Ions with Si, Ge, Sn, or Hg centres are more reactive than analogous carbonium ions. A simple 'sticking collision' mechanism is favoured.

Germanium. Processes occurring in the spectra of permethylcyclopolygermanes, $(Me_2Ge)_n$ ($n = 5$—7) include loss of Me, CH_4, $GeMe_3$, and $HGeMe_3$;[101] the most intense ion is $[Me_3Ge]^+$. Only with $n = 5$ is loss of $GeMe_2$ found from the P^+ ion. A general mode of breakdown is shown in Scheme 8.

$$[Me_{2n}Ge_n]^+ \xrightarrow{-Me} [Me_{2n-1}Ge_n]^+$$
$$\downarrow -Me_3Ge \qquad\qquad \downarrow -Me_3GeH$$
$$[Me_{2n-3}Ge_{n-1}]^+ \xrightarrow{-CH_4} [Me_{2n-5}Ge_{n-1}CH_2]^+$$

Scheme 8

Tin. Major ions in the spectra of $Me_3SnSC_6H_4X$-p have been listed;[102] transfer of X to tin occurs with X = F, Cl, or Br. In these cases the ions $[SnSC_6H_3]^+$ are also found. Preferential cleavage of the benzyl–tin bond occurs with compounds $R_nSn(CH_2Ph)_{4-n}$ ($n = 1$—3), except for $Bu^tSn(CH_2Ph)_3$, where both bonds are cleaved with the same probability.[103] The spectra of the corresponding mono- and di-chlorides are also recorded. The sequence for ease of cleavage $PhCH_2 \sim Bu^t > Pr^i > Et > Me$ was deduced.

The spectra of $(C_6F_5)_{4-n}SnR_n$ ($n = 1$—3; R = Ph, Bu^n, or Me), and of $(Me_3Sn)_2C_6X_4$ (X = F or Cl) have been studied, and detailed fragmentation

---

[99] S. Sloan, D. J. Harvey, and P. Vouros, *Org. Mass Spectrometry*, 1971, **5**, 789.
[100] D. J. Harvey, M. G. Horning, and P. Vouros, *Org. Mass Spectrometry*, 1971, **5**, 599.
[101] E. Carbeny, B. D. Dombek, and S. C. Cohen, *J. Organometallic Chem.*, 1972, **36**, 61.
[102] T. A. George, *J. Organometallic Chem.*, 1971, **31**, 233.
[103] M. Gielen, M. R. Barthels, M. de Clerq, and J. Nasielski, *Bull. Soc. chim. belges*, 1971, **80**, 189.

schemes deduced.[104] Cleavage of R groups occurs, whereas the pentafluorophenyl group fragments by elimination of C_6F_4, with migration of fluorine to tin. The common loss of a fluorine atom is only found in three cases, loss of RF from three co-ordinate stannonium ions being found instead. Loss of metal fluoride (as SnF_2) is found in two of the spectra. In the first series, $Me_3SnC_6F_5$ appears to be unique in exhibiting a metastable-supported loss of CF_4. This reaction is responsible for ions which carry about 16% of the total ion current.

In the disubstituted aromatics, of which the *ortho-* and *para-*isomers were studied, few differences were found between fluorine and chlorine compounds. Migration of chlorine to tin occurs to a larger extent than with fluorine. Between isomers, formation of $[P - Me]^+$ is more ready in the *ortho* series, which also lose CH_4. Cyclic ions are proposed to be formed (Scheme 9). In the spectra

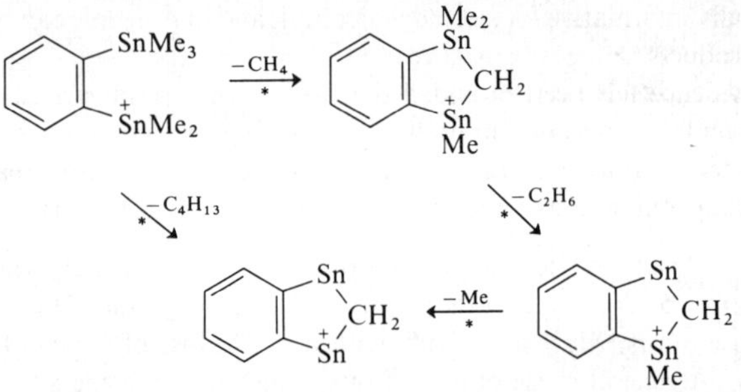

Scheme 9

of these and related compounds, the presence of the $[SnF]^+$ ion indicates a strong Sn—F affinity, as found chemically.

The formation of $[C_{18}F_n]^+$ ions occurs with $(C_6F_5)_3SnR$ derivatives, and of $[C_{12}F_n]^+$ with these and the $(C_6F_5)_2SnR_2$ compounds; these are probably polyphenylene ions, formed by elimination of a tin-containing fragment.

AP measurements have been used to evaluate heats of formation and bond dissociation energies in the compounds Ph_3SnR (R = Me, Et, Ph, I, SPh, Me_3Ge, Me_3Sn, or Ph_3Sn).[105] However, the values obtained by this method are often anomalous, and the authors urge that little significance should be attached to bond energies derived from electron-impact measurements, unless supported by independently determined values.

Nearly all ions in the spectrum of $SnBu^n_2(CH_2CH_2CN)_2$ contain tin; loss of butyl is slightly preferred over loss of cyanoethyl. Some evidence for the formation of cyclic ions containing Si—N bonds is presented.[24]

In the keto-organo-tins $RCO(CH_2)_nSnMe_3$ (R = Me or Ph; n = 2 or 3), the charge is localized on the tin atom, resulting in suppression of the McLafferty

---

[104] T. Chivers, G. F. Lanthier, and J. M. Miller, *J. Chem. Soc. (A)*, 1971, 2556.
[105] D. B. Chambers and F. Glockling, *Inorg. Chim. Acta*, 1970, **4**, 150.

rearrangement.[106] Fragmentation patterns of these compounds are dominated by ions retaining tin. Where a cyclic six-membered transition state can be formed, i.e. for $n = 3$, transfer of $SnMe_3$ from carbon to oxygen occurs:

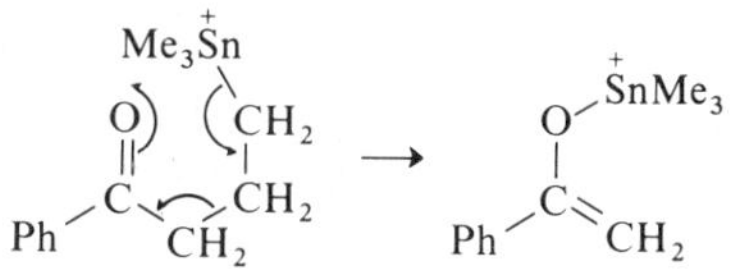

Skeletal rearrangements lead to the formation of $[PhSnMe_3]^+$ and $[PhSn]^+$ in the two benzoyl compounds; these ions are among the most abundant noted in these spectra.

The spectra of 10,10-disubstituted phenoxastannins have been reported,[107] and detailed fragmentation patterns were illustrated. A variety of related model compounds were used to establish the major features of the breakdown routes. Some modes are markedly dependent on the ligands attached to the metal. Generally the P^+ ions are either absent or have only very low abundances; the usual initial step is loss of a radical to form a three-co-ordinate tin ion. Where possible, alkyl is eliminated before aryl. Detailed charts of the breakdown of the resulting ions are given: neutral fragments lost include $C_{12}H_8O$ (dibenzofuran?), C_6H_4O, and benzyne. A feature of the breakdown is the formation of cyclic ions containing Sn—O bonds, such as (29). Metal-free ions found include most of those found in similar derivatives containing heteroatoms attached to at least two phenyl groups, such as (30) and (31).

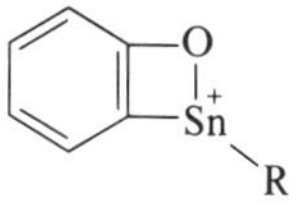

(29) 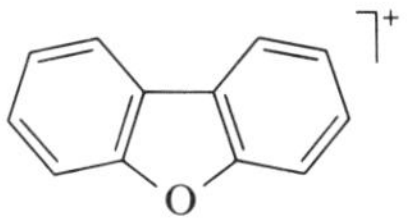 (30) 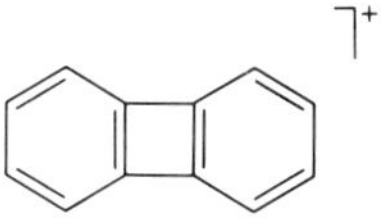 (31)

Lead. The spectra of the acetolysis products of hexa-aryldileads, Pb_2R_6, exhibit prominent $[R_2Pb(OAc)]^+$ and $[RPb(OAc)_2]^+$ ions; ions retaining both lead atoms are generally of very low abundance.[108] In Pb_2Ph_6 itself, the ions $[Pb_2Ph_n]^+$ ($n = 2$ or 3) have low intensities, and mononuclear ions predominate. The base peak is $[PbPh]^+$. In the putative $(Me_3Pb)_2CN_2$, the highest mass ion is $[C(PbMe_3)_2]^+$; the analogous silicon compound gave some evidence for association with a highest mass ion at m/e 357 $([\{(Me_3Si)_2CN_2\}_2 - Me]^+)$.[109]

[106] H. G. Kuivila, K.-H. Tsai, and D. G. I. Kingston, *J. Organometallic Chem.*, 1970, **23**, 129.
[107] I. Lengyel, M. J. Aaronson, and J. P. Dillon, *J. Organometallic Chem.*, 1970, **25**, 403.
[108] V. G. Kumar Das and P. R. Wells, *J. Organometallic Chem.*, 1970, **23**, 143.
[109] M. F. Lappert, J. Lorberth, and J. S. Poland, *J. Chem. Soc. (A),* 1970, 2954.

αα-Dihalogeno-organoleads show $[Ph_2PbCl]^+$ as a major fragment, indicating migration of halogen to metal.[110]

Group V.—The spectra of the tetraphenyl derivatives of Group IV elements, and the triphenyl compounds containing Group V elements, have been compared.[111] In the former, major ions are formed by loss of phenyl and $C_{12}H_{10}$ species. Further loss of phenyl is found for the Ge, Sn, and Pb compounds. This reaction is not accompanied by hydrogen scrambling, although further loss of hydrogen is a random process. The Group V derivatives show similar processes; loss of H_2 from $[EPh_2]^+$ is also random. In the negative-ion spectra of these compounds, the base peaks are $[P - Ph]^-$, and complete randomization precedes loss of hydrogen. These hydrogen-elimination reactions are assumed to involve cyclization before loss of hydrogen:

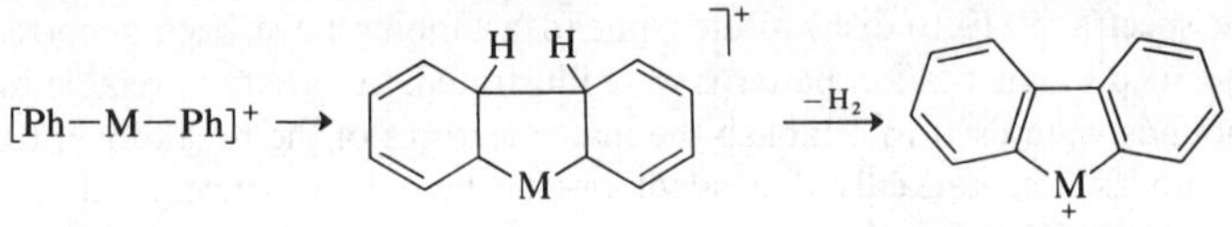

$[Ph-M-Ph]^+ \rightarrow$ [dihydro-dibenzometallole cation] $\xrightarrow{-H_2}$ [dibenzometallole cation]

Several polytertiary phosphine- and mixed phosphine–arsine ligands generally show features common to those previously noted.[112] Cleavage of P—C or As—C bonds predominated, e.g. loss of Ph_2P groups, and polyethylene-bridged compounds lost the bridge as olefin (usually C_2H_4). Some ions containing three phenyl groups were formed by phenyl-transfer reactions, although this reaction does not proceed from phosphorus to arsenic.

Features in the breakdown patterns in the series $(C_6F_5)_{3-n}EPh_n$ (n = 0—3; E = P or As), $SbPh_3$, $PhSb(C_6F_5)_2$, and $Sb(C_6F_5)_3$ have been correlated with the nature of the central atom.[113] There are few similarities between the spectra of these compounds and those of the neighbouring Group IV derivatives. For the triphenyl derivatives, loss of hydrogen from $[MPh_2]^+$ is followed by loss of metal; $[PhM]^+$, formed by loss of Ph_2, fragments further also by loss of metal. The abundances follow the order of bond strengths: P—C > As—C > Sb—C, with P^+ ions decreasing as the group is descended.

The tris(pentafluorophenyl) compounds exhibit $[P - F]^+$ ions, and loss of $C_{12}F_{10}$ is found. Migration of fluorine to the metal occurs by loss of MF, MF_2, and MF_3 species from $[M(C_6F_5)_2]^+$, generating fluorocarbon ions, and by elimination of C_6F_4 to give $[C_6F_5MF]^+$. Metal fluoride ions increase in abundance from phosphorus to antimony. In the bis(pentafluorophenyl) compounds, three competitive reactions predominate:

$$[(C_6F_5)_2MPh]^+ \rightarrow [PhC_6F_5]^+ + C_6F_5M$$
$$\rightarrow [PhMC_6F_5]^+ + C_6F_5^{\cdot}$$
$$\rightarrow [C_6F_5M]^+ + PhC_6F_5$$

[110] C. M. Warner and J. G. Noltes, *J. Organometallic Chem.*, 1970, **24**, C4.
[111] J. H. Bowie and B. Nussey, *Org. Mass Spectrometry*, 1970, **3**, 933.
[112] R. B. King and P. N. Kapoor, *J. Amer. Chem. Soc.*, 1971, **93**, 4158.
[113] A. T. Rake and J. M. Miller, *J. Chem. Soc. (A)*, 1970, 1881.

with analogous reactions found for $C_6F_5MPh_2$. In these compounds, the central metal prefers to remain attached to the more electronegative group. In contrast to the Group IV derivatives, both odd- and even-electron ions occur.

Although the familiar metallofluorene cation structure may be proposed for the ions containing at least two phenyl residues, migration of fluorine to metal with elimination of C_6F_4 suggests that discrete C_6F_5 groups are preserved, as in (32). However, an alternative process may account for the elimination of metal fluoride, as shown in (33). This also explains the high abundance of

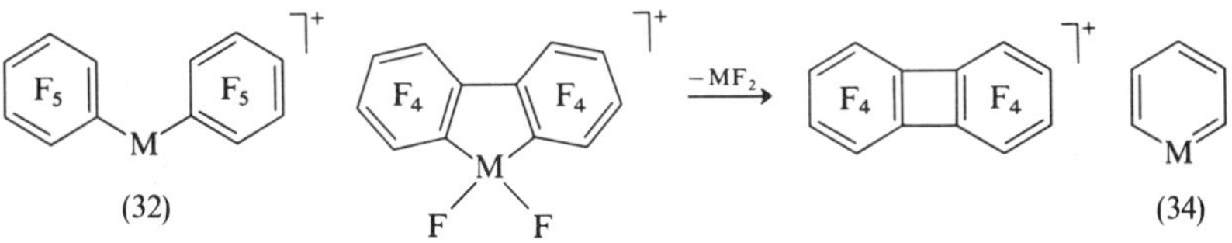

$[C_{12}F_8]^+$. Detailed fragmentation patterns have been proposed to rationalize the formation of the observed ions.

The spectra of the compounds C_5H_5M (M = N, P, As, or Sb) (34) have been reported briefly.[114,115] All show strong parent ions, as the base peaks; as the Group is descended, loss of C_2H_2 occurs to the exclusion of loss of HCM, the latter being entirely absent for the antimony compound. A strong Sb^+ ion was also noted in the latter spectrum.[115]

The spectra of several bis-2,2'-biphenylylenearsines[116] and similar antimony compounds[117] exhibit doubly charged molecular ions with relative intensities greater than the corresponding P^+ ion. Fragmentation of these ions gives a variety of cyclic ions, which may or may not retain the Group V atom.

Studies on several compounds of type (35) have been reported by various authors. Molecular ions of phenothiaphosphines are fairly stable, but undergo

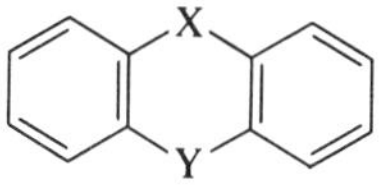

X = O,   Y = PR
X = S,   Y = PR, PO(OR)
X = SO_2, Y = PO(OR)
X = NH, Y = AsX, PX

a one-step elimination of the PR [or PO(OR)] group.[118] With fluorine substituents, loss of sulphur occurs to give (36); with methyl derivatives, ring-enlargement to tropylium ions takes place. With the sulphones, similar ring-expansion probably occurs to give (37). Phosphine oxides lose RO if R is aryl,

---

[114] A. J. Ashe, *J. Amer. Chem. Soc.*, 1971, **93**, 3293.
[115] A. J. Ashe, *J. Amer. Chem. Soc.*, 1971, **93**, 6690.
[116] D. Hellwinkel, B. Knabe, and G. Kilthau, *J. Organometallic Chem.*, 1970, **24**, 165.
[117] D. Hellwinkel and M. Bach, *J. Organometallic Chem.*, 1971, **28**, 349.
[118] I. Granoth, A. Kalir, Z. Pelah, and E. D. Bergmann, *Org. Mass Spectrometry*, 1970, **3**, 1359.

but only R· when R = Me. The effect of substituents on the breakdown of phenoxaphosphinic acids has been examined.[119] Generally, loss of OH and PO_2H from the P^+ ion was noted, although methyl-substituted compounds lost H_2O in a process characterized by an intense metastable ion, and probably involving a double rearrangement. Ethyl derivatives afforded intense $[P - Me]^+$ ions; in the 2,8-diethyl compound, $[P - 2Me]^{2+}$ has 17% of the abundance of $[P - Me]^+$, and is assigned structure (38). The spectra of some phenoxaphosphine derivatives have also been described.[120]

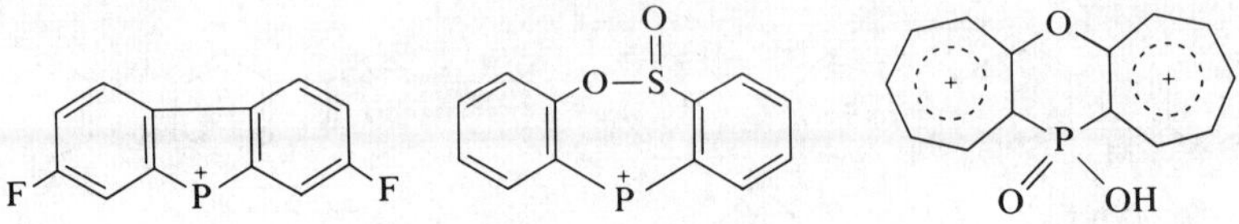

(36)   (37)   (38)

The phenarsazines[121] exhibit mass spectral fragmentation patterns which largely parallel pyrolytic decompositions; major pathways are loss of H and X to give (39), followed by loss of arsenic to give a heterocyclic nitrogen ion, possibly the diphenylenimine cation (40). The arsazinic acid (40a) shows ions at m/e 183 and 184, perhaps formed as shown.

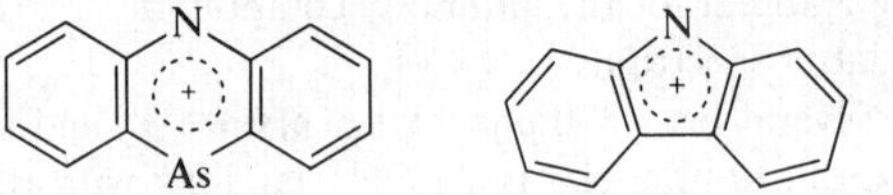

(39)   (40)

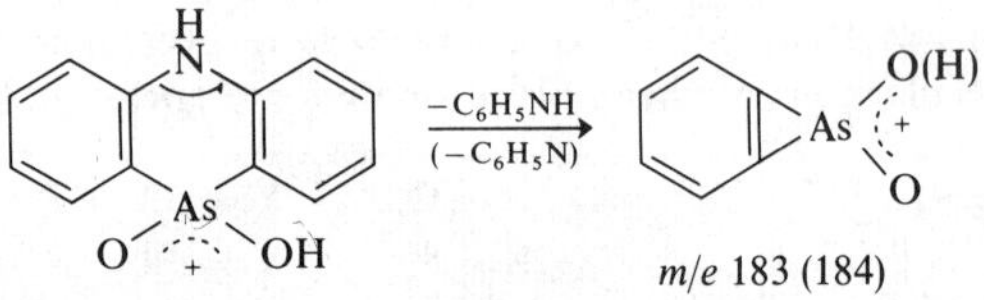

(40a) m/e 183 (184)

In the spectra of dihydrophenarsazines and their oxides, fragmentation leads to loss of exocyclic arsenic substituents, followed by loss of arsenic to give the carbazole ion.[122] Some differences from earlier results obtained with a chloro-compound are discussed in terms of As—Cl bond strengths. With the oxides, fragmentation patterns are related to those of the corresponding phosphorus compounds, although oxygen is lost more readily, in agreement with the relative chemical stabilities: P=O > As=O.

[119] I. Granoth and J. B. Levy, *J. Chem. Soc.* (*B*), 1971, 2391.
[120] I. Granoth, A. Kalir, Z. Pelah, and E. D. Bergmann, *Israel J. Chem.*, 1970, **8**, 621.
[121] J. C. Tou and C.-S. Wang, *Org. Mass Spectrometry*, 1970, **4**, 503.
[122] R. A. Earley and M. J. Gallagher, *Org. Mass Spectrometry*, 1970, **3**, 1287.

Phosphorus. The spectra of several acetylenes of the type $Ph_2PC\equiv CR$ have been recorded.[123] All except $R = P(NEt_2)_2$ give molecular ions. Complex decompositions were found for $Ph_2PC\equiv CCF_3$ and the analogous arsenic derivative, initial fragmentations involving loss of HF, CF_3 and C_2H_2, $PhC\equiv CCF_3$, or Ph groups. Simpler spectra were obtained for the non-fluorinated compounds (Scheme 10). The unsymmetrical $Ph_2PC\equiv CP(C_6F_5)_2$ gives $[Ph_2PC_2]^+$ as the most abundant peak indicating a preferential cleavage of the $C-P(C_6F_5)_2$ bond; with $Ph_2PC\equiv CAsPh_2$, cleavage of either bond occurs as

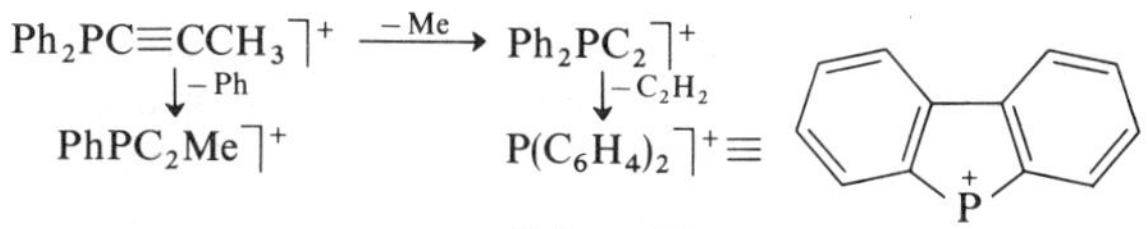

$$Ph_2PC\equiv CCH_3\,]^+ \xrightarrow{-Me} Ph_2PC_2\,]^+$$
$$\downarrow_{-Ph} \qquad\qquad \downarrow_{-C_2H_2}$$
$$PhPC_2Me\,]^+ \qquad P(C_6H_4)_2\,]^+\equiv$$

Scheme 10

a first step. In most cases, several high abundance ions are formed by rearrangement processes. The phosphino-acetylene $Ph_2PCH_2C\equiv CCH_2PPh_2$, and the corresponding dioxide and disulphide, exhibit P^+ ions.[124] These undergo P–C bond cleavage reactions, and all include the ion $[PPh_3]^+$, probably formed by pyrolysis. Other ions are related to those found in the spectra of similar compounds.

It is possible to obtain the mass spectra of $PhP(OH)_2$ and $PhPO(OH)_2$ under normal operating conditions between 50—85 °C.[125] The former shows characteristic ions corresponding to $[PhPO]^+$, $[PhPH_2]^+$, and $[C_6H_6O]^+$. The formation of phenylphosphine occurs *via* loss of oxygen from the charged ion, not *via* a thermal process. However, above 120 °C, peaks appear at higher m/e values, corresponding to fragments from the phosphonic acid formed by disproportionation. At lower temperatures, the latter gives a parent ion and associated fragment ions, but above 120 °C, dehydration occurs, and a highest m/e ion corresponding to $[(C_6H_5PO_2)_3]^+$ is observed. The degradation of this ion is discussed, and involves loss of PO, PO_2, and phenyl fragments.

Substituted *N*-phenyliminotriphenylphosphoranes exhibit little fragmentation, and spectra contain few ions other than the strong molecular ion.[126] The majority of fragment ions are analogous to those found with the corresponding ylides, with $[P - H]^+$ usually being of high abundance:

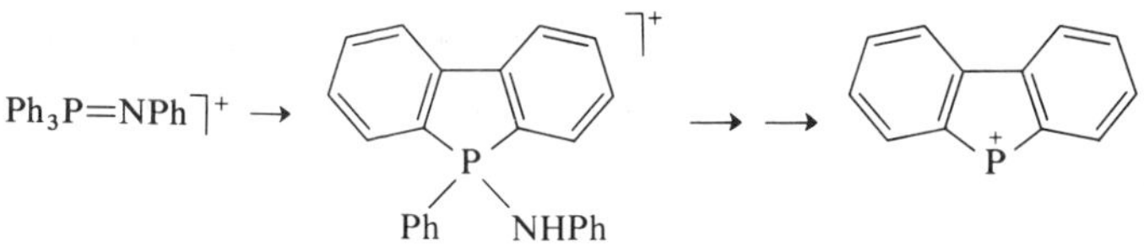

---

[123] A. J. Carty, N. K. Hota, T. W. Ng, H. A. Patel, and T. J. O'Connor, *Canad. J. Chem.*, 1971, **49**, 2706.
[124] R. B. King and A. Efraty, *Inorg. Chim. Acta*, 1970, **4**, 123.
[125] R. H. Wiley, *Org. Mass Spectrometry*, 1971, **5**, 675.
[126] T. Tökés and S. C. K. Wong, *Org. Mass Spectrometry*, 1970, **4**, 59.

The base peak in the spectrum of fluorenylidenetriphenylphosphorane is P^+; subsequent decomposition leads to many of the ions found in the spectra of other PPh_2 derivatives.[127] The annelated derivatives show much weaker P^+ ions, and $[PPh_3]^+$ is the most intense peak. Transfer of a phenyl group from phosphorus to the fluorene group gives the 9-phenylfluorenium ion (m/e 241).

The fragmentation of dimethylphosphonates derived from norbornane and related bi- and tri-cyclic hydrocarbons is directed by charge localization on phosphorus; the spectra of *endo*- and *exo*-isomers are similar.[128] Saturated derivatives tend to undergo fission of the P—C bond, whereas unsaturated compounds eliminate fragments *via* a retro-Diels–Alder reaction. Several processes can be formally rationalized by McLafferty-type rearrangements.

The spectra of acylphosphoranes, $Ph_3P=CR^1COR^2$, and related compounds have been studied with respect to P—O bond formation.[129] At higher temperatures (>200 °C) thermal decomposition occurs, and peaks due to

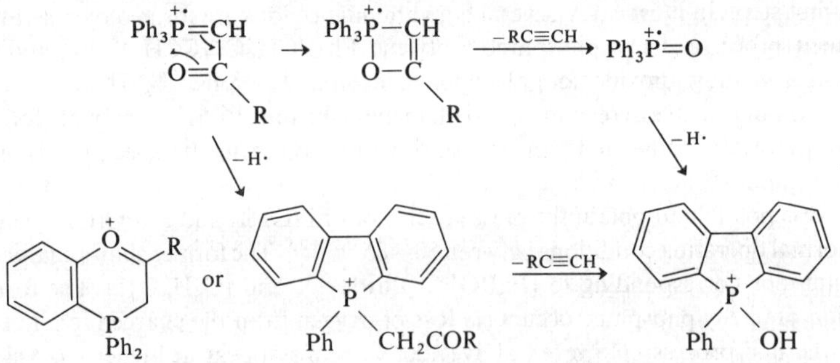

Scheme 11

PPh_3 and $OPPh_3$ predominate. At 70 °C, however, normal spectra are obtained, and a representative fragmentation is shown in Scheme 11. An isobutyl derivative (41) undergoes a rearrangement to the methylenetriphenylphosphorane ion:

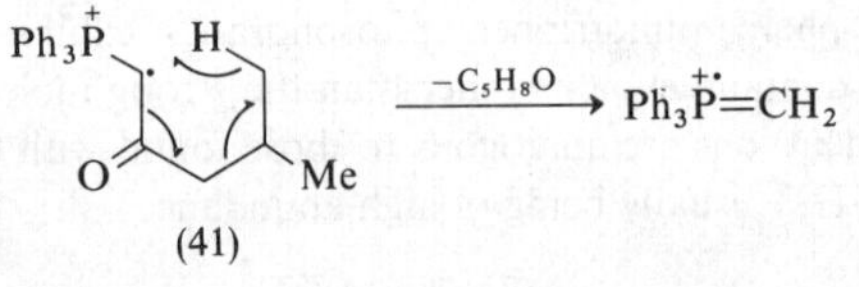

(41)

In the negative-ion spectra of diketophosphoranes such as (41a), the base peak is $[P - Ph]^-$, and other abundant ions include $[Ph_2PO]^-$ and $[C(COMe)-CO]^-$. Fairly extensive rearrangements are proposed to account for these ions.[130]

---

[127] E. D. Bergmann, M. Rabinowitz, C. Lifshitz, D. Shapiro, and I. Agranat, *Org. Mass Spectrometry*, 1970, **4**, 89.
[128] H. J. Callot and C. Benezra, *Org. Mass Spectrometry*, 1971, **5**, 343.
[129] A. P. Gara, R. A. Massy-Westrop, and J. H. Bowie, *Austral. J. Chem.*, 1970, **23**, 307.
[130] R. G. Alexander, D. B. Bigley, and J. F. J. Todd, *Chem. Comm.*, 1972, 553.

The parent ions of phosphine oxides appear to retain charge on oxygen, and the major process is formation of $[HP(O)R_2]^+$. In diethyl-*trans*-propenylphosphine oxide, relative intensities of P^+ and rearrangement ions indicate

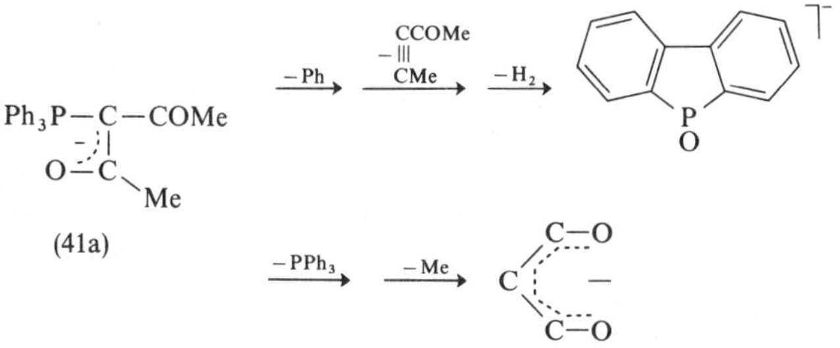

persistence of an intramolecular hydrogen bond.[131] In the spectra of 3-phospholene 1-oxides and 1-sulphides, high intensity P^+ ions are present, and loss of hydrocarbon occurs to only a small extent.[132] Instead, loss of butadiene gives ions such as $[P(O)(OPh)]^+$ and $[PSCl]^+$. Formation of $[C_4H_6]^+$ was correlated qualitatively with the reactivity of the compounds towards nucleophilic substitution.

Spectra of both $SPEt_3$ and $Et_4P_2S_2$ have been measured, and AP's determined; the P—P conjugation energy is about 25 kcal mol^{-1}. Relative intensities of ions have been correlated with electronic effects in the spectra of related oxides and sulphides.[133] Under high-resolution conditions,[134] the compositions of all ions were later confirmed; the P=S bond dissociation energy is about 85 kcal mol^{-1}. The spectra of Me_2PXMe (X = O or S) and Me_3PX are tabulated;[135] strong peaks are found for $[Me_2PO]^+$ and $[PO]^+$, and for $[Me_nPS]^+$ (n = 2 or 3) in the oxygen and sulphur compounds, respectively. In $Me_2P(O)H$, major ions are P^+, $[MePHO]^+$ (base peak), $[PO]^+$ and/or $[MePH]^+$, and $[PCH_2]^+$. Similar ions are found in the spectra of $Me_2P(O)OMe$, $(Me_2PO)_2O$, and Me_2POCl.[136] The spectrum of $[Me_2PPHMe_2]Cl$ is a superposition of those of Me_2PCl and Me_2PH, from which it is formed.[137]

The absence of peaks due to amine fragments and of the type $[CH_2=OPR_2]^+$, in the spectra of $ROP(NR^1_2)_2$ has been rationalized by $(p–d)\pi$ conjugation of nitrogen and oxygen lone pairs with phosphorus.[138] Principal fragmentation

[131] G. M. Bogolyubov, V. F. Plotnikov, V. M. Ignat'ev, and B. I. Ionin, *Zhur. obshchei Khim.*, 1971, **41**, 517.
[132] G. M. Bogolyubov, L. I. Zubtsova, N. N. Grishin, N. A. Razumova, and A. A. Petrov, *Zhur. obshchei Khim.*, 1971, **41**, 527.
[133] G. M. Bogolyubov, N. N. Grishin, and A. A. Petrov, *Zhur. obshchei Khim.*, 1971, **41**, 811.
[134] G. V. Fridlyanskii, V. A. Pavlenko, B. A. Vinogradov, N. N. Grishin, G. M. Bogolyubov, and A. A. Petrov, *Zhur. obshchei Khim.*, 1971, **41**, 1707.
[135] F. Seel and K.-D. Velleman, *Chem. Ber.*, 1972, **105**, 406.
[136] F. Seel and K.-D. Velleman, *Chem. Ber.*, 1971, **104**, 2972.
[137] F. Seel and K.-D. Velleman, *Chem. Ber.*, 1971, **104**, 2967.
[138] R. G. Kostyanovskii, I. A. Nuretidinov, N. P. Grechkin, and I. I. Chervin, *Izvest. Akad. Nauk S.S.S.R., Ser khim.*, 1969, 2588.

routes are summarized for these compounds, and also for $P(NR_2^1)_3$ compounds ($R_2^1 = Me_2$, Et_2, or $-CH_2CH_2-$).

In the spectra of cyanophosphines, the P—CN bond is cleaved before the P—R bond.[139] Thus with $PhP(CN)_2$, $[P - CN]^+$ and $[P - 2CN]^+$ are relatively intense ions. Loss of HCN is also found, giving $[C_6H_4PCN]^+$ (base peak). Phosphites show a strong tendency to form ions preserving the P—O bond, e.g. in $(MeO)_2PCN$, the base peak is $[P(OMe)_2]^+$; with $P(CN)_2(NMe_2)$, the base peak is $[P - CN]^+$, with $[PCN]^+$ having 20% relative abundance.

The spectra of $(CF_3)_2P(S)SP(CF_3)_2$ and $(CF_3)_2P(S)SSP(S)(CF_3)_2$ both show the expected fragmentation and rearrangement ions, although no P^+ ion occurs with the former compound.[140] The smallest fluorocarbon fragment retaining two sulphur atoms is $[(CF_3)_2PS_2]^+$. Neutral fragments lost include CF_3, CF_2, and sulphur in both cases. The spectra of $(CF_3)_2P(O)NHMe$ and the analogous sulphur compound have been tabulated.[141]

Cyclic polyphosphines exhibit P^+ ions corresponding to $[(C_2F_5P)_5]^+$, and $[(C_3F_7P)_n]^+$ ($n = 4$ or 5) have been distinguished;[142] the complex spectra also show ions corresponding to $[(C_2F_5P)_n]^+$ ($n = 3$ or 4); however, no trimeric product could be isolated.[143]

Spectra of $CF_3COP(C_6D_5)_2$, and of its product with oxygen $OP(C_6D_5)_2OCH$-$(CF_3)PO(C_6D_5)_2$ (41b) showed similar ions;[144] loss of $Ph_2PO(OH)$ occurred via the reaction shown.

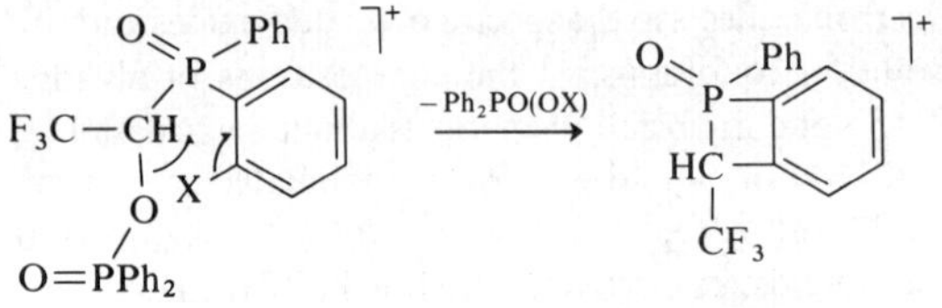

The parent ion forms the base peak in the spectrum of 1-methylphosphole (42), indicating the relative stability of this ring system.[145] Fragmentation is by loss of H or Me, and elimination of acetylene gives an ion corresponding to (43). As with N-, O-, and S-heterocycles, a prominent ion is also found at m/e 39. The spectrum of 1-phenylbenzophosphole similarly shows a strong P^+ ion.[146]

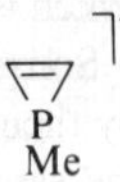

The spectrum of 1,6-diphosphatriptycene[147] contains strong multiply charged parent ions (P^+, P^{2+}, P^{3+}), together with ions at m/e 257 ($[C_{18}H_{10}P]^+$) and 183 ($[C_{12}H_8P]^+$).

[139] C. E. Jones and K. J. Coskran, Inorg. Chem., 1971, **10**, 1536.
[140] A. A. Pinkerton and R. G. Cavell, J. Amer. Chem. Soc., 1971, **93**, 2384.
[141] R. G. Cavell, T. L. Charlton, and W. Sim, J. Amer. Chem. Soc., 1971, **93**, 1130.
[142] P. S. Elmes, M. E. Redwood, and B. O. West, Chem. Comm., 1970, 1120.
[143] Cf. A. H. Cowley, T. A. Furtsch, and D. S. Dierdorf, Chem. Comm., 1970, 523.
[144] E. Lindner, H. D. Ebert, and A. Haag, Chem. Ber., 1970, **103**, 1872.
[145] L. D. Quin, J. G. Bryson, and C. G. Moreland, J. Amer. Chem. Soc., 1969, **91**, 3308.
[146] T. H. Chan and L. T. L. Wong, Canad. J. Chem., 1971, **49**, 530.
[147] K. G. Weinberg and E. B. Whipple, J. Amer. Chem. Soc., 1971, **93**, 1801.

Two major routes are followed in the breakdown of (44), each involving cleavage of a P—C bond.[148] Loss of Me gives an ion at m/e 115, and is followed by loss of H_2O. More important is loss of C_2H_4, followed by either loss of CO or

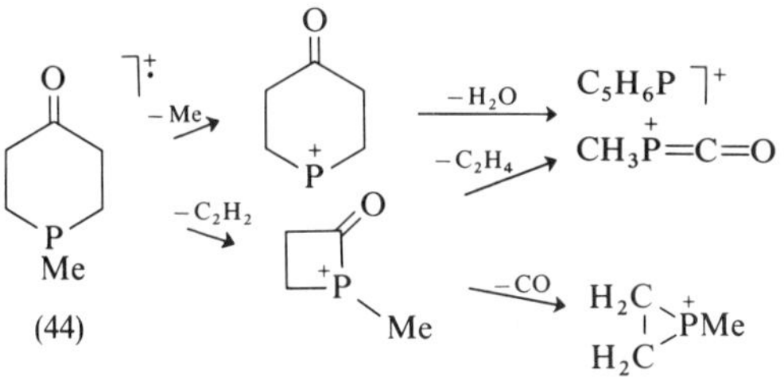

[Scheme showing fragmentation of (44): cyclic P-containing ketone losing Me, then C_2H_4 and H_2O to give $C_5H_6P^+$ which loses C_2H_4 to give $CH_3\overset{+}{P}=C=O$; alternative path through loss of C_2H_2 to a smaller cyclic cation, then $-CO$ to give $H_2C\text{-}\overset{+}{P}Me\text{-}H_2C$]

a second C_2H_4 group. Fragmentation of the nitrogen analogue (N-methyl-4-piperidone) is quite different, and initial cleavage of the C-2—C-3 bond occurs.

The partial mass spectrum[149] of $OCHBu^tCH_2=CBu^tP(O)Br$ contains a base peak formed by elimination of C_4H_8, and strong peaks corresponding to $[P - C_4H_8 - Me]^+$, $[P - Br]^+$, and $[P - C_4H_8 - Br]^+$.

Arsenic. The spectra of R_3As and R_4As_2 derivatives have been reported and show the expected features.[150] Loss of olefin from the intense P^+ peak, together with cleavage of one As—C bond, is observed. In the diarsine, As_2Et_4, the dissociation energy of the As—As bond was determined as 85 kcal mol^{-1}, a value which is compared with those of several similar Main Group derivatives.

Some vicinal bis(dimethylarsino) compounds have been partly characterized by their mass spectra.[151] Both *cis*- and *trans*-$(Me_2As)CH=CH(AsMe_2)$ exhibit the same spectrum. After ionization to (45), loss of Me is followed by loss of

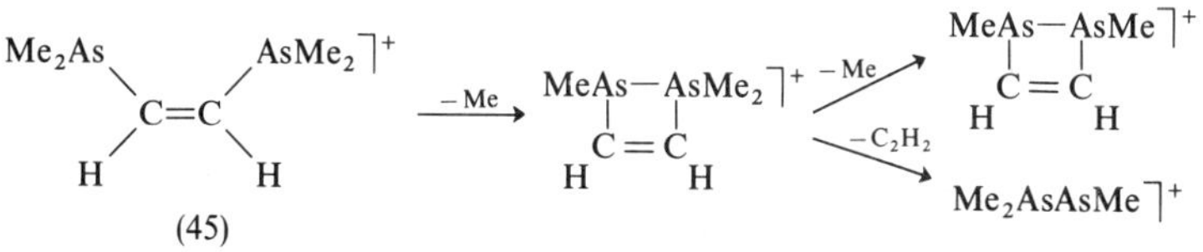

[Scheme: $(Me_2As)(H)C=C(H)(AsMe_2)$ (45) $\xrightarrow{-Me}$ $[MeAs(H)C=C(H)AsMe_2]^+$ $\xrightarrow{-Me}$ $[MeAs-AsMe\ HC=CH]^+$ and $\xrightarrow{-C_2H_2}$ $[Me_2AsAsMe]^+$]

acetylene or a second methyl group. Many ions at lower m/e values retain the two arsenic atoms. The spectrum of $Me_2AsC\equiv CAsMe_2$ was also recorded; major routes involve loss of all methyl groups, eventually giving $[As_2C_2]^+$. The formation of $[MeAs_2]^+$ by loss of C_2 from $[MeAs_2C_2]^+$ is also noted.

As with PPh_2 derivatives, compounds containing $AsPh_2$ groups give characteristic ions resulting from dehydrogenation of the $[AsPh_2]^+$ ion to the cyclic

---

[148] L. D. Quin and T. P. Toube, *J. Chem. Soc. (B)*, 1971, 832.
[149] R. S. Macomber, *J. Org. Chem.*, 1971, **36**, 2713.
[150] G. M. Bogolyubov, N. N. Grishin, and A. A. Petrov, *Zhur. obshchei Khim.*, 1971, **41**, 1710.
[151] R. D. Feltham and H. G. Metzger, *J. Organometallic Chem.*, 1971, **33**, 347.

arsenium ion.[152] In a variety of compounds, these characteristic ions are found at m/e 229, 227, 154, and 152, related as in Scheme 12. Comparisons of the various behaviours in the mass spectrometer with pyrolysis products enable few rationalizations to be made concerning these products and the mechanism of their formation.

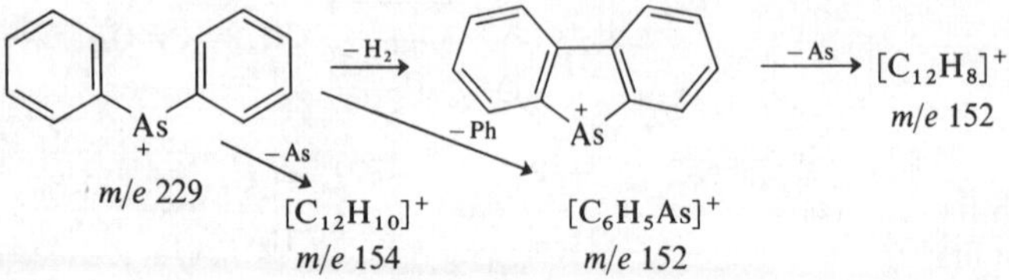

Scheme 12

Spectra of the cyclic arsines $(RAs)_5$ (R = Me, Et, or Pr^n) and $(PhAs)_6$ show P^+ peaks (often weak); the spectrum of the methyl derivative contains ions corresponding to all expected combinations $[As_mMe_n]^+$ except $[As_5Me_3]^+$. The base peak is $[AsMe_2]^+$, and neutral fragments lost include As_2 and $AsMe_2$. Longer chain alkyl groups also eliminate olefin. The phenyl derivative gives a strong $[As_3Ph_3]^+$ ion and the usual $[C_{12}H_8As]^+$ and $[C_{12}H_8]^+$ ions; at low ionizing energies, $[As_5Ph_5]^+$ and $[AsPh_3]^+$ were the only fragment ions.[153] The cyclohexyl and t-butyl compounds show highest m/e peaks corresponding to $n = 4$. Most ions retain three or four arsenic atoms.[154]

2,2'-Biphenylylene-arsenic derivatives give strong P^+ peaks,[155] and the spectra are characterized by loss of substituents to give the arsafluorene cation (45a). A strong parent ion also features in the spectrum of 10-phenyldibenz[be]arsenin.

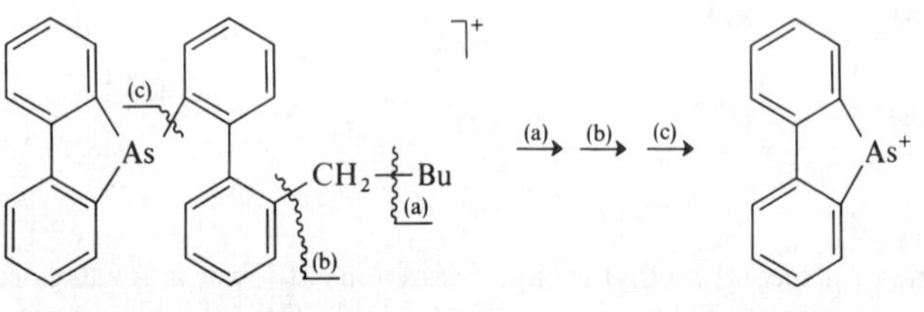

(45a)

In a series of arsenic compounds YCH_2CH_2XAsR (R = Ph, Et, NMe_2, etc.; X, Y = O, S, NMe) ready loss of the R group gives the base peak, assumed to be the cyclic arsenium ion (46), and much more stable than the corresponding boron

[152] J. C. Tou and C. S. Wang, *J. Organometallic Chem.*, 1972, **34**, 141.
[153] P. S. Elmes, S. Middleton, and B. O. West, *Austral. J. Chem.*, 1970, **23**, 1559.
[154] A. Tzschach and V. Kiesel, *J. prakt. Chem.*, 1971, **313**, 259.
[155] D. Hellwinkel and B. Knabe, *Chem. Ber.*, 1971, **104**, 1761.

cations.[156] In the phenyl derivatives, the formation of tropylium ion is found, as in other related heterocycles.

(46) 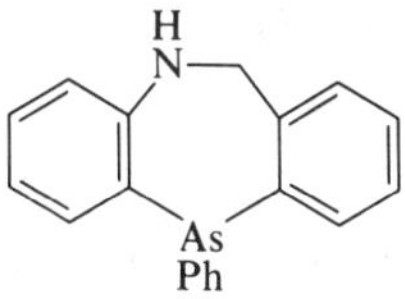 (47) 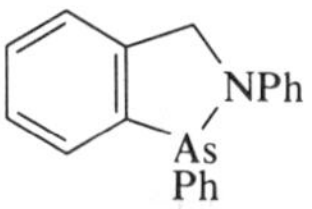 (48)

With the arsenic–nitrogen heterocycles (47)[122] and (48), arsenic substituents are lost first, followed by the arsenic atom.

Fragmentation in aza-arsatriptycene and its oxide have been reported.[157] Initial rearrangements of the P^+ ions give polycyclic systems, with preferential loss of arsenic. In addition to loss of oxygen, the oxide exhibits loss of CO and HCO from the P^+ ion; possible fragmentation pathways are discussed.

The compound $Me_4As_2S_2$ shows as prominent ions P^+, $[P - Me]^+$, $[P - S]^+$, and $[P - Me - S]^+$; other strong ions are $[AsMe_2]^+$ and $[AsS]^+$. From these data it was not possible to determine whether fission of As—S or As—C bonds had occurred more easily.[158]

Antimony and Bismuth. Mass spectra and AP's have been determined for $SbEt_3$ and Sb_2Et_4, and compared with data for the corresponding selenium and tellurium compounds.[159] In all cases, the parent peak is strong, and in the dinuclear compounds, the abundance of ions retaining the two metal atoms is high.

Complete analysis of the fragmentation of Me_2BiN_3 has been reported;[160] the P^+ ion loses Me or N_2 and N; some bismuth hydride ions were also observed.

3 Transition-metal Organometallics

Ion–Molecule Reactions.—The potential of ICR spectroscopy for the study of transition-metal complexes in the gas phase is illustrated in a report[161] describing ion–molecule reactions of $Fe(CO)_5$. At higher pressures, these reactions involve $[Fe(CO)_n]^+$ ($n = 0$—2) and $Fe(CO)_5$, to give $[Fe_2(CO)_n]^+$ ($n = 4$ or 5), and ligand substitution reactions in binary mixtures, for example, with MeF, H_2O, NH_3, and HCl:

$$[Fe(CO)_n]^+ + MeF \rightarrow [Fe(MeF)(CO)_{n-1}]^+ + CO \quad (n = 1\text{—}4)$$

[156] R. H. Anderson and R. H. Cragg, *Chem. Comm.*, 1971, 1414.
[157] R. A. Earley and M. J. Gallagher, *Org. Mass Spectrometry*, 1970, **3**, 1283.
[158] R. A. Zingaro, K. J. Irgolic, H. O'Brien, and L. J. Edmonson, *J. Amer. Chem. Soc.*, 1971, **93**, 5677.
[159] G. M. Bogolyubov, N. N. Grishin, and A. A. Petrov, *Zhur. obshchei Khim.*, 1969, **39**, 2244.
[160] J. Müller, *Z. anorg. Chem.*, 1971, **381**, 103.
[161] M. S. Foster and J. L. Beachamp, *J. Amer. Chem. Soc.*, 1971, **93**, 4924.

Other reactions studied include:

$$[Me_2F]^+ + Fe(CO)_5 \rightarrow [MeFe(CO)_4]^+ + CO + MeF$$

and extensive exchange of CO with NH_3 or H_2O. Using a π-bonding ligand, such as benzene, the reaction

$$[Fe(CO)_4]^+ + C_6H_6 \rightarrow [(C_6H_6)Fe(CO)_2]^+ + 2CO$$

was observed, whereas negative ion–molecule reactions include:

$$R^- + Fe(CO)_5 \rightarrow [RFe(CO)_3]^- + 2CO \quad (R = F \text{ or } OEt)$$

The wide range of information can be summarized thus: formation of dinuclear complexes, ligand replacement reactions, relative ligand binding energies, basicities of complexes, electrophilic and nucleophilic reactions, and formation of unusual π- and σ-bonded organometallic complexes.

At pressures between 10^{-5} and 10^{-6} Torr, dinuclear ions are formed in the mass spectrometer with a variety of organometallic complexes. With $M(CO)_6$, these are of the form $[M_2(CO)_n]^+$ ($n = 2$—11; M = Cr, Mo, or W), $[M_3(CO)_n]^+$ ($n = 6$—14), and $[W_4(CO)_n]^+$ ($n = 10$—14).[162] Mixtures of the hexacarbonyls give mixed ions, e.g. $[CrMo(CO)_n]^+$ ($n = 5$—7). Formation of these ions occurs by ion–molecule reactions; metastable peaks were observed for some decomposition processes:

$$[Cr_2(CO)_7]^+ \rightarrow [Cr_2(CO)_6]^+ + CO$$
$$[Cr_2(CO)_2]^+ \rightarrow Cr^+ + Cr(CO)_2$$

Secondary ions were also found in a variety of organochromium compounds.[163] With chromium carbonyl, the formation of the di- and tri-nuclear ions mentioned above was confirmed, and the major reactions responsible for their formation were determined to be:

$$[Cr(CO)_6]^+ + Cr(CO)_6 \rightarrow [Cr_2(CO)_{10}]^+ + 2CO$$
$$[Cr_2(CO)_{10}]^+ + Cr(CO)_6 \rightarrow [Cr_3(CO)_{14}]^+ + 2CO$$

With $Cr(C_6H_6)_2$, two ions are found, corresponding to $[Cr_2(C_6H_6)_n]^+$ ($n = 2$ and 3). Both di- and tri-nuclear ions $[Cr_2(C_6H_6)(CO)_n]^+$ ($n = 0$—3), $[Cr_2(C_6H_6)_2$-$(CO)_n]^+$ ($n = 0$—3), and $[Cr_3(C_6H_6)_2(CO)_n]^+$ ($n = 1$—6) are found in the spectrum of $(C_6H_6)Cr(CO)_3$.

In the cyclopentadienyl series, ions derived from $[(C_5H_5)_2Cr_2(NO)_2(CO)_2]^+$ by loss of CO and NO groups were found, and in the spectrum of $Cr(C_5H_5)_2$, dinuclear ions such as $[Cr_2C_{18}H_{16}]^+$, $[Cr_2C_{15}H_{13}]^+$, and $[Cr_2C_{10}H_9]^+$ were observed. These arise by collision of $Cr(C_5H_5)_2$ with $[Cr(C_5H_5)_2]^+$ or $[CrC_5H_5]^+$ ions.

Ions of greater mass than the P^+ ions in the spectra of (arene)$Cr(CO)_3$ complexes are of the type $[Ar_mCr_2(CO)_n]^+$ ($m = 1$ or 2, $n = 0$—3). Introduction of mixtures of complexes produced ions containing both arene groups,

[162] C. S. Kraihanzel, J. J. Conville, and J. E. Sturm, *Chem. Comm.*, 1971, 159.
[163] J. Müller and K. Fenderl, *Chem. Ber.*, 1971, **104**, 2199.

and this observation supports their formation by bimolecular gas-phase reactions.[164]

Secondary ions found in the spectra of the cyclopentadienyl carbonyls of vanadium, manganese, and cobalt decompose by loss of CO, as evidenced by the appropriate metastable peaks, and are formed by reactions such as:

$$[C_5H_5V(CO)_4]^+ + C_5H_5V(CO)_4 \rightarrow [(C_5H_5)_2V_2(CO)_4]^+ + 4CO$$
$$[C_5H_5Mn(CO)_3]^+ + C_5H_5Mn(CO)_3 \rightarrow [(C_5H_5)_2Mn_2(CO)_3]^+ + 3CO$$
$$[C_5H_5Co(CO)_2]^+ + C_5H_5Co(CO)_2 \rightarrow [(C_5H_5)_2Co_2(CO)_3]^+ + CO$$

However, other reactions also produce ions of this type, and a detailed discussion of possible ion–molecule reactions occurring with these complexes is given.[165] Ion–molecule reactions between the manganese complex and the fluorides EF_3 (E = P, As, Sb, but not N) and SF_4 (L) in the mass spectrometer, leading to secondary ions $[C_5H_5Mn(CO)L]^+$ or $[C_5H_5MnL]^+$, and expulsion of two or three CO groups, respectively, have been described.[166]

Metal Carbonyls.—Pyrolysis of dimeric manganese and cobalt carbonyls in graphite flow reactors connected to a mass spectrometer gave evidence for dissociation into the radical species ·$Mn(CO)_5$ and ·$Co(CO)_4$, respectively.[167] Equilibrium constants for these reactions were determined, and from these values for D(Mn—Mn) and D(Co—Co) of 21 ± 3 and 13 ± 3 kcal mol^{-1}, respectively, were obtained. The spectra of $M_2(CO)_{10}$ (M_2 = Mn_2, Re_2, or MnRe) have been measured again, together with the technetium complex.[168] The latter resembles the rhenium compound. IP and AP values for the various ions confirm trends found from other physical measurements in dissociation energies for the M—M and M—C bonds, and intensities of $[M(CO)_6]^+$ can be related to estimated activation energies for these rearrangements. Some predictions of relative rates for reactions involving M—M and M—CO bond ruptures are made.

The loss of CO groups from $[Ru_3(CO)_{12}]^+$ does not vary with ion-source temperature, and between 11—70 eV, fractional abundances accord with stepwise elimination reactions.[169] At 9 eV, predominant ions are $[Ru_3(CO)_n]^+$ (n = 7 or 8) formed by a different and unknown mechanism, probably involving ion-pair production.

Photo-ionization spectra have been described for $Fe(CO)_5$ and $Ni(CO)_4$, together with a discussion of the decomposition processes.[170]

Nitrosyls.—In the spectra of several cyclopentadienyl nitrosyl complexes,[171] loss of the elements of pyridine to give a metal oxide ion was found:

---

[164] J. R. Gilbert, W. P. Leach, and J. R. Miller, *J. Organometallic Chem.*, 1971, **30**, C41.
[165] J. Müller and K. Fenderl, *Chem. Ber.*, 1970, **103**, 3141.
[166] J. Müller and K. Fenderl, *Chem. Ber.*, 1971, **104**, 2207.
[167] D. R. Bidinosti and N. S. McIntyre, *Canad. J. Chem.*, 1970, **48**, 593.
[168] G. A. Junk and H. J. Svec, *J. Chem. Soc. (A)*, 1970, 2102.
[169] A. Herlan, *J. Organometallic Chem.*, 1971, **28**, 423.
[170] G. Distefano, *J. Res. NBS, Sect. A*, 1970, **74A**, 233.
[171] J. Müller, *J. Organometallic Chem.*, 1970, **23**, C38.

$[C_5H_5VNO]^+ \rightarrow [VO]^+ + (C_5H_5^- + N\cdot) \rightarrow$ [cyclopentadiene-NH structure] $\rightarrow$ [pyridine structure]

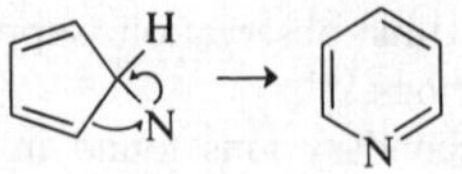

In other cases, transfer of a substituent group to nitrogen was found:

$[C_5H_5Cr(NO)C(OMe)Ph]^+ \xrightarrow{-Me} \xrightarrow{-CO} [C_5H_5-\underset{N}{\overset{|}{Cr}}-C_6H_5]^+ \rightarrow$
$\qquad\qquad\qquad\qquad\qquad\qquad\quad \underset{O}{\overset{|}{N}} \qquad [C_5H_5Cr:O]^+ + C_6H_5N$

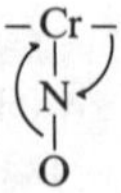

The complexes $M_3(CO)_{10}(NO)_2$ (M = Ru or Os) fragment by successive loss of all ligands to M_3^+, losing CO in preference to NO.[172] The ion $[Ru_2N]^+$, but not $[Ru_2C]^+$, is also found. For the ruthenium complex, loss of NO begins from the hexacarbonyl ion; with osmium, from the octacarbonyl. In the latter case loss of NO_2 from $[Os_3(CO)_9(NO)_2]^+$ gives $[Os_3(CO)_9N]^+$ or $[Os_3(CO)_8(NO)C]^+$.

Metal Carbonyl Hydrides and Halides.—The spectra of a variety of transition-metal hydrides have been discussed, and most show a P^+ ion of sufficient intensity to allow determination of molecular formulae.[173] Competitive loss of CO and hydrogen occurs, and analyses of the various ions formed allows an estimate of the relative importance of these two reactions. In general, terminal hydrides show $[P - H]^+$ ions, although this process may be subordinate to loss of HX from molecules such as $PtHX(PEt_3)_2$ (X = Cl, Br, CN, or CNO). Polynuclear complexes containing bridging hydride groups show no loss of hydrogen from the P^+ ion, and often retain hydrogen when the metal cluster fragments. In other cases, hydride is not lost until several CO groups have been eliminated.

The spectrum of $[Fe(CO)_3I]_2$ exhibits a P^+ ion, loss of CO groups to give eventually $[Fe_2I_2]^+$, and then decomposition by loss of iodine or iron.[174] In this case loss of halogen (assumed bridging) does not occur from the P^+ ion. Preservation of the Re_2Cl_2 nucleus is also found in the spectra of $[Re(CO)_3ClL]_2$ (L = Pr^iOH or Bu^iOH), giving strong ions centred on m/e 444.[175]

Transition-metal Cluster Compounds.—Parent ions are found for $H_2Ru_3(CO)_8X$ (X = S, Se, or Te) in high abundance;[176] detailed fragmentation studies revealed simultaneous loss of H and CO after the first two CO groups had been eliminated, to give $[Ru_3(CO)_nX]^+$ ($n = 0$—4). Doubly charged ions are present, and the percentages of trinuclear ions always exceed those of mono- and di-nuclear ions, although they decrease S > Se > Te.

The new polynuclear oxocarbonyl, $Os_6O_6(CO)_{16}$, has been identified by mass spectrometry,[177] and showed the expected ions $[Os_6O_6(CO)_n]^+$ ($n = 0$—16)

---

[172] J. R. Norton, J. P. Collman, G. Dolcetti, and W. T. Robinson, *Inorg. Chem.*, 1972, **11**, 382.
[173] B. F. G. Johnson, J. Lewis, and P. W. Robinson, *J. Chem. Soc.* (*A*), 1970, 1684.
[174] E. Koerner von Gustorf, J. C. Hogan, and R. Wagner, *Z. Naturforsch.*, 1972, **27b**, 140.
[175] W. Hieber and F. Stanner, *Chem. Ber.*, 1970, **103**, 2836.
[176] E. Sappa, O. Gambino, and G. Cetini, *J. Organometallic Chem.*, 1972, **35**, 375.
[177] C. W. Bradford and R. S. Nyholm, *J. Chem. Soc.* (*A*), 1971, 2038.

and $[Os_6O_6(CO)_n]^{2+}$ ($n = 0$—16). Breakup of the cluster gave ions such as $[Os_3O_n]^+$ ($n = 0$—2) and $[OsO]^+$.

After loss of CO from $[MeCCo_3(CO)_8(PPh_3)]^+$, the carbonyl-free ion fragments *via* a complex loss of C_6H_6, C_2H_2, and other neutrals.[178] No Co—Co bond cleavage is found, in contrast to the high abundance of dinuclear ions in the parent $MeCCo_3(CO)_9$ complex. In the arene derivatives, $YCCo_3(CO)_6$-(arene), which readily decomposed in the spectrometer, cobalt is eliminated from the carbonyl-free ion, with migration of Y to cobalt, and elimination of arene.[179] Random loss of hydrogen from the apical group and co-ordinated arene is found.

Besides the expected loss of CO groups, the spectrum of $Co_5(CO)_{15}C_3H$ shows the unusual series $[Co_4(CO)_nC_3H]^+$ ($n = 0$—10).[180] The predominant fragmentation of the $[Co_5C_3H]^+$ cluster ion is competitive loss of Co and H to give finally $[Co_2C_3]^+$. Doubly charged ions are present in reasonable abundance.

Compounds containing Bonds to Main Group Elements.—Spectra of $M[Re(CO_5]_2$ (M = Zn, Cd, or Hg) have been tabulated; apart from the expected P^+ ion and loss of CO groups, loss of an $Re(CO)_5$ group from the zinc and cadmium complexes was confirmed by a metastable peak.[181] The series $[Re_2(CO)_n]^+$ ($n = 0$—10) and $[Re(CO)_n]^+$ ($n = 0$—5) are also present. Metal complexes containing indium generally show fragment ions containing an intact metal-atom framework; $In[Re(CO)_5]_3$ is the only trisubstituted complex to show a P^+ ion.[182] Halogen-containing complexes, *e.g.* $Cl_2InRe(CO)_5$, give dimeric species such as $[Cl_4In_2Re_2(CO)_n]^+$ ($n = 2$ or 5).

The mass spectra of compounds $(\pi\text{-}C_5H_5)_2MClM'Ph_3$ (M = Zr or Hf; M' = Si or Ge) show weak P^+ peaks, and the only other ions retaining the metal–metal bond are $[(C_5H_5)_2MM']^+$. The base peak is always $[M'Ph_3]^+$; unusually, $[M'Ph_2]^+$ has appreciable intensity. Migration of the cyclopentadienyl group gives $[M'C_5H_5]^+$.[183]

The expected features are found in the spectra of the complexes $(\pi\text{-}C_5H_5)\text{-}(CO)_3MM'Me_3$ (M = Cr, Mo, or W; M' = Ge or Sn), and ions retaining the metal–metal bond carry 16 (Cr—Ge), 34 (Mo—Ge), 57 (W—Ge), 22 (Cr—Sn), 53 (Mo—Sn), and 80% (W—Sn) of the ion current.[184] AP measurements on suitable ions allowed a study of the energetics of the reaction:

$$(\pi\text{-}C_5H_5)(CO)_3M\text{—}M'Me_3 \rightarrow (\pi\text{-}C_5H_5)(CO)_3M\cdot + [M'Me_3]^+$$

and the metal–metal bond dissociation energies. Observed changes in these values are probably best rationalized in terms of steric effects, and 'relative stabilities' of the radicals $(\pi\text{-}C_5H_5)(CO)_3M\cdot$ and $Me_3M'\cdot$.

[178] T. W. Matheson, B. H. Robinson, and W. S. Tham, *J. Chem. Soc. (A)*, 1971, 1457.
[179] B. H. Robinson and J. L. Spencer, *J. Chem. Soc. (A)*, 1971, 2045.
[180] R. J. Dellaca, B. R. Penfold, B. H. Robinson, W. T. Robinson, and J. L. Spencer, *Inorg. Chem.*, 1970, **9**, 2197.
[181] A. T. T. Hsieh and M. J. Mays, *J. Chem. Soc. (A)*, 1971, 2648.
[182] A. T. T. Hsieh and M. J. Mays, *J. Organometallic Chem.*, 1972, **37**, 9.
[183] B. M. Kingston and M. F. Lappert, *J.C.S. Dalton*, 1972, 69.
[184] D. J. Cardin, S. A. Keppie, M. F. Lappert, M. R. Litzow, and T. R. Spalding, *J. Chem. Soc. (A)*, 1971, 2262.

The spectra of $H_3MM'(CO)_3(C_5H_5)$ (M = C or Si; M' = Cr, Mo, or W) have been briefly compared; loss of CO is followed by loss of MH_3, the charge remaining on the Group VI metal. Only for the Mo—Si complex were no ions found containing silicon.[185]

In the spectra of alkyltin-substituted arene-$Cr(CO)_3$ complexes, apart from the expected ions, abundant $[SnCr]^+$ ions are found.[186] Break-up of the ligand gives ions such as $[Me_2SnC_6H_4CH_2Cr]^+$, and elimination of benzene is found:

$$[(C_6H_5)_2SnCr]^+ \rightarrow [SnC_6H_4Cr]^+ + C_6H_6$$

The observed processes support the idea of some Sn—Cr interaction in the gas phase.

More than one-half the total ion current in the spectrum of $H_3GeCo(CO)_4$ is carried by ions retaining the Co—Ge bond.[187] In $H_3GeRe(CO)_5$, carbide ions $[ReC(CO)_n]^+$, and $[HRe(CO)_n]^+$, the latter formed by hydrogen migration, are found.[188]

In the spectra of $[Me_3SiFe(COSiMe_3)(CO)_3]_2$ and $[Me_3SiFe(COH)(CO)_3]_2$, singly and doubly charged ions retaining the C—C—Fe—Fe framework are found, and this favours a tetrahedral cage-type structure for these complexes.[189] The presence of $[P + CO]^+$ and $[P + 2CO]^+$ ions is also noted, perhaps arising from a gas-phase insertion of CO into the Si—Fe bond. Weak P^+ peaks are found in the spectra of tin–iron complexes containing the $(\pi\text{-}C_5H_5)Fe(CO)_2$ group; if chlorine is also present, as in $(\pi\text{-}C_5H_5)Fe(CO)_2SnPhCl_2$, peaks at higher m/e values were found, and correspond to $[P + O]^+$, $[P + Cl + CO]^+$, etc., suggesting catalytic effects on the ion source surfaces.[190] Thermal decomposition was unlikely at the temperatures used.

The nature of the Co—Si bond has been considered using a number of criteria. Mass spectral data[191] indicate that there is significant strengthening of this bond when three or four carbonyl groups are present, i.e. when there can be an interaction with three 'equatorial' carbonyl groups. In a later paper,[192] the mass spectra of $RCo(CO)_4$ (R = $SiCl_3$ or $SiMe_3$) were interpreted to support this idea. For example, in these and other complexes, ions of the type $[RCoL_n]^+$ (R = H, $SiCl_3$, or $SiMe_3$; L = CO or PF_3) are more abundant than $[CoL_{n-1}]^+$ for n = 3 or 4, which may be explicable in terms of this type of intramolecular interaction. Energetic measurements also indicated some strengthening of the Co—Si bond. Other features of the spectra include loss of CO before break-up of the SiX_3 group, and elimination of $SiCl_2$ from $[Cl_3SiCo(CO)_2]^+$.

[185] A. P. Hagen, C. R. Higgins, and P. J. Russo, *Inorg. Chem.*, 1971, **10**, 1657.
[186] T. P. Poeth, P. G. Harrison, T. V. Long, B. R. Willeford, and J. J. Zuckerman, *Inorg. Chem.*, 1971, **10**, 522.
[187] K. M. Mackay and R. D. George, *Inorg. Nuclear Chem. Letters*, 1970, **6**, 289.
[188] K. M. Mackay and S. R. Stobart, *Inorg. Nuclear Chem. Letters*, 1970, **6**, 687.
[189] M. A. Nasta, A. G. MacDiarmid, and F. E. Saalfeld, *J. Amer. Chem. Soc.*, 1972, **94**, 2449.
[190] D. S. Field and M. J. Newlands, *J. Organometallic Chem.*, 1971, **27**, 213.
[191] A. D. Berry, E. R. Corey, A. P. Hagen, A. G. MacDiarmid, F. E. Saalfeld, and B. B. Wayland, *J. Amer. Chem. Soc.*, 1970, **92**, 1940.
[192] F. E. Saalfeld, M. V. McDowell, and A. G. MacDiarmid, *J. Amer. Chem. Soc.*, 1970, **92**, 2324.

Complexes containing Pt—Si or Pt—Ge bonds, e.g. (diphos)Pt(MMe$_3$)$_2$, give observable P^+ ions, with a strong [(diphos)Pt]$^+$ ion cluster.[193] In the tin compound, elimination of SnMe$_2$, i.e. transfer of methyl from tin to platinum, and of methane, were found:

$$[(\text{diphos})\text{PtSnMe}_3]^+ \xrightarrow{-\text{SnMe}_2} [(\text{diphos})\text{PtMe}]^+ \xrightarrow{-\text{CH}_4} [\text{C}_{26}\text{H}_{23}\text{P}_2\text{Pt}]^+$$

Complexes containing antimony–transition-metal bonds include $\{[(C_5H_5)Fe(CO)_2]_3SbCl\}_2FeCl_4$[194] and $Co_4(CO)_{12}Sb_4$.[195] In the former, ready cleavage of the Fe—Sb bond occurs, and the highest mass ion is $[ClSb\{Fe(CO)_2(C_5H_5)\}_2]^+$; the ion $[SbFeC_5H_5]^+$ is also present. No mass spectrum could be obtained for the cobalt complex.

Complexes containing Metal–Carbon σ-Bonds.—Products from reactions between (π-C$_5$H$_5$)$_2$MoH$_2$ and various unsaturated molecules give spectra which confirm the assigned structures.[196] Hydrido-σ-alkyl complexes show low abundance P^+ ions, but hydrido-σ-alkenyl compounds show ions of greater intensity, perhaps because of conversion into olefin complexes. The diphenylacetylene complex ion, $[(C_5H_5)_2Mo(C_2Ph_2)]^+$, fragments to $[(C_5H_5)_2Mo]^+$, an ion found in high abundance in the spectra of all complexes, and also present as $[(C_5H_5)_2Mo]^{2+}$. Break-up of olefin complexes occurs by loss of substituent groups, e.g. of OMe and CO from the dimethyl fumarate complex.

Complexes of the type (π-C$_5$H$_5$)M(PEt$_3$)X (M = Pd or Pt; X = Br, I, or Ph)[197] fragment by loss of C$_5$H$_5$ and halide or phenyl; the metal–phosphorus bond is preserved to give $[M(PEt_3)]^+$. Similarly, in PtCl(C$_5$H$_5$)C$_6$H$_4$(AsEt$_2$)$_2$-o, loss of C$_5$H$_5$ or Cl occurs to give $[(C_6H_4As_2Et_2)Pt]^+$; the arsine ligand may also lose C$_4$H$_{10}$. A monomeric P^+ ion is present in the spectrum of (π-C$_5$H$_5$)PtMe$_3$, although in low abundance.[198] The dominant breakdown pathway is loss of methyl groups; the C$_5$H$_5$ group loses either C$_2$H$_2$ or a further methyl group. A fragmentation scheme is proposed, which also accounts for the formation of low abundance ions such as $[PtC_6H_5]^+$; loss of CH$_4$, C$_5$H$_6$, or CH$_3$C$_5$H$_5$ is also noted.

Phosphine–gold complexes, e.g. Me$_3$PAuMe and Me$_3$PAuMe$_3$, give $[AuPR_3]^+$ or the phosphine ligand ion as the base peak;[199] cleavage of the Au—C bonds occurs readily, and the trimethyl compound loses 30 m.u. (2CH$_3$ or C$_2$H$_6$) in accord with the thermal decomposition:

$$\text{Me}_3\text{AuPR}_3 \rightarrow \text{C}_2\text{H}_6 + \text{MeAuPr}_3$$

The complex Me$_3$PAuOSiMe$_3$[200] gives the expected P^+ and $[P - Me]^+$ ions; also abundant are $[AuPMe_3]^+$ and $[Me_3SiOSiMe_2]^+$.

[193] A. F. Clemmitt and F. Glockling, *J. Chem. Soc. (A)*, 1971, 1164.
[194] Trinh-Toan and L. F. Dahl, *J. Amer. Chem. Soc.*, 1971, **93**, 2654.
[195] A. S. Foust and L. F. Dahl, *J. Amer. Chem. Soc.*, 1970, **92**, 7337.
[196] A. Nakamura and S. Otsuka, *J. Amer. Chem. Soc.*, 1972, **94**, 1886.
[197] R. J. Cross and R. Wardle, *J. Chem. Soc. (A)*, 1971, 2000.
[198] K. W. Egger, *J. Organometallic Chem.*, 1970, **24**, 501.
[199] H. Schmidbaur and A. Shiotani, *Chem. Ber.*, 1971, **104**, 2821.
[200] A. Shiotani and H. Schmidbaur, *J. Amer. Chem. Soc.*, 1970, **92**, 7003.

Hydrocarbon–Metal π-Complexes.—*Olefins, Dienes, and Fluoro-olefins.* Loss of hydrogen from cyclohexadiene-Fe(CO)$_3$, a reaction which is competitive with loss of CO, occurs in a highly (but not 100%) stereospecific way.[201] Thus from the *endo*-D$_2$ species, loss of D$_2$ is preferred by a factor of 9:1. A reasonable explanation is aromatization by prior transfer of H$_2$ to metal, followed by cleavage of Fe—H bonds.

After loss of CO, further breakdown of [FeL]$^+$ (L = substituted 1,2-diazepin) occurs by loss of HCN, Me if present, and C$_2$H$_2$.[202] The analogous ruthenium complexes readily lose metal to give charged ligand ions. For methyl and phenyl derivatives, an additional route is loss of MeCN or PhCN, respectively. The spectra of other derivatives show essentially the same features, together with other ions resulting from the particular substituents.

Complexes of bicyclo[3,3,1]nona-2,6-diene with rhodium and platinum chlorides give P^+ ions, which lose Cl or HCl; fragmentation of the ligand occurs.[203] Many hydrocarbon ions are present, and indeed no metal-containing ions were found in the spectrum of the palladium complex.

A tetrafluorobenzobarrelene–rhodium complex (49) fragments by loss of C$_3$H$_4$ and (L − C$_3$H$_4$) groups to give eventually [C$_5$H$_5$Rh]$^+$; a charged ligand peak is also present.[204] The breakdown of a manganese complex shows no unusual features.

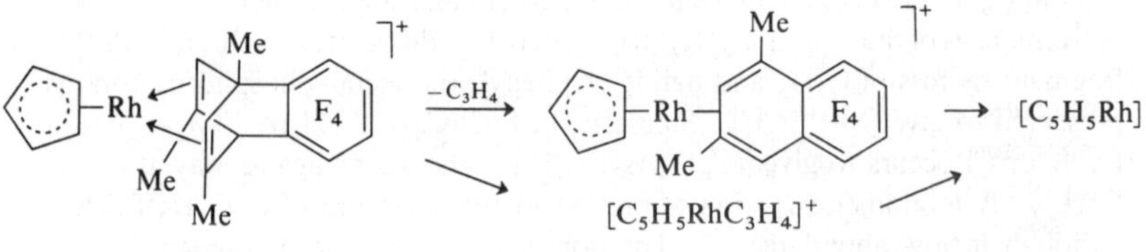

(49)

Fluoro-olefin-iron carbonyl complexes give no P^+ ions, major peaks arising by cleavage of the olefin, as found in the thermal decomposition of these complexes.[205] The complex C$_2$F$_3$ClFe(CO)$_4$ showed weak peaks [C$_2$F$_3$ClFe(CO)$_n$]$^+$ (n = 0 or 1) and [C$_2$F$_3$Fe(CO)$_n$]$^+$ (n = 1—4), suggested to arise from the formation of a σ-CF$_2$=CF complex in the spectrometer.

Acetylenes. Dinuclear acetylenic derivatives of the Fe–Ru–Os triad all give parent ions.[206] The iron complexes fragment to give mono- and di-nuclear ions, by loss of acetylene or iron. The ruthenium and osmium compounds are more robust, and show only dinuclear ions. Rearrangements involving these ions are

[201] T. H. Whitesides and R. W. Arhart, *Tetrahedron Letters*, 1972, 297.
[202] A. J. Carty, G. Kan, D. P. Madden, V. Snieckus, M. Stanton, and T. Birchall, *J. Organometallic Chem.*, 1971, **32**, 241.
[203] J. K. A. Clarke, E. McMahon, J. B. Thomson, and B. Zeeh, *J. Organometallic Chem.*, 1971, **31**, 283.
[204] D. M. Roe and A. G. Massey, *J. Organometallic Chem.*, 1971, **28**, 273.
[205] R. Fields, M. M. Germain, R. N. Haszeldine, and P. W. Wiggans, *J. Chem. Soc. (A)*, 1970, 1969.
[206] G. A. Vaglio, O. Gambino, R. P. Ferrari, and G. Cetini, *Org. Mass Spectrometry*, 1971, **5**, 493.

similar to those containing iron, however, although break-up of the ligand proceeds to give metal carbide ions such as $[Os_2C_4]^+$ and $[Os_2C_2]^+$. Loss of H was also noted. Doubly and triply charged ions are common in the spectra of phenyl-substituted compounds.

Allyls. Allyl derivatives of cyclopentadienylmetal carbonyls fragment by loss of CO, H_2, and C_2H_2; facile dehydrogenation of $[C_5H_5MC_3H_5]^+$ occurs, giving $[C_5H_5MC_3H_3]^+$, loss of C_2H_2 then affording $[C_6H_6M]^+$.[207]

π-Allylic iron carbonyl halides generally show weak P^+ peaks, except for the iodides; subsequent fragmentation involves the expected loss of CO and halogen.[208] Intense peaks $[RC_3H_3Fe]^+$, formed by elimination of H or HX, are found, which, with R = Me, may have butadiene- or trimethylenemethane-type structures. Formation of $[FeCH_3]^+$ by elimination of RC≡CH, or by methyl migration, is noted. A detailed breakdown scheme is presented.

The spectra of π-allylic palladium halides show P^+ ions which fragment by loss of halide and hydrocarbon groups.[209] Ions retaining the Pd_2X_2 nucleus readily lose halogen; variations in relative abundances are related to the various abilities of the allylic groups to stabilize the Pd—Pd bond in precursor ions.

The spectra of three isomers of $Fe_2(CO)_6$ complexes of allene trimer have been summarized;[210] the appearance of ions containing only one iron atom, *e.g.* $[FeC_9H_n]^+$ (n = 10 or 12), combined with other physical measurements, favours the absence of any metal–metal bond in these compounds.

Cyclobutadiene. IP's for cyclobutadiene and its iron carbonyl complex have been determined by photoelectron spectroscopy, and resolve the conflicting data obtained from mass spectrometric measurements.[211] For C_4H_4, the first and second IP's are estimated at 8.50 and 11.66 eV.

Loss of the tetraphenylcyclobutadiene ligand from some cobalt and iron derivatives occurs in two steps, with successive elimination of C_2Ph_2 fragments.[212] This may be contrasted with results reported[213] for the parent $C_4H_4Fe(CO)_3$ complex, in which absence of C_2H_2 losses was construed as evidence favouring the presence of a C_4 ring, and compared with the loss of two C_2H_2 fragments from $(π-C_5H_5)Co(π-C_4H_4)$.[214,215] No ion $[C_4Ph_4]^+$ is found in these spectra. Whereas loss of a phenyl group is only a minor process for the P^+ ion, for the P^{2+} ion, this is the major breakdown pathway.

Cyclopentadienyl. Further investigations[216] on $M(C_5H_5)_3$ (M = Nd, Sm, or Yb) and $Yb(C_5H_5)_2$ confirm the sequential loss of C_5H_5 groups; ions at lower m/e

[207] R. B. King and M. Ishaq, *Inorg. Chim. Acta*, 1970, **4**, 258.
[208] A. N. Nesmeyanov, Y. S. Nekrasov, N. P. Avakyan, and I. I. Kritskaya, *J. Organometallic Chem.*, 1971, **33**, 375.
[209] H. Lammens and G. Sartori, *Org. Mass Spectrometry*, 1971, **5**, 335.
[210] S. Otsuka, A. Nakamura, and K. Tani, *J. Chem. Soc. (A)*, 1971, 154.
[211] S. D. Worley, *Chem. Comm.*, 1970, 980.
[212] R. B. King and A. Efraty, *Org. Mass Spectrometry*, 1970, **3**, 1233.
[213] G. F. Emerson, L. Watts, and R. Pettit, *J. Amer. Chem. Soc.*, 1965, **87**, 131.
[214] R. Amiet and R. Pettit, *J. Amer. Chem. Soc.*, 1968, **90**, 1059.
[215] M. Rosenblum and B. North, *J. Amer. Chem. Soc.*, 1968, **90**, 1060.
[216] J. L. Thomas and R. G. Hayes, *J. Organometallic Chem.*, 1970, **23**, 487.

values include $[MC_3H_2]^+$ and $[MC_2H]^+$. Relative abundances of the ions indicate covalent bonding. IP and AP values indicate a first bond dissociation energy of $Yb(C_5H_5)_3$ to be $58 \pm 5\,kcal\,mol^{-1}$. Russian workers also have studied the spectra of the La, Pr, Sm, and Nd compounds, finding loss of C_5H_5 and C_2H_2 fragments from the parent ions.[217] IP and AP measurements were used to show that metal–ligand bond energies are higher than found for analogous transition-metal complexes, lying in the range 133—145 $kcal\,mol^{-1}$.

The results of an extensive investigation of cyclopentadienyl-titanium compounds include characterization of $C_{10}Me_{10}CH_2Ti$, showing P^+ and $[P - Me]^+$ ions, together with others derived from one of the rings.[218] Comparison of the spectra of $Ti(C_5Me_5)_2$ and $(C_5Me_5)_2TiMe_2$ show that these compounds give similar spectra, although the latter also exhibits P^+, $[P - Me]^+$, and $[P - CH_4]^+$ ions. The spectrum of $(C_{10}H_{10}Ti)_2$ differs substantially from those of the other metallocenes, with a dimeric P^+ ion, a very low abundance $[C_{10}H_{10}Ti]^+$ ion, and the striking loss of up to four H_2 molecules.[219] These results have been interpreted as indicating a double hydrogen bridge. The high-intensity $[C_5H_6]^+$ ion seems to be associated with the presence of a C_5H_4 group. In the spectrum of $(\pi\text{-}C_5H_5)_2TiMe_2$, loss of methyl groups affords the ion $[C_{10}H_{10}Ti]^+$ as a dominant species, but a rapid rearrangement is indicated by the presence of $[C_5H_6]^+$. Loss of hydrogen is not as important as with $(C_{10}H_{10}Ti)_2$.

Cyclopentadienyl-titanium and -zirconium azide complexes show P^+ ions, which undergo successive loss of N_3, C_5H_5, and N_2 groups.[220] The oxides $[(\pi\text{-}C_5H_5)_2M(N_3)]_2O$ show intense peaks retaining the M—O—M group, but fragment in the same manner.

The spectra of pyrrolyl-, indenyl-, and fluorenyl-manganese carbonyl complexes[221] show the usual features, i.e. loss of CO and C_2H_2 units; the EPh_3 (E = P, As, or Sb) derivatives show in addition the ions $[EPh_3]^+$, $[EPh]^+$, and $[C_{12}H_8E]^+$. Transfer of a phenyl group to manganese gives $[MnPh]^+$ in relatively high abundance. The pyrrolyl ligand loses HCN or C_2H_2N, as found for azaferrocene. The relative tendencies for loss of the ring or EPh_3 group are established, although some inconsistencies are apparent.

The mass spectra of the *endo*- and *exo*-isomers of $(C_5H_5)Rh(C_5HMe_5)$ (50) are considerably different.[222] The *exo*-isomer shows a weak P^+ peak, and the base peak is $[P - H]^+$. Loss of Me gave only a weak peak. In contrast, the base peak of the *endo*-isomer is $[P - Me]^+$, and both P^+ and $[P - H]^+$ have low intensity. In both cases, the stable ion corresponds to the methylated rhodicenium cation. The spectra of $(\pi\text{-}C_5Me_5)Ir(CO)_2$ and its oxidative addition products showed retention of the C_5Me_5Ir group;[223] with compounds of the type $(\pi\text{-}C_5Me_5)Ir(CO)RI$ (R = fluoroalkyl group), the ion $[(C_5Me_5)IrI_2]^+$ probably originates

[217] G. G. Devyatykh, S. G. Krasnova, G. K. Borisov, N. V. Larin, and P. E. Gaivoronskii, *Doklady Akad. Nauk S.S.S.R.*, 1970, **193**, 1069.
[218] J. E. Bercaw, R. H. Marvich, L. G. Bell, and H. H. Brintzinger, *J. Amer. Chem. Soc.*, 1972, **94**, 1219.
[219] H. H. Brintzinger and J. E. Bercaw, *J. Amer. Chem. Soc.*, 1970, **92**, 6182.
[220] R. S. P. Coutts and P. C. Wailes, *Austral. J. Chem.*, 1971, **24**, 1075.
[221] R. B. King and A. Efraty, *Org. Mass Spectrometry*, 1970, **3**, 1227.
[222] K. Moseley, J. W. Kang, and P. M. Maitlis, *J. Chem. Soc. (A)*, 1970, 2875.
[223] R. B. King and A. Efraty, *J. Organometallic Chem.*, 1971, **27**, 409.

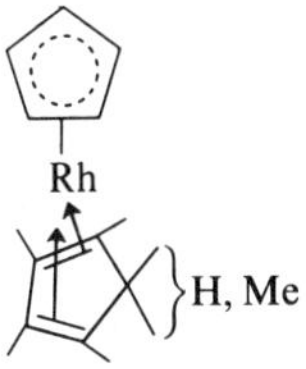

(50)

from $(\pi\text{-}C_5Me_5)Ir(CO)I_2$ produced by pyrolysis. Processes accounting for the majority of observed ions include loss of R, F, I, or CO, and migration of fluorine to metal, if present.

Arenes. Studies of styrene, α-[^{2}H]styrene, and the corresponding $Cr(CO)_3$ complexes have shown that randomization of H in $C_8H_8Cr^+$ is noticeably different from the complete scrambling of H in $C_8H_8^{\ddag}$ from styrene, before these ions decompose to $C_6H_5Cr^+$ and $C_6H_5^+$ respectively.[224]

The fragmentation of alkylbenzene-$Cr(CO)_3$ complexes proceeds *via* loss of CO, followed by fragmentation of the alkyl substituent, giving $[Cr(CH_2)_nC_6H_5]^+$ (*e.g.*, $n = 0$—3 for the butyl compound). The spectrum of this derivative is illustrated.[225]

Studies of isomeric substituted benzyl- and cycloheptatriene-$Cr(CO)_3$ complexes have shown that the benzyl–tropylium analogy found in the free ligands does not exist in these complexes.[226] With the unsubstituted compounds, the principal differences lie in the relative intensities of ions, especially those of $[CrC_7H_8]^+$, $[CrC_5H_6]^+$, and $[C_7H_7]^+$, all of which are higher in the cycloheptatriene complex. Indeed, the latter ion is present in high abundance in all the cycloheptatriene complexes. Differences in fragmentation of side-chains are also found, *e.g.* with $MeOC_7H_7Cr(CO)_3$, the ion $[CrC_7H_7O]^+$ is formed by loss of methyl, whereas the corresponding process is not found for the benzyl derivative. The *exo-* and *endo-*isomers of this complex can be distinguished on the basis of the intensities of the $[CrC_7H_7]^+$ ions. With the esters, RO_2CCH_2-$C_7H_7Cr(CO)_3$, the intermediate carbonyl-free ion loses C_6H_6 and $CH{=}CHCO_2R$ fragments, explained by the rearrangement:

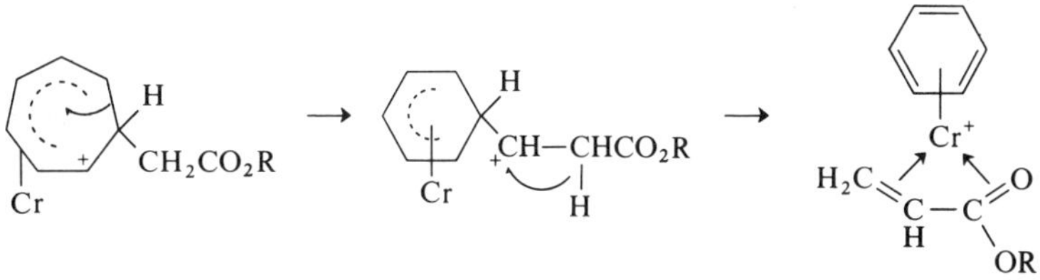

The spectrum of $C_7H_7C_7H_7Cr(CO)_3$ was also recorded.

[224] M. M. Bursey, F. E. Tibbetts, W. F. Little, M. D. Rausch, and G. A. Moser, *Tetrahedron Letters*, 1970, 1649.
[225] C. Segard, B. Roques, C. Pommier, and G. Guiochon, *Analyt. Chem.*, 1971, **43**, 1146.
[226] J. Müller and K. Fenderl, *Chem. Ber.*, 1970, **103**, 3128.

Complexes containing C_7 Hydrocarbons. The spectra of the complexes (π-C_5H_5)-$M(\pi$-C_7H_7) (M = Ti or V) are similar, and metal-containing ions of the type $[C_nH_nM]^+$ (n = 3, 5, 6, 7, 10, or 12) are present.[227] The characteristic fragmentation is loss of benzene; only the vanadium complex shows loss of C_7H_7, and also of C_2H_2, from the P^+ ion. The doubly charged titanium parent ion has 9.6% relative intensity. The spectrum of the vanadium compound has been studied by other workers, who also find the most interesting feature to be the formation of $[V(C_6H_6)]^+$.[228] Labelling studies show that this ion can be formed from either the C_5 or the C_7 ring, and originates by dissociation of $[V(C_6H_6)_2]^+$ produced by rearrangement of the P^+ ion (Scheme 13).

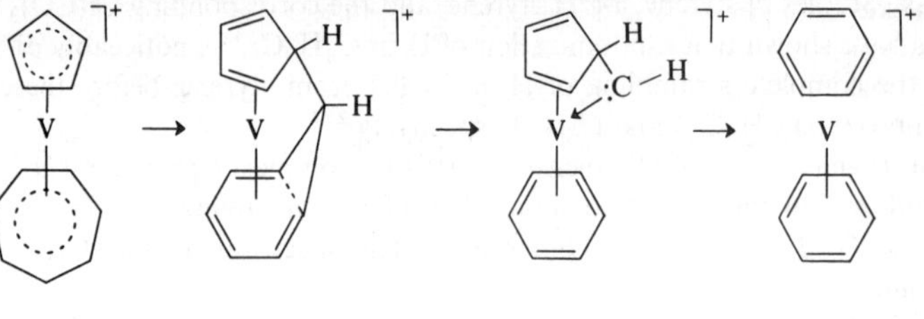

Scheme 13

Substituted cycloheptatrienyl-vanadium complexes (π-C_5H_5)V(π-C_7H_6X) or (π-C_7H_6X)V(CO)$_3$ show numerous rearrangements in the decomposition of the parent ions, and these are more complex than for the free ligands.[229] The cyclopentadienyl complexes exhibit the same rearrangements found for the parent complex; in particular, both ions $[VC_6H_6]^+$ and $[VC_6H_5X]^+$ are present. Where X = OR, ions $[C_5H_5VOR]^+$ are observed, suggesting migration of the alkoxy-group to the metal. More unusual processes are found in the decomposition of $[VC_7H_6X]^+$ from the tricarbonyl complexes. Three examples are illustrative:

$[VC_7H_6CH_3]^+ \xrightarrow{-C_2H_4} [VC_6H_5]^+ \xrightarrow{-C_6H_5} V^+$

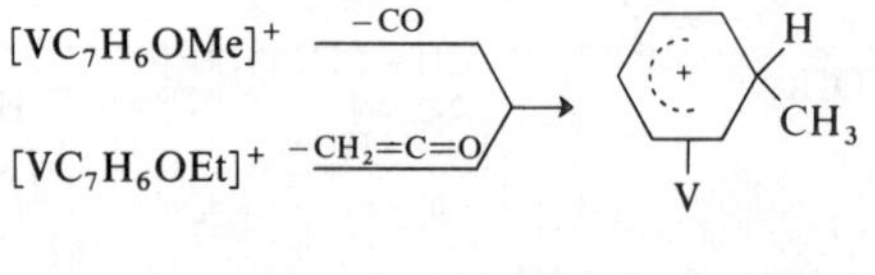

$[VC_7H_6OMe]^+ \xrightarrow{-CO}$

$[VC_7H_6OEt]^+ \xrightarrow{-CH_2=C=O}$

[227] H. O. van Oven and H. J. de Liefde Meijer, *J. Organometallic Chem.*, 1970, **23**, 159.
[228] M. F. Rettig, C. D. Stout, A. King, and P. Farnham, *J. Amer. Chem. Soc.*, 1970, **92**, 5100.
[229] J. Müller and B. Mertschenk, *J. Organometallic Chem.*, 1972, **34**, 165.

The paramagnetic borabenzene-cobalt complexes (51) and (52) have unusually low IP's [~6.5 eV for (51); ~7.1 eV for (52)], the extra electron being easily lost to form an 18-electron ion.[230]

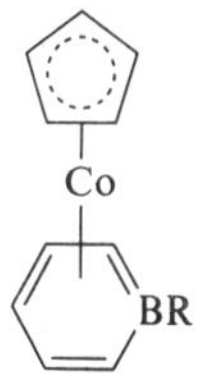

(51)

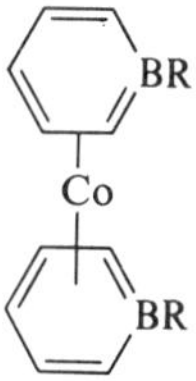

(52)

Complexes with Donor Ligands.—*Carbene Complexes.* The IP's of a series of ring-substituted carbene complexes $(RC_6H_4)(OMe)CCr(CO)_5$ have been determined and correlated with CO stretching force constants, and Jaffé σ-constants.[231,232] Similar relationships hold with the two constants, and the donor–acceptor properties of the ligands determine the energy of the highest occupied MO principally by charge transfer to the central metal atom.

In the spectra of chromium carbonyl carbene complexes, the abundance ratio parameter of selected ions does not necessarily bear an inverse linear relationship to the IP.[233] A wide variety of complexes has been studied, and several interesting rearrangements are suggested. In particular, the vinyl carbene complex shows ions derived from $(C_3H_5)Cr(CO)_4$:

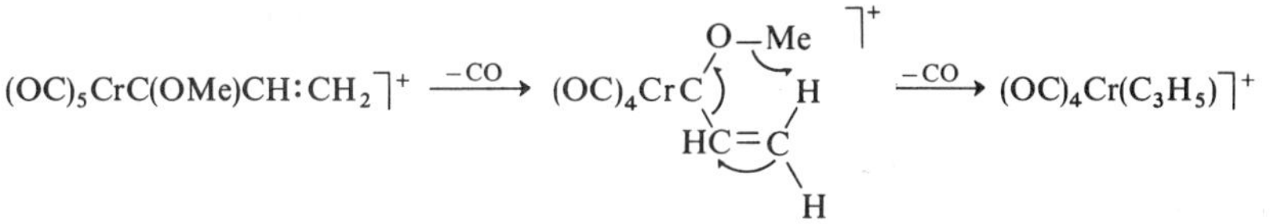

$$(OC)_5CrC(OMe)CH{:}CH_2]^+ \xrightarrow{-CO} (OC)_4CrC\genfrac{}{}{0pt}{}{O-Me}{HC=C}\genfrac{}{}{0pt}{}{}{H}]^+ \xrightarrow{-CO} (OC)_4Cr(C_3H_5)]^+$$

Fragmentation of carbonyl-free ions from mercapto-phenylcarbene chromium complexes[234] proceeds *via* elimination of Me, H_2S and styrene (Scheme 14). The diphenylcarbene ion (53) was found in the spectrum of $Ph(SPh)CCr(CO)_5$.

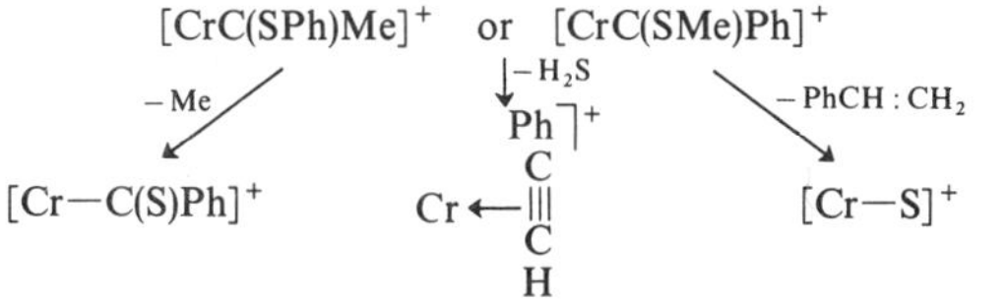

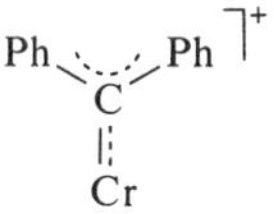

Scheme 14

[230] G. E. Herberich, G. Greiss, H. F. Heil, and J. Müller, *Chem. Comm.*, 1971, 1328.
[231] E. O. Fischer, C. G. Kreiter, H. J. Kollmeier, J. Müller, and R. D. Fischer, *J. Organometallic Chem.*, 1971, **28**, 237.
[232] E. O. Fischer, H. J. Kollmeier, C. G. Kreiter, J. Müller, and R. D. Fischer, *J. Organometallic Chem.*, 1970, **22**, C39.
[233] J. A. Connor and E. M. Jones, *J. Organometallic Chem.*, 1971, **31**, 389.
[234] E. O. Fischer, M. Leupold, C. G. Kreiter, and J. Müller, *Chem. Ber.*, 1972, **105**, 150.

Major fragmentation pathways for carbonyl-free ions derived from $Fe(CO)_4L$ (L = carbene ligand) show the previously established features.[235] Additional features were found in the spectra of the C_6F_5 complexes, e.g. $Fe(CO)_4C(OEt)$-C_6F_5, including the loss of iron as FeF_2, and the elimination of C_6F_4 and COC_6F_4 fragments. A derivative with two heteroatoms gave the pattern shown in Scheme 15.

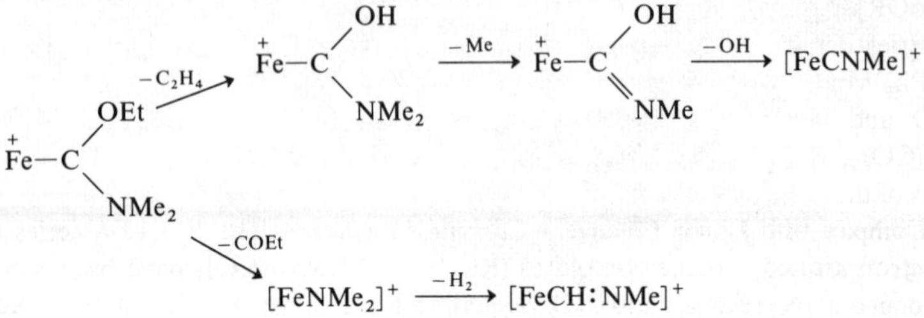

$$\overset{+}{Fe}-C\begin{matrix}OEt\\NMe_2\end{matrix} \xrightarrow{-C_2H_4} \overset{+}{Fe}-C\begin{matrix}OH\\NMe_2\end{matrix} \xrightarrow{-Me} \overset{+}{Fe}-C\begin{matrix}OH\\NMe\end{matrix} \xrightarrow{-OH} [FeCNMe]^+$$

$$\xrightarrow{-COEt} [FeNMe_2]^+ \xrightarrow{-H_2} [FeCH:NMe]^+$$

Scheme 15

Nitrogen Ligands. The spectra of diamine complexes, e.g. $(CHRNMe_2)_2Cr(CO)_4$, are characterized by the presence of ions derived from the ligands.[236] Only in the case of R = Ph, and the o-$C_6H_4(NMe_2)_2$ complex, were weak ions $[LCr(CO)_n]^+$ (n = 0—3) observed. This contrasts with the analogous tungsten complex containing $(CH_2NMe_2)_2$.

A series of Group VI carbonyl compounds containing the diphenylmethyleneamino ligand[237] show P^+ ions, loss of CO, followed by break-up of the ligand by successive loss of two PhCN groups, together with H or H_2. The highest mass organic ion is $[Ph_2CN]^+$. In the case of dinuclear complexes, ions retain two metal atoms until attached groups are lost. Similar features were found for related alkyl-aryl derivatives.[238]

Elucidation of the fragmentation patterns for $Mo(CO)_4(R_2NPPh_2)_2$ complexes was complicated by thermal decomposition to the ligand in the mass spectrometer, and by disproportion reactions.[239] However, P^+ ions were observed for the pentacarbonyls $(R_2NPPh_2)Mo(CO)_5$ (R = Et, Pr^n, or Bu^n). Loss of metal from $[P - 5CO]^+$ resulted in a breakdown similar to that found in the free ligand. No loss of CH_2NMe_2 with formation of ions of the type $[R_2NP(CH_2$-$NCH_2)Mo]^+$ was found.

The spectrum of $Re(CO)_5NO_3$ shows loss of CO and NO_2 to be the major routes; the features are interpreted in terms of the formation of six-co-ordinate ions.[240] The series $[Re(CO)_nO_2]^+$ (n = 0—3) is also present.

[235] E. O. Fischer, H.-J. Beck, C. G. Kreiter, J. Lynch, J. Müller, and E. Winkler, *Chem. Ber.*, 1972, **105**, 162.
[236] J. A. Connor and J. P. Lloyd, *J. Chem. Soc. (A)*, 1970, 3237.
[237] K. Farmery, M. Kilner, and C. Midcalf, *J. Chem. Soc. (A)*, 1970, 2279.
[238] M. Kilner and J. N. Pinkney, *J. Chem. Soc. (A)*, 1971, 2887.
[239] L. K. Atkinson and D. C. Smith, *J. Organometallic Chem.*, 1971, **33**, 189.
[240] C. C. Addison, R. Davis, and N. Logan, *J. Chem. Soc. (A)*, 1970, 3333.

Phosphorus Ligands. Characteristic ions in the mass spectra of $(\pi\text{-}C_5H_5)\text{Co-}$ $[P(OR)_3]_2$ (R = Me, Et, or Ph) are formed by loss of $P(OR)_3$ ligands, followed by the C_5 ring, with the base peak being $[P - P(OR)_3]^+$.[241] The alkyl phosphites undergo some cleavage of P—O and O—R bonds while attached to cobalt, and loss of CH_3 and OMe [from $P(OMe)_3$], and of CH_3CHO and C_2H_4 [from $P(OEt)_3$] are noted. Transfer of alkoxide to cobalt gives $[C_5H_5\text{CoOR}]^+$ and $[\text{CoOR}]^+$.

Brief details of the spectra of $(\pi\text{-}C_5H_5)\text{Fe(CO)}_2(\text{PPh}_2)\text{Fe(CO)}_4$, $\text{Fe(CO)}_4\text{-}$ $(\text{PPh}_2)\text{Mn(CO)}_4$, and $(\pi\text{-}C_5H_5)\text{Ni(PPh}_2)\text{Fe(CO)}_4$ include mention of loss of CO and benzene.[242,243] In $(\text{CO})_5\text{W(PMe}_2)\text{Mn(CO)}_5$ and $(\text{CO})_5\text{Cr(PMe}_2)\text{-}$ Re(CO)_5, the M—P—M′ group survives loss of CO groups; in both cases, loss of the lighter metal then occurs, giving $[\text{WPMe}_2]^+$ and Re^+, respectively.[244]

Comparison of PF_3 and CO complexes by a consideration of IP's, photoelectron spectra, and mass spectra, show that PF_3 is a weaker σ-donor and a stronger π-acceptor than CO.[245] More direct comparisons of this type are possible using IP values for complexes rather than ν(CO) frequencies and CO force constants.

Arsenic Ligands. The relatively high intensities of $[(C_5H_5)\text{Mn(CO)}_n]^+$ in the spectrum of $(\pi\text{-}C_5H_5)\text{Mn(CO)}_2\text{AsF}_3$ indicate that AsF_3 has similar donor properties to CO;[246] other ions include $[P - F]^+$ and $[\text{MnF}]^+$.

Group VI carbonyl derivatives containing $\text{As(NMe}_2)_3$ show a greater tendency to eliminate an NMe_2 group than the corresponding $\text{P(NMe}_2)_3$ complexes, so that the tungsten complex, for example, shows the series $[(\text{Me}_2\text{N})_m\text{AsW(CO)}_n]^+$ ($m = 1$—3; $n = 0$—5) as well as $[\text{AsW(CO)}_n]^+$ ($n = 0$—2).[247] However, the general features in the two series are similar, *e.g.* the iron complexes undergo the reaction:

$$[(\text{Me}_2\text{N})_2\text{EFeH}]^+ \rightarrow [(\text{MeNCH}_2)_2\text{Fe}]^+ + \text{EH}_3 \quad (\text{E = P or As})$$

The chromium complex eliminates neutral CrNMe_2 from $[(\text{Me}_2\text{N})_2\text{AsCrH}]^+$ in a process that may be related to elimination of metal fluoride and sulphide fragments. Other unusual processes include loss of As and NMe_2 groups, the former process giving $[\text{Cr(NMe}_2)_3]^+$.

In the spectra of $\text{Mn}_2(\text{CO})_8(\text{AsMe}_2)\text{X}$ [X = H, I, SCF_3, SMe, or $\text{P(CF}_3)_2$], loss of CO leading to a highly stable Mn_2AsX unit is observed.[248] From the fluorinated groups, loss of CSF_2 or CF_3PCF_2 results in formation of an Mn_2AsF bridge by fluorine migration to the metal.

---

[241] V. Harder, J. Müller, and H. Werner, *Helv. Chim. Acta*, 1971, **54**, 1.
[242] K. Yasufuku and H. Yamazaki, *J. Organometallic Chem.*, 1971, **28**, 415.
[243] K. Yasufuku and H. Yamazaki, *Bull. Chem. Soc. Japan*, 1970, **43**, 1588.
[244] W. Ehrl and H. Vahrenkamp, *Chem. Ber.*, 1971, **104**, 3261.
[245] J. Müller, K. Fenderl, and B. Mertschenk, *Chem. Ber.*, 1971, **104**, 700.
[246] J. Müller and K. Fenderl, *Angew. Chem.*, 1971, 83, 445; *Angew. Chem. Internat. Edn.*, 1971, **10**, 418.
[247] R. B. King and T. F. Korenowski, *Org. Mass Spectrometry*, 1971, **5**, 939.
[248] J. Grobe and F. Kober, *J. Organometallic Chem.*, 1971, **29**, 295.

Group VI Ligands. Major peaks in the spectra of $(\pi\text{-}C_5H_5)MX(ox)_2$ (M = Ti, Zr, or Hf; X = Cl or Br; ox = oxinate) are related by loss of C_5H_5, ox, and X.[249] Metal-containing ions carry about 85% of the ion current. Loss of H_2O from $[Ti(ox)_2]^+$ is found, but is not observed in the zirconium and hafnium complexes. All spectra show ions derived from putative redistribution products formed in the spectrometer, such as $(\pi\text{-}C_5H_5)_2MX_2$ or $(\pi\text{-}C_5H_5)_2M(ox)_2$.

IP's of several derivatives $M(CO)_5L$ (M = Cr or W) and $(\pi\text{-}C_5H_4Me)Mn(CO)_2L$ containing sulphur or nitrogen ligands show a linear correlation for the manganese complexes, but different correlation lines hold for different families of ligands.[250] In general, the fragmentation patterns show no unusual features. For manganese complexes containing sulphur-bonded ligands, also containing oxygen, the abundance of $[C_5H_4MeMnO]^+$ is always greater than that of $[C_5H_4MeMnS]^+$.

Reactions between $RRe(CO)_5$ (R = Me or Ph) and SO_3 gave the sulphonate complexes $Re(CO)_5SO_3R$;[251] the spectra are characterized by the series $[P - nCO]^+$ and $[RSORe(CO)_n]^+$ (n = 0—5). Other abundant ions include $[ORe(CO)_n]^+$ (n = 0—3), and $[PhCO]^+$.

The IP's of complexes $L_2Fe_2(CO)_6$ (L = SR or PR_2) depend on the IP of the ligand; less efficient σ-donors, *i.e.* SR, result in higher IP values for the complexes.[252] Phosphorus ligands reduce the value to below that of iron. From AP values for fragment ions, conclusions regarding Fe—CO bond strengths were derived. In phosphorus complexes, these bonds are stronger by *ca.* 0.2 eV than in sulphur compounds. When phenyl groups are present, a similar reduction in bond energy is found. The average bond dissociation energy for the first three CO groups is smaller than for the last three, as has been assumed previously.

Polynuclear iron carbonyl complexes containing SR groups show the expected loss of CO and alkene to give cluster ions such as $[Fe_3H(SH)]^+$ and $[Fe_3S]^+$ [from $Fe_3(CO)_9H(SH)$], $[Fe_3(SH)_2]^+$ and $[Fe_3S_2]^+$ [from $Fe_3(CO)_9(SR)_2$], and $[Fe_4S_3]^+$ [from $Fe_4(CO)_{12}S(SR)_2$]. The spectra of several $[Fe(CO)_3SR]_2$ complexes show similar features, resulting in the formation of the stable ions $[Fe_2(SR)_2]^+$ and $[Fe_2S_2]^+$.[253]

Comparison of the breakdown of the organic ligand derived from $M_2(CO)_6(SC_6F_5)_2$ (M = Fe or Co) and the cobalt-C_6Cl_5 derivative, show that the most abundant fragments from the cobalt complexes contain C_{12} units, together with some C_9 ions.[254] Sulphur remains bonded to cobalt, usually as an intense $[Co_2S_2]^+$ ion. The iron complex gives ions retaining the Fe—SC_6 unit, and fragmentation is similar to that of the ligand. These features are partly rationalized on the basis of the structures of these complexes.

---

[249] J. Charalambous, M. J. Frazer, and W. E. Newton, *J. Chem. Soc. (A)*, 1971, 2487.
[250] G. Distefano, G. Innorta, S. Pignataro, and A. Foffani, *Internat. J. Mass Spectrometry Ion Phys.*, 1971, **7**, 383.
[251] E. Lindner and R. Grimmer, *Chem. Ber.*, 1971, **104**, 544.
[252] G. A. Junk, F. J. Preston, H. J. Svec, and D. T. Thompson, *J. Chem. Soc. (A)*, 1970, 3171.
[253] J. A. de Beer and R. J. Haines, *J. Organometallic Chem.*, 1970, **24**, 757.
[254] G. Natile and G. Bor, *J. Organometallic Chem.*, 1972, **35**, 185.

Ferrocenes.—Predominant ions in the spectra of alkylferrocenes are formed by β-cleavage of the side-chain;[255] other ions are formed by ring-expansion reactions, e.g. $[FeC_7H_8]^+$. Some biferrocenyls are also reported. Some caution is urged in the implication of an unsubstituted C_5H_5 ring by the presence of the ion $[FeC_5H_5]^+$ (m/e 121), this ion being found in the spectra of 1,1'-dialkyl- and 1,1'-disilyl-ferrocenes. Ions present in the spectra of benzylferrocenes at m/e 208 and 152 are suggested to have benzopentalene structures.

In a series of ferrocenyl alcohols, distinction between exo- and endo-isomers could be made on the basis of transfer of OH to iron, which is aided by the formation of an Fe···H—O bond in the endo-isomer.[256] With this isomer, the ion $[FeC_5H_5]^+$ is weak, whereas m/e 138, $[C_5H_5FeOH]^+$, is strong. In the exo-isomer, it is the ion at m/e 202, $[C_5H_5FeC_5H_4OH]^+$, which is more intense.

Significant differences were found in the fragmentation of the syn- and anti-isomers of formylferrocene oxime at low (15—20 eV) energies, although no marked differences were found between 40 and 75 eV.[257] The major processes found with FcC(=NOH)R are loss of C_5H_5 and of ROH to give $[FeC_5H_4CN]^+$, and migration of OH to iron. Differences between the two isomers were confined to differences in intensity ratios $[C_5H_5FeOH]^+ : [C_5H_5Fe]^+$ (easier migration of OH with the syn form) and also $[P + 1]^+ : P^+$, where protonation of the oxime is easier for the syn-isomer. Other oximes, whose spectra are also recorded, do not show such significant differences.

Polychloroferrocenes, $(C_5H_{5-n}Cl_n)Fe(C_5H_5)$ and $Fe(C_5H_{5-n}Cl_n)_2$, fragment primarily by loss of $FeCl_2$ and the chlorinated ring; loss of C_5H_5, where present, is also found, as are $[P - HCl]^+$ ions, probably formed by a heteroannular process. As with other ferrocene derivatives, the base peaks are the P^+ ions; the elimination of $FeCl_2$ probably leads to a fulvalene cation. Metal-containing ions include $[FeCl]^+$ and Fe^+.[258]

A detailed description of the spectra of ferrocenylboron derivatives, e.g. $FcB(OH)_2$, includes proposed fragmentation schemes. Particularly stable are the Fc—B and B—O bonds, and the stability of $[FcB=O]^+$ is attributed to π-delocalization.[259]

The spectrum of azaferrocene shows strong P^+ and $[P - C_4H_4N]^+$ ions, but $[FeC_4H_4N]^+$ is very weak; IP and AP values lead to a value for $D[C_5H_5Fe—C_4H_4N] < 69.5$ kcal mol^{-1}, suggesting that the least firmly bound electron is in an antibonding orbital.[260]

4 Co-ordination Complexes

As in previous years the majority of reports relate to various aspects of the mass spectra of β-diketonato-complexes, and considerable discussion about the

[255] I. J. Spilners and J. G. Larson, *Org. Mass Spectrometry*, 1970, **3**, 915.
[256] B. Gautheron and R. Broussier, *Bull. Soc. chim. France*, 1971, 3636.
[257] K. Yamakawa and M. Hasatome, *Org. Mass Spectrometry*, 1972, **6**, 167.
[258] L. D. Smithson, A. K. Bhattacharya, and F. L. Hedberg, *Org. Mass Spectrometry*, 1970, **4**, 383.
[259] E. W. Post, R. G. Cooks, and J. C. Kotz, *Inorg. Chem.*, 1970, **9**, 1670.
[260] R. Cataliotti, A. Foffani, and S. Pignataro, *Inorg. Chem.*, 1970, **9**, 2594.

nature of the highest occupied MO's in these complexes has appeared. Other classes of co-ordination complex are also surveyed in the following pages, and a new section on alkoxides, alkylamides, and related compounds is also included.

β-Diketonato-complexes.—The continuing discussion on the nature of the orbital contributing the electron on ionization of β-diketone complexes, and often stated to involve the π-system of the ligand, has resulted in studies of the complexes $M(CO)_2(\beta$-diketonate) (M = Rh or Ir).[261] Determination of IP's, and comparisons between Co^{III} and Rh^{III} derivatives, lead to the conclusion that the differing values found can be explained by differing contributions of the metal AO's to the last occupied MO, although the maximum change in IP observed is only 0.35 eV. The following points can be accommodated: (i) the additive effect of different substituents on the chelate rings; (ii) the sensitivity of different metal systems to substituents should differ; (iii) differences should exist between the changes observed in IP's of H(β-diketonate) and M(β-diketonate)$_n$ as Me is replaced by CF_3; (iv) a linear correlation between IP and σ_p^+ for substituents on the central carbon of the ring; (v) the IP is sensitive to oxidation state.

Measurements of IP's of rhodium acetylacetonates progressively substituted with nitro-groups in the γ-positions show a linear progression of IP values, increasing with the number of nitro-groups.[262] This indicates that ionization should occur from a metal-dominated orbital, or any orbital in which each ligand influences the orbital energy equally.

Other evidence to be considered when discussing the nature of the highest occupied MO in these complexes includes the photoelectron spectra, which show no evidence for any inversion of orbital sequence, at least for the chromium, iron, and cobalt complexes.[263] It is considered that electron impact determinations fail to detect the first IP. However, a later communication[264] shows that

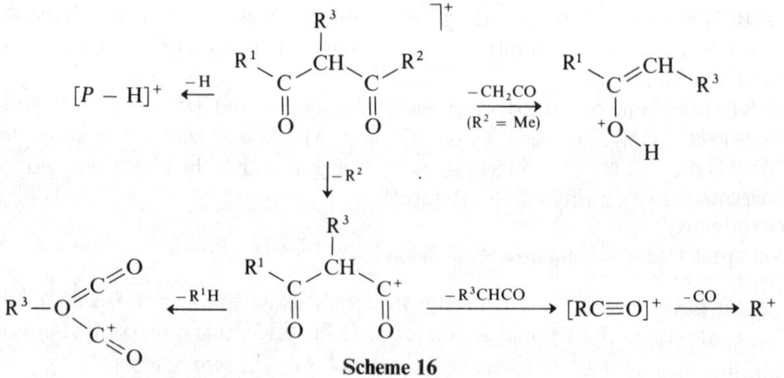

Scheme 16

[261] F. Bonati, G. Distefano, G. Innorta, G. Minghetti, and S. Pignataro, *Z. anorg. Chem.*, 1971, **386**, 107.
[262] M. M. Bursey and P. F. Rogerson, *Inorg. Chem.*, 1971, **10**, 1313.
[263] D. R. Lloyd, *Chem. Comm.*, 1970, 868.
[264] S. Evans, A. Hamnett, and A. F. Orchard, *Chem. Comm.*, 1970, 1282.

Organometallic and Co-ordination Compounds 241

weak bands in the photoelectron spectra are probably due to 'shadow' excitations. In addition, the difference in ionization energies cannot be identified with the ligand-field parameter Δ.

A comparative study of the fragmentation of β-diketones and their beryllium chelates, chosen because they are unlikely to change their oxidation state, has been described.[265] Major routes in the ligands are shown in Scheme 16, although the γ-phenyl ligand, *i.e.* R^3 = Ph, shows some different ions. For the metal complexes, major fragmentations include those set out in Scheme 17;

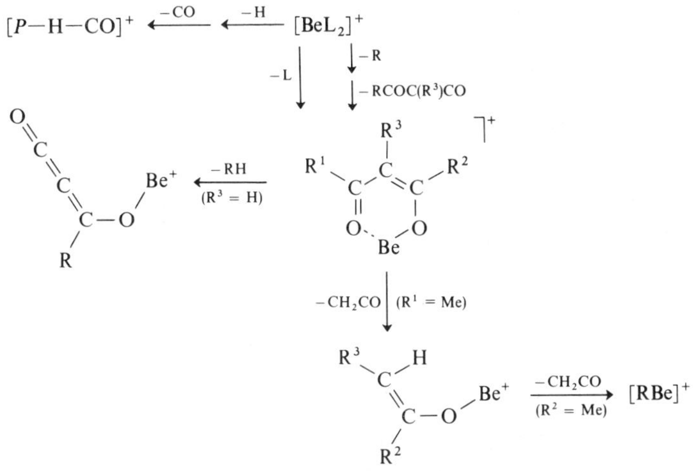

Scheme 17

again, there was some variation with R, particularly when CF_3 or Ph groups were present; some evidence for migration of fluorine to beryllium was obtained.

The spectra of a wide variety of tfac and hfac complexes of bi- and ter-valent metals have been compared[266] with the acac complexes; introduction of CF_3 groups generally leads to an increase in the proportion of fragment ions, owing to weaker $C-CF_3$ bonds, and stability of rearranged fragment ions containing M—F bonds. Features of the spectra were similar to those described earlier; energetic measurements on the major ions were reported. Neutral metal-containing fragments lost include AlF_3, CrF_3, FeF_2, and CuMe. The negative-ion spectra of these complexes are simple, *e.g.* the complex $Cu(tfac)_2$ gives only $[CuL_2]^-$ (base peak), L^-, and $[L - Me]^-$. The energetic data reported require some modifications to previously assumed ion structures, and these are discussed.

The extent of fragmentation in a series of copper chelates also depends on the substituents, and is larger if electron-withdrawing groups are present.[267] Other processes found include migration of an aryl group to copper, *e.g.* $[CuPh]^+$, and

---

[265] K. S. Patel, K. L. Rinehart, and J. C. Bailar, *Org. Mass Spectrometry*, 1970, **3**, 1239.
[266] C. Reichert, G. M. Bancroft, and J. B. Westmore, *Canad. J. Chem.*, 1970, **48**, 1362.
[267] C. Reichert and J. B. Westmore, *Canad. J. Chem.*, 1970, **48**, 3213.

hydrogen migration, *e.g.* to form [LCuH]$^+$ and [LCu − H]$^+$ ions. For these ions, possible involvement of an *ortho*-hydrogen on an aromatic ring (where present) is indicated. Some AP data are also given. Many of these results supplement and extend those reported in an earlier study.[10] In this paper, a series of β-diketonates of copper were found to eliminate alkyl groups in a manner dependent on the electronic characteristics of the substituents.

In some partially fluorinated iron β-diketonate chelates, loss of CO_2 has been reported,[268] and incorporation of this new route accounts for the major proportion of metal-containing ions (Scheme 18). This process seems well defined only for the iron and cobalt complexes of the fluorinated ligands.

$$[FeL_3]^{+\cdot} \xrightarrow{-L} [Fe^{III}L_2]^+ \leftrightarrow [Fe^{II}L_2]^{+\cdot} \xrightarrow{-CF_3} \xrightarrow{-(L-CF_3)} [Fe^{II}L]^+$$

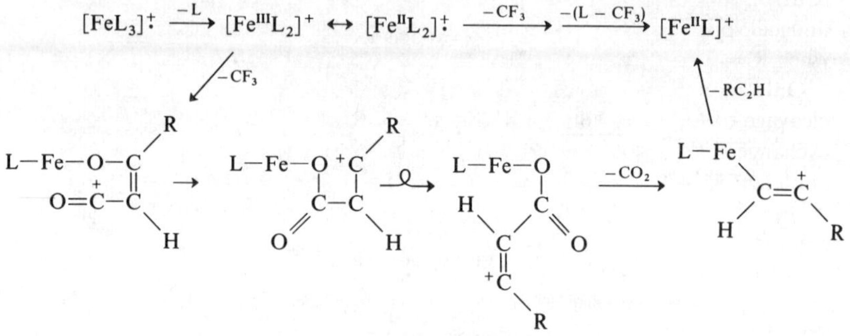

Scheme 18

The spectra of related Al, Cr, Fe, and Co tris-chelates have been discussed in terms of reactions arising from the ligand alone, and those attributed to the effect of the metal.[269] Generally, loss of a ligand is followed by fragmentation within the [ML$_2$]$^+$ ion by elimination of CF_2 groups, probably by migration of fluorine to metal. Loss of (L − CF$_3$) also occurs. Consideration of [ML$_2$]$^+$ as either [M^{III}L$_2$]$^+$ or [M^{II}L$_2$]$^{+\cdot}$ enables loss of the R group to be rationalized *via* the latter possibility, *i.e.* loss of a radical from an odd-electron ion. Consequently, the nature of the metal (relative stability of M^{III} *vs* M^{II}) determines whether a radical, an even-electron group, or both, are eliminated from the ion [ML$_2$]$^+$, while the nature of the ligand determines the nature of these fragments. In the special instance of the fluorinated ligands, loss of neutral metal fluorides includes the following species: AlF$_2$L, CrF$_3$, FeF$_2$, FeF$_3$, FeLF$_2$, and CoF$_2$. In the spectra of these complexes, certain metastable peaks are associated with consecutive decomposition reactions in the field-free region, *e.g.*

[AlL$_3$]$^+$ → [AlL$_2$(L − CF$_3$)]$^+$ → [AlL$_2$]$^+$
m/e 441   *m/e* 272   *m/e* 169
L = hfac   m* 64.8 [441 → 169]

[268] A. L. Clobes, M. L. Morris, and R. D. Koob, *Org. Mass Spectrometry*, 1970, **3**, 1255.
[269] A. L. Clobes, M. L. Morris, and R. D. Koob, *Org. Mass Spectrometry*, 1971, **5**, 633.

Features in the spectrum of Re(hfac)$_3$ indicate that breakdown is quite different from Cr(hfac)$_3$, and includes ions [ReOL$_3$]$^+$ (formed by elimination of C$_5$HF$_6$O ?).[270] A scheme based on the previous observation[271] that fluorinated acac derivatives fragment by fission of one metal–oxygen bond, followed by rotation of part of the ligand about the C-2—C-3 bond, is presented. This process brings halogen atoms close to the metal, and thus facilitates migration to the metal. In the present case, ions such as [ReFL$_n$]$^+$ and [ReOFL$_n$]$^+$ are formed. The spectra of M(acac)$_3$ (M = Re, Ti, V, Cr, or Mn) are also compared.

The dominant features found with Cu(MeCOCHRCOMe)$_2$ [R = (CH$_2$)$_2$Ph] are loss of one or both benzyl groups from [CuL$_2$]$^+$, and a similar process with [CuL]$^+$. In addition, loss of the complementary ligand fragments affords the unusual ions [CuL(C$_7$H$_7$)]$^+$ and [CuC$_7$H$_7$]$^+$. These reactions are also found when R = (CH$_2$)$_3$Ph, but not for R = benzyl.[272]

Only monomeric species Ti(acac)$_2$I$_2$ exist in the gas phase, although easy cleavage of the Ti—I bond gives [P − I]$^+$ as the highest mass ion.[273] Ligand exchange reactions occur readily in the spectrometer, and [Ti(acac)$_3$]$^+$ is an abundant ion; others include [Ti(acac)$_2$]$^+$ and [TiO(acac)]$^+$. The Group IV elements Ti, Zr, and Hf have been determined by the integrated ion current method, using the benzoyltrifluoroacetonates.[274] The spectra, which contain P^+ ions, together with [P − L]$^+$, and [MFL$_n$]$^+$ (n = 1 or 2), are illustrated. Particular ions used for the determination were [MF(btfa)$_2$]$^+$, with limits of detection: Ti, 10^{-11}; Zr, 5×10^{-14}; Hf, 5×10^{-12} g.

The spectrum of the lead chelate of heptafluorobutanoylpivaloylmethane shows a P^+ ion, which loses a Bu^t group; base peak is [P − L]$^+$. At higher m/e values, low-intensity peaks are present, corresponding to polymeric ions.[275] Using the base peak, the integrated ion-current method was used to detect lead to a lower limit of 10^{-14} g. Other similar reports refer to Mg, Ca, and Sr tetramethylheptanedionates, which contain as highest mass ions [M$_2$(thd)$_3$]$^+$; mixtures of two complexes contain ions such as [CaSr(thd)$_3$]$^+$ or [MgCa(thd)$_3$]$^+$;[276] and the thenoyltrifluoroacetonates of Zn, Ni, and Cr, used in their determination.[277]

In connection with preparative g.c. of a range of β-diketonates of Al, Cr, and Fe, eluates were characterized by abundant ions in the mass spectra. Trifluoroacetylpivaloylmethane (tpmH) derivatives were most suited to separation, and major features in the spectra of these compounds are tabulated. These include the expected loss of a tpm group, and elimination of Bu^t and CF$_3$ groups; the former is preferred.[278]

[270] W. D. Courrier, W. Forster, C. J. L. Lock, and G. Turner, *Canad. J. Chem.*, 1972, **50**, 8.
[271] G. M. Bancroft, C. Reichert, and J. B. Westmore, *Inorg. Chem.*, 1968, **7**, 870.
[272] J. A. Kemlo and T. M. Shepherd, *Inorg. Nuclear Chem. Letters*, 1972, **8**, 119.
[273] R. C. Fay and R. N. Lowry, *Inorg. Chem.*, 1970, **8**, 2048.
[274] M. G. Allcock, R. Belcher, J. R. Majer, and R. Perry, *Analyt. Chem.*, 1970, **42**, 776.
[275] R. Belcher, J. R. Majer, W. I. Stephen, I. J. Thompson, and P. C. Uden, *Analyt. Chim. Acta*, 1970, **50**, 423.
[276] J. E. Schwarberg, R. E. Sievers, and R. W. Moshier, *Analyt. Chem.*, 1970, **42**, 1828.
[277] J. K. Terlouw and W. Heerma, *Z. analyt. Chem.*, 1971, **257**, 177.
[278] R. Belcher, C. R. Jenkins, and W. I. Stephen, *Talanta*, 1970, **17**, 455.

Hexafluoromonothioacetylacetone, Hhfsacac, has been proposed as a new reagent for use in g.c. analysis of inorganic materials.[279] Complexes are volatile and readily identified mass spectrometrically, giving strong P^+ peaks and a base peak corresponding to loss of one ligand. Loss of F, CF_3, and a C_5HF_6O fragment also occurs. The spectra of the palladium and platinum chelates are illustrated. With the nickel complex, the P^+ ion loses ligand, followed by the metal atom; in contrast, the most prominent ion in $Ni(tfsac)_2$ is L^+, with only a weak P^+ peak.[280] Prominent ions in the spectra of chelates of $MeCSCH_2$-$COMe$ (sacacH) include P^+, $[P - L]^+$, and L^+; those of bivalent metals, except Pd, show loss of SH, and the Zn and Cd complexes also show $[P - Me]^+$. These complexes have been used to determine metals by the integrated ion-current technique; using the nickel parent ion, 10^{-7} to 10^{-8} g chelate can be measured.[281]

Schiff-base Complexes.—Comparisons of the spectra of NN'-bis(salicylidene)-ethylenediamine, and the corresponding isobutylenediamine compound, and of their metal chelates, have shown that the principal fragmentations are generally the same for both chelates and ligands.[282] In the former, P^+ ions are abundant, but few metal-free ions are found, except with the copper and palladium derivatives. In the ligands, two important decomposition routes are:

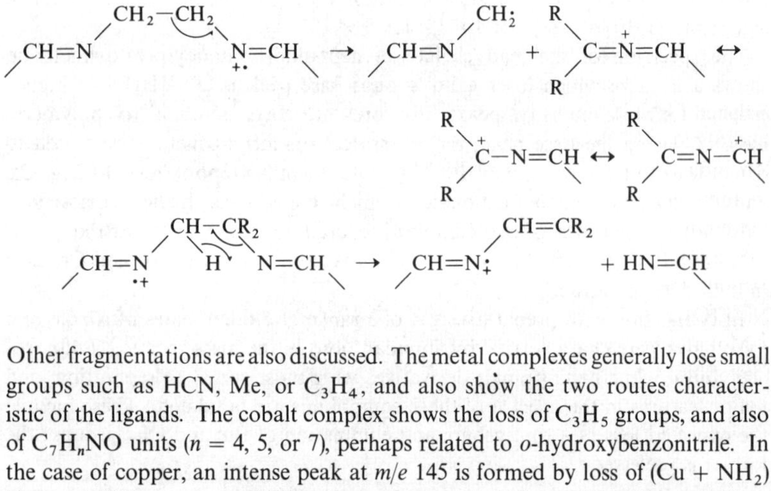

Other fragmentations are also discussed. The metal complexes generally lose small groups such as HCN, Me, or C_3H_4, and also show the two routes characteristic of the ligands. The cobalt complex shows the loss of C_3H_5 groups, and also of C_7H_nNO units ($n = 4$, 5, or 7), perhaps related to o-hydroxybenzonitrile. In the case of copper, an intense peak at m/e 145 is formed by loss of $(Cu + NH_2)$ from m/e 224, and loss of NH_3 is also observed (Scheme 19). The palladium complex is of interest, since a ready loss of H or H_2 is found, together with many metal-free ions. Other papers briefly document the spectra of some related

---

[279] E. Bayer, H. P. Müller, and R. Sievers, *Analyt. Chem.*, 1971, **43**, 2012.
[280] E. Bayer and H. P. Müller, *Tetrahedron Letters*, 1971, 533.
[281] R. Belcher, W. I. Stephen, I. J. Thomson, and P. C. Uden, *J. Inorg. Nuclear Chem.*, 1971, **33**, 1851.
[282] K. S. Patel, K. L. Rinehart, and J. C. Bailar, *Org. Mass Spectrometry*, 1970, **4**, 441.

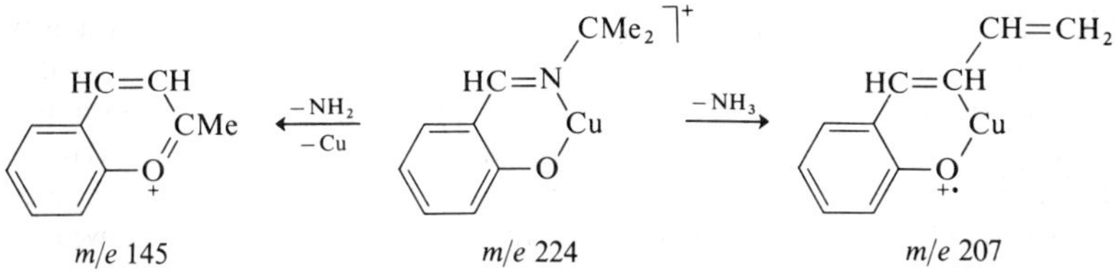

Scheme 19

nickel[283] and cobalt complexes.[284] The thallium complexes, Tl(sal—NR) give P^+ ions, although in some cases ions at higher m/e values are found.[285] High abundance Tl^+ and ligand ions, such as $[OC_6H_4CH=N]^+$ are present. With dithallium derivative, e.g. Tl_2(salen), $[TlL]^+$ can also occur.

The macrocyclic Schiff-base complexes (54) show $[P - Br]^+$ as highest mass ion; the doubly charged ion is also present.[286] Fragmentation involves loss of the backbone in a series of steps, until the fairly stable ion (55) remains.

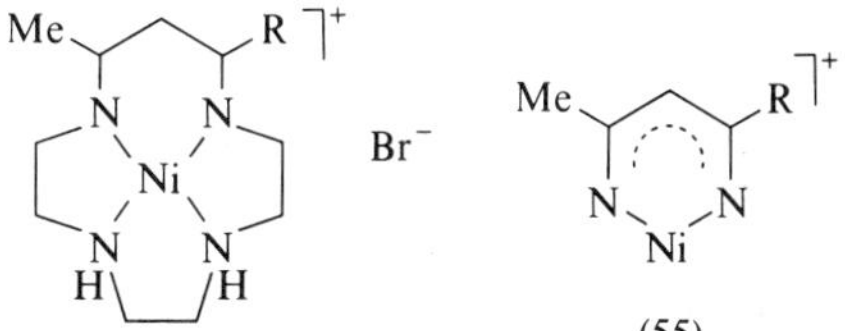

(54) R = Me or CF_3

(55)

Sulphur Ligands.—Mass spectral studies of nickel dithiobenzoates containing photolytically added sulphur (^{34}S) show an increase in intensity of the $[PhC^{34}S]^+$ ion, indicating that insertion of sulphur occurs adjacent to carbon; some sulphur scrambling occurs, probably by thermal decomposition.[287] In the zinc complex, scrambling occurs on addition of sulphur.

Spectra of nickel complexes containing dialkyldithiophosphate ligands exhibit a number of rearrangement processes.[288] Thus the base peak for $Ni[(EtO)_2PS_2]_2$ is P^+, and loss of OEt and C_2H_4 occurs, the latter probably by a McLafferty-type rearrangement; four such reactions leave an NiS_4-type ion (56). Similar elimination of olefin occurs from other alkyl compounds studied;

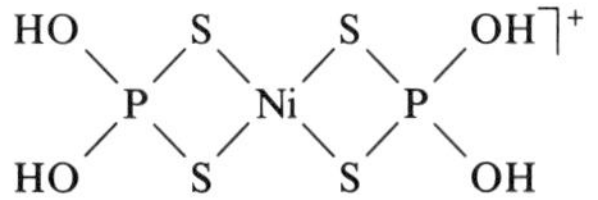

(56)

---

[283] C. J. Jones and J. A. McCleverty, *J. Chem. Soc.* (*A*), 1971, **38**, 1052.
[284] D. St. C. Black and M. J. Lane, *Austral. J. Chem.*, 1970, **23**, 2039.
[285] R. J. Cozens, K. S. Murray, and B. O. West, *Austral. J. Chem.*, 1970, **23**, 683.
[286] S. C. Cummings and R. E. Sievers, *Inorg. Chem.*, 1970, **9**, 1131.
[287] J. P. Fackler, J. A. Fetchin, and J. A. Smith, *J. Amer. Chem. Soc.*, 1970, **92**, 2910.
[288] S. E. Livingstone and A. E. Mihkelson, *Inorg. Chem.*, 1970, **9**, 2545.

the 2,2'-bipyridyl and phenanthroline adducts lose the nitrogen ligand as the first step in the breakdown of P^+, although this may also be a thermal process.

Mass spectra of the tropolone complexes VCl(trop)$_2$Htrop, VOF(trop)$_2$, and VOCl(trop)Htrop show monomeric ions, but a P^+ ion is only found for the former.[289] The base peaks for both vanadyl compounds is [VO(trop)$_2$]$^+$. The fluoro-complex also shows [VOF]$^+$. Dithiotropolonato (sst) complexes of several bivalent metals have been reported;[290] apart from P^+ ions, other peaks corresponding to [M(sst)]$^+$, [sst]$^+$, [(sst) − S]$^+$, [C$_7$H$_5$]$^+$, and [C$_6$H$_5$]$^+$ are present. The spectrum of dithiotropolone is also discussed.

The molybdenyl chelates containing diethyldithiocarbamate (edtc) ligands form two isomers.[291] The α form is monomeric whereas, in the vapour state, the β form is dimeric. Abundant parent ions lose oxygen; the monomer also gives [MoOS(edtc)]$^+$ by loss of S=CNEt, which also forms the base peak. The dimers give more complex spectra, containing as highest mass peak [Mo$_2$O$_2$S$_2$(edtc)$_2$]$^+$; fragments with two metal atoms always retain two oxygen atoms in addition. Similar features are found for [Mo$_2$O$_3$(edtc)$_4$], although in this case ions postulated to have a Mo—O—Mo bridge were found.

The spectrum of [Zn(edtc)$_2$]$_2$ shows weak peaks above [Zn(edtc)$_2$]$^+$ which arise by loss of Et$_2$NCS or Et$_2$NCS$_2$ from the dimeric P^+ ion.[292] These data demonstrate the existence of out-of-plane interactions in the gas phase similar to those found in the solid state.

Fragmentation of the copper complexes of α-diketonebis(thiosemicarbazone) ligands (57), related to some anticarcinogens, has been studied, and the complexes are of interest as containing only one, quadridentate, ligand molecule.[293] Monomeric parent ions are found, and observed ions can be relationalized in terms of the cleavages indicated. The p-bromophenyl derivative also shows the ion [BrC$_6$H$_4$C$_2$C$_6$H$_4$Br]$^+$, formed by cleavage of the two adjacent C—N bonds. Loss of metal occurs as CuS, with formation of an [L − S]$^+$ ion.

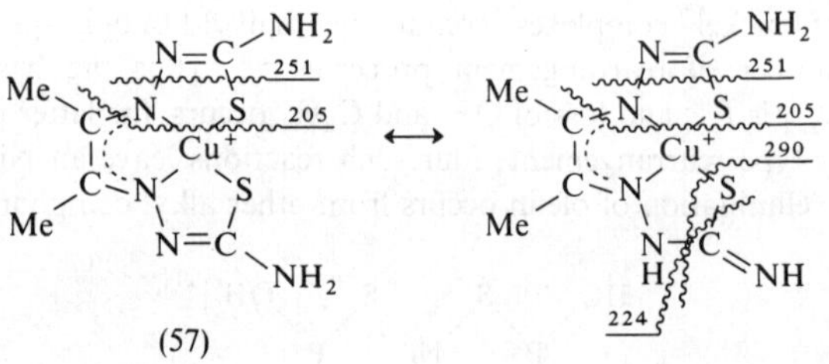

(57)

Complexes containing Nitrogen.—The major decomposition pathways of parent ions of oxine complexes of tervalent metals (Al, Ga, In, Sc, Cr, Fe) involve loss of

[289] J. B. White and J. Selbin, *J. Inorg. Nuclear Chem.*, 1970, **32**, 2434.
[290] C. E. Forbes and R. H. Holm, *J. Amer. Chem. Soc.*, 1970, **92**, 2297.
[291] R. Colton and G. G. Rose, *Austral. J. Chem.*, 1970, **23**, 1111.
[292] J. F. Villa, M. M. Bursey, and W. E. Hatfield, *Chem. Comm.*, 1971, 307.
[293] L. E. Warren, M. M. Bursey, and W. E. Hatfield, *Org. Mass Spectrometry*, 1971, **5**, 15.

ox radicals, and in the cases of $MX_{3-n}(ox)_3$ ($n = 1$ or 2), of X groups.[294] Relative abundances of fragment ions can be related to the oxidation states of the metals, and to the ease of reduction. For example, all compounds $M(ox)_3$ give $[M(ox)_2]^+$ ions, but with Al or Sc, M^+ is absent, and $[M(ox)]^+$ has only low abundance. Elimination of small molecules, e.g. CO, H_2O, HX, C_2H_2, from the co-ordinated ligands is facilitated when the metal can change its oxidation state by one unit. Several examples of redistribution reactions occurring in the spectrometer are described.

The spectrum of the copper chelate of 1-(2-pyridylazo)-2-naphthol (*o-β*-pan) shows nine copper-containing ions, including $[Cu_2L_2]^+$, $[CuL_2]^+$, and $[CuL]^+$. This result[295] does not agree with an earlier study,[296] which claimed that the complex had a 1 : 2 stoicheiometry, and suggests instead a dimeric 1 : 1 complex, forming $[CuL_2]^+$ by loss of copper.

Nickel and copper complexes of halogenated *o*-benzoquinone monoximes fragment by elimination of ligand, NO, CO, and oxygen, and where a halogen is present on the ligand, migration to the metal occurs.[297] Metal-containing ions constitute 60—70 and 10—15% of the total ion currents, respectively. Most complexes show a P^+ ion; for some, weak peaks above this ion are also found. The successive loss of radicals is interpreted in terms of valency changes of the metal in these ions, but another factor which is important is the stability of NO, the radical which is often lost.

Substituted nickel and palladium glyoximates revealed a fragmentation pattern common to a variety of complexes with varying substituents (Scheme 20).[298]

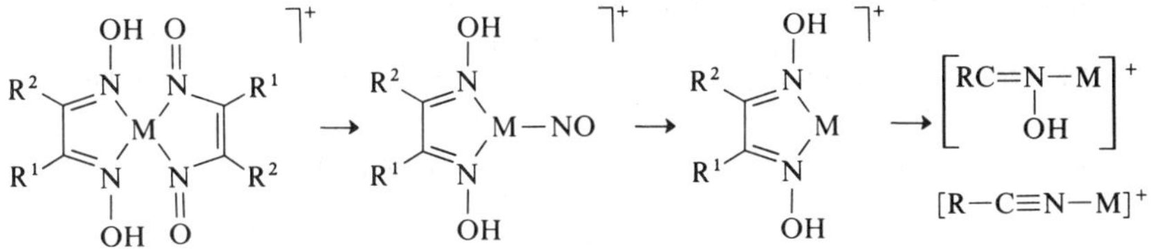

Scheme 20

Porphyrins.—Several mesoporphyrins and their metal complexes have been studied,[299] and the spectra are characterized by the presence of very stable parent ions, together with abundant doubly (18%) and triply charged (6%) fragments. In the porphyrin, the electron lost apparently comes from the conjugated π-system whereas, in the metal derivatives, it comes from a non-bonding

---

[294] J. Charalambous, M. J. Frazer, R. K. Lee, A. H. Qureshi, and F. B. Taylor, *Org. Mass Spectrometry*, 1971, **5**, 1169.
[295] J. K. Terlouw and J. J. de Ridder, *Talanta*, 1970, **17**, 891.
[296] D. Betteridge and D. John, *Talanta*, 1968, **15**, 1227.
[297] J. Charalambous and M. J. Frazer, *J. Chem. Soc. (A)*, 1970, 2645.
[298] A. V. Ablov, K. S. Kharitmov, and Z. Y. Vaisbein, *Doklady Akad. Nauk S.S.S.R.*, 1971, **201**, 345.
[299] A. D. Adler, J. H. Green, and M. Mautner, *Org. Mass Spectrometry*, 1970, **3**, 955.

metal orbital with the stable π-system being preserved. Several examples of ion–molecule reactions are noted, e.g. in the spectrum of tetraphenylporphin (TPP), weak peaks are found for $[P + 28]^+$, $[P + 77]^+$, and $[P + 90]^+$, whereas the tetrachloro-analogue shows intense peaks as high as $[P + 114]^+$. In the spectra of some metal derivatives, a peak was found corresponding to $[Re(TPP)]^+$, probably formed in the source by exchange with the filament.

Transalkylation of vanadyl octaethylporphin occurs on heating with alumina and related materials, as shown in the mass spectrum.[300] Above and below P^+, ions appear at intervals of 28 m.u., corresponding to the nona- and hepta-ethyl compounds. Complex spectra were obtained for the aluminium derivatives of octaethylporphin; loss of methyl or of OH or OMe groups was found for LAlOR (R = H or Me).[301] Loss of ethyl groups also occurred. In these compounds, and especially with LAlOAlL, multiply charged ions were abundant, including a strong P^{3+}.

Carboxylates.—The spectra of the mixed carboxylates $Be_4O(OAc)_n(CCl_3CO_2)_{6-n}$ (n = 5, 3, or 1) have been tabulated; in most ions the Be_4O core is retained, loss of chlorine or carboxylate occurring.[302] Some ions showed loss of one metal atom. The parent ions are weak or non-existent, the highest ion generally being $[P - (CX_3CO_2)]^+$ (X = H or Cl). Dimeric rhodium(II) trifluoroacetate gives $[Rh_2(CF_3CO_2)_4]^+$ as highest ion, which loses three carboxylate groups in stepwise fashion.[303] The base peak is $[Rh_2(CF_3CO_2)_2]^+$. Fluorine migration gives ions such as $[Rh_2F]^+$; other ions always contain two rhodium atoms, e.g. $[Rh_2OF]^+$ and $[Rh_2O]^+$.

Other Complexes.—The spectra of complexes $PtCl_2L_2$ (L = tertiary phosphine, PR_3) show ions formed by elimination of Cl or HCl, and R groups.[304] In the $PEtPhCH_2Ph$ complex, the source of the hydrogen appears to be the benzyl group. Replacement of benzyl by p-tolyl resulted in loss of Cl rather than of HCl. Minor peaks result from fragmentation of the free phosphine. The spectrum of the halogen-bridged complex $Pt_2Cl_4(PMeBu^tPh)_2$ also showed successive elimination of Cl and Bu^t groups, although no metal-containing ion free of chlorine was found.

Other authors[305] have considered the spectra of the cis- and trans-isomers PtX_2L_2 (X = Cl, Br, or I; L = NH_3, py, PEt_3, or PPh_3). Generally, parent ions can be observed, even for these relatively involatile complexes. The ligands fragment while attached to the metal, and some of these reactions involve the metal atom. Comparison of formally similar complexes with chelating and non-chelating ligands, e.g. $PtCl_2(py)_2$ and $PtCl_2(bipy)$, show considerable differences, such as in the ease of loss of Cl vs HCl. With the former complex and

---

[300] R. Bonnett, P. Brewer, K. Noro, and T. Noro, *Chem. Comm.*, 1972, 562.
[301] J. W. Buchler, L. Puppe, and H. H. Schneehage, *Annalen*, 1971, **749**, 134.
[302] K. J. Wynne and W. Bauder, *J. Amer. Chem. Soc.*, 1969, **91**, 5920; *Inorg. Chem.*, 1970, **9**, 1985.
[303] J. Kitchens and J. L. Bear, *Thermochim. Acta*, 1970, **1**, 537.
[304] T. H. Chan and E. Wong, *Org. Mass Spectrometry*, 1971, **5**, 1307.
[305] P. Haake and S. H. Mastin, *J. Amer. Chem. Soc.*, 1971, **93**, 6823.

$PtCl_2(NH_3)_2$, no difference is found in the spectra of the *cis*- and *trans*-isomers, a feature which may be explained by excitation to a tetrahedral stereochemistry in the spectrometer.

After ionization, the ease of bond cleavage is Pt—I > Pt—Br > Pt—Cl, in agreement with the respective Pt—X bond strengths. Rearrangement processes involving loss of H or C_6H_6 from the PPh_3 complexes, for example, suggest the structures (58) and (58a) for some ions. Differences in the spectra of *cis*- and *trans*-$PtX_2(PPh_3)_2$ include: more abundant P^+, $[P - X$ (or $HX)]^+$, and rearrangement ions, such as (58a), in the spectra of the *trans*-isomers.

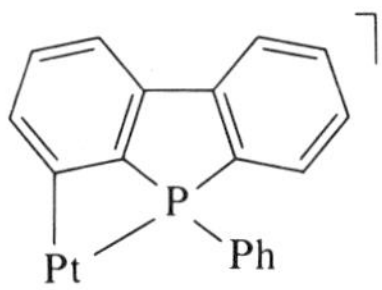

(58) 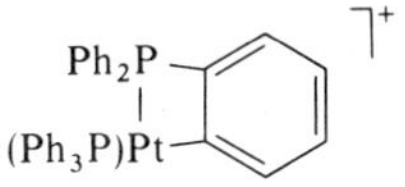   (58a)

Copper complexes of tricyclohexylphosphine show dimeric ions $[Cu_2(PCy_3)_2$-$X_2]^+$, which fragment by loss of halide or PCy_3.[306]

Alkoxides, Alkylamides, and Related Compounds.—In the vapour state, aluminium alkoxides generally form tetramers; for $Al(OEt)_3$, small peaks correspond to $[Al(OEt)_3]_5$ and derived ions.[307] Favoured decomposition routes include loss of OEt, Et, and AlO(OEt). With the corresponding gallium compounds, tetrameric ions are found for the isopropoxide, although most intense ions occur in the dimer region. Fragmentation occurs by loss of OPr^i, C_2H_4O, or C_3H_4O; elimination of $GaO(OPr^i)$ or $MeGaOPr^i$ is found for the higher mass ions. The ethoxide gives peaks arising from a hexamer, although major peaks again occur in the dimer region. The mixed compound $Al(OPr^i)(OBu^t)_2$ gives dimeric ions, but no P^+ ion; the two major decomposition paths are initial loss of OPr^i, or of an Me group.[308] Loss of olefin then occurs. The t-butoxide group is probably eliminated as CH_3, C_2H_6, and CO, although this has not been substantiated by any quantitative measurements.

The compound $LaAl_3(OPr^i)_{12}$ (59) shows tetranuclear ions, with the most abundant ion being $[P - OPr^i]^+$; some evidence for loss of the neutral $AlO(OPr^i)$ group (102 m.u.) is presented.[308]

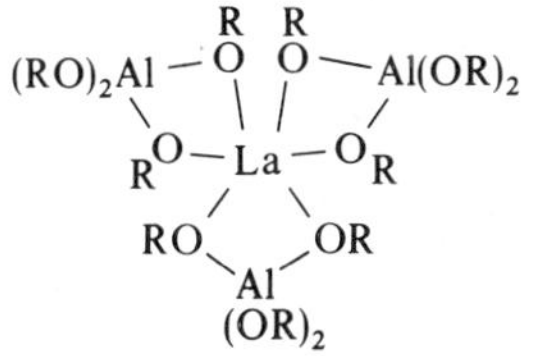

(59)

[306] F. G. Moers and P. H. op het Veld, *J. Inorg. Nuclear Chem.*, 1970, **32**, 3225.
[307] J. G. Oliver and I. J. Worrall, *J. Chem. Soc. (A)*, 1970, 2347.
[308] J. G. Oliver and I. J. Worrall, *J. Chem. Soc. (A)*, 1970, 845.

A series of tetra-alkoxysilanes, $Si(OR)_4$, containing alkyl groups up to five carbon atoms, all give prominent ions corresponding to $[Si(OR)_3]^+$ and $[Si(OR)_3OCH_2]^+$, formed by loss of alkoxy or alkene residues.[309] In many cases, loss of aldehyde occurs with concomitant hydrogen transfer to silicon:

$$\begin{array}{c} \diagdown \overset{+}{Si} \diagup \\ H \diagup \diagdown O \\ \diagdown C \diagup \\ H \quad Me \end{array} \rightarrow \begin{array}{c} \diagdown Si_+ \diagup \\ H \end{array} + \begin{array}{c} H \diagdown \\ C=O \\ Me \diagup \end{array}$$

This reaction has been previously reported for $SiMe_3(OEt)$.[310] With $Si(OMe)_4$, formation of $[Si(OMe)_2]^+$ is envisaged to occur as follows:

$$\begin{bmatrix} MeO \diagdown \diagup OMe \\ Si \\ MeO \diagup \diagdown OMe \end{bmatrix}^+ \xrightarrow{-MeOH} \begin{bmatrix} MeO \diagdown \\ Si \\ MeO \diagup \diagdown O-CH_2 \end{bmatrix}^+ \xrightarrow{-CH_2O} \begin{bmatrix} MeO \diagdown \\ Si \\ MeO \diagup \end{bmatrix}^+$$

Preliminary mass spectral data[311] for isopropoxides of some rare-earth elements (Nd, Er, Tb, and Lu) show ions at low m/e values for $[OPr^i]^+$ and derived ions; at higher values significantly abundant peaks correspond to dimeric and tetrameric species. A series of tungsten chloride–alkoxide derivatives is reported to give P^+ ions.[312]

The spectra of the hexafluoroisopropoxides of several Group IV elements (Si, Ge, Ti, Zr, and Hf) indicate that these compounds are monomeric in the vapour state.[313] Other ions result from loss of F, CF_2, CF_3, and C_2F_3 groups, although the important fragmentation is loss of $OCH(CF_3)_2$ groups. At 65 °C, polymeric ions containing between one and four sodium atoms have been found in the spectrum of $NaOC(CF_3)_3$, with varying numbers of $OC(CF_3)_3$ groups. At lower m/e values, the spectrum is very complex.[314] Mass spectra have been tabulated for boron, phosphorus, titanium, and chromium derivatives of hexafluoropinacol.[315]

In the spectrum of $Hg[SC(CF_3)_3]_2$, the parent ion fragments by loss of CF_3 and CF_2 groups, and also by cleavage of the Hg—S bond.[316]

Spectra for $[Ph_2C=NAlCl_2]_2$ and $Al[N=CBu^t_2]_3$ are tabulated;[317] a strong P^+ ion is found in the dimeric chloride, but the latter gives only weak metal-containing ions, with $[Bu^tCNH]^+$ as the base peak. Ions in the spectrum of

[309] G. Dube, *J. prakt. Chem.*, 1971, **313**, 357.
[310] G. Dube, *Z. Chem.*, 1970, **10**, 301.
[311] L. M. Brown and K. S. Mazdiyasni, *Inorg. Chem.*, 1970, **9**, 2783.
[312] W. J. Reagan and C. H. Brubaker, *Inorg. Chem.*, 1970, **9**, 827.
[313] K. S. Mazdiyasni, B. J. Schaper, and L. M. Brown, *Inorg. Chem.*, 1971, **10**, 889.
[314] R. E. A. Dear, W. B. Fox, R. J. Fredericks, E. E. Gilbert, and D. K. Huggins, *Inorg. Chem.*, 1970, **9**, 2590.
[315] A. P. Conroy and R. D. Dresdner, *Inorg. Chem.*, 1970, **9**, 2739.
[316] B. L. Dyatkin, S. R. Sterlin, B. I. Martynov, and I. L. Knunyants, *Tetrahedron Letters*, 1971, 345.
[317] R. Snaith, C. Summerford, K. Wade, and B. K. Wyatt, *J. Chem. Soc. (A)*, 1970, 2635.

[Pr^{i}NHGaD$_2$]$_2$ correspond to a dimeric parent; more intense ions result from loss of deuterium.[318]

The spectrum of Nb(NEt$_2$)$_4$ gives a molecular ion and many metal-containing ions;[319] metastable-characterized processes include loss of Me, Et, H$_2$, or ethylene, and in one case, MeCH=NEt. The chromium analogue, Cr(NEt$_2$)$_4$, gave a strong P^+ ion, and a fragmentation pattern which suggests considerable stability for the Cr—N bonds.[320] Some discussion of the origin of the electron lost, and comparison with the vanadium complex, is given. The strongest ions are [Cr{NEt(C$_2$H$_4$)}{N(C$_2$H$_4$)$_2$}]$^+$ and [Cr{NEt(C$_2$H$_4$)}$_2$]$^+$; neutral fragments include MeCH=NEt, Et$_2$NH, H$_2$, and C$_2$H$_4$. The parent ion is the base peak in the spectrum of Mo(NMe$_2$)$_4$; other intense ions were $[P - \text{Me}]^+$ and P^{2+}, but other ions, formed by loss of Me or hydrocarbon fragments, were only of low abundance.[321] Parent ions are given by both Ti[N(SiMe$_3$)$_2$]$_3$ and the corresponding vanadium compound.[322]

5 Ionization Potential Data

Tables 1, 2, and 3 summarize the IP's of organometallic compounds of the Main Group elements, of the transition metals, and of co-ordination complexes, respectively, determined since the last report. The data are reported as published, and have not been critically assessed. Standard deviations, if given, relate to the last figure.

Table 1 *Main-group organometallics*

Compound	IP (eV)	Ref.
GaMe$_3$	9.87 (2)	49
Ga(CH=CH$_2$)$_2$	10.81 (10)	49
SiMe$_4$	9.85 (16)	61
	9.86 (2) (PE)	66
SiHMe$_2$Ph	8.92 (3)	70
SiHMePh$_2$	8.75 (4)	70
SiMe$_3$Ph	8.81 (3)	70
SiHPh$_3$	8.80 (3)	70
SiPh$_4$	8.65 (1)	70
PhSiMe$_2$SiMe$_3$	7.82	68
PhC≡CSiMe$_3$	8.16	68
p-C$_6$H$_4$(SiMe$_3$)$_2$	8.25	68
p-C$_6$H$_4$(CH$_2$SiMe$_3$)$_2$	7.25	68
o-C$_6$H$_4$(CH$_2$SiMe$_3$)$_2$	7.74	68
p-C$_6$H$_4$(NMe$_2$)(SiMe$_3$)	6.73	68
Si$_2$Me$_6$	8.35 (12)	61
	8.46 (1)	70

[318] A. Storr and A. D. Penland, *J. Chem. Soc. (A)*, 1971, 1237.
[319] D. C. Bradley and M. H. Chisholm, *J. Chem. Soc. (A)*, 1971, 1511.
[320] J. S. Basi, D. C. Bradley, and M. H. Chisholm, *J. Chem. Soc. (A)*, 1971, 1433.
[321] D. C. Bradley and M. H. Chisholm, *J. Chem. Soc. (A)*, 1971, 2741.
[322] D. C. Bradley and R. G. Copperthwaite, *Chem. Comm.*, 1971, 764.

Table 1 (continued)

Compound	IP (eV)	Ref.
$SiMe_3SiMe_2Ph$	8.35 (3)	70
$SiMe_3SiMePh_2$	8.38 (4)	70
$SiMe_3SiPh_3$	8.30 (1)	70
$(SiMe_2Ph)_2$	8.11 (3)	70
$(SiMePh_2)_2$	8.05 (7)	70
Si_2Ph_6	8.16 (14)	70
$GeMe_4$	9.29 (14)	61
$GeMe_3CMe_3$	8.98 (12)	61
$GeMe_3SiMe_3$	8.31 (10)	61
Ge_2Me_6	8.18 (11)	61
$SnMe_4$	8.76 (12)	61
$SnMe_3CMe_3$	8.34 (11)	61
$SnMe_3SiMe_3$	8.18 (14)	61
$SnMe_3GeMe_3$	8.20 (10)	61
Sn_2Me_6	8.02 (15)	61
$PbMe_4$	8.26 (17)	61
$PbMe_3CMe_3$	7.99 (13)	61
Pb_2Me_6	7.41 (10)	61
PEt_3S	8.0	133
$P_2Et_4S_2$	7.8	133
$AsEt_3$	7.9	150
$AsPr_3^n$	7.8	150
$AsBu_3^n$	8.3	150
As_2Et_4	7.4	150
$As_2Pr_4^n$	7.0	150
$As_2Bu_4^n$	8.2	150

Table 2 Transition-metal organometallics

Compound	IP (eV)	Ref.
$LaCp_3$[a]	7.9 (3)	217
$PrCp_3$	8.2 (2)	217
$NdCp_3$	8.0 (2)	217
	8.3 (1)	216
$SmCp_3$	8.0 (1)	216
$YbCp_2$	7.62 (9)	216
$YbCp_3$	7.72 (9)	216
$CpVC_7H_7$	7.24	229
$CpVC_7H_6Me$	7.06	229
$CpVC_7H_6Ph$	7.01	229
$CpVC_7H_6OMe$	7.07	229
$CpVC_7H_6OEt$	6.92	229
$CpVC_7H_6CH_2CO_2Me$	7.05	229
$CpVC_7H_6CO_2Et$	7.32	229
$CpVC_7H_6CN$	7.78	229
$V(C_7H_8)_2$	6.79 (10)	474
$(CO)_3VC_7H_7$	6.60	229
$(CO)_3VC_7H_6Me$	6.73	229

Table 2 (continued)

Compound	IP (eV)	Ref.
$(CO)_3VC_7H_6Ph$	6.78	229
$(CO)_3VC_7H_6OMe$	6.49	229
$(CO)_3VC_7H_6OEt$	6.48	229
$(CO)_3VC_7H_6OPr^n$	6.50	229
$(CO)_3VC_7H_6CO_2Et$	6.58	229
$Cp(CO)_3CrGeMe_3$	9.06	184
$Cp(CO)_3CrSnMe_3$	9.09	184
$(CO)_3CrC_6H_5Me$	7.19	226
$(CO)_3CrC_7H_8$	7.10	226
$(CO)_3CrC_7H_6OMe$-endo	7.03	226
$(CO)_3CrC_7H_7OMe$-exo	7.16	226
$(CO)_3CrC_7H_7CH_2CO_2Me$-endo	7.12	226
$(CO)_3CrC_7H_7CH_2CO_2Me$-exo	7.21	226
$(CO)_5CrC(OMe)Me$	7.46	231
$(CO)_5CrC(OMe)Ph$	7.26	231, 232, 234
$(CO)_5CrC(OMe)C_6H_4Me$-p	7.13	231, 232
$(CO)_5CrC(OMe)C_6H_4F$-p	7.32	231, 232
$(CO)_5CrC(OMe)C_6H_4Cl$-p	7.34	231, 232
$(CO)_5CrC(OMe)C_6H_4CF_3$-p	7.42	231, 232
$(CO)_5CrC(OMe)C_6H_4OMe$-o	7.05	231, 232
$(CO)_5CrC(OMe)C_6H_4CF_3$-o	7.34	231, 232
$(CO)_5CrC(SMe)Me$	7.30	234
$(CO)_5CrC(SMe)Ph$	7.08	234
$(CO)_5CrC(SPh)Me$	7.17	234
$(CO)_5CrC(NHMe)Ph$	7.04	234
$Cr(CO)_5PF_3$	8.70	245
$Cr(CO)_5PCl_3$	8.26	245
$Cr(CO)_4(PF_3)_2$	8.85	245
$Cr(CO)_3(PF_3)_3$	8.90	245
$Cr(CO)_5S(CH_2)_4$	7.45	250
$Cr(CO)_5SOMe_2$	7.64	250
$Cr(CO)_5SO(OCH_2)_2$	7.80	250
$Cp(CO)_3MoGeMe_3$	9.63	184
$Cp(CO)_3MoSnMe_3$	9.85	184
$Cp(CO)_3WGeMe_3$	9.84	184
$Cp(CO)_3WSnMe_3$	10.05	184
$W(CO)_5NC_5H_5$	7.53	250
$W(CO)_5NC_5H_4CN$-2	7.65	250
$W(CO)_5NC_5H_4Me$-4	7.46	250
$W(CO)_5NC_5H_3Me_2$-2,6	7.43	250
$Mn_2(CO)_{10}$	8.46 (3)	168
$Tc_2(CO)_{10}$	8.30 (3)	168
$Re_2(CO)_{10}$	8.36 (3)	168
$MnRe(CO)_{10}$	8.14 (1)	168
$CpMn(CO)_2PH_3$	7.28 (5)	b
$CpMn(CO)_2PF_3$	8.24	245
$CpMn(CO)_2PCl_3$	8.12	245
$CpMn(CO)_2AsH_3$	7.16 (10)	b
$(MeC_5H_4)Mn(CO)_2SPh_2$	6.27	250
$(MeC_5H_4)Mn(CO)_2S(CH_2)_4$	6.45	250
$(MeC_5H_4)Mn(CO)_2SOPh_2$	6.76	250

Table 2 (*continued*)

Compound	IP (eV)	Ref.
$(MeC_5H_4)Mn(CO)_2SO(CH_2)_4$	6.79	250
$(MeC_5H_4)Mn(CO)_2SOMe_2$	7.19	250
$(MeC_5H_4)Mn(CO)_2SO(OCH_2)_2$	7.38	250
$Fe(CO)_4CN_4Me_2$	7.26	c
$Fe(CO)_4C(NHMe)_2$	6.84	c
$Fe(CO)_4C(NMeCH)_2$	6.83	c
$Fe(CO)_3(PF_3)_2$	8.47	245
$Fe(CO)_2(PF_3)_3$	8.61	245
$Fe(CO)(PF_3)_4$	8.62	245
$Fe(PF_3)_5$	8.83	245
$Fe_2(CO)_9$	7.91 (1)	252
$Fe_2(CO)_6(PMe_2)_2$	7.73 (1)	252
$Fe_2(CO)_6(PEt_2)_2$	7.67 (2)	252
$Fe_2(CO)_6(PPh_2)_2$	7.70 (3)	252
$Fe_2(CO)_6(SPh)(PPh_2)$	7.81 (5)	252
$Fe_2(CO)_6(SMe)_2$	8.07 (1)	252
$Fe_2(CO)_6(SPr^i)_2$	8.05 (1)	252
$Fe_2(CO)_6(SPh)_2$	7.90 (1)	252
azaferrocene	7.17 (10)	260
$Cl_3SiCo(CO)_4$	9.0 (1)	192
$CpCoC_5H_5BMe$	6.56	230
$CpCoC_5H_5BPh$	6.63	230
$Co(C_5H_5BMe)_2$	7.15	230
$Co(C_5H_5BPh)_2$	7.25	230
$Co(C_5H_5BOMe)_2$	7.02	230
$Ni(PF_3)_4$	8.81	245

[a] $Cp = C_5H_5$; [b] E. O. Fischer, W. Bathelt and J. Müller, *Chem. Ber.*, 1970, **103**, 1815; [c] K. Öfele and C. G. Kreiter, *ibid.*, 1972, **105**, 529.

Table 3 *Co-ordination compounds*

Compound	IP (eV)	Ref.
$Al(acac)_3$	7.95 (5)	266
$Cr(acac)_3$	8.10 (5)	266
$Fe(acac)_3$	8.45 (5)	266
$Fe(acac)_2$	8.10 (5)	266
$Cu(acac)_2$	7.75 (5)	10
	8.31 (5)	266, 267
$Zn(acac)_2$	8.62 (5)	266
$Cu(MeCOCHCOEt)_2$	7.68 (3)	10
$Cu(MeCOCHCOPr^i)_2$	7.61 (6)	10
$Cu(MeCOCHCOBu^t)$	7.59 (5)	10
$Cu(Meacac)_2$	7.97 (5)	267
$Cu(Phacac)_2$	8.05 (5)	267
$Cu(bzac)_2$	8.37 (5)	267
$Cu(dbzm)_2$	8.28 (5)	267
$Al(tfac)_3$	9.05 (10)	266
$Cr(tfac)_3$	9.09 (5)	266

Table 3 (continued)

Compound	IP (eV)	Ref.
Fe(tfac)$_3$	9.10 (5)	266
Fe(tfac)$_2$	8.75 (10)	266
Cu(tfac)$_2$	8.61 (5)	10
	9.05 (5)	266, 267
Zn(tfac)$_2$	9.4 (1)	266
Cu(bztfac)$_2$	9.06 (5)	267
Cu(furoyltfac)$_2$	8.89 (5)	267
Cu(ttac)$_2$	8.90 (5)	267
Cu(naphtfac)$_2$	8.39 (5)	267
Cu(acac)(hfac)	8.65 (1)	10
	9.03 (5)	267
Al(hfac)$_3$	9.8 (1)	266
Cr(hfac)$_3$	10.13 (5)	266
Fe(hfac)$_3$	10.13 (3) (PE)	264
	10.2 (1)	266
Fe(hfac)$_2$	9.7 (1)	266
Cu(hfac)$_2$	9.68 (1)	10
	9.86 (5)	266
Zn(hfac)$_2$	10.07 (5)	266
Rh(acac)$_3$	7.34 (1)	262
	7.75	261
Rh(acac)$_2$(NO$_2$acac)	7.65 (2)	262
Rh(acac)(NO$_2$acac)$_2$	7.97 (3)	262
Rh(NO$_2$acac)$_3$	8.39 (4)	262
(CO)$_2$Rh(acac)	8.6	261
(CO)$_2$Rh(bzac)	8.4	261
(CO)$_2$Rh(dbzm)	8.4	261
(CO)$_2$Rh(tfac)	8.85	261
(CO)$_2$Rh(hfac)	9.2	261
(CO)$_2$Ir(acac)	8.6	261
(CO)$_2$Ir(hfac)	8.85	261

6 Appendix

During the preparation of this review, approximately half the papers scanned were found to contain some mass spectral data. These usually consisted of a listing of ions, with or without assignments, and often only well-established features were noted. Consequently these accounts do not merit any discussion in the main text, but they are listed here for completeness, classified according to the headings used above.

Group II.—Perfluoroalkylmercury compounds;[323] fluorenyl-HgCl;[324] N-hydroxysuccinimidyl chloromercuriacetate;[325] chloromercury derivatives of cyclohexane and cyclopentanecarboxylic acid.[326]

[323] B. L. Dyatkin, L. G. Zhuravkova, B. I. Martynov, E. I. Mysov, S. R. Sterlin, and I. L. Knunyants, *J. Organometallic Chem.*, 1971, **31**, C15; *Tetrahedron*, 1971, **27**, 2843.
[324] E. Samuel and M. D. Rausch, *J. Organometallic Chem.*, 1972, **37**, 29.
[325] G. Folsch, *Acta Chem. Scand.*, 1970, **24**, 1115.
[326] H. J. Roth and B. Muhlenbruch, *Pharmazie*, 1970, **25**, 597.

Group III.—$C_5H_5BEt_2$;[327] 9-mesityl-9,10-dihydro-9-bora-anthracene;[328] di-isopropoxyborane;[329] [PhPSBMe$_2$]$_n$ (showing a monomeric P^+);[330] 1,2,4,5-tetra-aza-3,6-diborinane monomers;[331] $Ph_3B^-CN^+N=NPh$;[332] Me_2NCH_2-CO_2BH_2;[333] 1-pyrrolylboranes;[334] several B-substituted borazines;[335] $(CF_3)_2$-$C=NBPh_2$.[336]

Carboranes and Related Compounds. B-substituted *o*- and *m*-carboranes;[337] 2,3,4,5-tetracarba-*nido*-hexaborane(6);[338] 1-germa-, 1-stanna-, and 1-plumba-2,3-dicarba-*closo*-dodecarboranes(11);[339] germa- and stanna-undecarboranes;[340] silicon and germanium derivatives of the type $R_2MHC_2B_5H_6$;[341] 1,2-$C_2B_3H_7$;[342] 1-MeEC$_2$B$_4$H$_6$ (E = Ga or In);[343] $C_3B_5H_7$;[344] some benzocarborane derivatives.[345]

Group IV.—*Silicon.* Products from SiF_2 and $CF_3C\equiv CH$, perhaps CF_3-$\overline{C=CSiF_2SiF_2}$ and $H\overline{C=C(CF_3)C(CF_3)=CHSiF_2SiF_2}$;[346] non-fluorinated 1,2,3-trisilacycloheptane derivatives (methylated on silicon);[347] $MeSiH_2PH_2$;[348] $Ph_2PNMeSiMe_3$ and $Ph_3SiNMePPh_2$;[349] compounds such as $Ph_2SiHSiH_3$ and $PhGe(SiH_3)_3$;[350] Me_3SiONO_2;[351] Me_3SiR (together with PR_3, OPR_3, AsR_3, BR_3) [R = N=C(CF$_3$)$_2$];[352] $OS(NSiMe_3)_2$ and Me_3SiNSF_2O;[353] $PhBu^tC=NSiMe_3$;[354] fluorophosphine–sulphur compounds;[355] $S(NSiMe_3)_3$ and SF_2-

---

[327] H. Grundke and P. Paetzold, *Chem. Ber.*, 1971, **104**, 1136.
[328] R. van Veen and F. Bickelhaupt, *J. Organometallic Chem.*, 1971, **30**, C51.
[329] T. P. Fehlner, *Inorg. Chem.*, 1972, **11**, 252.
[330] H. Vahrenkamp, *J. Organometallic Chem.*, 1971, **28**, 167.
[331] J. J. Miller and F. A. Johnson, *Inorg. Chem.*, 1970, **9**, 69.
[332] E. Brehm, A. Hage, and G. Hesse, *Annalen*, 1970, **737**, 80.
[333] N. E. Miller, *J. Amer. Chem. Soc.*, 1970, **92**, 4564.
[334] H. Bellut and R. Koster, *Annalen*, 1970, **738**, 86.
[335] O. T. Beachley, *J. Amer. Chem. Soc.*, 1970, **92**, 5372.
[336] K. Niedenze, C. D. Miller, and F. C. Nahm, *Tetrahedron Letters*, 1970, 2441.
[337] J. S. Roscoe, S. Kongpricha, and S. Papetti, *Inorg. Chem.*, 1970, **9**, 1561.
[338] E. Groszek, J. B. Leach, G. T. F. Wong, C. Ungermann, and T. Onak, *Inorg. Chem.*, 1971, **10**, 2770.
[339] R. W. Rudolph, R. L. Voorhees, and R. E. Cochry, *J. Amer. Chem. Soc.*, 1970, **92**, 3351.
[340] R. E. Loffredo and A. D. Norman, *J. Amer. Chem. Soc.*, 1971, **93**, 5587.
[341] W. A. Ledoux and R. N. Grimes, *J. Organometallic Chem.*, 1971, **28**, 37.
[342] D. A. Franz, V. R. Miller, and R. N. Grimes, *J. Amer. Chem. Soc.*, 1972, **94**, 412.
[343] R. N. Grimes, W. J. Rademaker, M. L. Denniston, R. F. Bryan, and P. T. Greene, *J. Amer. Chem. Soc.*, 1972, **94**, 1865.
[344] M. L. Thompson and R. N. Grimes, *J. Amer. Chem. Soc.*, 1971, **93**, 6677.
[345] D. S. Matteson and N. K. Hota, *J. Amer. Chem. Soc.*, 1971, **93**, 2893.
[346] C. S. Liu and J. C. Thompson, *Inorg. Chem.*, 1971, **10**, 1100.
[347] H. Sakurai, Y. Kobayashi, and Y. Nakadaira, *J. Amer. Chem. Soc.*, 1971, **93**, 5272.
[348] J. W. Anderson and J. E. Drake, *J. Chem. Soc. (A)*, 1971, 1424.
[349] R. Keat, *J. Chem. Soc. (A)*, 1970, 1795.
[350] F. Feher, P. Plichta, and R. Guillery, *Tetrahedron Letters*, 1970, 4443.
[351] L. Birkofer and M. Franz, *Chem. Ber.*, 1972, **105**, 470.
[352] R. F. Swinkell, D. P. Babb, T. J. Ouellette, and J. M. Shreeve, *Inorg. Chem.*, 1972, **11**, 242.
[353] O. Glemser, M. F. Feser, S. P. von Halasz, and H. Saran, *Inorg. Nuclear Chem. Letters*, 1972, **8**, 321.
[354] M. Kilner and J. N. Pinkney, *J. Chem. Soc. (A)*, 1971, 2887.
[355] H. W. Roesky and L. F. Grimm, *Chem. Ber.*, 1970, **103**, 1664.

(NSiMe$_3$)$_2$;[356] cyclic silicon–nitrogen compounds;[357] products from BF$_3$ and (Me$_3$Si)$_2$NH;[358] trimethylsilylamine;[359] R$_2$SiCH$_2$CHMeCH$_2$S(O)O;[360] compounds containing the MeSiHF group;[361] chlorinated methyldisilanes, e.g. MeClSiHSiH$_2$Cl;[362] 1-aza-2-silacyclopentanes and related compounds;[363] products from polylithiated aromatic hydrocarbons treated with Me$_3$SiCl;[364] derivatives of H$_3$SiCH$_2$GeH$_3$;[365] alkylpolysilanes, used in the identification of pyrolysis products of MeSi$_2$H$_5$ and (Me$_2$SiH)$_2$;[366] various methyl-chloro polysilanes;[367] MeSi$_2$H$_5$ and deuteriated derivatives;[368] Et$_3$SiCHPhNC$_4$H$_8$;[369] ClC≡CSiCl$_3$;[370] 1,2-bis(trimethylsilyl)indene;[371] some trialkoxysilanes;[372] products from photolytic addition of Me$_3$SiH to C$_2$F$_4$, of the type Me$_3$Si-(CF$_2$)$_n$H;[373] several dimethylaminosilanes.[374]

Germanium. Methyltrigermanes;[375] digermylmethane;[376] Me$_3$SiGeH$_3$, Ge$_2$H$_3$-Me$_3$, and related compounds;[377] fluorocarbon–GeEt$_3$ compounds;[378] the Ge—P adamantane analogue P$_4$(GeMe$_2$)$_6$;[379] (C$_6$F$_5$)$_3$GeF and (C$_6$F$_5$)$_3$-GeOH.[380]

Tin. Organotin fluorides, R$_3$SnF;[381] Me$_3$SnSO$_2$Me;[382] Ph$_3$SnSO$_2$Ph;[383] Me$_3$SnPH$_2$;[384] (Bu$_3^n$Sn)$_2$SO$_4$;[385] trimethyltin derivatives containing a variety of fluorinated P—O, S—O, and P—S groups.[386]

[356] O. Glemser and J. Wegener, *Angew. Chem.*, 1970, **82**, 324; *Angew. Chem. Internat. Edn.*, 1970, **9**, 309.
[357] U. Wannagat, E. Bogusch, and F. Rabet, *Z. anorg. Chem.*, 1971, **385**, 261.
[358] G. Elter, O. Glemser, and W. Herzog, *Chem. Ber.*, 1972, **105**, 115.
[359] N. Wiberg and W. Uhlenbrock, *Chem. Ber.*, 1971, **104**, 2643.
[360] J. Dubac, P. Mazerolles, M. Joly, W. Kitching, C. W. Fong, and W. H. Atwell, *J. Organometallic Chem.*, 1972, **34**, 17.
[361] E. W. Kifer and C. H. van Dyke, *Inorg. Chem.*, 1972, **11**, 404.
[362] A. J. van der Wielen and M. A. Ring, *Inorg. Chem.*, 1972, **11**, 246.
[363] T. Tsai and C. J. Marshall, *J. Org. Chem.*, 1972, **37**, 596.
[364] A. Halasa, *J. Organometallic Chem.*, 1971, **31**, 369.
[365] C. H. van Dyke, E. W. Kifer, and G. A. Gibbon, *Inorg. Chem.*, 1972, **11**, 408.
[366] R. B. Baird, M. D. Sefick, and M. A. Ring, *Inorg. Chem.*, 1970, **10**, 883.
[367] C. R. Bettler, J. C. Sendra, and G. Urry, *Inorg. Chem.*, 1970, **9**, 1060.
[368] P. Estacio, M. D. Sefick, E. K. Chan, and M. A. Ring, *Inorg. Chem.*, 1970, **9**, 1068.
[369] J. A. Connor and P. D. Rose, *J. Organometallic Chem.*, 1970, **24**, C45.
[370] J. Binenboym and R. Schaeffer, *Inorg. Chem.*, 1970, **9**, 1578.
[371] A. Davison and P. E. Rakita, *J. Organometallic Chem.*, 1970, **23**, 407.
[372] W. E. Newton and E. G. Rochow, *Inorg. Chem.*, 1970, **9**, 1071.
[373] R. N. Haszeldine, S. Lythgoe, and P. J. Robinson, *J. Chem. Soc. (B)*, 1970, 1634.
[374] J. Grobe and U. Möller, *J. Organometallic Chem.*, 1971, **33**, 13.
[375] S. T. Hosfield and W. M. Mackay, *J. Organometallic Chem.*, 1970, **24**, 107.
[376] R. M. Dreyfuss and W. L. Jolly, *Inorg. Chem.*, 1971, **10**, 2567.
[377] R. D. George, K. M. Mackay, and S. R. Stobart, *J. Chem. Soc. (A)*, 1970, 3250.
[378] B. I. Petrov, O. A. Kruglaya, N. S. Vyazankin, B. I. Martynov, S. R. Sterlin, and B. L. Dyatkin, *J. Organometallic Chem.*, 1972, **34**, 299.
[379] A. R. Dahl and A. D. Norman, *J. Amer. Chem. Soc.*, 1970, **92**, 5525.
[380] G. B. Deacon and J. C. Parrott, *J. Organometallic Chem.*, 1970, **22**, 287.
[381] K. Licht, H. Geissler, P. Kochler, K. Hottmann, H. Schnorr, and H. Kriegsmann, *Z. anorg. Chem.*, 1971, **385**, 271.
[382] E. Lindner, U. Kunze, G. Ritter, and A. Haag, *J. Organometallic Chem.*, 1970, **24**, 119.
[383] E. Lindner, U. Kunze, G. Vitzhum, G. Ritter, and A. Haag, *J. Organometallic Chem.*, 1970, **24**, 131.
[384] A. D. Norman, *J. Organometallic Chem.*, 1971, **28**, 81.
[385] R. H. Herber and C. H. Stapfer, *Inorg. Nuclear Chem. Letters*, 1971, **7**, 617.
[386] H. W. Roesky and H. Wiezer, *Chem. Ber.*, 1971, **104**, 2258.

Group V.—$(CF_3)_2P(S)OH$ and $[(CF_3)_2P(S)]_2O$;[387] trialkylalkylidenephosphoranes and their BPh_3 adducts;[388] $PhEtC(CN)P(O)(OEt)_2$;[389] $(OPPh_2)_2OCHCF_3$;[390] adducts of tetracyclone with diphenylphosphinous acid;[391] alkylchloramination products of $(Ph_2P)_2NMe$;[392] $MePX_2$ (X = Br or CN);[393] products of reactions between olefins and Me_2PH or $(CF_3)_2PH$;[394] further products from phosphines and $C_2(CO_2Me)_2$;[395] $F_3SiCH_2CH_2PMe_2$;[396] phosphabarrelene compounds;[397] $(CF_3)_3PPF_2$ and its BH_3 adduct;[398] $(CF_3S)_2PCF_3$;[399] MePFCl and MePFBr;[400] products obtained from PF_3 and primary amines, *e.g.* $RNHPF_2$;[401] phosphonitrilic derivatives obtained from $(PNCl_2)_4$ and PhMgBr;[402] $Ph_2PCH_2CH_2NC$;[403] 10,11-dihydro-5-phenyl-5H-dibenzo-[bf]phosphepin;[404] the phenylethynyl-fluorophosphonitrilic derivatives $N_3P_3F_5(C_2Ph)$ and $N_3P_3F_4(C_2Ph)_2$, and the cobalt carbonyl complexes obtained from these;[405] some derivatives of $[\overline{CH_2(CH_2)_3P(CH_2)_3CH_2}]I$;[406] and the new As—S heterocycle $(MeAs)_2S_2N_4$.[407]

Carbonyls.—$Os_2(CO)_9$.[408]

Nitrosyls.—Cyclopentadienylmolybdenum nitrosyl complexes containing tertiary phosphines, arsines, or isocyanides;[409,410] cyclopentadienylmanganese nitrosyl complexes with sulphur-containing ligands;[411] similar cobalt complexes containing bridging NO and EPh groups (E = S, Se, or PPh) and the related halides $(\pi\text{-}C_5H_5)Co(NO)X$ (X = Br or I) (the latter showing dinuclear ions).[412]

---

[387] A. A. Pinkerton and R. G. Cavell, *J. Amer. Chem. Soc.*, 1972, **94**, 1870.
[388] R. Koster, D. Simic, and M. A. Grassberger, *Annalen*, 1970, **739**, 211.
[389] J. H. Boyer and R. Selvarajan, *J. Org. Chem.*, 1970, **35**, 1224.
[390] E. Lindner, H. D. Ebert, and P. Junkes, *Chem. Ber.*, 1970, **103**, 1364.
[391] J. A. Miller, *Tetrahedron Letters*, 1969, 4335.
[392] D. F. Clemens, M. L. Caspar, D. Rosenthal, and R. Peluso, *Inorg. Chem.*, 1970, **9**, 960.
[393] R. Foester and K. Cohn, *Inorg. Chem.*, 1970, **9**, 1571.
[394] R. Fields, R. N. Haszeldine, and N. F. Wood, *J. Chem. Soc. (A)*, 1970, 1370.
[395] N. E. Waite, J. C. Tebby, R. S. Ward, M. A. Shaw, and D. H. Williams, *J. Chem. Soc. (C)*, 1971, 1620.
[396] J. Grobe and U. Möller, *J. Organometallic Chem.*, 1971, **31**, 157.
[397] G. Märkl, F. Lieb, and C. Martin, *Tetrahedron Letters*, 1971, 1249.
[398] H. W. Schiller and R. W. Rudolph, *Inorg. Chem.*, 1971, **10**, 2500.
[399] I. B. Mishra and A. B. Burg, *Inorg. Chem.*, 1972, **11**, 664.
[400] H. W. Schiller and R. W. Rudolph, *Inorg. Chem.*, 1972, **11**, 187.
[401] J. S. Harman and D. W. A. Sharp, *J. Chem. Soc. (A)*, 1970, 1935.
[402] M. Biddlestone and R. A. Shaw, *J. Chem. Soc. (A)*, 1970, 1750.
[403] R. B. King and A. Efraty, *J. Amer. Chem. Soc.*, 1971, **93**, 564.
[404] J. L. Suggs and L. D. Freedman, *J. Org. Chem.*, 1971, **36**, 2566.
[405] T. Chivers, *Inorg. Nuclear Chem. Letters*, 1971, **7**, 827.
[406] B. D. Cuddy, J. C. F. Murray, and B. J. Walker, *Tetrahedron Letters*, 1971, 2397.
[407] O. J. Scherer and R. Wies, *Angew. Chem.*, 1971, **83**, 882; *Angew. Chem. Internat. Edn.*, 1971, **10**, 812.
[408] J. R. Moss and W. A. G. Graham, *Chem. Comm.*, 1970, 835.
[409] T. A. James and J. A. McCleverty, *J. Chem. Soc. (A)*, 1971, 1596.
[410] T. A. James and J. A. McCleverty, *J. Chem. Soc. (A)*, 1971, 1068.
[411] P. Hydes, J. A. McCleverty, and D. G. Orchard, *J. Chem. Soc. (A)*, 1971, 3660.
[412] H. Brunner and S. Loskot, *Z. Naturforsch.*, 1971, **26b**, 757.

Organometallic and Co-ordination Compounds

Hydrides, Halides, and Pseudohalides.—$H_4Re_4(CO)_{12}$;[413] ruthenium carbonyl fluoride;[414] manganese and rhenium carbonyl-phosphine isocyanates.[415]

Compounds with Bonds to Main Group Elements.—$Zn_2Co_4(CO)_{15}$;[416] Ti-$(SnPh_3)_3$;[417] $(\pi\text{-}C_6H_6)Cr(CO)_2H(SiCl_3)$;[418] $Bu_2^nSnCr(CO)_5$;[419] $Ge_2H_5Mn(CO)_5$;[420] organo-silicon and -tin(carbonyl)osmium complexes;[421] $Et_2Si[Co(CO)_4]OCCo_3(CO)_9$;[422] $PhGeCo_3(CO)_{11}$;[423] $(\pi\text{-}C_5H_5)Rh(CO)HSi(CH_2\text{-}Ph)_3$ and $(\pi\text{-}C_5H_5)Rh(CO)H(SiPh_3)$;[424] $(1,5\text{-}C_8H_{12})RhCl(SiCl_3)_2$ and $(\pi\text{-}C_5H_5)Rh(C_8H_{14})Rh(SiCl_3)_2$;[425] $Fe_5C(CO)_{15}$;[426] derivatives containing the $CCo_3(CO)_9$ group;[427] $Me_3SiOCCo_3(CO)_9$;[428] $(\pi\text{-}C_5H_5)_3Ni_3CR$ (R = Ph or p-tolyl) and $(\pi\text{-}C_5H_5)_2(\pi\text{-}C_5H_4CH_2Ph)Ni_3CPh$.[429]

σ-Bonded Complexes.—$Hg[C_3F_7Fe(CO)_4]_2$ and $HFe(CO)_4(C_3F_7)$;[430] octafluoronickelacyclopentane complexes;[431] two titanium metallocycles;[432] carboxylate complexes of rhenium carbonyl.[433,434]

Acetylene Complexes and σ-Acetylides.—A bis(t-butyl)acetylene-iron complex;[435] diphenylacetylene complexes of $Os_3(CO)_{12}$;[436] alkyne-cobalt and -rhodium carbonyl complexes;[437,438] the tetrafluorobenzyne derivatives C_6F_4-

[413] R. Saillant, G. Barcelo, and H. Kaesz, *J. Amer. Chem. Soc.*, 1970, **92**, 5739.
[414] C. J. Marshall, R. D. Peacock, D. R. Russell, and I. L. Wilson, *Chem. Comm.*, 1970, 1643.
[415] J. T. Moelwyn-Hughes, A. W. B. Garner, and A. S. Howard, *J. Chem. Soc. (A)*, 1971, 2370, 2361.
[416] J. M. Burlitch and S. E. Hayes, *J. Organometallic Chem.*, 1971, **29**, C1.
[417] M. F. Lappert and A. R. Sanger, *J. Chem. Soc. (A)*, 1971, 1314.
[418] W. Jetz and W. A. G. Graham, *Inorg. Chem.*, 1971, **10**, 4.
[419] T. J. Marks, *J. Amer. Chem. Soc.*, 1971, **93**, 7090.
[420] S. R. Stobart, *Chem. Comm.*, 1970, 999.
[421] S. A. R. Knox and F. G. A. Stone, *J. Chem. Soc. (A)*, 1970, 3147.
[422] S. A. Fieldhouse, A. J. Cleland, B. H. Freeland, C. D. M. Mann, and R. J. O'Brien, *J. Chem. Soc. (A)*, 1971, 2536.
[423] R. Ball, M. J. Bennett, E. H. Brooks, W. A. G. Graham, J. Hoyano, and S. M. Illingworth, *Chem. Comm.*, 1970, 592.
[424] A. J. Oliver and W. A. G. Graham, *Inorg. Chem.*, 1971, **10**, 1.
[425] F. Glockling and G. C. Hill, *J. Chem. Soc. (A)*, 1971, 2137.
[426] R. P. Stewart, U. Anders, and W. A. G. Graham, *J. Organometallic Chem.*, 1971, **32**, C49.
[427] R. Dolby and B. H. Robinson, *Chem. Comm.*, 1970, 1058.
[428] C. D. M. Mann, A. J. Clelland, S. A. Fieldhouse, B. H. Freeland, and R. J. O'Brien, *J. Organometallic Chem.*, 1970, **24**, C61.
[429] T. I. Voyevodskaya, I. M. Pribytkova, and Y. A. Ustynyuk, *J. Organometallic Chem.*, 1972, **37**, 187.
[430] F. Seel and G.-V. Röschenthaler, *Z. anorg. Chem.*, 1970, **373**, 183.
[431] C. S. Cundy, M. Green, and F. G. A. Stone, *J. Chem. Soc. (A)*, 1970, 1647.
[432] M. D. Rausch and L. P. Klemann, *Chem. Comm.*, 1971, 354.
[433] E. Lindner and R. Grimmer, *J. Organometallic Chem.*, 1971, **31**, 249.
[434] A. M. Brodie, G. Hulley, B. F. G. Johnson, and J. Lewis, *J. Organometallic Chem.*, 1970, **24**, 201.
[435] K. Nicholas, L. S. Bray, R. E. Davis, and R. Pettit, *Chem. Comm.*, 1971, 608.
[436] O. Gambino, G. A. Vaglio, R. P. Ferrari, and G. Cetini, *J. Organometallic Chem.*, 1971, **30**, 381.
[437] R. S. Dickson and P. J. Fraser, *Austral. J. Chem.*, 1970, **23**, 475, 2403.
[438] R. S. Dickson and G. R. Tailby, *Austral. J. Chem.*, 1970, **23**, 229, 1531.

Co$_4$(CO)$_{10}$ and C$_6$F$_4$Fe$_2$(CO)$_8$;[439] (π-C$_5$H$_5$)$_2$M(C$_2$Ph)$_2$ (M = Ti or Zr) and (π-C$_5$H$_4$Me)$_2$Zr(C$_2$Ph)$_2$;[440] MnBr(CO)$_4$(C$_2$PPh$_3$).[441]

Olefin Complexes.—[C$_2$(OMe)$_4$]Fe(CO)$_4$;[442] PtX(acac)(C$_2$H$_4$) (X = Cl or Br);[443] butadiene-, cyclobutadiene-, and butatriene-iron carbonyl complexes;[444-446] fluoro-olefin-phosphine and -phosphite complexes of iron and ruthenium carbonyls, and similar fluoroalkyne complexes;[447] thiophen-1,1-dioxide-Fe(CO)$_3$;[448] substituted cyclo-octatetraene-Fe(CO)$_3$ complexes;[449] iron carbonyl complexes of various C$_8$ hydrocarbons[450] and of heptafulvene;[451] C$_8$H$_8$FeRu(CO)$_5$;[452] bullvalene-Cr(CO)$_3$;[453] cyclopentadienylrhodium complexes of pentamethylenecyclodecane.[454]

Allyl Complexes.—A keto-π-cyclobutenyl-cobalt carbonyl.[455]

Cyclopentadienyl Complexes.—(π-C$_5$H$_5$)$_4$Sm$_2$Cl$_2$;[456] (π-C$_5$H$_5$)$_3$UL (L = THF, C$_6$H$_{11}$NC, or l-nicotine);[457] (π-C$_5$H$_5$)$_2$Ti[OP(CF$_3$)$_2$]$_2$;[458] (π-C$_5$H$_5$)$_2$V$_2$(CO)$_5$;[459] (π-C$_5$H$_5$)$_2$NbCl$_2$;[460] the hydrides (π-C$_5$H$_5$)$_2$TaH$_3$ and (π-C$_5$H$_5$)$_2$MoH$_2$;[461] [Mo(π-C$_5$H$_5$)$_2$]$_n$ (giving a dimeric P^+), the monocarbonyl derivative, and the decamethyl-substituted complex;[462] π-trimethylsilylcyclopentadienyl-molybdenum and iron complexes;[463] indenyl-molybdenum complexes (π-ind)Mo(CO)$_3$X (X = Cl, Br, or I) and phosphine derivatives;[464] complexes derived from 2-benzylpyrrole containing π-bonded Mn(CO)$_3$ and Cr(CO)$_3$ groups;[465]

[439] D. M. Roe and A. G. Massey, *J. Organometallic Chem.*, 1970, **23**, 547.
[440] A. D. Jenkins, M. F. Lappert, and R. C. Srivastava, *J. Organometallic Chem.*, 1970, **23**, 165.
[441] D. K. Mitchell, W. D. Korte, and W. C. Kaska, *Chem. Comm.*, 1970, 1384.
[442] M. Herberhold and H. Brabetz, *Z. Naturforsch.*, 1971, **26b**, 656.
[443] G. Hulley, B. F. G. Johnson, and J. Lewis, *J. Chem. Soc. (A)*, 1970, 1732.
[444] H. A. Brune, W. Schwab, and H. P. Wolf, *Z. Naturforsch.*, 1970, **25b**, 982.
[445] H. A. Brune, H. Hanebeck, and H. Hüther, *Tetrahedron*, 1970, **26**, 3099.
[446] H. A. Brune and W. Schwab, *Tetrahedron*, 1970, **26**, 1357.
[447] R. Burt, M. Cooke, and M. Green, *J. Chem. Soc. (A)*, 1970, 2975, 2981.
[448] Y. L. Chow, J. Fossey, and R. A. Perry, *J.C.S.Chem. Comm.*, 1972, 501.
[449] B. F. G. Johnson, J. Lewis, and G. L. P. Randall, *J. Chem. Soc. (A)*, 1971, 422.
[450] P. Janse van Vuuren, R. J. Fletterick, J. Meinwald, and R. E. Hughes, *Chem. Comm.*, 1970, 883.
[451] D. J. Ehntholt and R. C. Kerber, *Chem. Comm.*, 1970, 1451.
[452] E. W. Abel and S. Moorhouse, *Inorg. Nuclear Chem. Letters*, 1970, **6**, 621.
[453] R. Aumann, *Angew. Chem.*, 1970, **82**, 810; *Angew. Chem. Internat. Edn.*, 1970, **9**, 800.
[454] R. B. King and P. N. Kapoor, *J. Organometallic Chem.*, 1971, **33**, 383.
[455] R. B. King and A. Efraty, *J. Organometallic Chem.*, 1970, **24**, 241.
[456] P. G. Laubereau, *Inorg. Nuclear Chem. Letters*, 1970, **6**, 611.
[457] B. Kanellakopulos, E. O. Fischer, E. Dornberger, and F. Baumgärtner, *J. Organometallic Chem.*, 1970, **24**, 507.
[458] W. J. Reagan and A. B. Burg, *Inorg. Nuclear Chem. Letters*, 1971, **7**, 741.
[459] E. O. Fischer and R. J. J. Schneider, *Chem. Ber.*, 1970, **103**, 3684.
[460] F. W. Siegert and H. J. de Liefde Meijer, *J. Organometallic Chem.*, 1970, **23**, 177.
[461] F. Macasek, V. Mikulaj, and P. Drienovsky, *Coll. Czech. Chem. Comm.*, 1970, **35**, 993.
[462] J. L. Thomas and H. H. Brintzinger, *J. Amer. Chem. Soc.*, 1972, **94**, 1386.
[463] C. S. Kraihanzel and J. Conville, *J. Organometallic Chem.*, 1970, **23**, 357.
[464] A. J. Hart-Davis, C. White, and R. J. Mawby, *Inorg. Chim. Acta*, 1970, **4**, 431.
[465] K. J. Coleman, C. S. Davies, and N. J. Gogan, *Chem. Comm.*, 1970, 1414.

several $(\pi\text{-}C_5H_5)Ru(CO)_2X$ complexes e.g. $X = CONHNH_2$, NCO, or NCS.[466]

Arene Complexes.—$(\pi\text{-Mesitylene})V(CO)_3X$ ($X = H$ or I);[467] $Cr(PhSiMe_3)_2$;[468] triptycene-$Cr(CO)_3$;[469] $(\pi\text{-}C_6H_6)Mo[P(OMe)_3]_3$;[470] bis(hexamethylbenzene)-ruthenium;[471] $(\pi\text{-}2,4,6\text{-triphenylphosphorin})Cr(CO)_3$.[472]

Complexes containing Larger Rings.—$(\pi\text{-}C_7H_7)Ti(\pi\text{-}C_7H_9)$;[473] the paramagnetic complex $V(C_7H_8)_2$;[474] $(\pi\text{-}C_5H_5)M(\pi\text{-}C_7H_7)$ ($M = Mo$ or W);[475] azulene-$Ru_3(CO)_7$.[476]

Complexes containing Donor Ligands.—*Boron Ligands.* $PhCH_2OB(CH{=}CH_2)_2$-$Fe(CO)_3$;[477] $(\pi\text{-}C_3H_5)Ni(Me_2BNMe_2)_2$;[478] $(\pi\text{-}2\text{-}MeC_3B_3H_5)Mn(CO)_3$;[479] bis-$[\pi\text{-}(3)\text{-}1,2\text{-dicarbollyl}]$-metallates of Ni, Pd, and Pt.[480]

Carbene and Related Complexes. $(EtO)(R_2N)CCr(CO)_5$;[481,482] bis-(1,3-dimethyl-4-imidazolin-2-ylidene)$Cr(CO)_4$;[483] $HFe_3(CO)_{10}CNMe_2$;[484] $PtCl_2(PEt_3)$-$\overline{CNPh(CH_2)_2NPh}$.[485]

Nitrogen Donors. $(Me_2N)_2M[NArC(NMe_2){=}NAr]_2$ ($M = Ti$ or Zr);[486] molybdenum and tungsten complexes containing the $N{\equiv}CBu^t_2$ ligand;[487] meso-

[466] A. E. Kruse and R. J. Angelici, *J. Organometallic Chem.*, 1970, **24**, 231.
[467] A. Davison and D. L. Reger, *J. Organometallic Chem.*, 1970, **23**, 491.
[468] C. Elschenbroich, *J. Organometallic Chem.*, 1970, **22**, 677.
[469] R. L. Pohl and B. R. Willeford, *J. Organometallic Chem.*, 1970, **23**, C45.
[470] M. L. H. Green, L. C. Mitchard, and W. E. Silverthorn, *J. Chem. Soc. (A)*, 1971, 2929.
[471] E. O. Fischer and C. Elschenbroich, *Chem. Ber.*, 1970, **103**, 162.
[472] J. Deberitz and H. Nöth, *Chem. Ber.*, 1970, **103**, 2541.
[473] H. O. van Oven and H. J. de Liefde Meijer, *J. Organometallic Chem.*, 1971, **31**, 71.
[474] J. Müller and B. Mertschenk, *J. Organometallic Chem.*, 1972, **34**, C41.
[475] H. W. Wehner, E. O. Fischer, and J. Müller, *Chem. Ber.*, 1970, **103**, 2258.
[476] M. R. Churchill, F. R. Scholer, and J. Wormald, *J. Organometallic Chem.*, 1971, **28**, C21.
[477] G. E. Herberich and H. Müller, *Angew. Chem.*, 1971, **83**, 1020; *Angew. Chem. Internat. Edn.*, 1971, **10**, 937.
[478] G. Schmid, *Chem. Ber.*, 1970, **103**, 528.
[479] J. W. Howard and R. N. Grimes, *Inorg. Chem.*, 1972, **11**, 263.
[480] L. F. Warren and M. F. Hawthorne, *J. Amer. Chem. Soc.*, 1970, **92**, 1157.
[481] E. O. Fischer and H. J. Kollmeier, *Angew. Chem.*, 1970, **82**, 325; *Angew. Chem. Internat. Edn.*, 1970, **9**, 309.
[482] E. O. Fischer, E. Winkler, C. G. Kreiter, G. Huttner, and B. Krieg, *Chem.*, 1971, **83**, 1021; *Angew. Chem. Internat. Edn.*, 1971, **10**, 922.
[483] K. Öfele and M. Herberhold, *Angew. Chem.*, 1970, **82**, 775; *Angew. Chem. Internat. Edn.*, 1970, **9**, 739.
[484] R. Greatrex, N. N. Greenwood, I. Rhee, M. Ryang, and S. Tsutsumi, *Chem. Comm.*, 1970, 1193.
[485] D. J. Cardin, B. Cetinkaya, M. F. Lappert, L. Manojlovic-Muir, and K. W. Muir, *Chem. Comm.*, 1971, 400.
[486] G. Chandra, A. D. Jenkins, M. F. Lappert, and R. C. Srivastava, *J. Chem. Soc. (A)*, 1970, 2550.
[487] M. Kilner and C. Midcalf, *J. Chem. Soc. (A)*, 1971, 292.

porphyrin IX dimethyl ester complexes of rhenium carbonyl[488] and nitrosylruthenium;[489] a dehydroguanidine-$Fe_2(CO)_6$ complex.[490]

Phosphorus Donors. $(\pi\text{-}C_5H_5)Mn(CO)_2(PPhH_2)$;[491] tertiary phosphine derivatives of $(\pi\text{-}C_5H_5)Mn(CO)_3$;[492] $C_3H_7COMn(CO)_3[P(OC_3H_7)_3]_2$;[493] iron and ruthenium carbonyl complexes containing olefinic tertiary phosphines;[494] iron carbonyl complexes of $Me_2P(CH_2)_2SiR_3$ (R = Me or F);[495] complexes of the type $(\pi\text{-}C_5H_5)Ni(RC_2R^1)(PPh_2)Fe(CO)_3$ (R, R^1 = H, Me, Ph, CO_2Me);[496] $Ni(PMe_3)_4$;[497] $H_2Fe_2(CO)_6[P(CF_3)_2]_2$;[498] $(\pi\text{-}C_5H_5)_2Fe_3(CO)_2[OP(CF_3)_2]_4$;[499] $Fe_2(PF_2)_2(PF_3)_6$ and $Ni(PF_3)_3(PH_3)$, $Ni(PF_2Cl)_4$;[500] phosphorus trifluoride substitution products of cyclohexadiene-$Fe(CO)_3$;[501] fluorocarbon phosphine or arsine complexes of ruthenium carbonyl,[502] and some analogous iron complexes.[503,504]

Arsenic, Antimony, and Bismuth Donors. Group VI carbonyl complexes of AsH_3 and SbH_3;[505] pentamethylcyclopenta-arsine complexes of Group VI carbonyls;[506] polymethylcyclopolyarsine complexes of manganese and iron carbonyls;[507] $Fe(CO)_4(AsC_6F_5)_2$, $Fe_3(CO)_9(PC_6F_5)_2$, and $Fe_2(CO)_6(PC_6F_5)_4$;[508] $Ph_3BiM(CO)_5$ (M = Cr, Mo, or W).[509]

Sulphur Donors. Cyclopentadienylmolybdenum dithiolene complexes;[510] $(\pi\text{-}C_5H_5)_2M(SPh)_2M'(CO)_4$ (M = Mo or W; M' = Cr, Mo, or W);[511] (thiazolidine-

[488] D. Ostfeld, M. Tsutsui, C. P. Hrung, and D. C. Conway, *J. Amer. Chem. Soc.*, 1971, **93**, 2548.
[489] T. S. Srivastava, L. Hoffman, and M. Tsutsui, *J. Amer. Chem. Soc.*, 1972, **94**, 1385.
[490] N. J. Bremer, A. B. Cutcliffe, and M. F. Farona, *Chem. Comm.*, 1970, 932; *J. Chem. Soc. (A)*, 1971, 3264.
[491] M. Höfer and M. Schnitzler, *Chem. Ber.*, 1971, **104**, 3117.
[492] A. J. Hart-Davis and W. A. G. Graham, *J. Amer. Chem. Soc.*, 1971, **93**, 4388.
[493] E. P. Ross, R. T. Jernigan, and G. R. Dobson, *J. Inorg. Nuclear Chem.*, 1971, **33**, 3375.
[494] M. A. Bennett, G. B. Robertson, I. B. Tomkins, and P. O. Whimp, *Chem. Comm.*, 1971, 341.
[495] J. Grobe and U. Möller, *J. Organometallic Chem.*, 1972, **36**, 335.
[496] K. Yasufuku and H. Yamazaki, *J. Organometallic Chem.*, 1972, **35**, 367.
[497] H. F. Klein and H. Schmidbaur, *Angew. Chem.*, 1970, **82**, 885; *Angew. Chem. Internat. Edn.*, 1970, **9**, 903.
[498] R. C. Dobbie and D. Whittaker, *Chem. Comm.*, 1970, 796.
[499] R. C. Dobbie, P. R. Mason, and R. J. Porter, *J.C.S. Chem. Comm.*, 1972, 612.
[500] P. L. Timms, *J. Chem. Soc. (A)*, 1970, 2526.
[501] J. D. Warren, M. A. Busch, and R. J. Clark, *Inorg. Chem.*, 1972, **11**, 452.
[502] W. R. Cullen and D. A. Harbourne, *Inorg. Chem.*, 1970, **9**, 1839.
[503] W. R. Cullen, D. A. Harbourne, B. V. Liengme, and J. R. Sams, *Inorg. Chem.*, 1970, **9**, 707.
[504] J. P. Crow, W. R. Cullen, J. R. Sams, and J. E. H. Ward, *J. Organometallic Chem.*, 1970, **22**, C29.
[505] E. O. Fischer, W. Bathelt, and J. Müller, *Chem. Ber.*, 1970, **103**, 1815; 1971, **104**, 986.
[506] P. S. Elmes and B. O. West, *Austral. J. Chem.*, 1970, **23**, 2247.
[507] P. S. Elmes and B. O. West, *J. Organometallic Chem.*, 1971, **32**, 365.
[508] P. S. Elmes, P. Leverett, and B. O. West, *Chem. Comm.*, 1971, 747.
[509] R. A. Brown and G. R. Dobson, *J. Inorg. Nuclear Chem.*, 1971, **33**, 892.
[510] T. A. James and J. A. McCleverty, *J. Chem. Soc. (A)*, 1970, 3308, 3318.
[511] A. R. Dias and M. L. H. Green, *Rev. Port. Quim.*, 1969, **11**, 61.

2-thione)M(CO)$_5$ (M = Cr, Mo, or W);[512] PhCS$_2$Mn(CO)$_4$;[513] MeCS$_2$Re-(CO)$_4$;[514] Et$_2$PS$_2$Re(CO)$_4$ and related compounds;[515,516] [Fe(CO)$_3$EPh]$_2$ (E = S, Se, or Te);[517] H$_2$Ru$_3$(CO)$_9$S;[518] Fe$_2$(CO)$_8$SO$_2$;[519] (C$_6$F$_5$S)$_2$Co$_2$-(CO)$_6$.[520]

Ferrocenes.—Fluoroferrocene;[521] ferrocenylfuran derivatives;[522] the supposed ferrocenylketimine, shown to be FcC(CN)=C(NH$_2$)CH$_2$Fc;[523] 1,2-terferrocenyl and related compounds.[524]

---

[512] D. DeFilippo, F. Devillanova, C. Preti, E. F. Trogu, and P. Viglino, *Inorg. Chim. Acta*, 1972, **6**, 23.
[513] E. Lindner and R. Grimmer, *J. Organometallic Chem.*, 1970, **25**, 493.
[514] E. Lindner, R. Grimmer, and H. Weber, *J. Organometallic Chem.*, 1970, **23**, 209.
[515] E. Lindner and K.-M. Matejcek, *J. Organometallic Chem.*, 1970, **24**, C57.
[516] E. Lindner and K.-M. Matejcek, *J. Organometallic Chem.*, 1971, **29**, 283.
[517] E. D. Schermer and W. H. Braddley, *J. Organometallic Chem.*, 1971, **30**, 67.
[518] A. J. Deeming, R. Ettore, B. F. G. Johnson, and J. Lewis, *J. Chem. Soc. (A)*, 1971, 1797.
[519] D. S. Field and M. J. Newlands, *J. Organometallic Chem.*, 1971, **27**, 221.
[520] G. Bor and G. Natile, *J. Organometallic Chem.*, 1971, **26**, C33.
[521] F. L. Hedberg and H. Rosenberg, *J. Organometallic Chem.*, 1971, **28**, C14.
[522] F. D. Popp and E. B. Moynahan, *J. Heterocyclic Chem.*, 1970, **7**, 351.
[523] P. L. Pauson and S. Toma, *Tetrahedron Letters*, 1971, 3367.
[524] S. I. Goldberg and J. G. Breland, *J. Org. Chem.*, 1971, **36**, 1499.

6
Computerized Data Acquisition and Handling

BY S. D. WARD

1 Introduction

In the two-year period covered by this report the use of computers in mass spectrometry has expanded considerably. Computerized acquisition of data is becoming the rule rather than the exception. Because the principles of on-line computerized data acquisition are well established and earlier problems with interfaces disposed of, it can be expected that this expansion will continue at an increasing rate. There is now ample evidence from the literature to show that the advantages of using a computer can out-weigh the additional expenditure necessary. As the hardware of mass spectrometer–computer systems becomes more standardized (for conventional mass spectrometer instrumentation) and the software[1] becomes readily available, it can also be expected that the cost of basic data acquisition packages will stabilize (or even fall!). The past two years have seen relatively few innovations in the instrumentation–data-acquisition field as covered by this report but rather a widening of the use of well-founded methods. In consequence, the section dealing with data acquisition, especially for photographic recording, contains rather fewer references than in Volume 1.

It was emphasized in the previous report that the vastly increased data-flow created by the use of computers for data acquisition from mass spectrometers could not in most cases be handled efficiently by conventional methods. The computer itself should be the answer and, indeed, work in this field has continued at a rapidly increasing rate.

Once again many papers referenced below are taken from the proceedings of important conferences on mass spectrometry (American Society for Testing and Materials meetings and the Triennial International Conference in Brussels in September 1970) and it may be of interest to note that 25% of all the references came from these sources, 24% came from the single journal *Analytical Chemistry* and the remainder came from other sources. Some references omitted in the last report have been included.

Broadly based review papers and books covering the use of computers in mass spectrometry have appeared.[2-9] Real-time computer applications in high-resolution mass spectrometry have been reviewed.[10] Specialized reviews and conference reports covering applications of computerized mass spectrometry

---

[1] The term 'software' refers to the programs which are required by the computer in order that it can be told the task it has to perform. The programs are usually fed into the computer initially on punched cards, paper, or magnetic tape.

in natural product research,[11] biochemical applications,[12] gas chromatography[13,14] polymer research,[15] and general organic chemistry[16—20] have been published. The future design of computerized instruments[21] and coupling of computers and mass spectrometers[22] have been discussed. A recent survey of computer applications in the laboratory by Perone[23] contains some useful references to mass spectrometric applications. The use of computerized mass spectrometry in geological research[24,25] and for investigation of planetary surfaces[26] has been reported.

2 Data Acquisition

The number of papers merely reporting conventional configurations of mass spectrometer–computer link-ups has decreased somewhat in the current two-year period relative to that covered by the last report. The causes of this have been

[2] A. L. Burlingame and G. A. Johanson, *Analyt. Chem.*, 1972, **44**, 337R; D. C. De Jongh, *ibid.*, 1970, **42**, 169R.
[3] F. Erni and J. T. Clerc, *Chimia*, 1970, **24**, 388.
[4] 'Spectral Analysis: Methods and Techniques', ed. J. A. Blackburn, Marcel Dekker, New York, 1970, Chap. 7, p. 235.
[5] F. W. McLafferty, *Adv. Mass Spectrometry*, ed. A Quayle, Institute of Petroleum, London, 1971, Vol. 5.
[6] 'Recent Topics in Mass Spectrometry', ed. R. I. Reed, Gordon and Breach, New York, 1971.
[7] 'Mass Spectrometry: Techniques and Applications', ed. G. W. A. Milne, Wiley–Interscience, New York, 1971.
[8] 'Topics in Organic Mass Spectrometry', *Adv. Analyt. Chem. and Instr.*, ed. A. L. Burlingame, Wiley–Interscience, New York, 1970, Vol. 8.
[9] 'Recent Developments in Mass Spectroscopy', Proceedings of International Conference, Kyoto, Japan, 1969, ed. K. Ogata and T. Hayakawa, Univ. Park Press, Baltimore, London, and Tokyo, 1970.
[10] T. O. Merren and R. W. Olsen, in 'Computers in Analytical Chemistry', *Progr. Analyt. Chem.*, Plenum Press, New York and London, 1970, Vol. 4, p. 17.
[11] P. Vallon and R. Erickson, *Drug Cosmetic Ind.*, 1970, **107**, 44; *Soap Chem. Specialities*, 1970, **46**, 44.
[12] 'Biochemical Applications of Mass Spectrometry', ed. G. R. Waller, Wiley, New York, 1972, p. 1136.
[13] 'Ancillary Techniques of Gas Chromatography', ed. L. S. Ettre and W. H. McFadden, Wiley–Interscience, New York, 1969, p. 145.
[14] Proceedings of International Symposium on Chromato-Mass Spectrometry, Moscow, May 1968, Inst. Chem. Phys. Akad. Nauk SSR, Moscow, 1969.
[15] J. Mitchell and Jen Chiu, *Analyt. Chem.*, 1971, **43**, 267R.
[16] R. Masot and J. Ulrick, *Inter-Electronique*, 1970, **6**, 34.
[17] A. Tatematsu, *Japan Analyst*, 1970, **19**, 983.
[18] W. Koegler, *Glas-Instrum. Tech. Fachzlab*, 1970, **14**, 6.
[19] A. Herlan, *Z. analyt. Chem.*, 1971, **253**, 1.
[20] H. A. Vant Klooster, *Chem. Weekblad*, 1971, **67**, 9.
[21] J. W. Fraser, Proceedings of the 1968 IBM Scientific Computer Symposium on Computing in Chemistry, 1969, p. 265.
[22] D. Stahl, *Chimia*, 1971, **25**, 149.
[23] S. P. Perone, *Analyt. Chem.*, 1971, **43**, 1288.
[24] R. D. Russell, J. Blenkinsop, R. D. Meldrum, and D. L. Mitchell, *Mass Spectrometry (Japan)*, 1971, **19**, 19.
[25] J. S. Stacey, E. E. Wilson, Z. E. Peterman, and R. Terrazas, *Canad. J. Earth Sci.*, 1971, **8**, 371.
[26] K. Bieman, Chemistry Institute of Canada and American Chemical Society Joint Conference, Toronto, May 1970, Abstr. ANAL 17.

mentioned in the Introduction. There have been some reports of larger system configurations using time-shared computers which are either general purpose computers or smaller systems dedicated to the laboratory environment. Despite the often greater complexity of new systems it is still possible to assign them to one or more of the basic classes of configuration described in this section of the previous report and reproduced in Figure 1 for reference.

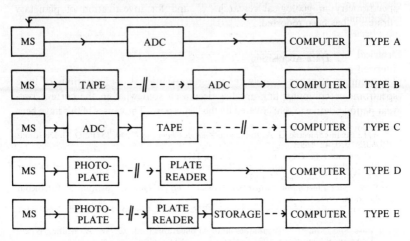

Figure 1 *Configurations of mass spectrometer–computer systems*

A computer–machine interface unit suitable for mass spectrometric data acquisition systems has been described by Page and Nowak.[27] The interface in any system is of critical importance and can often be the cause of long periods of delay in commissioning of new mass spectrometer data acquisition systems. The interface in a system is very often a special piece of electronic engineering designed for a particular mass spectrometer and computer link-up and as such it cannot be expected to be as reliable or fault-free as the computer hardware itself which has benefited from extensive research, experimentation, technical experience, and, above all, extensive use under real-run conditions. It is important therefore that some effort is directed towards standardizing where possible the features of mass spectrometer–computer interfaces. The paper[27] referred to above describes an interface for real-time acquisition of either electron multiplier signals or the output signals from a photoplate reader for high-resolution mass spectrometry. The design philosophy of the interface is such that it is machine independent. A crystal-controlled oscillator and frequency divider are used to trigger a high-speed 14-bit + sign analog-to-digital converter (ADC) with a sample-and-hold amplifier at rates from 312 Hz to 80 kHz for electron multiplier signals. For the photoplate reader, triggering is provided at 0.5 micron intervals

[27] R. P. Page and A. V. Nowak, Proceedings of the 19th Annual Conference on Mass Spectrometry and Allied Topics, Atlanta, 1971, p. 84.

by a shaft encoder. A further design for a mass spectrometer–computer interface has appeared.[28]

A significant advance in the data acquisition and instrumentation field in the past two years has been the advent of the commercially available double-beam electrical recording mass spectrometer.[29] This may have wide implications in several aspects of data acquisition and is discussed further in the g.c.–m.s., electrical recording, and reference compound sections.

Optimum sampling and smoothing conditions for digitally recorded spectra have been discussed.[30]

Electrical Recording.—*System Configurations and Design.* The first computerized mass spectrometer systems were more often than not concerned with high-resolution data, but it was soon realized that routine acquisition of low-resolution data, especially when in conjunction with gas chromatography, is potentially just as important if not more so in certain circumstances. The techniques and the software[1] required for low-resolution data acquisition differ from those required at high resolution, and it is for these reasons that development of low-resolution software has often lagged behind.

An extensive low-resolution system has been described by Binks and co-workers.[31] The system configuration is a combination of types A and B (see Figure 2), on-line operation being supplemented by the inclusion of a tape

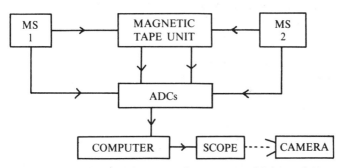

Figure 2 *A combination of system configurations A and C from ref. 31*

recorder. This enables the system to handle data from two mass spectrometers (which can be dealt with separately on-line) simultaneously, one on-line and the other by recording the data on magnetic tape. Another feature of the system is the oscilloscope, which can be used for rapid presentation of results and other information. The computer used in this system has available an eight-channel ADC of 9-bit precision, and by dividing the ion-current output from the mass

---

[28] D. L. Mitchell and J. Blenkinsop, *Trans. Amer. Geophys. Union U.S.A.*, 1971, **52**, 433.
[29] M. Barber, J. R. Chapman, B. N. Green, and T. O. Merren, *Adv. Mass Spectrometry*, ed. A. Quayle, Institute of Petroleum, London, 1971, Vol. 5.
[30] J. P. Porchet and Hs. H. Gunthard, *J. Sci. Instr.*, 1970, Ser. 2, **3**, 261.
[31] R. Binks, R. L. Cleaver, J. S. Littler, and J. MacMillan, *Chem. in Britain*, 1971, **7**, 8.

spectrometers into six channels with an intensity ratio of four between each and then digitizing each channel separately, intensity measurements could be made to better than 1% over a dynamic range of 4096:1. The digitization is under program control and genuine mass peaks are distinguished from noise by application of a threshold, together with the fact that peaks must be within certain limits of width and must be a minimum number of samples apart. The signal maxima and the time at which they occur are stored, the time reference being obtained from a crystal clock. The more satisfactory methods of calculation of peak areas and centroids could not be used because of the time-consuming software routine required for the necessary division operation. At the end of a mass spectrometer scan the data for the scan (maximum 1000 peaks) are transferred to digital magnetic tape while the magnet is resetting. Up to 63 further scans can be made immediately following each other, thus allowing rapid collection of spectra during g.c.–m.s. runs. The conversion of the time $vs.$ intensity matrix into the mass spectrum is achieved by the standard look-up table method in which the peak times are compared with values in a calibration table containing the time intervals between all consecutive mass numbers. Such tables usually hold good for a particular mass spectrometer for several weeks or even months without recalibration. Mass numbers cannot be assigned, however, until a starting point has been identified. It is usually possible to apply some unambiguous check to identify m/e 17, 18, 28, 32, 40, and 44 which are always present as background, and this method was used in this case. Once the mass spectra have been determined and normalized they are returned to magnetic tape and can be incorporated into a library system. Overall, this system illustrates how a powerful data acquisition configuration can be built up round a very small (12-bit, 4K words of core store) computer.

Another system designed for the rapid scanning required for g.c.–m.s. combinations and built around a PDP 8/I computer with 8K of 12-bit word core store has been described by Sweeley et al.[32] An ever-increasing number of systems are taking advantage of the greater flexibility offered by the inclusion of magnetic disc and digital magnetic tape backing stores and this system[32] is a good example. The configuration is basically of type A but includes several additional items of equipment which are connected to the computer via a common input–output (I/O) highway[33] (see Figure 3). As in the previous system described,[31] unprocessed scan data are transferred to a backing store immediately after the end of the scan thus enabling repetitive rapid scanning. According to the number of peaks in each scan, 31—62 scans could be held on the backing store (in this case a magnetic disc). In this system the magnetic field of the mass spectrometer was also sampled during the scan by use of a Hall effect probe. The scans were initiated by computer control and the output of a selected intensity channel of the ion current from the mass spectrometer digitized at 25 kHz. When the summit of a peak was detected

[32] C. C. Sweeley, B. D. Ray, W. I. Wood, J. F. Holland, and M. I. Krichevesky, *Analyt. Chem.*, 1970, **42**, 1505.

[33] 'Highway' is a term for a special type of connection line to a computer. Transfer of control signals and data between the computer and several items of peripheral equipment attached to a single highway connection is possible.

the input multiplexor was switched under computer control so that the output of the Hall effect probe was digitized. Five samples were taken and the median of these values stored along with the maximum peak height before the multiplexor was switched back to the ion-current output channel. The scan matrix stored on the disc at the end of the scan was thus one of intensity *vs.* Hall effect voltages which are proportional to the magnetic field. At this stage the operator could initiate a new scan, or process the current or previous scans. To convert the scan matrix into the mass spectrum a relationship between the Hall effect

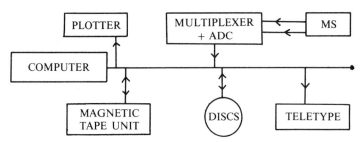

Figure 3 *System configuration of type A with computer input/output highway, from ref. 32*

voltages and exact mass was required. This in turn required a calibration procedure in which the computer was given the exact masses of about 100 peaks in the spectrum of a reference compound (*e.g.* perfluorokerosine). The matrix of Hall effect voltages and corresponding masses was then used to construct an interpolation table and a table of cubic coefficients for use in the interpolation. In processing scan data stored on the disc, the Hall effect voltages were brought down into core one at a time and converted into masses using the cubic coefficients by an interpolation procedure due to Pennington.[34] The next stages involved conversion of exact masses into nominal integer masses and storage of the processed spectrum on magnetic tape. Various options were then available to the operator including spectrum averaging and spectrum subtraction, both important in the processing of spectra from g.c.–m.s. runs. The system was capable of scan speeds up to 0.8 s per mass decade at the 25 kHz sampling rate and processing the scan matrix for about 300 peaks took about 8 s. The mass spectra could be printed on the printer or plotted on an incremental plotter.

Using medium-sized process control oriented computers, in a time-shared environment, for acquiring and processing mass spectrometric data is often an attractive alternative proposition to the dedicated computer. Such a system in which an IBM 1800 computer is used to acquire data from X-ray diffraction, i.r., and Raman spectrometers as well as a mass spectrometer, and also to handle batch processing schedules, has been reported.[35] By utilization of the interrupt

[34] R. H. Pennington, 'Introductory Computer Methods and Numerical Analysis', Macmillan Co., New York, 1965, p. 404.
[35] W. F. Haddon, D. R. Black, and R. H. Elsken, Proceedings of the 18th Annual Conference on Mass Spectrometry and Allied Topics, San Francisco, 1970, p. B450.

and priority structures available with this and similar computers, simultaneous and asynchronous operation of the various spectrometers was achieved. Although the computer was remote from the mass spectrometer, it appeared to the operator of the mass spectrometer that the computer was dedicated to his use. High-resolution mass spectra were obtained at a scan speed of 100 per decade and resolution of 10,—20 000 utilizing the maximum 18.5 kHz digitization rate of the IBM analog input system. The mass spectrometer ion-current was digitized, transferred to the computer core store, and after being thresholded was transferred to magnetic tape. A few seconds after the end of the scan the peak centres had been calculated and the computer fed back information to the user *via* a cathode ray tube (CRT) display giving details of the number of valid peaks in certain mass ranges and the number of peaks exceeding the dynamic range of the ADC. The advantage of rapid feedback which is available with dedicated on-line systems was therefore still maintained with this time-shared system. One of the advantages of using a larger computer in a time-shared mode is that more costly items of equipment such as line-printers and large disc backing stores, which are often included with these computer systems, become directly accessible to the mass spectrometrist. In this system[35] the disc was used to store various programs used to process the raw peak centroids *vs*. intensity data. Perfluorokerosine (PFK) was used as an internal standard and standard masses were taken six at a time and fitted using polynomial regression routines. The first six reference peaks were recognized by their standard time differences and intensities. The masses of non-reference peaks were found by interpolation. Other programs were available on the disc for magnetic tape handling and element-mapping with output on the line-printer.

Use of an IBM 1800 computer in a time-shared mode with a quadrupole mass spectrometer has also been reported.[36] Sampling rates of 20 kHz down to 20 Hz and routines for spectrum plotting, data smoothing, peak location, and mass identification were available.

A further step up in the complexity of time-shared systems is illustrated[37] by a system in which a mini-computer acted as a time-shared interface and pre-processing unit between a mass spectrometer, a photoplate comparator, several items of I/O equipment, and a large IBM 360/50 computer which itself operated in a time-shared environment. The large computer was mainly used for transferring data between disc files and the mini-computer and for running applications programs.

Another large system[38] using a PDP10 computer with 32K of 36-bit words of core store as the centre of a time-shared laboratory computer system could handle up to 32 'slow' instruments and 8 'fast' instruments, the instrument 'speed' relating to the digitization rate required. Despite its complexity (see Figure 4)

---

[36] D. L. Raimondi, H. F. Winters, P. M. Grant, and D. C. Clarke, *IBM J. Res. Development*, 1971, **15**, No. 4.

[37] R. N. Stillwell, J. G. Leferink, N. R. Earle, and L. S. Wascho, Proceedings of the 19th Annual Conference on Mass Spectrometry and Allied Topics, Atlanta, 1971, p. 93.

[38] E. Ziegler, D. Henneberg, and G. Schomberg, *Analyt. Chem.*, 1970, **42**, 51A.

the system is still basically of type A. The 'slow' ADC could handle instruments requiring digitization rates up to 20 kHz such as fast-scan low-resolution or slow-scan high-resolution mass spectrometers. Although a multiplexor connected eight channels to the 'fast' ADC, the time-sharing software of the system dictated that only one fast instrument, such as a fast-scanning high-resolution mass spectrometer, be allowed to transmit data at any given time. This is, in fact, quite a common restriction in even the largest multi-instrument handling systems. The organization of the software was such that up to 15 jobs could be run concurrently by having the 15 relevant programs in the core store at the same time. When core-store space was insufficient programs could be shuffled in core and transferred to and from the disc as required. One 'job program' handled data

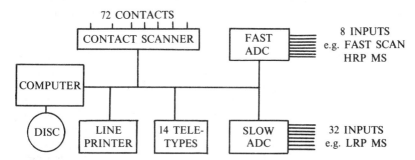

Figure 4 *Configuration of a multi-experiment laboratory computer system from ref. 38*

acquisition from all 'slow' instruments but a separate program was used for each fast instrument. When a fast instrument scan request was made the relevant program was brought into core to accept the data, and this program stayed in core as long as the instrument was running (a few seconds), and during this time no other fast instrument scan requests were allowed. Even with a fast instrument requiring a 5 kHz data rate, 97% of central processor time was still available for other jobs. The time between samples was 200 μs (rate $\equiv$ 5 kHz), the transfer time from the ADC to core was 6 μs leaving 194 μs for other tasks. A problem arose, however, in acquiring data from a g.c.–m.s. combination where the response time from a request for a scan and commencement of the scan was too long for the fast narrow peaks of the gas chromatogram produced from a high-resolution capilliary column. This problem was solved by a modification to the time-sharing monitor software. The total ion-current monitor of the mass spectrometer (representing the gas chromatogram) was sampled *via* one channel of the slow ADC and analysis of this data was used to initiate successive mass spectral scans between the beginning and the end of a g.c. peak. This represents an example of real-time data processing as well as real-time data acquisition. The system was also used for off-line computational work for an analytical chemistry laboratory. A further advantage of larger computer systems is emphasized by the authors.[38] High-level languages (such as FORTRAN) can be used for producing new programs and the editing, debugging, and testing facilities available on larger systems make program development a much simpler task. When it is considered that manufacturers' special software development costs can be extremely high it becomes very desirable that users can develop their own software. Program development on this system could be carried out as one of the time-shared jobs without interrupting data collection.

Other workers have reported[39] on a simple method of acquiring data from a quadrupole mass filter using two conventional audio tape decks. This system of type C is in complete contrast to the large time-shared system described above and represents the other end of the scale of complexity. Spectra were digitized by a voltage-to-frequency converter and recorded on one track of the stereo tape deck while on the other track timing pulses from a counter were recorded. The tape was played back at twice the recording speed on an identical tape deck interfaced to a small computer. Spectra could be averaged, smoothed, and printed on a teletype or displayed on an oscilloscope.

The trend towards completely integrated data acquisition and data processing systems is illustrated by an on-line system[40] of type A which has been used successively for over two years for analysis of multi-component mixtures of groups of compounds encountered in the petroleum industry. Low-resolution data were acquired by a medium-size computer in an on-line environment. The computer sampled the ion current and the accelerating voltage of the mass spectrometer at 250 Hz and also the magnetic field of the main magnet at the beginning and the end of each scan. Voltage scanning of the spectrum was used. By use of a central control panel the system operator could initiate various functions, including calibration of accelerating voltage with mass number, scan acquisition, printing of integral mass spectra, or analysis routines for mixtures of different groups of compounds such as petroleum naphthas. A low electron-beam voltage was used for recording the spectra of some of the aromatic mixtures. The software of the system was designed so that as more analysis routines were developed they could be integrated into the system and called into use by the operator as required. A disc backing store was used to store all the programs.

In an extensive paper, McLafferty and co-workers[41] have described an on-line system using a small (4K) computer, with a disc backing store and a line-printer, for use with a high resolving-power double-focusing mass spectrometer. The paper also includes an enlightening discussion of practical as well as theoretical considerations in the choice of such a system.

The usefulness of CRT display facilities in a mass spectrometer–computer system has been demonstrated for a low-resolution g.c.–m.s. combination.[42] A spectrum could be displayed after completion of the scan, and the displays were expandable so that small regions of the spectra could be examined in more detail.

[39] N. A. Jones, R. D. Friesen, and J. W. Pyper, *Rev. Sci. Instr.*, 1970, **41**, 1828.
[40] H. E. Lumpkin and J. H. Harding, Proceedings of the 18th Annual Conference on Mass Spectrometry and Allied Topics, San Francisco, 1970, p. B458.
[41] R. J. Klimowski, R. Venkataraghaven, and F. W. McLafferty, *Org. Mass Spectrometry*, 1970, **4**, 17.
[42] W. F. Holmes, W. H. Holland, and J. A. Parker, *Analyt. Chem.*, 1971, **43**, 1806.

The system used a Hall effect probe to establish the mass scale and allowed magnetic tape storage of spectra which could then be retrieved for display, plotting, or further processing.

It has been shown[43] that in a simple quadrupole mass spectrometer–computer system, by suitable design of the software so that the computer and the mass spectrometer operator could assist each other, operations corresponding to those carried out by much larger computer systems with far more complex programs could be performed.

Other electrical recording systems have been reported, including a total laboratory system,[44] low-resolution systems,[45–48] a space-craft system,[49] and combined high- and low-resolution systems.[50,51] In one of the latter papers[51] a detailed description of a unified approach to the system software is also given.

A method for the digitization of mass spectra due to Carrick[52] was reported in Volume 1. The instrumentation required for this approach, which is based on the identification of peak maxima by real-time differentiation of the ion current, is now commercially available. The advantages of the system remain its relative cheapness and the fact that its digitized output can be accepted by any available device such as a paper tape punch, a digital magnetic tape recorder, or a computer. The method has now been extended to deal with intensity measurements and a new report on its performance in various systems at both high and low resolving power has appeared.[53]

A method for peak-matching by computer in order to obtain atomic mass differences has been reported,[54] and peak-area integration using a small computer in conjunction with the peak-switching device available with double-focusing mass spectrometers has been described.[55]

Accuracy of Mass and Intensity Measurement. A detailed examination of sources of error in high-resolution data acquisition systems has been presented by McLafferty and co-workers.[41] Errors are identified as arising from three sources, the mass spectrometer, the analog system and interface, and the computer. Mass spectrometric errors may be due to mechanical vibrations or electronic

[43] P. Ronteix, *Scientific and Technical Aerospace Reports*, 1970, **8**, 1254.
[44] H. Guenzler, *Chem. Ing. Tech.*, 1970, **42**, 877.
[45] R. Knutti and R. E. Bühler, *Chimia*, 1970, **24**, 437.
[46] G. R. Waller, H.-Y. Li, K. Kinneberg, R. Saunders, D. Simpson, and L. Mills, *Proc. Okla. Acad. Sci.*, 1970, **50**, 19.
[47] D. Bell and R. B. Stewart, *Scientific and Technical Aerospace Reports*, 1971, **9**, 684.
[48] S. W. Downer, G. R. Smith, and B. R. Willoughby, *Appl. Spectroscopy*, 1971, **25**, 149.
[49] R. J. Leite, B. B. Hinton, and C. J. Mason, *Scientific and Technical Aerospace Reports*, 1970, **8**, 674.
[50] P. Powers and M. J. Wallington, Proceedings of the 19th Annual Conference on Mass Spectrometry and Allied Topics, Atlanta, 1971, p. 88.
[51] D. H. Smith, R. W. Olsen, E. C. Walls, and A. L. Burlingame, *Analyt. Chem.*, 1971, **43**, 1796.
[52] A. Carrick, *Internat. J. Mass Spectrometry Ion Phys.*, 1969, **2**, 333.
[53] A. Carrick, *Adv. Mass Spectrometry*, ed. A. Quayle, Institute of Petroleum, London, 1971, Vol. 5, p. 330.
[54] J. O. Meredith, R. C. Barber, and H. E. Duckworth, *Bull. Amer. Phys. Soc.*, 1970, **15**, 769.
[55] N. M. Frew and T. L. Isenhour, *Analyt. Chem.*, 1972, **44**, 659.

ripple or noise, all of which affect the shape and stability of the mass peaks at the collector. In the analog system the bandwidth of the signal is all important. The dynamic range and the extent to which digitization distorts the peak shape are both dependent on the saturation limit of the ADC and the amplification system used to present the signal to the ADC. Although digital calculations can be exact, in processing the data from the mass spectrometer various algorithms, which define the digital calculations, are followed in the compilation and these algorithms are often approximations. Calculation of peak positions by the centroid method, the contribution of noise to observed errors, and an extrapolation procedure for reference peak identification are also considered. The overall mass measurement accuracy of the system studied in this paper[41] was estimated at about 12 p.p.m. at 10 000 resolving power, and it was concluded in general that, for scan speeds giving equivalent signal bandwidths, increased instrument resolving power gives increased accuracy of peak-centre determination for the same sample flow rate.

Smith et al.[51] have also made a detailed study of sources of error in a high-resolution system, questioning the conclusions of the previously mentioned paper[41] on the grounds that the standard deviation of the mass measurements found by McLafferty (38.5 p.p.m. at 10 000 resolving power) is an order of magnitude larger than that obtained by them under similar conditions. These workers[51] conclude that, in the resolving power range 2500—30 000, mass measurement accuracy is a function of the product of resolution and sensitivity and not dependent on resolution *per se* to any important extent.

One of the standard procedures in the data acquisition of low-resolution mass spectral data from an electrical recording magnetic deflection instrument is to convert a time *vs.* intensity matrix into a mass spectrum, using data obtained from a calibration spectrum of a known compound to relate the time and mass scales. Early workers in this field assumed that a continuously cycling magnet sweep of about 5 s period was required to establish the necessary reproducible time *vs.* mass relationship.[56] Recently, however, other workers have reported[57] that idling the magnet at full current produces conditions resulting in reproducible scans from high current (high m/e) to low current (low m/e) for an AEI MS12 mass spectrometer. The data acquisition system was of a conventional type A configuration. The scans and digitization were started synchronously and a signal-averaging program was used to test the scan reproducibility. A doublet requiring a resolution of 2000 to separate the components was repetitively scanned 100 times and the signal averaged without significantly destroying the resolution. The scan law of the mass spectrometer was taken as

$$t = K_1 \ln(m/e^{\frac{1}{2}}) + K_2$$

The mass scale was calibrated using perfluoroalkane (PFA), which has regularly spaced peaks of an intensity pattern which could be automatically recognized by the computer. The above scan law was not found to deviate from linearity when

[56] See, for instance, R. A. Hites and K. Biemann, *Analyt. Chem.*, 1968, **40**, 1217.
[57] J. R. Plattner and S. P. Markey, *Org. Mass Spectrometry*, 1971, **5**, 463.

a variety of scan conditions were used. In the calibration procedure all reference peaks were identified by an extrapolation procedure and assigned an accurate $\ln(m/e^{\frac{1}{2}})$ value. Scans of unknown compounds were then directly interpolated against the run of the calibration compound to assign the m/e values. When ten 15 s scans of PFA were acquired and interpolated against an eleventh scan, the standard deviation of the measured mass against actual mass of a variety of peaks averaged about 0.03 a.m.u., varying from 0.006 a.m.u. for m/e 31 to 0.051 for m/e 793. The results were equally encouraging for the repetitive scan mode with a variety of cycle times. By using an internal standard of PFA with an unknown compound in the mass spectrometer and scanning the spectrum at both 8 kV accelerating voltage and a lower accelerating voltage, these workers[57] were able to interpolate the scan of the mixture obtained at low accelerating voltage against that obtained at 8 kV for PFA and then, by applying an accurate multiplication factor, assign m/e values up to m/e 4000 with an accuracy of 500 p.p.m. (*e.g.* 1.5 a.m.u. at m/e 3628). The problems here are with the mass spectrometer sensitivity not the data system. A practical useful limit of m/e 2000 was given by the authors. For mass value interpolation within the mass range covered by the reference compound as an external standard the mass scale accuracy was consistently about 200 p.p.m.

In another low-resolution system[32] (see p. 268) in which Hall effect voltages were used in an interpolation procedure to assign m/e values, it was found that variation in exact mass over ranges of the mass scale was up to 0.3 a.m.u. The exact mass minus nominal mass relationship was not as smooth nor its amplitude as small as would be expected solely on the basis of the mass defects for the compounds under study. This was partly attributed to drift of voltages and the magnetic field of the system. If calibration runs for the Hall effect voltages were not carried out frequently drift errors could accumulate to 1 or 2 a.m.u. at m/e 500. Algorithms for applying corrections to the calculated masses to avoid these cumulative errors were investigated. A method in which the median of the divergence of a set of successive mass peaks was used to calculate the average rate of drift was considered to be the most successful.

McMurray and Lipsky have investigated[58] in detail the levels of accuracy to which mass peak abundances can be measured by a data acquisition system when recording mass spectra of samples introduced *via* a direct inlet system into the ion source of a magnetic scanning mass spectrometer.

A method for improving the signal-to-noise ratio in mass spectrometry and thereby improving the accuracy of measurement of both mass and intensity has been proposed.[59] The method is based on a signal multiplexing technique previously described for optical spectroscopic methods. One of the problems of obtaining precise mass measurements from very small quantities of material using a data acquisition system is that an instrument resolution at least sufficient to

---

[58] W. J. McMurray and S. R. Lipsky, Proceedings of the 19th Annual Conference on Mass Spectrometry and Allied Topics, Atlanta, 1971, p. 74.
[59] C. J. Eckhardt and M. L. Gross, *Internat. J. Mass Spectrometry Ion Phys.*, 1970, **5**, 223.

resolve the internal reference compound peaks (usually PFK) from the mass peaks of the sample must be used, despite the fact that the spectroscopist may wish to reduce the resolution to obtain better sensitivity needed for samples at the nanogram level. This can be particularly evident in g.c.–m.s. work. Campbell and Halliday have shown previously[60] that mass measurements from electrical recording mass spectrometers can still be of useful accuracy at resolutions of less than 10 000 and even down to 1000. The advent of the double-beam mass spectrometer allows a solution to the above problem. Workers from the mass spectrometer manufacturers[61] have shown[62] that a scan of a reference compound in one source/beam of the mass spectrometer can be used to calibrate a simultaneous scan of an unknown compound (which could be a g.c. effluent) in the second source/beam of the mass spectrometer. The mass scales of the two beams were off-set by 2000 p.p.m., and the signals from both beams were added and fed to the data acquisition system. Since the mass scales of the reference compound spectrum and the sample spectrum were artificially separated by a known amount, there was no problem about resolving the mass peaks of the reference compound from those of the sample when they occurred at the same nominal mass. Any resolution could be used for each of the two beams. As an alternative to the off-set method, a two-channel data acquisition system could be used. Six 10 s per decade scans of PFK and hexachlorobutadiene were digitized at 12 kHz and 1000 resolution. Two scans were used to determine the mass off-set between the two channels, and for the other four the PFK scan was used as a reference compound for the hexachlorobutadiene. The mean r.m.s. error in the measured masses of eight of the largest peaks was 0.8 m.m.u. and the individual errors varied from −3.2 m.m.u. to +3.1 m.m.u., with 68% of the results better than 1.5 m.m.u. The accuracies achievable at 10 000 and 1000 resolution for various amounts of sample were calculated and given as shown in Table 1. It can thus be seen that half the

Table 1 *Some results from ref. 62 illustrating the relationship between mass measurement accuracy, sample quantity, and resolution*

Resolution	Sample quantity/ng	Mass measurement accuracy/p.p.m.	Intensity range
10 000	400	<5	10:1
1000	200	<5—50	1000:1
1000	20	5—15	10:1

amount of sample is required at 1000 resolution relative to 10 000 resolution to achieve essentially the same mass measurement accuracy and furthermore the intensity range of observable peaks at the lower resolution is 100 times greater.

---

[60] A. J. Campbell and J. S. Halliday, Proceedings of the 13th Annual Conference on Mass Spectrometry and Allied Topics, St. Louis, 1965.
[61] AEI Scientific Apparatus Ltd., Urmston, Manchester, England.
[62] M. Barber, J. R. Chapman, B. N. Green, T. O. Merren, and A. Riddock, Proceedings of the 18th Annual Conference on Mass Spectrometry and Allied Topics, San Francisco, 1970, p. B299.

An inherent disadvantage in computerized data acquisition from electrical recording mass spectrometers is the fact that, during the scan, data relevant to the actual mass peaks is collected only during a fraction of the total scan time. This problem can be overcome and scan times thus reduced in the case of computer-controlled low-resolution quadrupole mass spectrometers (see p. 293), but in the case of high-resolution magnetic-scanning mass spectrometers the same approach is not possible. McLafferty[63] has proposed, however, that the time between peaks could be utilized to rescan the peak (or group of peaks) just scanned and thus increase the signal-to-noise ratio by averaging. The accuracy of mass and intensity measurement would also thereby be improved. To perform this rescanning it would be necessary to change by a small increment the ion-accelerating voltage or the electrostatic analyser voltage, or to introduce a small opposing magnetic field by an auxiliary scan coil (as is used for peak scanning in the peak matching process). It is suggested[63] that changes to the electrostatic analyser voltage under on-line computer control is the most practical approach.

Photographic Recording.—*System Configuration and Design.* Nearly all the papers published in the two-year period under review and concerning data acquisition from photoplates have dealt with spark source mass spectrometry, and this perhaps is tending to reflect the relative importance of electrical and photographic recording in modern organic mass spectrometry. Many of the principles relevant to spark source data acquisition, however, are also relevant to data acquisition from photoplates produced by conventional electron-impact ion sources, and therefore some general points will be discussed here.

One of the drawbacks of photographic recording is the long data turn-round time and it is here where computer techniques have helped. A system of type D developed with this problem in mind has been reported.[64] The system used a small computer on-line to a microdensitometer. The computer was also equipped with a disc backing store, a magnetic tape unit, and a storage display oscilloscope. The output of the photomultiplier of the microdensitometer, after being fed through an operational amplifier for impedance matching, was fed *via* 40 feet of cable to the ADC of the computer. The signal was digitized at 450 μs intervals under program control using the interrupt facility of the computer. The digital signals were stored sequentially in one of two 512 word buffers in the core store of the computer. When one buffer was full, values were stored in the other buffer while the contents of the full buffer were written to magnetic tape. The end of the scan could be determined either by the operator or by a preset limit switch on the microdensitometer. The photoplates (15 in) were scanned in 75 s and the digitization rate represented a resolution of 2.5 μm at this speed. Files holding spectra on the magnetic tape could be labelled by the operator. One tape could hold data from six complete 15 in scans and required 7.5 min of operating time. In an automatic data-reduction phase the tape was rewound and the data processed

[63] F. W. McLafferty, R. Venkataraghaven, J. E. Coutant, and B. Giessner, *Analyt. Chem.*, 1971, **43**, 967.
[64] G. Lauer, R. A. Meyer, and R. S. Carpenter, Proceedings of the 18th Annual Conference on Mass Spectrometry and Allied Topics, San Francisco, 1970, p. B445.

to give line positions and intensities as output on paper tape, as a printout on a teletype, or as a display on an oscilloscope. The paper tape had to be reloaded for a further program to calculate masses. As an alternative to the automatic data-reduction program, another program giving a much greater degree of operator control was developed. A 25 000 data-point window was read from the tape to the disc. Any 512 successive points could be displayed on the oscilloscope thus enabling the operator to examine in detail any portion of any scan and determine parameters for the data reduction. The peak-searching algorithm was based on the arithmetic difference between successive incremental density data points. This difference represents a measure of the slope of the distance vs. intensity curve and should change from zero to +ve to −ve and back to zero on proceeding through a peak. Peaks were detected by comparing this measure of slope with preset parameters representing the minimum values expected at the beginning and the end of a peak.

By way of contrast, in a system[65] where neither an automatic microdensitometer nor an on-line computer was available, distance vs. intensity data measured from a photoplate with a microscope arrangement were subsequently coded on to punched cards and then processed by an off-line computer (type E system) into high-resolution mass measurements and elemental compositions. Ten hours were required for plate reading, three hours for punching the data on to cards, and a further six hours for data turn-round on the computer.

Another completely automated photoplate analysis system has been described by Franzen et al.[66] A photometer was equipped with x- and y-direction scanning motors and a photoelectric distance measuring system which initiated transfers of the digitized transparency values into the core store of an IBM 1130 computer at 10 μm intervals. With the photoplate being continuously scanned at 1 cm s^{-1} this represented a data rate of 1 kHz. The data transfers were made by the direct store access (cycle-steal[67]) facility of the computer and not under program control. A suite of 16 self-linking programs stored on a disc backing store handled the data acquisition and subsequent analysis, including determination of calibration functions for ion-abundance calculations and average background photoplate transmissions, determination of masses, and printout of analysis reports. Computer–operator interaction via an oscilloscope is also emphasized in a paper by Perone et al.[68] which deals with a photoplate analysis system. There have been other reports dealing with photoplate systems.[44,69—71]

[65] P. Haug, D. Bonnell, B. Phillips, J. Runyan, and C. Becker, Proceedings of the 18th Annual Conference on Mass Spectrometry and Allied Topics, San Francisco, 1970, p. B447.
[66] J. Franzen, W. Shonfield, and D. Stüwer, Adv. Mass Spectrometry, ed. A Quayle, Institute of Petroleum, London, 1971, Vol. 5, p. 322.
[67] 'Cycle-steal' is a term describing the technique whereby information may be transferred to a computer core store without the intervention of a program by utilization of the computer hardware address logic during the part of the cycle of instruction execution. The rate of execution of program instructions is usually slowed down slightly.
[68] J. W. Frazer, L. R. Carlson, A. M. Kray, M. R. Bertoglio, and S. P. Perone, Analyt. Chem., 1971, 43, 1479.

Accuracy of Mass and Intensity Measurement. The problem of calibration functions for intensity measurements from photoplate data has remained a centre of attention.[72] A double-exposure method using a Mattauch–Herzog type double-focusing mass spectrometer, and precise mass measurements for peaks greater than m/e 1000 with an automated data analyser have been described.[73]

Reference Compounds for Mass Calibration.—Unfortunately, progress in this important field is still slow as far as high mass markers are concerned. However, the double-beam mass spectrometer[29] could provide useful assistance in getting a little bit more out of existing reference compounds such as PFK. It has already been described how one source/beam of the double-beam mass spectrometer can be used to reference the other beam and so assist in the case where very small amounts of sample are available (see p. 276). In many cases, when obtaining precise mass measurements for high molecular weight compounds, a real practical problem is ion-source saturation resulting in suppression of the PFK reference compound peaks. This effect, coupled with the fact that the high mass peaks of the sample are very often of very low intensity and require a considerable total ion current from the compound in order that they can be detected, often frustrates mass measurements much above m/e 800 using PFK. It is possible that, by using a double-beam mass spectrometer with PFK in one source/beam and the compound in the other source/beam thus removing the suppression of reference peaks problem and allowing a lower resolution to be used, existing mass reference compounds might be used over their full mass range.

Other Techniques.—*Metastable Ion Detection.* McLafferty[5,74] has reported on the use and computerized acquisition of metastable ion data by methods which were fully described in this section of Volume 1.

Detection of metastable ions by a data system operating with a mass spectrometer under high-resolution conditions has been reported by Carrick.[75] The data system is described elsewhere (see p. 273). It was previously thought that metastable ions from the second field-free region of a double-focusing mass spectrometer cease to be detectable under high-resolution conditions.[76] However, using an analog system with suitably high bandwidth these metastable ions have been detected, each one as a series of single-ion or ion-packet peaks of which the integral makes up the familiar shape of the low-resolution metastable peak.

---

[69] J. Pellet, *Inter-Electronique*, 1970, **25**, 36.
[70] C. A. Bailey, R. D. Carver, R. A. Thomas, and R. J. Dupzyk, in 'Developments in Applied Spectroscopy', ed. E. L. Grove and A. J. Perkins, Plenum Press, New York and London, 1969, p. 294.
[71] M. S. Chupakhia and B. N. Khusainova, *Zhur. analit. Khim.*, 1971, **26**, 1083.
[72] S. R. Taylor, *Geochim. Cosmochim. Acta*, 1971, **35**, 1187.
[73] T. Komori, T. Kawasaki, T. Aoyama, M. Shino, M. Togashi, and M. Arai, ref. 9, p. 281.
[74] J. E. Coutant and F. W. McLafferty, *Internat. J. Mass Spectrometry Ion Phys.*, 1972, **8**, 323.
[75] A. Carrick and H. M. Paisley, Fifth Mass Spectroscopy Group Meeting, Bristol, England, July 1971.
[76] J. H. Beynon, R. A. Saunders, and A. E. Williams, 'Mass Spectrometry of Organic Molecules', Elsevier, Amsterdam, 1968, p. 56.

Measurements on mass, intensity, and shape for narrow and broad metastable ions were collected at 10 000 resolution and 10 s per decade scan speed.

Another system has been described[77] for the detection of metastable ions under low-resolution conditions. The signal from the electron multiplier of the mass spectrometer was transformed logarithmically, and conventional computer identification of m/e values was followed by a program to suppress normal m/e peaks and detect low-intensity metastable ions. The stored normal and metastable ion data were then analysed according to the standard parent–daughter–metastable ion relationship, and metastable ions so recognized were superimposed on a computer-drawn bar graph of the low-resolution mass spectrum.

Spark Source Mass Spectrometry. Some spark source systems have already been mentioned[64,66] in the section dealing with photographic recording. In connection with the automation of spark source mass spectrometry, the advantages and disadvantages of electrical and photographic recording have been discussed objectively[78] and this author is, on balance, clearly in favour of electrical detection on the grounds of speed of availability of results and ease of interfacing with a computer, coupled with the fact that the accuracy achievable with electrical recording can be at least equivalent to that achievable with photographic recording.

There have been further reports of data systems used with electrical recording spark source mass spectrometers.[79]

Time of Flight Mass Spectrometry. Data acquisition systems for time of flight mass spectrometers have been described[80,81] and techniques for the analysis of data from fast-scanning time of flight mass spectrometers have been presented.[82,83]

3 Data Processing

As emphasized in the preceding sections, this aspect of the use of computers in mass spectrometry is of rapidly growing importance because of the much increased data flow rates achievable with computerized acquisition techniques associated with modern fast-scanning mass spectrometers. Applications of data processing will be categorized as those dealing with direct comparison of mass

---

[77] A. M. Duffield, W. E. Reynolds, D. A. Anderson, R. A. Stillman, and C. A. Carrol, Proceedings of the 19th Annual Conference on Mass Spectrometry and Allied Topics, Atlanta, 1971, p. 63.

[78] F. D. Leipziger, Proceedings of the 18th Annual Conference on Mass Spectrometry and Allied Topics, San Francisco, 1970, p. 332.

[79] R. Brown and P. Powers, *Analyt. Chem.*, 1971, **43**, 1079; R. A. Bingham, R. Brown, and P. Powers, Proceedings of the 18th Annual Conference on Mass Spectrometry and Allied Topics, San Francisco, 1970, p. B227.

[80] M. A. Grayson and R. J. Conrads, *Analyt. Chem.*, 1970, **42**, 456.

[81] M. B. Fallgatter and R. J. Hanrahan, *Scientific and Technical Aerospace Reports*, 1971, **9**, 486.

[82] R. W. Loser, C. A. Chambers, and E. D. Ruby, *Scientific and Technical Aerospace Reports*, 1970, **8**, 1210.

[83] J. A. Benek, M. R. Busby, and H. M. Powell, *U.S. Govt. Res. Dev. Reports*, 1970, **70**, 173.

spectra by library-searching techniques, and those dealing with the analysis of single spectra or learning machine[84] methods.

Biemann[85] has discussed methods such as computer-controlled microfilming of graphic displays, retrieval of mass spectra according to compound type from spectral collections, and retrieval of pertinent literature references.[86]

The advantages and disadvantages of the computer compared with the mass spectrometrist for current awareness data retrieval have been considered.[87] A generalized theoretical approach to the processing of isotopic mass spectra has been given by Bir.[88]

Single Spectra and Learning Machine Methods.—*Low Resolution.* The group at Stanford University have actively pursued their work on the use of the DENDRAL[89] program for the analysis of low-resolution mass spectra. The program has been extended to deal with the case of saturated amines.[90,91] This represents a more significant advance than may at first be apparent, since the number of possible isomers of a given empirical formula is considerably larger for amines than for the previously considered cases of ketones and ethers. Furthermore, simple applications of the methods used previously would not have been sufficient to handle the case of saturated amines. The heuristic DENDRAL program consists of essentially three main subprograms called PRELIMINARY INFERENCE MAKER, STRUCTURE GENERATOR, and PREDICTOR. The first subprogram finds structural features consistent with the mass spectrum and the elemental composition of the compound. The second subprogram uses this information to generate a list of all the possible structures for the compound. Finally, PREDICTOR then predicts the mass spectrum of each of these structures, compares it with the recorded mass spectrum, and reports on the best fits. N.m.r. data could also be used to assist in the last stage of the sequence. In order to reduce the number of possible structures which would be generated for amines, the PRELIMINARY INFERENCE MAKER was improved by incorporating much more mass spectrometric theory about fragmentation mechanisms, and by the use of n.m.r. data (if available) at this early stage rather than at the last stage as in the earlier version. Structural subunits ('superatoms') with at least one free valence are entered on a GOODLIST for use by STRUCTURE GENERATOR. After the program has decided that it is dealing with an amine (from the empirical formula), the rules

[84] N. J. Nilsson, 'Learning Machines', McGraw-Hill, New York, 1965.
[85] J. Biller, H. S. Hertz, and K. Biemann, Proceedings of the 19th Annual Conference on Mass Spectrometry and Allied Topics, Atlanta, 1971, p. 85.
[86] H. S. Hertz, D. A. Evans, and K. Biemann, *Org. Mass Spectrometry*, 1970, **4**, 453.
[87] R. H. Searle, *J. Documentation*, 1970, **26**, 221.
[88] R. Bir, *Adv. Mass Spectrometry*, ed. A. Quayle, Institute of Petroleum, London, 1971, Vol. 5, p. 339.
[89] J. Lederberg, G. L. Sutherland, B. G. Buchanan, and E. A. Feigenbaum, *U.S. Govt. Res. Dev. Reports*, 1970, **70**, 73; B. Buchanan and G. Sutherland, *ibid.*, 1968, **68**, 50.
[90] A. Buchs, A. M. Duffield, G. Schroll, C. Djerassi, A. B. Delfino, B. G. Buchanan, G. L. Sutherland, E. A. Feigenbaum, and J. Lederberg, *J. Amer. Chem. Soc.*, 1970, **92**, 6831.
[91] A. M. Duffield, A. Buchs, A. B. Delfino, and C. Djerassi, Proceedings of the 18th Annual Conference on Mass Spectrometry and Allied Topics, San Francisco, 1970, p. B135.

for amines are used and 31 possible superatoms initially placed on the GOODLIST. The mass spectrum and the n.m.r. data are then accepted by the program and used for validation tests on each superatom which is then either retained or discarded from the GOODLIST. The use of n.m.r. data was optional but, as an example, reduced the number of possible isomers from 26 to 1 in the case of N-methyl-n-propyl-n-hexylamine. A typical validation test on the superatoms for amines used the well-known α-fission fragmentation process (Figure 5). Other successive tests

$$\left[R^1\text{—}\underset{\underset{R^2}{|}}{N}\text{—}CH_2R^3 \right]^{\ddag} \rightarrow [R^1(R^2)NCH_2]^+ + R^3_{\cdot}$$

Figure 5 *Fragmentation process in aliphatic amines used in the* DENDRAL *program*

were used to detect branching and rearrangement peaks in the mass spectrum. In general, PRELIMINARY INFERENCE MAKER was so efficient that where n.m.r. data was available only the correct superatom remained on the GOODLIST in the 93 test cases. Often only one structure was compatible with this superatom. Even without n.m.r. data the number of possible isomers was remarkably reduced. For example, for the empirical formula of tri-n-hexylamine there are 38 649 142 possible amine isomers and this was reduced to 1938. In most cases the first subprogram of DENDRAL was sufficient to identify the mass spectrum, especially if n.m.r. data were available. Indeed, at the time of publication the second and third subprograms were unable to deal with the GOODLIST generated for amines.[90]

The original DENDRAL program is able to deal with only saturated acyclic structures, but an approach to the inclusion of cyclic structures and unsaturation has been reported.[92] To investigate the problems involved in this development of the program the specific problem of the ketonic isomers of $C_6H_{10}O$ was studied. A special purpose program generated all possible single, simple, ring structures. Major problems immediately recognized were firstly that of redundancy in the structures generated (symmetry effects) and secondly the computer time needed owing to the vastly increased number of isomers. Considering possible ketones, aldehydes, ketens, alcohols, and ethers there are 150 acyclic and 558 cyclic isomers for $C_6H_{10}O$ and the time required to generate all the cyclic isomers was proportionately larger (66 s for C_5H_8O). It was feasible therefore to deal only with relatively low molecular weight compounds. The BADLIST[93] was used to exclude structures with a double bond in a three-membered ring, triple bonds in rings with less than eight atoms and allenes in rings with less than nine atoms. The possible isomers for $C_6H_{10}O$ (see Table 2) include many more cyclic ethers and alcohols than ketones. A feasible scheme for distinguishing between the 27 cyclic carbonyl isomers (after excluding cyclopropanones on the grounds of limited

---

[92] Y. M. Sheikh, A. Buchs, A. B. Delfino, G. Schroll, A. M. Duffield, C. Djerassi, B. G. Buchanan, G. L. Sutherland, E. A. Feigenbaum, and J. Lederberg, *Org. Mass Spectrometry*, 1970, **4**, 493.

[93] J. Lederberg, G. L. Sutherland, B. G. Buchanan, E. A. Feigenbaum, A. V. Robertson, A. M. Duffield, and C. Djerassi, *J. Amer. Chem. Soc.*, 1969, **91**, 2973.

Table 2 *Data from ref. 92 showing isomer distribution of* $C_6H_{10}O$

C_6H_{10}	Non-cyclic	Cyclic
Ketones	13	19
Aldehydes	21	13
Ketens	7	0
Alcohols	47	215
Ethers	62	310
Totals	150	558

stability) based on mass, n.m.r., i.r., and u.v. spectrometry was devised. Rules ('heuristics') based on such a scheme could be programmed for a computer, but it was accepted that in the case where a much greater number of candidate structures are available, such as the 525 cyclic ethers/alcohols for $C_6H_{10}O$, general heuristics would be required. Since there is little or no spectrometric information on many of the candidate structures (many as yet unsynthesized) on which to base such programming, the problem assumes unmanageable proportions. The investigation[92] thus indicated that although it was possible to generate candidate structures satisfactorily it was not feasible to program general heuristics to distinguish between them.

A further paper[94] reviews improvements and extension of the DENDRAL program to deal with thio-ethers and thiols and saturated acyclic ethers of any structure. The major improvement is that the empirical formula of the compound is no longer required as input data. Improvements to the INFERENCE MAKER are such that the other subprograms need not be used for any of these three classes of compound. Although it is accepted that these compounds represent only a small proportion of known compounds it is stated that the program in general performs better than an experienced mass spectrometrist.

It is apparent that one of the problems of the Stanford group's approach to the application of artificial intelligence in mass spectrometry is the degree to which PRELIMINARY INFERENCE MAKER and PREDICTOR depend on the mass spectrometrist himself analysing and programming the information he finds in the mass spectra either completely empirically or according to simple theories (*e.g.* charge localization theory of mass spectra). As was found in the paper investigating cyclic structures,[92] the large volumes of data required to deduce the general rules which would be required for such programs is often unavailable. Extension of the method to larger molecules and more complex structures it seems will be fraught with this difficulty.

The application in mass spectrometry of artificial intelligence based on learning machine theory[84] is fundamentally similar to the Stanford group's approach, except that here the computer 'looks' at the mass spectra and builds up its own knowledge about the nature of the mass spectra of the varying types of compound instead of the mass spectrometrists and programmers producing programs

---

[94] A. Buchs, A. B. Delfino, A. M. Duffield, C. Djerassi, B. G. Buchanan, E. A. Feigenbaum, and J. Lederberg, *Helv. Chim. Acta*, 1970, **53**, 1394.

based on the available data. Such a method has the obvious disadvantage that accepted hypotheses or theories cannot be used, but it is likely that this is far outweighed by the simultaneous advantage that 'learning' is totally unrestricted. The theories of current learning machine applications, particularly in mass spectrometry, have recently been reviewed by Isenhour and Jurs.[95] Further papers[96-101] from these workers have extended their earlier work in this field (see Vol. 1, pp. 278, 279).

Briefly, the learning machine method depends on representing mass spectra as points in a d-dimensional hyperspace (x_i) where x_i $(i = 1 \rightarrow d)$ represent the intensities of mass spectral peaks. Such points are represented by d-dimensional vectors (X) with d components. Points corresponding to mass spectra with similar features are expected to cluster in the hyperspace. Positioning a 'decision surface' between clusters gives a method of assigning a new point to one cluster or another depending on which side of the decision surface it is determined to fall. Feedback methods are used to determine the location of these decision surfaces in the hyperspace. It is advantageous to reduce the dimensionality of the d-dimensional pattern vectors in order to reduce the computational effort required to calculate the position of the points relative to decision surfaces. This in fact means that as few peaks as possible in the mass spectrum under study should be utilized, and the technique of selecting the most significant peaks is called feature selection. Using 630 low-resolution mass spectra of compounds with compositions in the range $C_{1-10}H_{1-24}O_{0-4}N_{0-2}$, Jurs[98] has shown that information relevant to the presence or absence of nitrogen or oxygen is centred in restricted portions of the individual mass spectra. Methods of feature selection for this group of compounds are discussed and are illustrated by satisfactory pattern recognition and prediction using classifiers based on only 20% of the original m/e values in the mass spectra.

The answer as to which side of a decision surface a particular point falls is normally made by calculating the dot product of the pattern X of the point with the normal vector between the point and the decision surface, the latter vector usually being called the weight factor or pattern classifier W. The sign of the scalar result determines which side of the decision surface the point lies. There are four important properties of pattern classifiers: recognition ability, reliability, convergence rate, and predictive ability. The first of these concerns the ability of a 'trained' pattern classifier to classify correctly data from the training set; reliability refers to the ability of the same 'trained' classifier to correctly classify data from the training set which has been distorted; the convergence rate relates to the number of iterations required to train the classifier to the required degree;

---

[95] T. L. Isenhour and P. C. Jurs, *Analyt. Chem.*, 1971, **43**, N10, 20A.
[96] P. C. Jurs, B. R. Kowalski, T. L. Isenhour, and C. N. Reilly, *Analyt. Chem.*, 1970, **42**, 1387.
[97] P. C. Jurs, *Appl. Spectroscopy*, 1971, **25**, 483.
[98] P. C. Jurs, *Analyt. Chem.*, 1970, **42**, 1633.
[99] P. C. Jurs, *Analyt. Chem.*, 1971, **43**, 22.
[100] P. C. Jurs, *Analyt. Chem.*, 1971, **43**, 1812.
[101] L. E. Wangen, N. M. Frew, and T. L. Isenhour, *Analyt. Chem.*, 1971, **43**, 845.

Computerized Data Acquisition and Handling 285

perhaps most important of all, the predictive ability refers to the success rate of the classifier with data patterns which were not part of the training set. A 50% predictive ability for binary classification (yes or no) would be expected with random guessing, but with trained binary pattern classifiers used for low-resolution mass spectra abilities are often >98%. Techniques for improving reliability and predictive abilities of such binary pattern classifiers for low-resolution mass spectra have been discussed.[99] These techniques include various transformations of the original data (*e.g.* using the logarithm or square root of normalized peak intensities), the use of thresholds against which the scalar result of the vector product $X.W$ is compared in the decision process, and the effects of using layered systems of classifiers in which several classifiers are trained by the same pattern simultaneously and then a second layer classifier determines a majority decision from the output of the first layer.[99]

A further paper by Jurs[100] has shown that preprocessing of mass spectral data by forming Fourier transform representations of the mass spectra can improve both the convergence rate and the reliability of the pattern classifiers trained with this data. Optimization of pattern classifiers has also been discussed.[101]

Other workers have reported on the use of binary pattern classifiers,[102–105] and Robertson and Reed have described the application of set theory for the determination of molecular structure from mass spectra.[106,107] Alkanes, alkenes, alkynes, and cycloalkanes have been included in this latter method.

A different approach to the computer analysis of low-resolution mass spectra has been suggested by Crawford and Morrison.[108] They have adopted an *ab initio* approach limited to C, H, N, and O compounds of mol. wt. less than 200. Following basically the steps followed by the mass spectrometrist, the program examines the mass spectrum and searches for the molecular mass peak, evidence for functional groups and groups adjacent to functional groups, and finally attempts to draw the molecular skeleton. The method differs from the systematic approach of the Stanford group[90–94] in that exhaustive lists of possible structures are not generated at an early stage of the program, but rather a less systematic interrogation of the mass spectrum is followed. The program is written in FORTRAN IV and structured in the well-established form of a relatively short control program which can call a number of service programs as required. It is important that complex programs such as these can easily be modified and added to, and the use of a well-established high-level language together with the above-mentioned program structure greatly assists in this task. The program is 26 000 words long

---

[102] Kain-Sze Kwok, R. Venkataraghaven, and F. W. McLafferty, Proceedings of the 19th Annual Conference on Mass Spectrometry and Allied Topics, Atlanta, 1971, p. 70.
[103] J. R. Chapman, Fifth Mass Spectrometry Group Meeting, Bristol, England, July, 1971.
[104] D. H. Smith, Fifth Mass Spectrometry Group Meeting, Bristol, England, July, 1971.
[105] D. H. Smith, *Analyt. Chem.*, 1972, **44**, 536.
[106] R. I. Reed and D. H. Robertson, ref. 7, p. 1281.
[107] D. H. Robertson and R. I. Reed, Proceedings of the 19th Annual Conference on Mass Spectrometry and Allied Topics, Atlanta, 1971, p. 68.
[108] L. R. Crawford and J. D. Morrison, *Analyt. Chem.*, 1971, **43**, 1790.

and can be run on small computers by the use of chaining methods.[109] The main program determines a molecular weight from the mass spectrum by taking the highest m/e value and then performing various isotope peak checks. Various group analysis routines are then called and these form probability tables which list the probabilities of the compound belonging to any of twelve chosen groups: aromatic, ester, ether, acid, ketone, aldehyde, alkene, alkane, alcohol, cycloalkane, diene, and amine. On the basis of this information the program calls other subroutines to determine the empirical formula of the compound, test for special chemical classes of compound, or perform a general interrogation of the spectrum. As the program proceeds it builds up possible structures and substructures and holds the information in the form of nearest neighbour tables according to a method previously described.[110] Finally, a structure-drawing routine converts the nearest neighbour information into recognizable structures and prints them in a character format. The program was tested with 76 low-resolution mass spectra of widely differing compound types. Each unknown was processed by the program in about 6 s. Typical output gave the molecular weight, main fragmentation peaks, the most probable functional groups or types, an estimation of the empirical formula and a most probable structure drawing. The results obtained were compared with those achievable by fourth year students after 16 lectures on organic mass spectrometry and are summarized in Table 3.

Table 3 *Data from ref. 108 comparing the performance of humans and a computer in the analysis of 76 mass spectra*

	Human	*Computer*
Correct molecular ion	59	62
Correct formula in highest rating	48	53
Correct or nearly correct structure	50	45

It can be seen that the performances were about the same, but the significant improvement of the computer was in the time of 6 s required to analyse a spectrum compared to the 10–15 min required by the students. The method is encouraging because it can deal with a relatively large range of functional groups and the program is small enough to be run on small computers, but the success rate is not so impressive as for the Stanford group's approach (see p. 281). As the authors state, mainly monofunctional compounds were tested and it remains to be seen whether polyfunctional molecules can be dealt with; if so, the method should prove very useful.

A method for the reduction of low-resolution mass spectra of compounds containing C, H, N, and O into complete or partial mono-isotopic mass spectra

---

[109] 'Chaining' is a method of segmenting a program into small sections so that while the whole program is stored on a backing store, individual sections can be brought into the computer core store and run one at a time.

[110] L. R. Crawford and J. D. Morrison, *Analyt. Chem.*, 1969, **41**, 994.

using a short computer program has been described.[111] Any approach to the classification of compounds from their mass spectra is important in the field of computerized identification of mass spectra if it is sufficiently logical to enable it to be programmed for a computer. A new classification of alkyl benzenes which satisfies these requirements has been proposed.[112]

High Resolution. In view of the greatly improved facilities now available for obtaining rapidly complete high-resolution mass spectra, it is surprising that more work on the data processing of these spectra has not been published in the period under review.

The interest in peptide mass spectra has been maintained. In a purely theoretical paper, Dayhoff and Eck[113] have proposed a new method for the analysis of proteins by mass spectrometry. The method relies on obtaining from the unknown protein a mixture of all the possible smaller peptides which could be formed by random cleavage of the intact molecule, and contain up to about eight or nine amino-acid residues.

Thus, considering a small portion of a protein –A–B–C–D– the following peptides should be present in the mixture:

A B C D	B C D	C D	D
A B C	B C	C	
A B	B		
A			

This mixture would then be subjected to mass spectrometry without separation. The important features required from the mass spectrum are the molecular ions of all the peptide molecules in the mixture. Using this high-resolution data, a computer program then reconstructs the original molecule by building up from the small peptide units found by comparing and searching for overlapping sections. It was estimated that the mass measurement accuracy required to distinguish between different peptide molecular ions at the same mass, allowing for some errors of distinction, was probably less than that required in standard sequencing methods by mass spectrometry. An artificial mass spectrum was generated in order to test the method for a protein. The idea is important because it is applicable to large peptides and to unseparated mixtures, but it is doubtful whether the two requirements of random hydrolytic cleavage of the intact peptides and essentially molecular-ion type mass spectra are within the realms of present experimental techniques, especially when it is considered that very often only micrograms of unknown peptide material are available. Another method using peptide mixtures has in part been demonstrated by Perrine

---

[111] B. Boone, R. K. Mitchum, and S. E. Scheppele, *Internat. J. Mass. Spectrometry Ion Phys.*, 1970, **5**, 21.
[112] R. G. Gillis, *Org. Mass Spectrometry*, 1971, **5**, 79.
[113] M. O. Dayhoff and R. V. Eck, *Comput. Biol. Medicine*, 1970, **1**, 5.

et al.[114] The method depends on sequential Edman degradation of peptide mixtures followed by quantitative identification of the methyl thiohydantoin derivatives of the N-terminal amino-acids of the peptides present in the mixture by mass spectrometry. This sequential degradation is repeated up to 15 times thus yielding the first 15 N-terminal amino-acids in each peptide. By using several different mixtures of peptides obtained from various selective hydrolyses of a larger peptide and analysing the above data by means of a computer program the original peptide could be sequenced.

The DENDRAL program (see p. 281) has recently been applied[115] to the analysis of high-resolution mass spectra of 43 oestrogenic steroids.

It could be argued that a drawback in the use of complete high-resolution mass spectra is that, for large molecules, at the first stage of data processing when elemental compositions are determined, a large amount of redundant data can be generated. This arises because, in order to allow for all possible ion compositions and a degree of error in mass measurements, the chemist must allow the elemental compositions program a fairly wide search area. For example, an unknown protected peptide has a molecular ion with a measured mass of 548.286 25. If the compositions program is limited to 40-^{12}C, 1-^{13}C, 82-^{1}H, 8-^{14}N, and 10-^{16}O, with an error tolerance of 5 p.p.m., then there are 11 possible formulae for the molecule (data from ref. 116):

$C_{36}{}^{13}CH_{39}O_4$ $C_{29}H_{42}NO_9$

$C_{32}H_{40}N_2O_6$ $C_{34}{}^{13}CH_{37}N_3O_3$

$C_{27}H_{40}N_4O_8$ $C_{30}H_{38}N_5O_5$

$C_{22}{}^{13}CH_{41}N_5O_{10}$ $C_{32}{}^{13}CH_{35}N_6O_2$

$C_{25}H_{38}N_7O_7$ $C_{28}H_{36}N_8O_4$

$C_{20}{}^{13}CH_{39}N_8O_9$

A considerable number of these formulae can be discarded at once by the chemist; the ^{13}C peaks because there is no abundant $M - 1$ peak, $C_{28}H_{36}N_8O_4$ because there should be more oxygens than nitrogens in the peptide, *etc.* However, it is unfortunate that the mass spectrometrist is left to sift such data, which when summed over the whole spectrum can be very voluminous. A technique, which must have been previously used in various forms in many laboratories, to alleviate this problem has been described by Kunderd et al.[116] Instead of giving the elemental compositions program the task of searching for combinations of atomic masses, it is asked to search for ions containing combinations of sub-molecular groups of atoms, such as Me, C_6H_6, CO_2, CONH, *etc.* The approach is obviously more especially useful when the chemist has some idea of the structure

[114] C. Perrine, E. Cannon, C. Cavender, T. Fairwell, and R. E. Lovins, *Adv. Mass Spectrometry*, ed. A. Quayle, Institute of Petroleum, London, 1971, Vol. 5, p. 728.
[115] D. H. Smith, B. G. Buchanan, R. S. Engelmore, A. M. Duffield, A. Yeo, E. A. Feigenbaum, J. Lederberg, and C. Djerassi, *J. Amer. Chem. Soc.*, 1972, in press.
[116] A. Kunderd, R. B. Spencer, and W. L. Budde, *Analyt. Chem.*, 1971, **43**, 1086.

of the compound under study and can select suitable groups for the search procedure, but even in the case of completely unknown compounds, judicious choice of groups can vastly reduce the data output. The program described[116] allowed up to 50 nuclide or submolecular group masses in the searching process and its use was demonstrated for a tetra-peptide and another compound containing acetate, phenyl, and ether groups. The high-resolution spectra were obtained with a photoplate-recording mass spectrometer, microdensitometer, and computer system.

Computer aided mass spectrometric identification of stereoisomeric monosaccharides has been reported.[117]

Library Searching and Spectrum Comparison Techniques.—Investigations into the identification of mass spectra by comparison with a large number of other spectra in the form of a computer compatible library have continued. In the majority of cases the aim is to be able to process the large number of low-resolution spectra produced by a g.c.–m.s. computer combination on a realistic time scale.

It is important that suitable mass spectral libraries are available for this type of work and the Mass Spectrometry Data Centre, Aldermaston, England has reported on progress in this direction particularly with a view to providing a library on computer compatible magnetic tape.[118] A new eight-peak index of more than 17 000 mass spectra is now available.[119]

There are three important factors concerning library search methods which should be considered in relation to any proposed method. (i) File searching time. Large amounts of computer time may be prohibitively expensive. (ii) Success of matching criteria for spectra present in the library after taking into consideration random and systematic changes in the mass spectrum of a compound due to varying operating conditions, and the recording stability of the mass spectrometer(s) used to determine the spectra. (iii) Information gained after a search for a compound which is not present in the library.

In a series of papers[120—123] Grotch has investigated the effect of different methods of encoding the intensity of peaks in mass spectra which are subsequently used in library searches. In one paper[120] he studies from the viewpoint of information theory and statistics the encoding and subsequent matching of mass spectra with peak heights encoded to one bit only (0 for absence, 1 for presence of a peak above a predetermined threshold). A library of 3000 spectra was divided into three groups and all spectra in each group compared pairwise

[117] J. Vink, J. H. W. Bruins Slot, J. J. de Ridder, J. P. Kamerling, and J. F. G. Vliegenthart, *J. Amer. Chem. Soc.*, 1972, **94**, 2542.
[118] R. G. Ridley and W. M. Scott, Proceedings of the 19th Annual Conference on Mass Spectrometry and Allied Topics, Atlanta, 1971, p. 58.
[119] 'An eight peak Index of Mass Spectra', compiled by ICI Ltd., Dyestuffs Division, with MSDC, Aldermaston, England, Vols. I, II, and III.
[120] S. L. Grotch, *Analyt. Chem.*, 1970, **42**, 1214.
[121] S. L. Grotch, Proceedings of the 18th Annual Conference on Mass Spectrometry and Allied Topics, San Francisco, 1970, p. B453.
[122] S. L. Grotch, Proceedings of the 19th Annual Conference on Mass Spectrometry and Allied Topics, Atlanta, 1971, p. 72.
[123] S. L. Grotch, *Analyt. Chem.*, 1971, **43**, 1362.

with each spectrum in the other two groups. The resulting distribution of mismatches was analysed using the statistics of the individual groups, and the results indicated that mass spectral patterns are still highly specific even when the peak heights are encoded to only one bit. The information content of the mass spectral pattern is dependent on the choice of the transition threshold. For the spectra in this library it was found that peaks less than 0.3% of the base peak contribute little to the uniqueness of the spectra but that information content of the spectra falls rapidly for transition levels above 1%. (The transition level is the intensity above which the peak is encoded as 1 rather than 0.) In practice, however, noise in the recording system often makes the measurement of meaningful intensity values at these low levels impossible and therefore some compromise must be reached. Generalization of the analysis of the one-bit case to multi-bit encoding showed that the optimum transition levels for multi-bit encoding should be on a logarithmetric scale of peak intensity.

A computer program for performing library searches with mass spectra encoded to one bit has been developed.[121] In the method, a criterion of mismatch, C, was determined for each spectrum in the library L_i, when compared with the unknown X, such that

$$C_i = A - NB_i$$

where A is the total number of channels which disagree in the comparison of X and L_i, and B is the number of channels which match minus the total number of peaks (i.e. ones) in the spectrum of the unknown X; N is a weighting factor usually 2—5. The search produced the ten library spectra with the lowest values of C. Using a library of 7000 spectra and 168 newly measured test 'unknown' spectra, 156 were correctly identified in the top ten. Tests with mixtures of compounds proved promising. The advantages of one-bit encoding are in the much greater efficiency of comparison, and the economy, of core store required for the computation, and backing store required for the library. The overall result is that a much larger number of spectra can be compared in a given time than with fully encoded spectral library search methods.

In further papers[122,123] Grotch has compared one-bit encoding to encoding at 2, 8, and 10^4 levels. Two methods of spectrum comparison based on the disagreement criterion C_j for the unknown and the jth library spectrum were used. In the first

$$C_j = \sum_1^M |x_i - l_{ij}|$$

where x_i is the level of intensity of the ith peak (i = mass number $1 \rightarrow M$) in the unknown spectrum and l_{ij} is the level of intensity of the ith peak in the jth library spectrum. In the second method the N_1 most intense peaks in the unknown spectrum were compared with the N_2 most intense peaks in the library, and those library members which maximized the coincidence of the mass numbers of these peaks were determined. As before, the top ten spectra were summarized at the end of the search. The methods were tested using a library of

6800 low-resolution mass spectra and 125 unknown spectra obtained from five independent sources. The program consisted of about 500 FORTRAN statements and was run and tested on an IBM 360/44 computer. Both unrestricted and prefiltered searches were carried out. In the prefiltered approach only library members which satisfied certain preconditions were compared with the unknown. These conditions were (i) the mass corresponding to the base peak of the unknown must be among the five most intense peaks of the library spectrum and *vice-versa*; (ii) The mol. wt. of the library member must be greater than a specified fraction (usually 0.7) of the mass of the highest unknown peak with an intensity greater than 0.5% of the total ion current. The prefiltering could increase the search speed by a factor of 2—3, and several compounds which may otherwise have appeared in the best ten list were often removed from the list thus making it less ambiguous. A variety of thresholds for the transitions between one level and the next, based on intensities relative to the base peak or the total ion current, were investigated. As may be expected, selection of the correct spectrum from the library in the best ten fits improves as the number of levels considered for the intensity encoding is increased. However, encoding to one or three bits was found to be surprisingly successful and in some applications the greatly reduced computer time and storage requirements may be of overriding importance. If only one unknown is searched at a time it is likely that the reading of the library from magnetic tape into core will be the time-limiting stage in the search. In this case a '3-bit search' takes approximately four times as long as the '1-bit search' and a library fully encoded to 10^4 levels (intensities 0.1—100%) takes roughly an order of magnitude longer.

Wangen *et al.*[124] have also investigated one-bit encoding and further reduction of the data space required by the mass spectrum. By reducing to 48 the number of bits required to encode a whole spectrum without loss of pertinent information a library of over 6000 spectra was searched in 2 s. This paper also considers other methods of minimizing search times and data storage requirements so that the searching can be carried out with the minimum hardware of a moderate speed CPU, 4K of core memory, a 25 i.p.s. 9-track tape drive, and standard input/output facilities.

The use of hash coding[125] in searching information files has been discussed by Jurs,[126] and another paper[127] comments on his findings with reference to mass spectrometry.

Knock *et al.*[128] have investigated, with reference to library searching techniques, the variation of the mass spectrum of one compound owing to instrumental and other factors. Four matching routines were used. In the first, the m/e values of the n most intense peaks in the unknown and the library spectrum,

[124] L. E. Wangen, W. S. Woodward, and T. L. Isenhour, *Analyt. Chem.*, 1971, **43**, 1605.
[125] I. Flores, 'Data Structure and Management', Prentice-Hall Inc., Englewood Cliffs, N.J., 1970.
[126] P. C. Jurs, *Analyt. Chem.*, 1971, **43**, 364.
[127] F. E. Lytle, *Analyt. Chem.*, 1971, **43**, 1335.
[128] B. A. Knock, I. C. Smith, D. E. Wright, and R. G. Ridley, *Analyt. Chem.*, 1970, **42**, 1516.

irrespective of order, and the number of agreements A were determined. The degree of matching was given by

$$P_1 = A/n$$

In the second method, the n strongest peaks in order of decreasing intensity were used. If the corresponding peaks were not at the same m/e value, P_2 was reduced in proportion to the difference in position, thus

$$P_2 = (1/n^2) \sum_{k=1}^{A} (n - |i - j|_k)$$

where A is the number of agreements irrespective of order and i and j are the positions in the respective sets of the kth pair of equal m/e values.

In order to take account of mass discrimination effects due to type of instrument two further methods were used. These latter matching methods were essentially the same as the first two, but performing the determination of P over R equal mass ranges in the spectra and then averaging the results to obtain a single P value for comparison of the spectrum with each library spectrum. In the search the ten library members with the highest P values were determined and ranked. The effect of instrument variability was tested by using the spectra of three compounds recorded on 12 different mass spectrometers of four different types, together with a special library of compounds with the same molecular weight as the unknown. In the searches, methods 1 and 3 retrieved the correct compound as first choice in 93% of the trials and methods 2 and 4 were 100% successful. Other studies of the general groups of compounds, terpenes, hydrocarbons, and non-hydrocarbons, were made and counting similar isomers as successes the retrieval was successful in 97% of the cases. Even when the compound was not present in the library, if closely similar compounds were present, useful information would be retrieved. The program was written in FORTRAN for an IBM machine and required 3—30 s for a search.

Biemann and co-workers have reported[129] more fully on their library searching system for use with a mass spectrometer–gas chromatograph–computer system. For the searching procedure the mass spectra were abbreviated by taking only the two most intense peaks in each 14 unit mass range from m/e 9 upwards. This considerably reduced storage and computer time requirements. The library used contained 7600 spectra of a wide variety of compounds of mol. wt. up to 1000. Spectra from the library to be compared with the unknown were again preselected (see ref. 128 and p. 291) according to several broadly based criteria, such as the requirement that the mass range of the unknown and library member should not differ by more than a factor of three. It was unusual for more than a few hundred spectra to be preselected from the library for any one unknown. In the comparison of the spectra a similarity index was determined. This was based on a weighted ratio of the known to the unknown abbreviated spectra taken mass for mass. The ratio was normalized so that a value of one was equivalent to a perfect match. Each search with preselection required on average about 1 min of

[129] H. S. Hertz, R. A. Hites, and K. Biemann, *Analyt. Chem.*, 1971, **43**, 681.

computer time on an IBM 1800 machine. The search system was evaluated by using the actual mass spectra obtained from g.c.–m.s. runs of two mixtures, one a test synthetic mixture and the other from an actual problem being investigated in the laboratory. In the majority of cases the correct compound was retrieved as the first search find. When one g.c. peak contained more than one compound, spectrum subtraction could be employed and using the resulting spectra the search also proved useful. Even when the unknowns were not present in the library useful information in the form of closely similar compounds was retrieved, and it was usually possible to distinguish between correct finds and 'closely similar' finds from the magnitudes of the similarity indices.

Apart from dealing with the general case of unknown mass spectra, searching techniques can be extremely valuable when used in conjunction with libraries of mass spectra of a strictly limited range of compounds. For instance, a particular laboratory working a great deal with a single compound class could rapidly build up its own limited library of compounds encountered in this research. Spiteller has described[130] such a search system for steroids based on mol. wt., key fragments, and key mass differences. The search is possible on a very small computer or even with a punched card system.

Searching techniques utilizing i.r., u.v., and n.m.r. data as well as mass spectral data have been described by Erni and Clerc.[131]

4 Computer Control

A sophisticated example of the use of a computer to control the running of a mass spectrometer has been described by Crawford and co-workers.[132] The system was used for gas analysis and the inlet system could be controlled by the computer, the pneumatically operated valves being driven by relays activated by digital outputs from the computer. A pressure transducer monitored the pressure in the sample inlet and provided feedback information to the computer to tell it how to admit the samples into the reservoir. During the mass spectrometer scan initiated by the computer, the ion current, accelerating voltage, magnetic field, sample inlet pressure, and sample reservoir pressure readings from the mass spectrometer were digitized and fed to the computer. As the spectrum was scanned peak tops were located and, after the scan, mass and intensity information was printed immediately on a teletype printer near the mass spectrometer. This information was also written to magnetic tape to enable further processing by a larger computer. This system (see Figure 6) used a medium-size 18-bit wordlength computer which was time-shared with seven other instruments.

Other examples[133,134] of computer control in mass spectrometry have been mainly concerned with data acquisition from quadrupole mass analysers.

[130] G. V. Unruk, M. Spiteller-Friedmann, and G. Spiteller, *Tetrahedron*, 1970, **26**, 3039.
[131] Von F. Erni and J. T. Clerc, *Chimia*, 1970, **24**, 388.
[132] R. W. Crawford, J. W. Frazer, and R. K. Stump, Proceedings of the 18th Annual Conference on Mass Spectrometry and Allied Topics, San Francisco, 1970, p. 449.
[133] W. E. Reynolds, V. A. Bacon, J. C. Bridges, T. C. Coburn, B. Halpern, J. Lederberg, E. C. Levinthal, E. Steed, and R. B. Tucker, *Analyt. Chem.*, 1970, **42**, 1122.
[134] J. Houseman and F. W. Hafner, *J. Sci. Instr.*, 1971, **4**, 46.

In this application the computer is programmed to generate control voltages for the rods of the quadrupole and so determine itself which mass peaks are transmitted by the analyser and detected. The control voltages required to transmit each mass peak are usually determined previously by a calibration run with a reference compound. The advantages of this approach may be summarized as follows: (i) Since control voltages for particular mass peaks are predetermined, the data collected can be stored directly as an m/e vs. intensity matrix within the computer thus reducing processing time required during the scan. (ii) In conventional scanning mass spectrometers operating with computers in real-time,

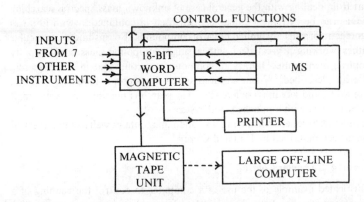

Figure 6 *System configuration with feedback control facilities, from ref. 132*

much of the scan time, processor time, and in some cases core storage is wasted by the need to digitize and process the background signal between all peaks (sometimes $>90\%$ of the scan time). In computer controlled systems this does not occur since only data relevant to mass peaks are digitized. (iii) The time saving resulting from (i) and (ii) allow a greater dwell-time on each mass peak for equivalent scan times thus resulting in collection of a greater number of ions and thus a better signal-to-noise ratio and a better dynamic intensity range. (iv) Because processor and storage requirements are minimized, the computer hardware required for these systems can often be less rather than more expensive than that required for conventional data systems. (v) A more flexible approach to operator/mass spectrometer interfaces is possible.

The system described by Reynolds et al.[133] was operated with a gas chromatograph coupled to the mass spectrometer via a 50–50 splitter, (50% to flame ionization detector), a Biemann separator, and a solenoid-actuated valve used to prevent the initial solvent peak from the chromatograph entering the spectrometer. The computer could control all valves including that from the fluorine reference compound reservoir into the spectrometer. All control functions were entered via a teletype keyboard with no need for the operator to adjust the spectrometer controls. The actual mass peaks sampled and the dwell-time on each peak could be predetermined. Since measurements were taken on the ion current when the signal was in an essentially stationary state, full integration of the signal to enhance the signal-to-noise ratio could be employed. Standard commercial FET operational amplifiers were used for both integration and the electrometer head amplifiers. The hardware system was designed to be essentially independent of individual computer characteristics and had been used with a 12-bit 2K machine and a time-shared IBM 360/50 buffered via an IBM 1800. The computer was connected to the rest of the instrumentation by a 500 feet cable. It was stated that most modern small general purpose computers would be able to handle the computational requirements of the system, although some form of magnetic backing store would be desirable for program and data storage.

Another quadrupole mass spectrometer data system based on the same control philosophy has been designed for the specific purpose of analysis of gases in a rocket engine combustion chamber.[134] Here the workers were interested in only five mass peaks and required minimal operator intervention. To overcome the problem of drift of mass peaks on the control voltage scale, nine digital samples were taken in a pattern around the calibrated voltages according to experimentally determined drift characteristics of the instrument. The largest of these samples was then taken as the peak height. After a calibration phase the system could be run almost independent of the operator, acquiring nine samples on each peak in 9 ms and transferring the digital data to magnetic tape for later processing on a larger computer.

Other similar quadrupole systems have been reported[135] and they include such facilities as selection of electron-beam energy with a resolution of 10 mV[136] and pulse counting of electron multiplier output.[136]

McLafferty[63] has proposed a method for rescanning peaks during a normal magnetic scan of a double-focusing electrical-recording mass spectrometer by computer control of the electrostatic analyser voltage (see p. 277).

Direct digital control in mass spectrometry in general,[137] computer control in molecular beam mass spectrometry,[138] control of the accelerating voltage in spark source mass spectrometry,[139] and other applications of control[140,141] have been discussed.

---

[135] A. Kupperman, Proceedings of the 19th Annual Conference on Mass Spectrometry and Allied Topics, Atlanta, 1971, p. 225.
[136] M. A. Frisch, Proceedings of the 19th Annual Conference on Mass Spectrometry and Allied Topics, Atlanta, 1971, p. 227.
[137] J. W. Amy, R. B. Spencer, A. W. Kunderd, and J. H. Beynon, 21st Mid-American Symposium on Spectroscopy, Chicago, June, 1970, Abstr. 84, p. 94.
[138] J. R. Wyatt, G. A. Pressley, and F. E. Stafford, *High Temp. Sci.*, 1971, **3**, 130.
[139] J. P. Boulloud, C. Jonnet, J. Maragnon, and P. Tauveron, *Adv. Mass Spectrometry*, ed. A Quayle, Institute of Petroleum, London, 1971, Vol. 5, p. 344.
[140] E. J. Bonelli, *American Lab.*, 1971, **3**, 27.
[141] L. E. Tighe, *Nuclear Sci. Abs.*, 1971, **25**, 819.

5 Miscellaneous Applications

Gas Chromatography and Mass Spectrometry.—The use of computer techniques in conjunction with g.c.–m.s.[142,143] and new instrumentation[144] in this field have been reviewed. A great many mass spectrometer–computer systems are now designed to operate with a g.c. inlet to the spectrometer and several are mentioned elsewhere in this chapter. For greater detail of applications and techniques in this general field the reader is referred to Chapter 7. In a general review of g.c.–m.s., Junk[145] has considered computer applications and future trends in this field.

Biemann and co-workers have continued to report on their system for continuously scanned mass spectra of g.c. effluents.[146] In one paper[147] they report on the technique of displaying the change in abundance of certain ions during the running of the gas chromatogram. The resulting spectra are called mass chromatograms. A particular mass value was chosen and then the abundance of this ion was plotted against spectrum number (from a series of continuously scanned spectra) by the computer. Several such mass chromatograms for different mass numbers could be plotted in different colours on the same plot with the same axes. Choice of the mass values was made by the chemist from a knowledge of the sample or the mass spectra of the compound groups under study, or from suggestions presented by the computer. These suggestions consisted of a list of the most common mass peaks in all the mass spectra from a single gas chromatogram (often several hundred spectra). A detailed discussion with examples of use of this technique is given in the paper.[147]

Other g.c.–m.s. systems with repetitive scanning have been reported.[57,148] A data system capable of rapid change between high- and low-resolution operation for recording mass spectra from a g.c.–m.s. combination with full facilities for spectrum storage and spectrum subtraction has been described.[50] As pointed out by Bonelli,[149] spectrum subtraction facilities are particularly useful for the analysis of g.c. multiplets and also for the removal of column bleed background spectra from the spectra of the g.c. peaks due to components of the mixture under analysis.

In a paper[150] describing analysis of steroids by off-line computerized g.c.–m.s., a method is reported whereby the g.c. effluent is scanned at steadily increasing time intervals instead of the more normal constant time intervals. The spectra were recorded on magnetic tape and processed off-line by a medium-size computer.

---

[142] D. Henneberg and G. Schomberg, *Adv. Mass Spectrometry*, ed. A. Quayle, Institute of Petroleum, London, 1971, Vol. 5, p. 253.
[143] F. W. Karasek, *Analyt. Chem.*, 1972, **44**, 32A.
[144] Pittsburgh Conference Review, *Chem. and Eng. News*, 1970, 44.
[145] G. A. Junk, *Internat. J. Mass Spectrometry Ion Phys.*, 1972, **8**, 1.
[146] R. A. Hites and K. Biemann, *Analyt. Chem.*, 1968, **40**, 1217.
[147] R. A. Hites and K. Biemann, *Analyt. Chem.*, 1970, **42**, 855.
[148] L. Bergstedt and G. Widmark, *Chromatographia*, 1970, **2**, 59.
[149] E. J. Bonelli, *Analyt. Chem.*, 1972, **44**, 603.
[150] R. Reimendal and J. Sjovall, *Analyt. Chem.*, 1972, **44**, 21.

Many examples of the application of computerized g.c.–m.s.,[151] and the computer processing of g.c.–m.s. data have been published.[152]

Ionization and Appearance Potential Measurement.—There have been several reports of the use of computers for both acquisition and processing of ionization efficiency data as obtained from electron impact mass spectrometers.

Apart from the problems of determining ionization and appearance potentials from ionization efficiency (IE) data, acquiring the data itself is often a time-consuming process. Manual measurements of the ion current at 0.2—0.5 eV intervals up and down the ionization efficiency curve must be made for at least two mass positions while other instrumental conditions are kept stable. Semi-automatic methods have been previously proposed[153] but now several groups have reported computerized acquisition of IE data.[154–156]

In the method described by Johnstone et al.,[154] using a conventional magnetic scanning double-focusing mass spectrometer the electron-beam energy was decreased at a constant rate of about 0.1 eV per second from a preselected start position (nominally 14 eV) while the ion current was sampled by the computer at about 10 kHz. The digital samples were averaged over a predetermined number in real-time and then stored within the computer ready for output at the end of the scan. The end of the scan was detected by the computer when the ion current reached zero for a large number of samples. A single scan required only 50 s to record and 10 s to output on paper tape. Typically, ten scans of each IE curve would be recorded, and subsequently these would be averaged before being processed by another program to obtain the appearance potential data. The averaging at three stages of the data acquisition resulted in much improved signal-to-noise ratios and, in consequence, more reliable data for the foot of the IE curve. Furthermore, because of the greatly reduced time scale for obtaining the data, problems of instrument stability were reduced and it was even possible to obtain meaningful AP data for involatile compounds by using a direct insertion probe and recording the IE curve in about 30 s. An important consideration for this method was that a conventional ion source with *no* modification should be used so that normal routine use of the mass spectrometer should not be unnecessarily interrupted either before or after using the instrument for energy

---

[151] J. R. Althaus, K. Biemann, J. Biller, P. F. Donaghue, D. A. Evans, H. J. Forster, H. S. Hertz, C. E. Hignite, R. C. Murphy, G. Preti, and V. Reinhold, *Experientia*, 1970, **26**, 714; N. C. Law, V. Aandahl, H. M. Fales, and G. W. A. Milne, *Clin. Chim. Acta*, 1971, **32**, 221; D. S. Millington and K. L. Rinehart, Proceedings of the 19th Annual Conference on Mass Spectrometry and Allied Topics, Atlanta, 1971, p. 96; J. Roboz, L. Sarkozi, A. Ruhig, and F. Hutterer, *ibid.*, p. 106.

[152] S. Oshima, T. Nishishita, T. Konishi, and Y. Sugiyama, *Mass Spectrometry (Japan)*, 1970, **18**, 1254; P.-A. Jansson, S. Melkersson, R. Ryhage, and S. Wikstrom, *Arkiv Kemi*, 1970, **31**, 565.

[153] G. G. Meisels, J. Y. Park, and B. G. Giessner, *J. Amer. Chem. Soc.*, 1970, **92**, 254.

[154] R. A. W. Johnstone, F. A. Mellon, and S. D. Ward, *Internat. J. Mass Spectrometry Ion Phys.*, 1970, **5**, 241; *Adv. Mass Spectrometry*, ed. A. Quayle, Institute of Petroleum, London, 1971, Vol. 5, p. 334.

[155] R. G. Dromey, J. D. Morrison, and J. C. Traeger, *Internat. J. Mass Spectrometry Ion Phys.*, 1971, **6**, 57.

[156] T. P. Georgobiani and N. I. Starkovskii, *Russ. J. Phys. Chem.*, 1971, **45**, 732.

measurements. Because of the much improved signal-to-noise ratio obtained, fine structure in the IE curves was clearly visible and IP's measured for a range of compounds were in good agreement with photo-ionization values and improved relative to previous electron impact values. The method due to Winters[157] was used to process the IE curves.

Automation of the measurement of IE curves has also been described using a quadrupole mass spectrometer with an electron impact ion source.[155] In this method, during sweeps of the IE curve consisting of 1024 voltage steps, the ion current was digitized four times at each step, the values summed and stored. A sweep required 5 s and was repeated many times, each new curve being added to the resultant time-averaged curve. The system produced IE curves with very good signal-to-noise ratios. After further mathematical smoothing the curves were processed by a Fourier transform method. Results for N_2 and CO_2 are discussed in the paper.[155] By the nature of the instrumentation, however, this method is limited to the study of gases or liquids with high vapour pressures. The computer system was used both to digitize and process the ion current and to sweep the electron energy by providing an analog signal *via* a digital-to-analog converter (see Figure 7). The IE and processed curves could also be displayed on an oscilloscope which was connected to the computer.

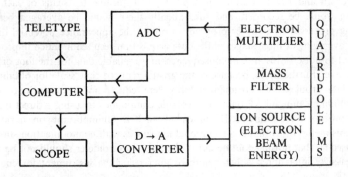

Figure 7 *Quadrupole mass spectrometer–computer system for energy measurements, from ref. 155*

Automatic determination of IE curves has been discussed by other workers[156,158] using conventional mass spectrometer types, and Gross and Wilkins[159] have described the coupling of an ion cyclotron resonance spectrometer and a computer with particular reference to AP measurements. These workers

[157] R. E. Winters and J. H. Collins, *J. Chem. Phys.*, 1966, **45**, 1931.
[158] G. G. Meisels and B. G. Giessner, *Adv. Mass Spectrometry*, ed. A. Quayle, Institute of Petroleum, London, 1971, Vol. 5, p. 352.
[159] M. L. Gross and C. L. Wilkins, *Analyt. Chem.*, 1971, **43**, 1624.

also aimed for a routine, rapid, and accurate method of measurement but with less emphasis on obtaining fine structure in the curves. It was stressed by the authors that once the tedious nature of obtaining AP data by conventional manual techniques has been overcome, use of this data should become much more widespread and an example concerning the identification of C_5H_{10} isomers was given.

Computers playing the role of 'number crunchers' have long been important in the processing of IE data to obtain IP's and AP's. A major practical problem in many laboratories is that the same mass spectrometer must often be used for a multitude of purposes (high-resolution scanning, low-resolution scanning, metastable defocusing, energy measurements, g.c.–m.s., *etc.*) and modification of the electronics or, in particular, the ion source to suit energy measurements is undesirable. For AP measurements with conventional electron-impact ion sources this leaves the problem of energy-spread in the incident electron beam and the resultant loss of fine structure in the IE curve, as well as the difficulty in making an objective determination of the ionization on-set. In the past, two important mathematical approaches to the removal of the effect of the energy spread of the electron beam have been used, namely, the Fourier transform method of Morrison[160] and the electron distribution difference method due to Winters *et al.*[157] These two methods have been carefully compared by Giessner and Meisels.[161] As these workers point out, part of the importance which must be attached to these methods is that an improvement in the data is often possible without instrumental modification. The use of computers for these calculations is invaluable if not essential.

Deconvolution[162] and inverse convolution[163] techniques for processing IE data have been discussed by Morrison and Dromey.

Mixture Analysis.—Several workers[164–166] have described the use of computers to aid in the problem of mixture analysis by mass spectrometry. Hites[166] has used semi-automatic data acquisition of the mass spectra of triglyceride mixtures followed by off-line processing of the resultant punched paper tape on a small computer. The approach was to record spectra during molecular distillation of the mixture from a direct insertion probe. Other papers[164] describe a similar method using a temperature-programmed direct insertion probe. A new procedure for petroleum fraction analysis has been described by Robinson,[165] and

[160] J. D. Morrison, *J. Chem. Phys.*, 1963, **39**, 200.
[161] B. G. Giessner and G. G. Meisels, Proceedings of the 18th Annual Conference on Mass Spectrometry and Allied Topics, San Francisco, 1970, p. B431.
[162] J. D. Morrison and R. G. Dromey, *Internat. J. Mass Spectrometry Ion Phys.*, 1970, **4**, 475.
[163] R. G. Dromey and J. D. Morrison, *Internat. J. Mass Spectrometry Ion Phys.*, 1971, **6**, 256.
[164] R. D. Grigsby, C. O. Hansen, D. G. Mannering, W. G. Fox, and R. H. Cole, *Analyt. Chem.*, 1971, **43**, 1135; R. D. Grigsby, J. S. Bird, and K. E. Bird, Proceedings of the 19th Annual Conference on Mass Spectrometry and Allied Topics, Atlanta, 1971, p. 110.
[165] C. J. Robinson, *Analyt. Chem.*, 1971, **43**, 1425.
[166] R. A. Hites, *Analyt. Chem.*, 1970, **42**, 1736.

Edwards[167] has approached the problem of mixture analysis by using library search methods.

Computer Programs and General Data Processing.—Computers can be used to simulate ion sources or other mass spectrometer features by the use of mathematical models. Such studies can aid the design and use of instruments. Such computer techniques have been used in a study of field ionization sources with the aim of identifying factors which could improve sensitivity.[168] A similar study has been reported for a quadrupole mass spectrometer.[169] Mathematical models have been used to examine ion–polar molecule collisions relevant in mass spectrometry.[170]

Reports have appeared of various utility programs providing such facilities as generation of accurate mass tables for C, H, N, O, and P containing compounds,[171] aiding the determination of unique molecular formulae from mass spectral data,[172] and a general computer language for a chemical laboratory.[173] Applications of Fourier transform techniques which may be useful in mass spectrometry have been described.[174]

6 Future Trends

Standard computer hardware has become relatively cheaper over the past two years whereas software costs have risen. This situation would be expected to lead to users developing their own specialized software often more cheaply than manufacturers or outside software consultants. For example, a language particularly designed for chemical applications has been developed.[173] Falling hardware costs can be exploited in two ways; sophisticated hardware can be used to replace programmable general purpose computer hardware,[175] and larger computer systems capable of controlling and handling data from mass spectrometers and other instruments at the same time can be used. Such a total laboratory system has recently been described.[176] The alternative to the dedicated computer for total laboratory automation is a hierarchical system of a group of small computers, one for each application, but communicating with a large time-shared computer for use of bulk storage devices, efficient software for program development, line-printers *etc.*

[167] T. R. Edwards, *Adv. Mass Spectrometry*, ed. A. Quayle, Institute of Petroleum, London, 1971, Vol. 5, p. 319.
[168] J. P. Pfiefer, A. M. Falick, and A. L. Burlingame, Proceedings of the 19th Annual Conference on Mass Spectrometry and Allied Topics, Atlanta, 1971, p. 48.
[169] M. Mosharrafa and G. M. Wood, Proceedings of the 19th Annual Conference on Mass Spectrometry and Allied Topics, Atlanta, 1971, p. 229.
[170] J. V. Dugan and J. L. Magee, Proceedings of the 18th Annual Conference on Mass Spectrometry and Allied Topics, San Francisco, 1970, p. B395.
[171] R. C. Belsky and L. P. Gordon, *U.S. Govt. Res. Dev. Reports*, 1970, **70**, 69.
[172] R. L. Anderson and J. R. Boal, *U.S. Govt. Res. Dev. Reports*, 1970, **70**, 69.
[173] E. C. Toren, R. N. Carey, A. E. Sherry, and J. E. Davis, *Analyt. Chem.*, 1972, **44**, 339.
[174] G. Horlick, *Analyt. Chem.*, 1972, **44**, 943.
[175] See, for instance, the section on Hardware Design Philosophy in ref. 41.
[176] J. Farren, R. G. Houghton, and R. F. Warren, 'Trends in On-line Computer Control Systems', Institute of Electrical Engineers, Conference Publication 85, 1972, p. 20.

Whichever trend is followed more actively, there seems little doubt that applications of computer techniques in mass spectrometry will again grow considerably during the next two years.

7
Gas Chromatography–Mass Spectrometry

BY C. J. W. BROOKS AND B. S. MIDDLEDITCH

1 General Considerations

Introduction.—The technique of combined gas chromatography–mass spectrometry (g.c.–m.s.) has now diffused into very many laboratories: its broadened scope is perhaps associated with some diminution in its sharpness of impact. This Report follows, in the main, the pattern of the first: however, the availability of extra space, and the reduced need for introductory notes, have led us to include a more comprehensive list of references here. While we regard this trend towards annotated bibliography as undesirable in principle, we think it may be of practical value, pending the emergence of a more substantial review literature on g.c.–m.s., which would engender a more selective Report.

A survey of the literature indicates that only rarely are the current capabilities of g.c.–m.s. fully deployed. The power of accurately determined retention data in structural elucidation[1] tends still to be disregarded by some workers, though its value in certain areas (*e.g.* lipid and steroid analyses) has long been established. Greater enlistment of gas chromatographic evidence is, however, likely to ensue from the recent development of high-resolution glass columns,[2,3,4a] their successful adaptation for g.c.–m.s.,[4b,5] and their application to biological samples.[4b] It seems probable also that the advantages manifest in 'reaction gas chromatography', *i.e.* the controlled alteration of chemical character of samples *in transitu*, as elegantly developed by Beroza,[6] will eventually promote more general use of this approach in g.c.–m.s. In the meantime, technical advances have largely centred on mass spectrometry and on computer data acquisition and reduction. References especially relevant to g.c.–m.s. are included here: others are in Chapter 6.

Pertinent books and reviews are cited in the Table, with brief indications of their scope.

Practical Aspects of G.C.–M.S.—The most significant advances in the instrumentation of g.c.–m.s. have been concerned with automated data handling: this work is reviewed in Chapter 6.

[1] A. J. P. Martin, in 'Gas Chromatography in Biology and Medicine' (Ciba Foundation Symp.), ed. R. Porter, Churchill, London, 1969, p. 197.
[2] K. Grob, *Helv. Chim. Acta*, 1968, **51**, 718.
[3] M. Novotny and A. Zlatkis, *Chromatog. Rev.*, 1971, **14**, 1.
[4] (*a*) J. A. Völlmin, *Chromatographia*, 1970, **3**, 233; (*b*) *ibid.*, p. 238.
[5] P. Schulze and K. H. Kaiser, *Chromatographia*, 1971, **4**, 381.
[6] M. Beroza, *Accounts Chem. Res.*, 1970, **3**, 33.

Gas Chromatography–Mass Spectrometry

Table Books and reviews pertaining to g.c.–m.s.

Principal topic	References
Advances in m.s.	a, b
Techniques and applications of m.s.	c
Biochemical applications of m.s. (and g.c.–m.s.)	d—g
Analytical applications of m.s.	e
M.s. of organic compounds	h
Techniques of g.c.–m.s.	i—m
G.c.–m.s./computer techniques	n, o
Mass fragmentography	p
Multiple detectors for g.c.	q
Biochemical applications of preparative g.c.	r
M.s. of fatty acid derivatives	s—u
M.s. of steroids	s
M.s. of hormones	v
M.s. of heterocyclic compounds	w
M.s. of carbohydrates	x

(a) 'Advances in Mass Spectrometry', ed. A. Quayle, Institute of Petroleum, London, 1970, vol. 5 (763 pages); (b) 'Recent Topics in Mass Spectrometry', ed. R. I. Reed, Gordon and Breach, London, 1971 (355 pages); (c) 'Mass Spectrometry, Techniques and Applications', ed. G. W. A. Milne, Wiley–Interscience, New York, 1971 (521 pages); (d) 'Biochemical Applications of Mass Spectrometry', ed. G. R. Waller, Wiley–Interscience, New York, 1972 (872 pages); (e) A. L. Burlingame and G. A. Johanson, *Analyt. Chem.*, 1972, **44**, 337R—378R (1743 references); (f) J. A. McCloskey, in ref. a, pp. 715—719 (351 references); (g) G. W. A. Milne, in ref. c, pp. 327—371 (232 references); (h) D. H. Williams, in ref. a, pp. 569—588 (65 references); (i) C. Merritt, *Appl. Spectroscopy Reviews*, 1970, **3**, 263—325 (97 references); also ref. b, pp. 195—211 (45 references); (j) R. Ryhage and S. Wikström, in ref. c, pp. 91—119 (65 references); (k) R. A. Flath, in 'Guide to Modern Methods of Instrumental Analysis', ed. T. H. Gouw, Wiley–Interscience, 1972, pp. 323—350 (68 references); (l) G. A. Junk, *Internat. J. Mass Spectrometry Ion Phys.*, 1972, **8**, 1—71 (777 references); (m) D. Henneberg and G. Schomburg, in ref. a, pp. 605—614 (33 references); (n) E. J. Bonelli, M. S. Story, and J. B. Knight, in 'Dynamic Mass Spectrometry', Heyden and Son, London, 1971, Vol. 2, pp. 177—202 (12 references); (o) F. W. Karasek, *Analyt. Chem.*, 1972, **44**, 32A—42A (21 references); (p) C. G. Hammar, *Acta Pharm. Suecica*, 1971, **8**, 129—152; (q) H. Brandenberger, *Pharm. Acta Helv.*, 1970, **45**, 394—413; (r) W. J. A. VandenHeuvel and G. W. Kuron, in 'Preparative Gas Chromatography', ed. A. Zlatkis and V. Pretorius, Wiley–Interscience, New York, 1971, pp. 277—324 (174 references); (s) J. A. McCloskey, *Methods in Enzymology*, 1969, **14**, 382—450 (117 references); (t) J. A. McCloskey, in 'Topics in Lipid Chemistry', ed. F. D. Gunstone, Logos, London, 1970, Vol. 1, pp. 369—440 (95 references); (u) B. H. Kennett, in 'Biochemistry and Methodology of Lipids', ed. A. R. Johnson, Interscience, New York, 1971, pp. 251—263 (36 references); (v) L. L. Engel and J. C. Orr, in ref. d, pp. 537—572 (130 references); (w) G. Spiteller, in 'Physical Methods in Heterocyclic Chemistry', ed. A. R. Katritzky, Academic Press, New York, 1971, pp. 223—296 (327 references); (x) S. Hanessian, *Methods Biochem. Analysis*, 1971, **19**, 105—228 (116 references).

Excellent urinary steroid profiles have been obtained by using glass capillary columns for g.c.–m.s.,[7,8] although the problems of sample introduction and g.c.–m.s. coupling[9,10] are greater for capillary columns than for packed columns.

[7] J. A. Völlmin and H.-C. Curtius, *Z. klin. Chem. klin. Biochem.*, 1971, **9**, 43.
[8] A. Ros and I. F. Sommerville, *J. Obstet. Gynaecol.*, 1971, **78**, 1096.
[9] J. Eyem, in 'Column Chromatography', ed. E. sz. Kovats, Swiss Chemists' Association, 1970, p. 224.
[10] G. Nota, G. Marino, and A. Malorni, *Chem. and Ind.*, 1970, 1294.

A system for the controlled introduction of mass calibrant in capillary g.c.–m.s. has been devised.[11] The need for moderately polar stationary phases of higher thermal stability has yet to be fulfilled. The poly-m-carboranylene-siloxane phase Dexsil 300GC was expected to be capable of high-temperature operation,[12,13] but has proved somewhat disappointing in this respect.[14] Dexsil 400GC phases, which are similar to Dexsil 300GC but with some methyl groups replaced with phenyl groups, appear to be promising.[15] Applications of sample modification during g.c.–m.s. include hydrogenolysis of monoterpenoids[16] and deuterium–hydrogen exchange in terpenoids and steroids.[17]

The dynamics of molecular separators are now better understood, and it is realized that maintenance of optimum gas flow is essential for satisfactory operation.[18,19] It has been found that more satisfactory spectra of minor components of mixtures are obtained if the major components are not introduced into the ion source. A convenient instrumental modification is the inclusion of a pumped splitter between the g.c. column and a Watson–Biemann-type separator (Figure 1).[20] A method for determining trace constituents, using a membrane separator and incorporating a peak 'cutting' system, has been applied to volatile

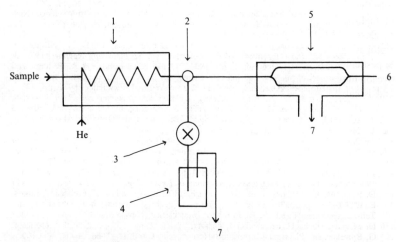

Figure 1 *1 Gas chromatograph; 2 Splitter; 3 Needle valve; 4 Cold trap; 5 Watson–Biemann separator; 6 Ion source; 7 To pumps*

[11] D. M. Schoengold and W. H. Stewart, *Analyt. Chem.*, 1972, **44**, 884.
[12] M. Novotny and A. Zlatkis, *J. Chromatog.*, 1971, **56**, 353.
[13] R. W. Finch, *Analabs Res. Notes*, 1970, **10**, 1.
[14] C. J. W. Brooks and B. S. Middleditch, *Clinica Chim. Acta*, 1971, **34**, 145.
[15] G. E. Pollock, *Analyt. Chem.*, 1972, **44**, 634.
[16] R. E. Kepner and H. Maarse, *J. Chromatog.*, 1972, **66**, 229.
[17] G. M. Anthony and C. J. W. Brooks, in 'Gas Chromatography 1970', ed. R. Stock, Institute of Petroleum, London, 1971, p. 70.
[18] F. W. Karasek and W. H. McFadden, *Res./Develop.*, 1971, **22**, No. 9, p. 52.
[19] R. Kaiser, *Z. analyt. Chem.*, 1970, **252**, 119.
[20] S. Goto and M. Noshiro, *Reports Res. Lab. Asahi Glass Co.*, 1970, **20**, 113.

nitrosamines.[21] A porous nickel molecular enricher[22] and a versatile gas chromatograph–mass spectrometer interface system employing a silver membrane separator[23] have been described. Further details have appeared of the palladium separator mentioned in the last Report.[24] A design has been proposed for a palladium alloy electrolytic separator. It is claimed that 100% of the sample is retained, while as much as 99.9999% of the carrier gas (hydrogen) is removed in one separation stage.[25] A novel trapping system and technique for indirect g.c.–m.s. has been described.[26]

A tandem thermogravimetric analyser–gas chromatograph–high-resolution mass spectrometer system has been described.[27] Chemical ionization (CI),[28] dual CI–electron impact,[29] and field ionization (FI)[30] sources appear to have great potential for g.c.–m.s.: in CI systems, the carrier gas may also serve as the reactant gas, thus obviating the need for a molecule separator.[31] (CI and FI methods are discussed more fully in Chapter 1.)

The technique of selective ion monitoring, which comprises single-ion monitoring, multiple-ion detection, and mass fragmentography (but is not to be confused with mass chromatography[32]), has been used extensively. When operated in this mode, g.c. with low-resolution m.s. can provide a detection method of great selectivity, superior in many practical applications to procedures relying wholly on high-resolution m.s.[33] G.c.–m.s. may be essential, for example, when mixtures of closely related compounds containing several relatively abundant natural isotopes are to be examined[34] or when single components of complex mixtures of pesticides[35] or steroid hormones[36] are to be quantitatively determined. Selective ion monitoring is particularly useful in the location of drug metabolites,[37–39] and is capable of greater sensitivity than is usually obtained by use of flame

[21] T. A. Gough and K. S. Webb, *J. Chromatog.*, 1972, **64**, 201.
[22] L. Charpenet, M. Eudier, H. Eustache, P. Jacquot, and A. Jacquot, *Chim. analyt.*, 1970, **52**, 1407.
[23] A. J. Luchte, jun. and D. C. Damoth, *Amer. Lab.*, 1970, No. 9, p. 33.
[24] J. E. Lovelock, P. G. Simmonds, G. R. Shoemake, and S. Rich, *J. Chromatog. Sci.*, 1970, **8**, 452; P. G. Simmonds, G. R. Shoemake, and J. E. Lovelock, *Analyt. Chem.*, 1970, **42**, 881.
[25] D. P. Lucero, *J. Chromatog. Sci.*, 1971, **9**, 105.
[26] F. W. Karasek and R. J. Smythe, *Analyt. Chem.*, 1971, **43**, 2008.
[27] Teh-Liang Chang and T. E. Mead, *Analyt. Chem.*, 1971, **43**, 534.
[28] D. M. Schoengold and B. Munson, *Analyt. Chem.*, 1970, **42**, 1811.
[29] G. P. Arsenault, J. J. Dolhun, and K. Biemann, *Analyt. Chem.*, 1971, **43**, 1720.
[30] J. N. Damico and R. P. Barron, *Analyt. Chem.*, 1971, **43**, 17.
[31] G. P. Arsenault, J. J. Dolhun, and K. Biemann, *Chem. Comm.*, 1970, 1542.
[32] C. E. Bennett, L. W. DiCave, jun., D. G. Paul, J. A. Wegener, and L. J. Levase, *Amer. Lab.*, 1971, No. 5, p. 67.
[33] A. G. Sharkey, in 'Recent Topics in Mass Spectrometry', ed. R. I. Reed, Gordon and Breach, New York, 1971, 127.
[34] B. Johansson, R. Ryhage, and G. Westöö, *Acta Chem. Scand.*, 1970, **24**, 2349.
[35] E. J. Bonelli, *Analyt. Chem.*, 1972, **44**, 603.
[36] L. Siekmann, H.-O. Hoppen, and H. Breuer, *Z. analyt. Chem.*, 1970, **252**, 294.
[37] C.-G. Hammar, B. Alexanderson, B. Holmstedt, and F. Sjöqvist, *Clin. Pharmacol. Ther.*, 1971, **12**, 496.
[38] C. J. W. Brooks, A. R. Thawley, P. Rocher, B. S. Middleditch, G. M. Anthony, and W. G. Stillwell, *J. Chromatog. Sci.*, 1971, **9**, 35.
[39] A. M. Lawson and C. J. W. Brooks, *Biochem. J.*, 1971, **123**, 25P.

ionization or (in g.c.–m.s.) total ion-current detection.[14,40,41] A range of sensitivities may be obtained even for different derivatives of the same compound,[42] and suitable internal standards must be employed.[43,44] Deuterium-labelled analogues of the components to be measured provide convenient standards for quantitation – a procedure first applied to prostaglandin derivatives.[45,46] A further example of the use of the method is the use of ^{2}H- and ^{13}C-labelled precursors in the investigation of metabolic processes *in vivo*. Thus, the potential hazards of using radioactive tracers are avoided.[47–51] A combined multiple ion detector–peak matcher device has been described,[52] and methods for recording a 'normal' chromatogram and one[14] or more[53] traces of ion intensities at particular m/e values on a multi-channel recorder have been used. Elegant computer-assisted display methods for selective ion monitoring are available[35,54] (see Chapter 6). Compounds containing polyisotopic elements such as Cl, Br, Hg, and Pb can be determined in ng quantities by repetitive scanning over a narrow range (3—20 a.m.u.) of the mass spectrum, in conjunction with the use of an on-line multi-channel analyser for data processing.[55]

Selection of Functional Derivatives for G.C.–M.S.—Trimethylsilyl (TMS) ethers and related derivatives have retained their dominant position in g.c.–m.s. Among their manifold and peculiar virtues are the following. They are readily preparable, mostly as stable compounds, from practically all classes of hydroxygroup and from many other functional types; they protect various groupings against thermal decomposition during g.c.; their low polarity renders them almost ideal for g.c., compensating for column imperfections and improving peak resolution; their steric properties often confer distinctive differences between diastereomers in respect of gas-chromatographic retention, and sometimes also of fragmentation pattern; they permit the volatilization of polyols of high molecular weight; they are chemically unreactive, suffer little adsorption during g.c.–m.s., and are rapidly removed from the ion source; their mass spectra frequently show clear molecular ions or indirect evidence thereof; their fragmen-

[40] B. Holmstedt, in 'Psychotomimetic Drugs', ed. D. H. Efron, Raven Press, New York, 1971, p. 151.
[41] R. Bonnichsen, A. C. Maehly, Y. Marde, R. Ryhage, and B. Schubert, *Zacchia*, 1970, **6**, 1.
[42] T. A. Baillie, C. J. W. Brooks, and B. S. Middleditch, *Analyt. Chem.*, 1972, **44**, 30.
[43] R. W. Kelly, *J. Chromatog.*, 1971, **54**, 345.
[44] S. H. Koslow, F. Cattabeni, and E. Costa, *Science*, 1972, **176**, 177.
[45] B. Samuelsson, M. Hamberg, and C. C. Sweeley, *Analyt. Biochem.*, 1970, **38**, 301.
[46] U. Axen, K. Gréen, D. Hörlin, and B. Samuelsson, *Biochem. Biophys. Res. Comm.*, 1971, **45**, 519.
[47] P. D. Klein, J. R. Haumann, and W. J. Eisler, *Analyt. Chem.*, 1972, **44**, 490.
[48] P. D. Klein, J. R. Haumann, and W. J. Eisler, *Clinical Chem.*, 1971, **17**, 735.
[49] T. E. Gaffney, C.-G. Hammar, B. Holmstedt, and R. E. McMahon, *Analyt. Chem.*, 1971, **43**, 307.
[50] J. L. Pinkus, D. Charles, and S. C. Chattoraj, *J. Biol. Chem.*, 1971, **246**, 633.
[51] E. Usdin, *Psychopharmacol. Bull.*, 1972, **8**, 14.
[52] C.-G. Hammar and R. Hessling, *Analyt. Chem.*, 1971, **43**, 298.
[53] R. D. McCoy, G. T. Porter, and B. O. Ayers, *Amer. Lab.*, 1971, No. 9, p. 17.
[54] R. Reimendal and J. Sjövall, *Analyt. Chem.*, 1972, **44**, 21.
[55] L. Bergstedt and G. Widmark, *Chromatographia*, 1970, **3**, 59.

tations, usually *via* facile α-cleavage, are structurally informative; in certain instances, ions relating to the silyl group and its environs are formed in very high relative abundance and serve as sensitive, specific indicators of particular structural features; fragment ions observed in low-resolution g.c.–m.s. can conveniently be interpreted with the aid of parallel results for 'labelled' TMS derivatives – notably [^{2}H$_9$]TMS, but also dimethylsilyl and especially halogenomethyl(dimethyl)silyl derivatives, which afford in addition distinctive chromatographic properties and isotopic patterns; finally, TMS ethers can, in general, be converted back into their parent compounds, or directly into other derivatives, in micro-scale reactions.

Disadvantages of TMS derivatives include their high mass increment (72 a.m.u. per reactive group), the tendency for partial conversion of unprotected ketones into enol TMS ethers, the ease of hydrolysis of certain classes (TMS esters, TMS amides), and the incidence in di- and multi-functional compounds of ion rearrangements and ion–molecule interactions which may complicate mass spectrometric interpretations. However, the compounds most affected are those for which the TMS derivatives aid the elucidation of stereochemistry: moreover, the abundance of certain 'rearrangement ions' has been found to be markedly dependent on structural (and in some instances stereochemical) factors, and may thus be turned into an advantage, as exemplified briefly below.

$$HC\begin{matrix}OR\\+\\OR\end{matrix} \quad \begin{matrix}a; R = SiMe_3 : m/e\ 191\\b; R = Me : m/e\ 75\end{matrix}$$

(1)

The ion of m/e 191 (1a), arising in high yield from vicinal TMS ether groups in certain cyclic polyol derivatives, serves as a characteristic ion for the determination of oestetrol (see p. 320). Formation of the corresponding ion of m/e 75 (1b) from analogous methyl ethers has been studied in detail for a range of cycloalkane-diols and -triols and found to be insensitive to configurations.[56a] (Stereochemically controlled elimination of methoxy-groups was, however, observed for ethers in which these groups were non-vicinal.[56b]) The ion of m/e 147, due to [Me$_2$SiOSiMe$_3$]$^+$, which appears in the mass spectra of many diol di-TMS ethers, has been studied in several trimethylsilylated pyranoses, in which its relative intensity (in 20 eV spectra) was related to the stereochemistry.[57] (Intensities of other ions in sugar TMS ether spectra also permit identification of diastereomers.[58]) Reference to the ion of m/e 147 in the spectra of steroid diol di-TMS ethers is made on p. 317.

[56] (a) J. Winkler and H. F. Grützmacher, *Org. Mass Spectrometry*, 1970, **3**, 1117; (b) *ibid.*, p. 1139.
[57] S. C. Havlicek, M. R. Brennan, and P. J. Scheuer, *Org. Mass Spectrometry*, 1971, **5**, 1273.
[58] J. Vink, J. H. W. Bruins Slot, J. J. de Ridder, J. P. Kamerling, and J. F. G. Vliegenthart, *J. Amer. Chem. Soc.*, 1972, **94**, 2542.

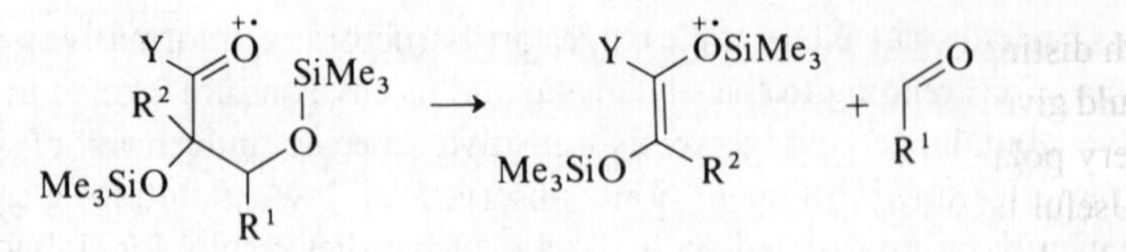

$$Y \underset{R^2}{\overset{\overset{+\cdot}{O}}{\diagdown}} SiMe_3 \longrightarrow Y \underset{Me_3SiO}{\overset{\overset{+\cdot}{OSiMe_3}}{\diagup}} R^2 + \underset{R^1}{\overset{O}{\diagup}} \quad (1)$$

$$\underset{Me_3SiO}{Y \overset{\overset{+\cdot}{O}}{\underset{R^2}{\diagdown}} \underset{R^1}{\overset{SiMe_3}{\underset{O}{|}}}}$$

(2)

Petersson[59] has explored the diagnostic utility of the McLafferty-type rearrangement [equation (1)] in a study by g.c.–m.s. of TMS derivatives of a large number of hydroxy-acids, hydroxy-ketones, and hydroxy-lactones. The rearrangement ions were found to be of great value in structural analysis, affording the base peak at 20 eV for many compounds. Among many other investigations on rearrangements of TMS derivatives under electron impact may be cited those on halogeno-acids and -alcohols[60] and steroidal phosphates[61] (see also p. 314). Related observations have been reported on the occurrence of ion–molecule reaction products, attributable to the intermolecular transfer of TMS groups, in a variety of substrates.[62]

As the above citations indicate, the rearrangements to which TMS derivatives are prone have added to, rather than detracted from, their interest.

Spiteller[63] has advocated the oxidation of hydroxylic steroids to the corresponding ketones for indirect characterization by m.s., and has interpreted the mass spectra of a number of hydroxy-keto-steroids.[64] Such correlations of the fragmentations of free steroids are of great intrinsic interest, but their applications in g.c.–m.s. are more limited than those of derivatives. For example, fluorinated corticosteroids may conveniently be identified in pharmaceutical preparations by mass spectrometry,[65] but where circumstances require g.c.–m.s., the use of derivatives is necessary.[66] Moreover, the conversion of alcohols into ketones merges many individual structures: the sixteen pregnane-3,16,20-triols (3),

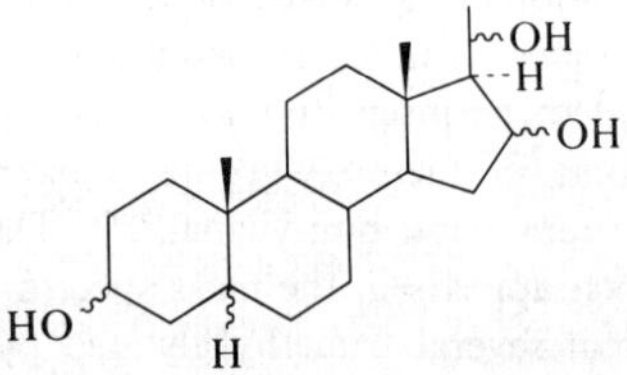

(3) 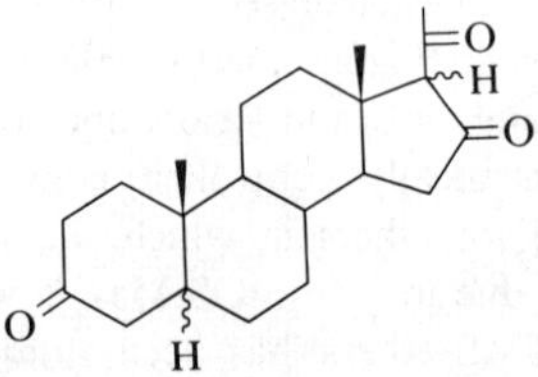   (4)

[59] G. Petersson, *Org. Mass Spectrometry*, 1972, **6**, 565, 577.
[60] E. White and J. A. McCloskey, *J. Org. Chem.*, 1970, **35**, 4241.
[61] D. J. Harvey, M. G. Horning, and P. Vouros, *Tetrahedron*, 1971, **27**, 4231.
[62] D. J. Harvey, M. G. Horning, and P. Vouros, *Chem. Comm.*, 1970, 898; *Analyt. Letters*, 1970, **3**, 489.
[63] H. Obermann, M. Spiteller-Friedmann, and G. Spiteller, *Chem. Ber.*, 1970, **103**, 1497.
[64] M. Ende and G. Spiteller, *Monatsh.*, 1971, **102**, 929 and references there cited.
[65] B. A. Lodge and P. Toft, *J. Pharm. Pharmacol.*, 1971, **23**, 196.
[66] C. J. W. Brooks and A. M. Lawson, *Excerpta Med. Int. Congr. Ser.*, 1972, No. 219, p. 238.

each distinguished by the retention time of its TMS ether (on SE-30 and QF-1),[67] would give rise to only two pregnanetriones (4), with the complications of forming a very polar β-diketone system and of epimerization at C-17.

Useful tables of key fragments and common mass differences in steroid mass spectra have been compiled[68] for free steroids and derivatives. Examples of old and new derivatives are cited in Section 2.

Current Trends.—As already mentioned, there has been great activity in data processing: reports from 16 laboratories on this topic form a timely and instructive section of G. R. Waller's book.[69] In respect of the technique of g.c.–m.s., the potentialities of alternative ionization methods are attracting much attention, and the use of combined EI–CI sources[70] seems certain to develop fruitfully.

2 Applications

Hydrocarbons.—The deuterolysis of neopentane over iron films has been studied by g.c.–m.s.[71] Detailed analyses of complex hydrocarbon mixtures from walnut leaf wax[72] and petroleum[73] have been reported, and illustrate well the value of g.c.–m.s. in the characterization of minor components. The non-conjugated heneicosahexaene mentioned in Volume 1 (p. 292) has been isolated from marine plankton.[74] A sex attractant pheromone of the housefly (*Musca domestica*) has been identified as (Z)-9-tricosene.[75] Branched mono-olefins of *Sarcina lutea* and *S. flava* have been characterized.[76] An interesting application of Beroza's micro-techniques for hydrogenolysis[16] has been made to a number of monoterpenes: cyclopropane rings were cleaved, but cyclobutane and cyclopentane rings remained intact.[16] Important evidence for the enzymic conversion of 12-methyl- and 12'-methyl-farnesyl pyrophosphate (5) into bishomosqualenes

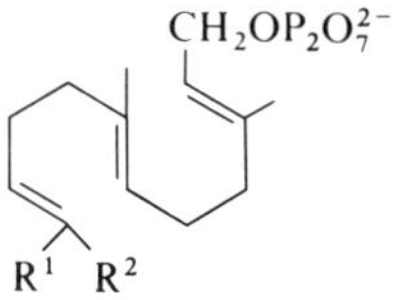

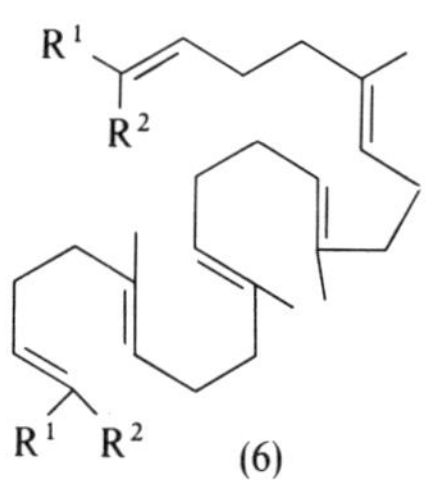

a; $R^1 = Et, R^2 = H$
b; $R^1 = H, R^2 = Et$

---

[67] J.-Å. Gustafsson and A. Stenberg, *European J. Biochem.*, 1971, **22**, 246.
[68] G. von Unruh and G. Spiteller, *Tetrahedron*, 1970, **26**, 3289, 3303, 3329.
[69] Ref. *d* (Table, p. 303), pp. 51—132.
[70] G. P. Arsenault, in ref. *d* (Table, p. 303), p. 817.
[71] R. S. Dowie, C. Kemball, J. C. Kempling, and D. A. Whan, *Proc. Roy. Soc.*, 1972, A**327**, 491.
[72] K. Stransky, M. Streibl, and V. Kubelka, *Coll. Czech. Chem. Comm.*, 1970, **35**, 882.
[73] M. Kuraš and S. Hála, *J. Chromatog.*, 1970, **51**, 45.
[74] M. Blumer, M. M. Mullin, and R. R. L. Guillard, *Marine Biol.*, 1970, **6**, 226.
[75] D. A. Carlson, M. S. Mayer, D. L. Silhacek, J. D. James, M. Beroza, and B. A. Bierl, *Science*, 1971, **174**, 76.
[76] T. G. Tornabene and S. P. Markey, *Lipids*, 1971, **6**, 190.

(6), and for the absence of monomethylsqualenes from the products, was obtained by g.c.–m.s.[77] Squalene and cholest-5-ene have been identified in human atherosclerotic plaques,[78] and hop-22(29)-ene has been found in prokaryotic organisms.[79–81]

Long-chain Compounds.—G.c.–m.s. has been used in model procedures, based on reduction of olefinic bonds with N_2D_4, for the characterization of polyenoic acids[82] and wax esters.[83] Location of double-bond positions in 16 isomeric octadecadienoic acids, by g.c.–m.s. (20 eV) of the derived tetra(trimethylsilyloxy)-octadecanoates, has been discussed.[84] A briefly described method for double-bond location, involving oxidative cleavage of the epoxide with HIO_4–ether, is suggested to be suitable for g.c.–m.s.[85] The resolution of enantiomeric ω2-hydroxy-alkanoic acids[86] as their diastereomeric N-(R)-1-phenylethylurethanes (7)[87,88] has been elegantly applied to demonstrate the stereospecificity of the

$$\begin{array}{c} Me \quad H \quad O \quad R^1 \quad R^2 \\ \diagdown \diagup \diagdown \diagup \diagdown \diagup \\ C \quad C \quad C \\ \diagup \diagdown \diagup \diagdown \diagup \diagdown \\ Ph \quad N \quad O \quad H \\ \quad H \end{array} \qquad \begin{array}{l} R^1 = Me \\ R^2 = (CH_2)_nCO_2Me \end{array} \} (S)\text{-Hydroxy-acid}$$

$$\begin{array}{l} R^1 = (CH_2)_nCO_2Me \\ R^2 = Me \end{array} \} (R)\text{-Hydroxy-acid}$$

(7)

microsomal ω2-hydroxylation of decanoic acid,[89] and to study the stereoselectivity of the oxido-reduction of 17-hydroxystearic acid by enzymes present in microsomal and soluble fractions of liver homogenates.[90] G.c.–m.s. has also allowed identification of oxo-acids obtained by microsomal dehydrogenation of ten isomeric hydroxystearic acids.[91]

Detailed analyses of straight- and branched-chain acids in human milk lipids[92] have required a wide range of physical and chemical methods, including

[77] K. Ogura, T. Koyama, and S. Seto, *J. Amer. Chem. Soc.*, 1972, **94**, 307.
[78] C. J. W. Brooks, G. Steel, and W. A. Harland, *Lipids*, 1971, **5**, 818.
[79] C. W. Bird, J. M. Lynch, S. J. Pirt, W. W. Reid, C. J. W. Brooks, and B. S. Middleditch, *Nature*, 1971, **230**, 473.
[80] M. de Rosa, A. Gambacorta, L. Minale, and J. D. Bu'Lock, *Chem. Comm.*, 1971, 619.
[81] C. W. Bird, J. M. Lynch, S. J. Pirt, and W. W. Reid, *Tetrahedron Letters*, 1971, 3189.
[82] K. K. Sun, H. W. Hayes, and R. T. Holman, *Org. Mass Spectrometry*, 1970, **3**, 1035.
[83] A. J. Aasen, H. H. Hofstetter, B. T. R. Iyengar, and R. T. Holman, *Lipids*, 1971, **6**, 502.
[84] U. Murawski, H. Egge, and F. Zilliken, *Z. Naturforsch.*, 1971, **26b**, 1241.
[85] K. Kusumran and N. Polgar, *Lipids*, 1971, **6**, 961.
[86] M. Hamberg, *Chem. and Phys. Lipids*, 1971, **6**, 152.
[87] W. Freytag and K. H. Ney, *J. Chromatog.*, 1969, **41**, 473.
[88] W. Pereira, V. A. Bacon, W. Patton, B. Halpern, and G. E. Pollock, *Analyt. Letters*, 1970, **3**, 23.
[89] M. Hamberg and I. Björkhem, *J. Biol. Chem.*, 1971, **246**, 7411.
[90] I. Björkhem and M. Hamberg, *J. Biol. Chem.*, 1971, **246**, 7417.
[91] I. Björkhem, *Biochim. Biophys. Acta*, 1972, **260**, 178.
[92] (a) H. Egge, U. Murawski, R. Ryhage, P. György, and F. Zilliken, *F.E.B.S. Letters*, 1970, **11**, 113; (b) H. Egge, U. Murawski, R. Ryhage, F. Zilliken, and P. György, *Z. analyt. Chem.*, 1970, **252**, 123; (c) U. Murawski, H. Egge, P. György, and F. Zilliken, *F.E.B.S. Letters*, 1971, **18**, 290; (d) H. Egge, U. Murawski, R. Ryhage, P. György, W. Chatranon, and F. Zilliken, *Chem. and Phys. Lipids*, 1972, **8**, 42.

g.c.–m.s. with packed and open tubular columns. Double bonds were located by oxidative cleavage[92b] or by hydroxylation and study of TMS ethers.[84,92c] Branched-chain acids from three sources – vernix caseosa,[93] *Staphylococcus aureus*,[94] and *Nocardia erythropolis*[95] – are exceedingly complex mixtures. 10-Methylpalmitic acid and 10-methylstearic acid were major components of the fatty acids of *Microspora parva*[96] as indicated, *inter alia*, by fragment ions at m/e 199, *viz.* $[\text{MeCH}(\text{CH}_2)_8\text{CO}_2\text{Me}]^+$. (Z)-6-Hexadecenoic acid was characterized in a seed oil[97] by the methoxymercuration–demercuration procedure of P. Abley *et al.* (Vol. 1, p. 294), and was among several alkenoic acids identified *via* hydroxylation–trimethylsilylation in other seed oils.[98] Hydroxy-acids and dicarboxylic acids of plant cutins have been characterized[99] and the formation of methyl ethers as artefacts during esterification has been noted.[100] Evidence for the occurrence of 9,10-epoxy-18-hydroxystearic acid in cutin has been obtained by g.c.–m.s. of the TMS ether of the trideuteriated triol formed by reduction with LiAlD_4,[101] though a keto-acid precursor is not rigorously excluded. Hydroperoxides of methyl oleate have been characterized by g.c.–m.s. of the hydroxy-esters obtained by SnCl_2 reduction, the corresponding hydrogenation products, and TMS ethers.[102] Preen-gland waxes of certain species of birds comprise complex mixtures of esters, branched in both acyl and alkyl moieties;[103] others contain dialkanoic esters of alkanediols such as octadecane-2,3-diol.[104] Sixteen compounds formed by autoxidation of 2,4-decadienal have been identified with the aid of g.c.–m.s.[105] *O*-Methyloximes of long-chain aldehydes and ketones have been studied.[106]

Prostaglandins.—The problem of determining these compounds in nanogram amounts was ingeniously solved by a technique[45] based on a generally applicable principle. Dilution of the material containing isotopes in natural abundance is effected with a much larger amount of an isotope-labelled sample of the

[93] N. Nicolaides, *Lipids*, 1971, **6**, 901.
[94] H. Egge, U. Murawski, W. Chatranon, and F. Zilliken, *Z. Naturforsch.*, 1971, **26b**, 893.
[95] I. Yano, K. Saito, Y. Furukawa, and M. Kusunose, *F.E.B.S. Letters*, 1972, **21**, 215.
[96] A. Ballio and S. Barcellona, *Gazzetta*, 1971, **101**, 635.
[97] G. F. Spencer, R. Kleiman, R. W. Miller, and F. R. Earle, *Lipids*, 1971, **6**, 712.
[98] C. M. Zorzut and P. Capella, *Riv. Ital. Sostanze Grasse*, 1969, **46**, 66.
[99] P. J. Holloway and A. H. B. Deas, *Phytochemistry*, 1971, **10**, 2781; P. J. Holloway, A. H. B. Deas, and A. M. Kabaara, *ibid.*, 1972, **11**, 1443; R. Croteau and I. S. Fagerson, *ibid.*, p. 353; P. M. Loveland and M. L. Laver, *ibid.*, p. 430; T. J. Walton and P. E. Kolattukudy, *Biochemistry*, 1972, **11**, 1885.
[100] P. J. Holloway and A. H. B. Deas, *Chem. and Ind.*, 1971, 1140.
[101] P. E. Kolattukudy, T. J. Walton, and R. P. S. Kushwaha, *Biochem. Biophys. Res. Comm.*, 1971, **42**, 739.
[102] M. Piretti, P. Capella, and U. Pallotta, *Riv. Ital. Sostanze Grasse*, 1969, **46**, 652; P. Capella, M. Piretti, and A. Strocchi, *ibid.*, p. 659.
[103] J. Jacob and A. Zeman, *Z. Naturforsch.*, 1970, **25b**, 984, 1438; 1971, **26b**, 33, 1344, 1352.
[104] K. Saito and M. Gamo, *J. Biochem.*, 1970, **67**, 841; M. Gamo, *Internat. J. Biochem.*, 1971, **2**, 574.
[105] R. F. Matthews, R. A. Scanlan, and L. M. Libbey, *J. Amer. Oil Chemists' Soc.*, 1971, **48**, 745.
[106] B. S. Middleditch and B. A. Knights, *Org. Mass Spectrometry*, 1972, **6**, 179.

compound to be determined. G.c.–m.s., and measurement of the ratios of natural and isotopic species in terms of the relative abundance of appropriate ions, then follows. The 'carrier' material largely obviates difficulties (adsorption, *etc.*) encountered in gas chromatography of nanogram quantities. [Corrections may be necessary for 'isotope fractionation' (*i.e.* different retention times of substrate and isotope analogue) and for isotope effects on the mass spectrometric fragmentation.] In the original example, prostaglandin E_1, as its O-methyloxime methyl ester, was added to the O-[$^2H_3$]methyloxime analogue, the ratios of ^1H to ^2H forms being from $1:10^3$ to $8:10^3$. Trimethylsilylation and g.c.–m.s., with monitoring of the base peaks at m/e 470 and 473 ($[M-71]^+$) afforded a satisfactory linear relation between the peak area ratios and the sample weight ratios in the range 3—10 ng of (unlabelled) prostaglandin E_1. In a later procedure, [3,3,4,4-$^2H_4$]prostaglandin E_2 and [3,3,4,4-$^2H_4$]prostaglandin $F_{2\alpha}$ (8) were used

Me ester triacetate: base peak m/e 318 $[M-60]^+$

(8)

as carriers.[46] Detection and/or estimation of prostaglandins in biological material have also been achieved with TMS ether methyl esters and trifluoroacetate methyl esters, by precise correlation of the points of scanning with the retention times of reference samples.[107] The choice of derivatives has been discussed,[108] and the advantages of using a multiple-ion detector in conjunction with a small analogue computer have been outlined.[109]

Studies of the metabolism of prostaglandins have used g.c.–m.s. extensively.[110,111] The major urinary metabolite in man[111] after intravenous administration of prostaglandin E_1 or E_2 was 7α-hydroxy-5,11-dioxotetranorprostane-1,16-dioic acid, also produced from 11α-hydroxy-9,15-dioxoprosta-5,13-dienoic acid (9a). The O-methyloxime TMS ether methyl ester (9b) afforded an interesting base peak at m/e 180: as the assignment is obscured by an error,[111] it may be noted that an ion of type (10) would have the correct mass.

G.c.–m.s. of cyclic boronate esters[112] has been applied to the analytical characterization of prostaglandins E_1, E_2, $F_{1\alpha}$, and $F_{2\alpha}$.[113]

[107] C. J. Thompson, M. Los, and E. W. Horton, *Life Sci., Part I*, 1970, **9**, 983; K. P. Bland, E. W. Horton, and N. L. Poyser, *ibid.*, 1971, **10**, 509; H. A. Davis, E. W. Horton, K. B. Jones, and J. P. Quilliam, *Brit. J. Pharmacol.*, 1971, **42**, 569.
[108] J. T. Watson and B. J. Sweetman, *J. Amer. Oil Chemists' Soc.*, 1971, **48**, 333A.
[109] R. W. Kelly, *Acta Endocrinol. Suppl.*, 1971, **155**, 221.
[110] E. Granström, *Biochim. Biophys. Acta*, 1971, **239**, 120; E. Granström and B. Samuelsson, *J. Biol. Chem.*, 1971, **246**, 7470; E. Granström, *European J. Biochem.*, 1972, **25**, 581.
[111] M. Hamberg and B. Samuelsson, *J. Biol. Chem.*, 1971, **246**, 6713.
[112] C. J. W. Brooks and J. Watson, *Chem. Comm.*, 1967, 952.
[113] C. Pace-Asciak and L. S. Wolfe, *J. Chromatog.*, 1971, **56**, 129.

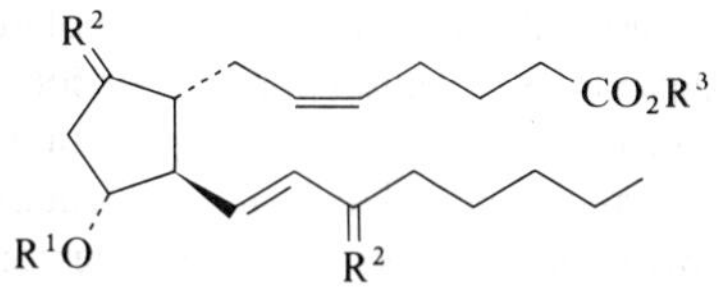

(9)

a; $R^1 = R^3 = H; R^2 = O$
b; $R^1 = TMS; R^2 = NOMe; R^3 = Me$

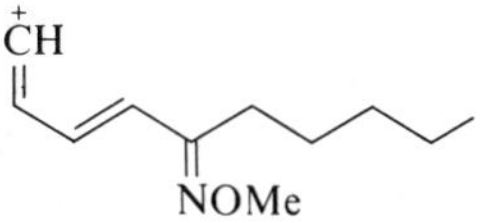

(10): m/e 180

Sphingosine Derivatives.—Cerebroside TMS ethers have mol. wt. exceeding 1000, but g.c.–m.s. continues to be fruitfully applied.[114–116] Two pathways of biosynthesis have been demonstrated by means of deuterium-labelled substrates: both ceramides[117] and glycosylsphinganines[118] underwent conversion into cerebrosides, in the presence, respectively, of mouse- and rat-brain microsomes. Deuterium incorporation was indicated by the occurrence of fragment ions corresponding in mass to relevant portions of the labelled substrate, as for example the ion (at [M − 467] in the unlabelled compounds) resulting from cleavage of the sugar moiety (Scheme 1).

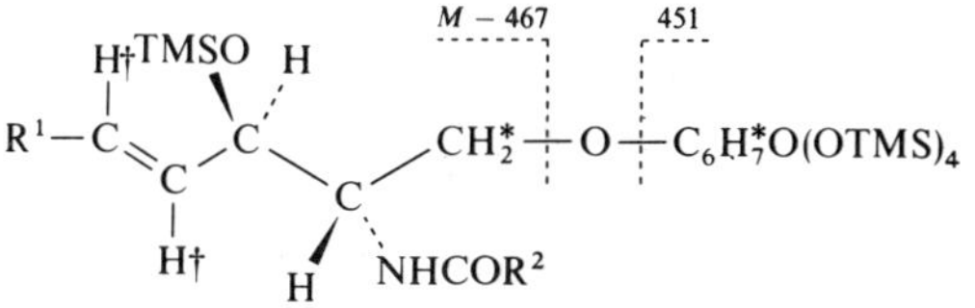

* Positions labelled in glycosylsphinganine substrates.
† Positions labelled in ceramide substrates.

Scheme 1

Stereospecific incorporation has been observed[119] of (S)-[2-$^2H_1$]palmitic acid into 4-hydroxysphinganine in the yeast *Hansenula ciferri*. G.c.–m.s. has been used to identify ceramides of human aortal lipids,[120] to secure evidence for a novel natural cerebroside containing xylose,[121] and to determine the structure of sphinga-4,8-dienine from oyster glycolipids.[122]

---

[114] K. Samuelsson and B. Samuelsson, *Chem. and Phys. Lipids*, 1970, **5**, 44.
[115] S. Hammarström, *European J. Biochem.*, 1971, **21**, 388.
[116] M. G. Horning, S. Murakami, and E. C. Horning, *Amer. J. Clin. Nutr.*, 1971, **24**, 1086.
[117] S. Hammarström, *Biochem. Biophys. Res. Comm.*, 1971, **45**, 468; S. Hammarström and B. Samuelsson, *J. Biol. Chem.*, 1972, **247**, 1001.
[118] S. Hammarström, *Biochem. Biophys. Res. Comm.*, 1971, **45**, 459.
[119] A. J. Polito and C. C. Sweeley, *J. Biol. Chem.*, 1971, **246**, 4178.
[120] M. Royer and J. L. Foote, *Chem. and Phys. Lipids*, 1971, **7**, 266.
[121] K.-A. Karlsson, B. E. Samuelsson, and G. O. Steen, *J. Lipid Res.*, 1972, **13**, 169.
[122] A. Hayashi and T. Matsubara, *Biochim. Biophys. Acta*, 1971, **248**, 306.

Carbohydrates.—Aldonitrile acetates, prepared by a simple procedure, have been studied by g.c.–m.s. as possible alternatives to alditol acetates for the analysis of polysaccharides. Excellent separations were obtained by g.l.c. on polyester phases, and the mass spectra of partially methylated aldonitrile acetates were informative.[123] Stereoisomeric pyranose TMS ethers have been distinguished,[57,58] and hexose epoxides analysed[124] by g.c.–m.s. The 40 eV mass spectrometric fragmentation of glucose O-methyloxime penta-TMS ether has been verified by comparison of labelled and unlabelled compounds.[125] Pyranose and furanose forms of 2-oxo-L-gulonic acid tetra-TMS ether TMS ester have been analysed.[126] Mass spectra of sugar phosphate[127] and glycerophosphate TMS derivatives,[128] which are notable for the abundance of rearrangement ions, have been further investigated.

Kärkkäinen[129] has developed useful procedures for di- and tri-saccharide analyses, based on borodeuteride reduction, methylation (Hakomori reagent), and g.c.–m.s. N-Acetylamino-groups were converted into N-acetyl-N-methylamino-groups. The mass spectra indicated the mol. wt. of the monosaccharide units and the position of the glycosidic bond next to the alditol unit (Scheme 2).

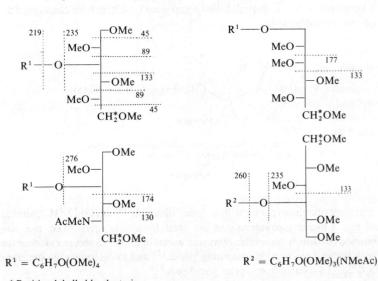

$R^1 = C_6H_7O(OMe)_4$ $R^2 = C_6H_7O(OMe)_3(NMeAc)$

* Position labelled by deuterium.

Scheme 2 *Major fragmentations of disaccharide alditol methyl ethers*

[123] B. A. Dmitriev, L. V. Backinowsky, O. S. Chizhov, B. M. Zolotarev, and N. K. Kochetkov, *Carbohydrate Res.*, 1971, **19**, 432.
[124] K. Capek and J. Jary, *Coll. Czech. Chem. Comm.*, 1970, **35**, 1727.
[125] R. A. Laine and C. C. Sweeley, *Analyt. Biochem.*, 1971, **43**, 533.
[126] H. G. J. de Wilt, *J. Chromatog.*, 1971, **63**, 379.
[127] D. J. Harvey, P. Vouros, and M. G. Horning, *Appl. Spectroscopy*, 1971, **25**, 139.
[128] J. H. Duncan, W. J. Lennarz, and C. C. Fenselau, *Biochemistry*, 1971, **10**, 927.
[129] J. Kärkkäinen, *Carbohydrate Res.*, 1970, **14**, 27; 1971, **17**, 11.

Permethylated methyl glycosides of trisaccharides have also been successfully studied[130] by g.c.–m.s., but the alditols yield simpler chromatograms (anomers being absent) and more readily interpretable spectra. G.c.–m.s. in the analysis of polysaccharides by methylation procedures has been reviewed.[131]

Research on natural carbohydrates has included the identification of α-ethylglucoside in sake,[132] analyses of carbohydrates in mushrooms,[133] and methylation analyses of polysaccharides of soya beans[134] and fungi (*Polyporus* spp.).[135]

A detailed survey of TMS ethers and acetates of the inositols indicated that although the TMS ethers yielded mass spectra consisting of the same ions, each diastereomer showed a characteristic intensity pattern.[136] The ion of m/e 507 is formed from all the isomers (by loss of CH_3 and trimethylsilanol) and was used in selective ion monitoring of inositols isolated from the American cockroach: *chiro*-inositol was identified for the first time in animal tissues.[137] An important application of g.c.–m.s. has been to elucidate structures of various esterified forms of indole-3-acetic acid (IAA) in maize kernels: three classes have been characterized as IAA myoinositols, IAA myoinositol arabinosides, and IAA myoinositol galactosides.[138]

The potential value of g.c.–m.s. in the analysis of cytokinins is signalized by the separation of the Δ^2- and Δ^3-isomers of N^6-(isopentenyl)adenosine by g.c. and their further distinctive characterization (in this instance, of samples collected from g.c.) by m.s.[139]

Oxygenated Terpenoids.—Separation and characterization of menthol stereomers and menthoglycols as TMS ethers, and of the glycols as cyclic n-butylboronates, has been reported.[140] Mass spectra of 24 monoterpenoid alcohol trifluoroacetates, recorded by g.c.–m.s., have been tabulated and discussed.[141]

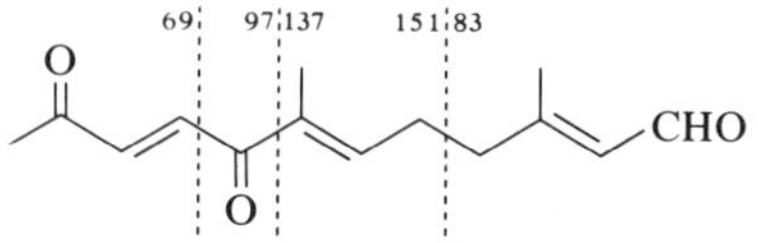

(11)

[130] J. Kärkkäinen, *Carbohydrate Res.*, 1971, **17**, 1.
[131] H. Björndal, C. G. Hellerqvist, B. Lindberg, and S. Svensson, *Angew. Chem. Internat. Edn.*, 1970, **9**, 610.
[132] T. Imanari and Z. Tamura, *Agric. and Biol. Chem. (Japan)*, 1971, **35**, 321.
[133] R. B. Holtz, *J. Agric. Food Chem.*, 1971, **19**, 1272.
[134] G. O. Aspinall and I. W. Cottrell, *Canad. J. Chem.*, 1971, **49**, 1019.
[135] H. Björndal and B. Lindberg, *Carbohydrate Res.*, 1970, **12**, 29.
[136] W. R. Sherman, N. C. Eilers, and S. L. Goodwin, *Org. Mass Spectrometry*, 1970, **3**, 829.
[137] P. P. Hipps, W. H. Holland, and W. R. Sherman, *Biochem. Biophys. Res. Comm.*, 1972, **46**, 1903.
[138] M. Ueda, A. Ehmann, and R. S. Bandurski, *Plant Physiol.*, 1970, **46**, 715.
[139] S. N. Alam and R. H. Hall, *Analyt. Biochem.*, 1971, **40**, 424.
[140] P. Bournot, B. F. Maume, and C. Baron, *J. Chromatog.*, 1971, **57**, 55.
[141] W. Ebbinghausen, E. Breitmaier, G. Jung, and W. Voelter, *Z. Naturforsch.*, 1970, **25b**, 1239.

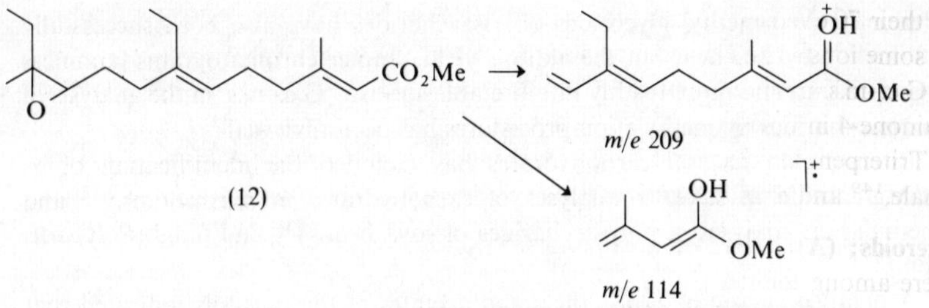

Scheme 3 *Juvenile hormone*

A nor-sesquiterpene aldehyde (11), isolated from the defence glands of gyrinid beetles, was identified by n.m.r., high-resolution m.s., and g.c.–m.s. of ozonolysis products.[142] An important method (Scheme 3) for the determination of insect juvenile hormone (12) from *Hyalophora cecropia* is based on the use of [^{2}H$_3$]-(12) (prepared by transesterification with CD$_3$OH) as a carrier and internal standard[143] (*cf.* p. 311): the authors estimate that 200 ng of (12) should be detectable in a lipid extract when 75 µg of [^{2}H$_3$]-(12) is added. Acyclic acids of isoprenoid origin occurring in shales may be partially identified and distinguished from associated 'pseudo'-isoprenoid acids.[144]

G.c.–m.s. remains an essential technique for the characterization of gibberellins and accompanying compounds in plant materials,[145] and has also been applied in a one-stage method for estimation of tocopherols and tocotrienols (as TMS ethers) in vegetable oils.[146] A notable example of the limitations of m.s. (also u.v., i.r., and n.m.r.) is afforded by the six possible structural isomers of dimethyl-phytyl-1-acetoxy-4-methoxybenzene (13). All of these, as well as α-tocopherol acetate (14) and '*ortho*-tocopherol acetate' (15), yielded the same fragment ions

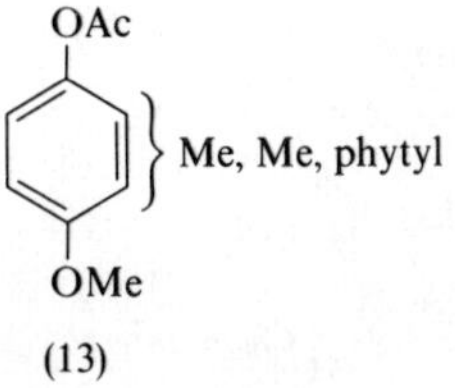

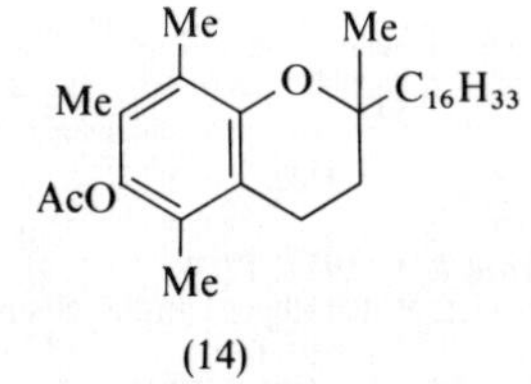

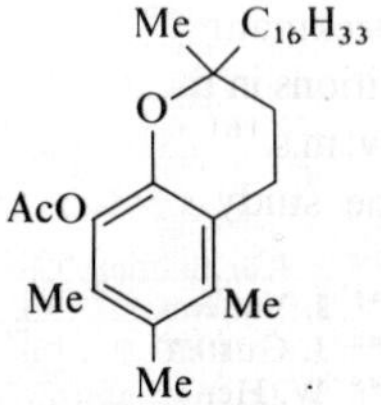

[142] J. Meinwald, K. Opheim, and T. Eisner, *Proc. Nat. Acad. Sci. U.S.A.*, 1972, **69**, 1208.
[143] M. A. Bieber, C. C. Sweeley, D. J. Faulkner, and M. R. Petersen, *Analyt. Biochem.*, 1972, **47**, 264.
[144] A. G. Douglas, M. Blumer, G. Eglinton, and K. Douraghi-Zadeh, *Tetrahedron*, 1971, **27**, 1071.
[145] R. C. Durley, J. MacMillan, and R. J. Pryce, *Phytochemistry*, 1971, **10**, 1891; A. Crozier, D. H. Bowen, J. MacMillan, D. M. Reid, and B. H. Most, *Planta (Berlin)*, 1971, **97**, 142.
[146] M. K. Govind Rao and E. G. Perkins, *J. Agric. Food Chem.*, 1972, **20**, 240.

in their 70 eV mass spectra, the only differences being in the relative intensities of some ions. The isomers appeared to be well distinguished by g.c.[147]

G.c.–m.s. has been applied to the analysis of minute amounts of hexahydroubiquinone-4 in lipid extracts.[148] Triterpenoids identified by g.c.–m.s. include tetrahymanol from Green River Shale,[149] and α- and β-amyrin from the blue-green alga *Spirulina platensis*.[150]

Steroids: (A) Reference Compounds.—*Alcohols*. Sterol methyl ethers, which were among the derivatives earliest used for g.c. of sterols, show characteristic mass spectra[151,152] and are potentially useful for g.c.–m.s. Methoxymethyl ethers[153] yield intense molecular ions. Sterically hindered 17α-hydroxy-groups are silylated by *N*-TMS-imidazole at 100 °C: where O-methyloximation of ketogroups has preceded this reaction, the excess of reagents (MeONH$_3^+$ Cl$^-$, pyH$^+$ Cl$^-$) catalyses silylation.[154] Oestr-4-en-17β-ol and its 17α-ethynyl, 17α-ethyl, and 17α-allyl analogues are detectable in subnanogram amounts by selective ion monitoring of free alcohols and chloromethyl(dimethyl)silyl ethers.[14] The incidence, in mass spectra of steroid diol di-TMS ethers, of the 'rearrangement ion' of *m/e* 147 ([Me$_3$SiOSiMe$_2$]$^+$), is such as to suggest that this results from a stereoselective interaction of the two substituents, not normally dependent on ring cleavage.[155] Related observations have been made on the loss of D$_2$O from steroid diols labelled in both hydroxy-groups.[156]

Ketones. A simple microchemical conversion of 3-oxo-Δ^4-steroids into 3β-hydroxy-Δ^5-steroids is achieved by reduction of the dienol TMS ethers with NaBH$_4$–Pr^{i}OH.[157] 'Persilylation' of steroids with hexamethyldisilazane, catalysed by Me$_3$SiBr at room temperature, effects trimethylsilylation even of 11β-hydroxy-groups and converts most ketones into enol TMS ethers: incomplete reaction was noted for 11-ketones and 20,17-ketols. The 'persilylated' steroids are useful for g.c.–m.s.[158] Satisfactory preparations of enol heptafluorobutyrates (HFB) of 3-oxo-Δ^4-steroids have been reported, with g.c.–m.s. data.[159,160] Beckmann fission of 17-oxo-steroid oxime derivatives occurs under certain conditions in the injector zone of a gas chromatograph, yielding nitriles, as confirmed by m.s.[161] *O*-Benzyloximes have been introduced as valuable derivatives for the study of ketonic steroids: their long retention times allow a complete

[147] J. Vance and R. Bentley, *Bio-organic Chem.*, 1971, **1**, 329, 345.
[148] J. Gürtler and R. Blomstrand, *Internat. J. Vitamin Nutr. Res.*, 1971, **41**, 204.
[149] W. Henderson and G. Steel, *Chem. Comm.*, 1971, 1331.
[150] M.-C. Forin, B. Maume, and C. Baron, *Compt. rend.*, 1972, **274**, *D*, 133.
[151] D. R. Idler, L. M. Safe, and S. H. Safe, *Steroids*, 1970, **16**, 251.
[152] C. R. Narayanan and A. K. Lala, *Org. Mass Spectrometry*, 1972, **6**, 119.
[153] E. S. Waight, J. E. Herz, and Y. Santoyo, *Org. Mass Spectrometry*, 1971, **5**, 359.
[154] J. P. Thenot and E. C. Horning, *Analyt. Letters*, 1972, **5**, 21.
[155] S. Sloan, D. J. Harvey, and P. Vouros, *Org. Mass Spectrometry*, 1971, **5**, 789.
[156] C. C. Fenselau and C. H. Robinson, *J. Amer. Chem. Soc.*, 1971, **93**, 3070.
[157] L. Aringer and P. Eneroth, *Steroids*, 1971, **18**, 381.
[158] L. Aringer, P. Eneroth, and J.-Å. Gustafsson, *Steroids*, 1971, **17**, 377.
[159] J. R. G. Challis and R. B. Heap, *J. Chromatog.*, 1970, **50**, 228.
[160] L. Dehennin, A. Reiffsteck, and R. Scholler, *J. Chromatog. Sci.*, 1972, **10**, 224.
[161] J. P. Thenot and E. C. Horning, *Analyt. Letters*, 1971, **4**, 683.

separation from hydroxy-steroids and unreactive ketones.[162] Dehydrogenation of [7β-²H]androst-4-ene-3,17-dione with chloranil gave the 4,6-dienedione, largely retaining the deuterium, as shown by g.c.–m.s.[163]

Corticosteroids.[154,158] Derivatives of fluorinated corticosteroids and their side-chain degradation products have been characterized by g.c.–m.s.:[66] mass spectra of the free steroids have been discussed.[65] Several classes of corticosteroid derivative have been compared by g.c.–m.s.,[42] and cyclic boronates of various corticosteroids have been studied.[164]

Steroids: (B) in Biological Material.—*Sterols.* G.c.–m.s. is now regularly used in many laboratories concerned with phyto- and zoo-sterols. 4α,14α-Dimethyl-5α-cholest-8-en-3β-ol (31-nordihydrolanosterol), a possible biosynthetic precursor of lophenol, has been identified in dandelion pollen,[165] and 4α-methyl-sterols have been found (only in esterified form) in the pitcher plant.[166] Changes in sterol and sterol ester composition during greening of etiolated barley shoots have been studied.[167] Sterols of algae (red[168] and green[169]) have been characterized, though the mass spectrum reported for the TMS ether of 'sterol 2' in ref. 169 is not consistent with the proposed structure (5α-ergosta-7,22-dien-3β-ol).[170]* Cholesterol occurs in cigarette tobacco and cigarette smoke.[171] G.c.–m.s. of TMS ethers of steryl glucosides has been effected with a column 1 m long (1% SE-30): glucosides of stigmasterol and β-sitosterol were identified in *Phaseolus aureus*.[172]

Among many interesting marine sterols,[173,174] the C_{26}-sterols[175] are detected particularly readily by g.l.c.,[174—176] while m.s.[175,176] and g.c.–m.s.[174] have afforded important structural evidence. Another new natural sterol, cholesta-5,(Z)-22-dien-3β-ol, has been found in the scallop.[177]

24-Methylenecholesterol has been shown to be an intermediate in the dealkylation of campesterol by the tobacco hornworm.[178] Sterols of milk fat

[162] P. G. Devaux, M. G. Horning, and E. C. Horning, *Analyt. Letters*, 1971, **4**, 151.
[163] J. C. Orr and J. M. Broughton, *J. Org. Chem.*, 1970, **35**, 1126.
[164] C. J. W. Brooks, B. S. Middleditch, and D. J. Harvey, *Org. Mass Spectrometry*, 1971, **5**, 1429.
[165] A. M. Atallah and H. J. Nicholas, *Steroids*, 1971, **17**, 611.
[166] A. S. Wan, R. T. Aexel, R. B. Ramsey, and H. J. Nicholas, *Phytochemistry*, 1972, **11**, 456.
[167] P. B. Bush, C. Grunwald, and D. L. Davis, *Plant. Physiol.*, 1971, **47**, 745.
[168] G. H. Beastall, H. H. Rees, and T. W. Goodwin, *Tetrahedron Letters*, 1971, 4935.
[169] D. M. Orcutt and B. Richardson, *Steroids*, 1970, **16**, 429.
[170] C. Djerassi, also G. Steel and C. J. W. Brooks, unpublished observations.
[171] C. Grunwald, D. L. Davis, and L. P. Bush, *J. Agric. Food Chem.*, 1971, **19**, 138.
[172] R. A. Laine and A. D. Elbein, *Biochemistry*, 1971, **10**, 2547.
[173] D. R. Idler and P. Wiseman, *Internat. J. Biochem.*, 1971, **2**, 91.
[174] L. J. Goad, I. Rubinstein, and A. G. Smith, *Proc. Roy. Soc.*, 1972, **B180**, 223.
[175] D. R. Idler, P. M. Wiseman, and L. M. Safe, *Steroids*, 1970, **16**, 451.
[176] S.-I. Teshima, A. Kanazawa, and T. Ando, *Comp. Biochem. Physiol.*, 1972, **B41**, 121.
[177] D. R. Idler and P. Wiseman, *Comp. Biochem. Physiol.*, 1971, **A38**, 581.
[178] J. A. Svoboda, M. J. Thompson, and W. E. Robbins, *Lipids*, 1972, **7**, 156.

* Dr. Orcutt has informed us that the peak at *m/e* 400 was apparently spurious.

have been analysed,[179] and cholesterol α-oxide has been identified in human serum.[180] Lanosterol and several later intermediates in the biogenesis of cholesterol have been identified in meconium and faeces from newborn infants, together with 22-, 23-, and 24-hydroxy- and 20,22-dihydroxy-cholesterol.[181]

Bile Acids. Very informative reviews have appeared of the application of g.c.–m.s.[182] and m.s.[183,184] to the analytical characterization of bile acids. Marked incorporation of deuterium into bile acids in rats with a bile fistula given $C^1H_3C^2H_2O^1H$ has been demonstrated: deuterium is transferred *via* NADH to the NADPH that is used in the biosynthesis of cholesterol and bile acids.[185]

Hormonal Steroids and Metabolites in the Human. Systematic studies, by the Helsinki school, of oestrogens in various biological fluids, using g.c.–m.s. in conjunction with other methods, convincingly show the power of the technique in identifying and determining individual phenolic steroids in these complex mixtures.[186—188] In some instances (*e.g.* 6α- and 6β-hydroxyoestradiols), epimers can be distinguished by the mass spectra of their TMS ethers at low electron energy.[188] Oestrogen production rates during pregnancy were estimated with the aid of deuterium-labelled oestradiol, isotope dilution data being obtained by g.c.–m.s.[50] Oestrone and oestradiol in plasma were determined by adding [6,7-$^2H_2$]-labelled material as internal standard and measuring the ratios of molecular ion peaks of the TMS ethers (m/e 342/344; 416/418 respectively).[36] G.c.–m.s. confirmed the specificity of a gas-chromatographic estimation of 15α-hydroxyoestriol ['oestetrol' (16)] in pregnancy urine:[189] another method

[179] C. R. Brewington, E. A. Caress, and D. P. Schwartz, *J. Lipid Res.*, 1970, **11**, 355.
[180] M. F. Gray, T. D. V. Lawrie, and C. J. W. Brooks, *Lipids*, 1971, **6**, 836.
[181] J.-Å. Gustafsson and P. Eneroth, *Proc. Roy. Soc.*, 1972, **B180**, 179.
[182] J. Sjövall, in 'Bile Salt Metabolism', ed. L. Schiff, J. B. Carey, and J. M. Dietschy, C. C. Thomas, Springfield, Ill., 1969, p. 205.
[183] J. Sjövall, P. Eneroth, and R. Ryhage, in 'The Bile Acids: Chemistry, Physiology, and Metabolism', ed. P. P. Nair and D. Kritchevsky, Plenum Press, New York, 1971, Chap. 7.
[184] W. H. Elliott, in 'Biochemical Applications of Mass Spectrometry', ed. G. R. Waller, Wiley–Interscience, New York, 1971, p. 291.
[185] T. Cronholm, I. Makino, and J. Sjövall, *European J. Biochem.*, 1972, **24**, 507.
[186] H. Adlercreutz, M. Ikonen, and T. Luukkainen, *Proc. 7th Int. Congr. Clin. Chem.*, Geneva/Evian, 1970, **3**, 14; A. L. Siegel, H. Adlercreutz, and T. Luukkainen, *Ann. Med. Exp. Fenn.*, 1969, **47**, 22.
[187] H. Adlercreutz, *Clinica Chim. Acta*, 1971, **34**, 231.
[188] H. Adlercreutz and T. Luukkainen, *Z. klin. Chem., klin. Biochem.*, 1971, **9**, 421.
[189] J. Heikkilä and H. Adlercreutz, *J. Steroid Biochem.*, 1970, **1**, 243.

of determination was based on selective monitoring of the fragment ion (1a) of m/e 191 formed from oestetrol tetra-TMS ether and from the tetra-TMS ether of the analogue (17) used as an internal standard.[43]

Investigations have been made of hormonal metabolism *in utero*, as evidenced by steroids in foetal tissues[190] and in amniotic fluid from normal[191] and anencephalic[192] pregnancies. Among a large number of C_{19} and C_{21} steroids[193] occurring as sulphates in pregnancy urine is the first reported example[194] in the human of a neutral 15β-hydroxy-steroid [(18); pregn-5-ene-3β,15β,17α,20α-tetraol]: it is suggested that this may be an indicator of steroid endocrine activity of the foetus.

```
                    ┌─OTMS    117 (98%)
                    │··OH     449 (100%)
                    
                         OTMS
     TMSO              H      261 (25%)
              (18)
```

The Karolinska group have continued their detailed explorations of steroid metabolism by newborn and young infants, through analyses of faecal,[195] meconial,[67] and urinary[196] steroid compositions. Among several Δ⁵- and saturated C_{19}- and C_{21}-steroids newly identified in infant urine were 18-hydroxy-DHEA (3β,18-dihydroxyandrost-5-en-17-one), 3β-hydroxypregna-5,16-dien-20-one (also reported from a case of adrenogenital syndrome),[197] and a group of 16-hydroxy-pregnanes and -pregnenes. A 1,3,16-trihydroxypregnan-20-one preponderated in the urine of an anencephalic newborn infant.[198] Groups of 7-hydroxylated C_{19}-steroids,[199] and 16-hydroxylated C_{19}-steroids[200] have been identified in sulphate fractions of adult urine.

Studies of infant and adult urinary steroids by 'steroid profile' procedures developed in Houston have been usefully extended by the introduction of O-benzyloxime TMS ethers[201] to complement the O-methyloxime TMS

[190] I. Huhtaniemi, M. Ikonen, and R. Vihko, *Biochem. Biophys. Res. Comm.*, 1970, **38**, 715; I. Huhtaniemi, T. Luukkainen, and R. Vihko, *Acta Endocrinol.*, 1970, **64**, 273.
[191] O. Jänne and R. Vihko, *J. Steroid Biochem.*, 1970, **1**, 279.
[192] T. Luukkainen, E. A. Michie, and R. Vihko, *J. Endocrinol.*, 1971, **51**, 109.
[193] O. Jänne and R. Vihko, *Acta Endocrinol.*, 1970, **65**, 50; H. Eriksson and J.-Å. Gustafsson, *European J. Biochem.*, 1970, **16**, 268.
[194] O. A. Jänne and R. Vihko, *Excerpta Med. Int. Congr. Ser.*, 1970, No. 219, p. 219.
[195] J.-Å. Gustafsson, C. H. L. Shackleton, and J. Sjövall, *Acta Endocrinol.*, 1970, **65**, 18.
[196] C. H. L. Shackleton, J.-Å. Gustafsson, and J. Sjövall, *Steroids*, 1971, **17**, 265.
[197] N. E. Brandstrup and L. R. Trieber, *J. Steroid Biochem.*, 1971, **2**, 133.
[198] P. Eneroth, H. Ferngren, J.-Å. Gustafsson, B. Ivemark, and A. Stenberg, *Acta Endocrinol.*, 1972, **70**, 113.
[199] O. Jänne and R. Vihko, *J. Steroid Biochem.*, 1970, **1**, 177.
[200] O. Jänne, *J. Steroid Biochem.*, 1971, **2**, 33.
[201] P. G. Devaux, M. G. Horning, R. M. Hill, and E. C. Horning, *Analyt. Biochem.*, 1971, **41**, 70.

ethers[202,203] originally introduced. The latter derivatives have been used in a capillary g.c. method for estimating urinary 'tetrahydro S' and 'tetrahydro DOC', the identity of the compounds being confirmed by g.c.–m.s.[204]

Hormonal Steroids of Animals or Plants. Urinary steroids of a newborn chimpanzee[205] and a pregnant macaque monkey[206] include many metabolites found in analogous human subjects: three androstane-3,16,17-triols,[206] well separated by g.c., await stereochemical assignments. Interest in the pheromonal Δ^{16}-C_{19}-steroids of the boar has continued.[207] A group of 15-hydroxy-14β-H-C_{21}-steroids has been partially characterized in faeces and urine from female rats and the effects of intestinal microflora have been studied.[208] 15α-Hydroxylation of 3β-hydroxy-5α-androstan-17-one, during incubation *in vitro* with rat liver microsomes, has been observed,[209] among numerous other modes of hydroxylation by such preparations.[210]

Ikekawa and co-workers[211] have reported a promising method for the determination of phytoecdysones by g.l.c. of TMS ethers and HFB esters. G.c.–m.s. confirmed the structures of the derivatives.

Amines: (A) Reference Compounds and Reaction Products.—Picogram quantities of primary amines may be detected by g.c. as their Schiff bases formed by reaction with C_6F_5CHO,[212] and *N*-pentafluorobenzamides formed from primary and secondary amines are also good derivatives for electron-capture detection:[213] the structures have been verified by g.c.–m.s. Primary amines can be conveniently characterized as the corresponding isothiocyanates, formed by reaction with CS_2:[214] the mass spectra of phenethyl isothiocyanates showed the expected benzylic ions due to loss of ·CHR·NCS, while the compounds (19)

(19)   a; R = H
       b; R = Me

[202] E. C. Horning and M. G. Horning, *J. Chromatog. Sci.*, 1971, **9**, 129.
[203] G. M. Maume, R. J. Bégué, B. F. Maume, and P. Padieu, *Semaine Hôp. Paris*, 1970, **46**, 2938.
[204] P. Koepp, J. A. Völlmin, M. Zachmann, and H.-C. Curtius, *Acta Endocrinol.*, 1971, **66**, 756.
[205] C. H. L. Shackleton and J.-Å. Gustafsson, *Steroids*, 1971, **18**, 175.
[206] M. E. Manson, C. H. L. Shackleton, F. L. Mitchell, J.-Å. Gustafsson, and J. Sjövall, *Steroids*, 1971, **18**, 51.
[207] D. B. Gower, F. A. Harrison, and R. B. Heap, *J. Endocrinol.*, 1970, **47**, 357; K. E. Beery, J. D. Sink, S. Patton, and J. H. Ziegler, *J. Food Sci.*, 1971, **36**, 1086.
[208] H. Eriksson, J.-Å. Gustafsson, and J. Sjövall, *European J. Biochem.*, 1971, **19**, 433.
[209] J.-Å. Gustafsson and B. P. Lisboa, *Biochim. Biophys. Acta*, 1970, **210**, 199.
[210] J.-Å. Gustafsson and B. P. Lisboa, *European J. Biochem.*, 1970, **12**, 369; 1970, **16**, 475; *Acta Endocrinol.*, 1970, **65**, 84.
[211] N. Ikekawa, F. Hattori, J. Rubio-Lightbourn, H. Miyazaki, M. Ishibashi, and C. Mori, *J. Chromatog. Sci.*, 1972, **10**, 233.
[212] A. C. Moffat and E. C. Horning, *Biochim. Biophys. Acta*, 1970, **222**, 248.
[213] A. C. Moffat, E. C. Horning, S. B. Matin, and M. Rowland, *J. Chromatog.*, 1972, **66**, 255.
[214] N. Narasimhachari and P. Vouros, *Analyt. Biochem.*, 1972, **45**, 154.

yielded strong peaks at m/e 117 and 131 ascribed to the loss of ·NCS. TMS derivatives of NN-dimethyltryptamine (DMT) and its 5-hydroxy- and 5-methoxy-analogues have been studied,[215] and DMT (produced by *in vitro* methylation of the monomethyl amine) has been analysed, with addition of a [$^2H_2$]-DMT as internal standard, by g.c.–m.s.[216] TMS derivatives of natural and synthetic pterins have been separated and characterized.[217]

G.c.–m.s. has been applied to the analysis of commercial oligoethylene oligoamines,[218] of products arising from hydroboration of ω-dimethylaminoalkenes,[219] and of reaction products from amines and CCl_4.[220] Successful g.c. of ergot alkaloids[221] suggests that g.c.–m.s. could be of value in studies of these compounds.

A gas-chromatographic procedure for the analysis of aqueous mixtures of lower fatty acids, amides, and N-acylamides has been checked by g.c.–m.s.,[222] which has also provided effective identification of methylvinylmaleimide and formylethylmaleimide formed by oxidation of haem and of chlorophylls *a* and *b*.[223] Pyrrolidides of carboxylic acids may be useful as an indicator of structural features, by virtue of their induction of electron-impact fragmentation at the β- and γ-bonds.[224]

Amines: (B) in Biological Material.—Monitoring of characteristic mass spectrometric fragments during g.c. ('mass fragmentography') has been used to detect nuclear-substituted phenethylamines in extracts of cacti. Ions at m/e 137 and 138 are typical of nuclear OH + OMe (as in 3-methoxytyramine), whereas ions at m/e 167 and 168 serve to indicate OH + 2OMe.[225] A similar technique, based on pentafluoropropionyl derivatives, permits the concurrent determination of norepinephrine and dopamine in amounts below 1 ng in tissue samples.[44] Melatonin (N-acetyl-5-methoxytryptamine) has been identified in chicken blood[226] and estimated in rat pineal glands[227] by g.c.–m.s. Quantitative determination of the antidepressant drug nortriptyline (20), and some of its metabolites, in body fluids has been described.[228,37] The specificity of a gas-

---

[215] N. Narasimhachari, J. Spaide, and B. Heller, *J. Chromatog. Sci.*, 1971, **9**, 502.
[216] R. W. Walker, Ho Sam Ahn, L. R. Mandel, and W. J. A. VandenHeuvel, *Analyt. Biochem.*, 1972, **47**, 228.
[217] T. Lloyd, S. Markey, and N. Weiner, *Analyt. Biochem.*, 1971, **42**, 108.
[218] L. Bergstedt and G. Widmark, *Acta Chem. Scand.*, 1970, **24**, 2713.
[219] Z. Polivka, V. Kubelka, N. Holubova, and M. Ferles, *Coll. Czech. Chem. Comm.*, 1970, **35**, 1131.
[220] J. R. Lindsay Smith and Z. A. Malik, *J. Chem. Soc. (B)*, 1970, 617.
[221] S. Agurell and A. Ohlsson, *J. Chromatog.*, 1971, **61**, 339.
[222] T. Nakagawa, J. H. Vermeer, and J. R. Dean, *J. Chromatog. Sci.*, 1971, **9**, 293.
[223] A. L. Williamson and R. K. Ellsworth, *Analyt. Biochem.*, 1971, **43**, 633.
[224] W. Vetter, W. Walther, and M. Vecchi, *Helv. Chim. Acta*, 1971, **54**, 1599.
[225] J. E. Lindgren, S. Agurell, J. Lundström, and U. Svensson, *F.E.B.S. Letters*, 1971, **13**, 21.
[226] R. W. Pelham, C. L. Ralph, and I. M. Campbell, *Biochem. Biophys. Res. Comm.*, 1972, **46**, 1236.
[227] P. H. Degen, J. R. DoAmaral, and J. D. Barchas, *Analyt. Biochem.*, 1972, **45**, 634.
[228] O. Borga, L. Palmér, A. Linnarsson, and B. Holmstedt, *Analyt. Letters*, 1971, **4**, 837.

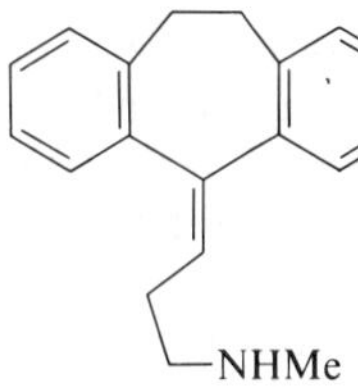

chromatographic determination of pyrimethamine (21) was readily checked by repetitive scanning over the molecular ion region, since the ions $M^{+\cdot}$ (248, 250) and $[M - 1]^+$ (247, 249) accounted for practically the entire ion current.[229]

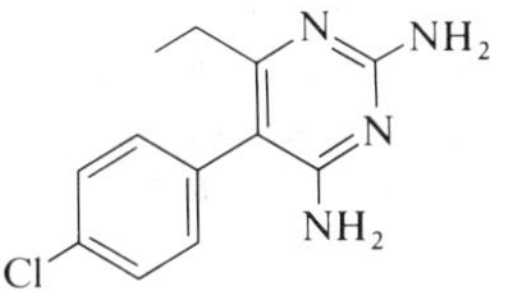

Amino-acids and Peptides.—G.c.–m.s. affords an excellent method of determining relative amounts of enantiomeric amino-acids, exemplified by model analyses of ($\pm$)-threonines and ($\pm$)-allothreonines as TMS ethers of the diastereomeric amides formed with (R)- and (S)-isomers of N-trifluoroacetylprolyl chloride: for example, the retention times of (S)-prolyl amides of (R)- and (S)-threonine were 7.2 and 9.2 min respectively on a 1.8 m column at 200 °C (10% diethyleneglycol succinate packing).[230] Separations of enantiomeric N-trifluoroacetyl-amino-acid isopropyl esters on chiral phases, such as N-trifluoroacetyl-(S)-phenylalanyl-(S)-leucine cyclohexyl ester,[231] are best achieved on capillary columns. A survey of mass spectra of N-trifluoroacetyl n-butyl esters of α-, β-, and γ-amino-C_3 and -C_4 acids (including N-methylamino-acids) has yielded useful diagnostic data.[232] Complex results were reported from a study of TMS derivatives of α- and ω-amino-acids.[233] Phenylthiohydantoin (PTH) derivatives

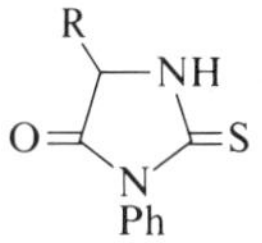

[(22) and (23)] of 21 amino-acids usually found in proteins can be well identified by g.c.–m.s., and the method is stated to be sufficiently sensitive to be applied to the products of automated Edman degradations of polypeptides:[234] mass spectra of phenylhydantoins were indeed applied to elucidate the structure of the decapeptide LRF (luteinizing-hormone-releasing factor) of ovine origin.[235]

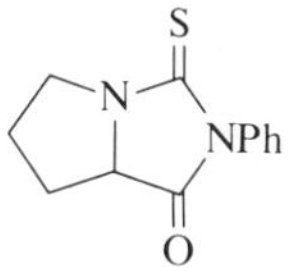

[229] P. C. Cala, N. R. Trenner, R. P. Buhs, G. V. Downing, J. L. Smith, and W. J. A. VandenHeuvel, *J. Agric. Food Chem.*, 1972, **20**, 337.
[230] J. C. Dabrowiak and D. W. Cooke, *Analyt. Chem.*, 1971, **43**, 791.
[231] W. A. Koenig, W. Parr, H. A. Lichtenstein, E. Bayer, and J. Oró, *J. Chromatog. Sci.*, 1970, **8**, 183.
[232] J. G. Lawless and M. S. Chadha, *Analyt. Biochem.*, 1971, **44**, 473.
[233] K. Bergström and J. Gürtler, *Acta Chem. Scand.*, 1971, **25**, 175.
[234] H. Hagenmaier, W. Ebbinghausen, G. Nicholson, and W. Voetsch, *Z. Naturforsch.*, 1970, **25b**, 681.
[235] R. Burgus, M. Butcher, M. Amoss, N. Ling, M. Monahan, J. Rivier, R. Fellows, R. Blackwell, W. Vale, and R. Guillemin, *Proc. Nat. Acad. Sci. U.S.A.*, 1972, **69**, 278.

Further observations on TMS derivatives of deuteriated amino-acids have been discussed with reference to fractionation in g.l.c. and mass spectrometric fragmentations.[236] The synthesis of 19 amino-acids by the action of electric discharges on a mixture of methane, nitrogen, water, and traces of ammonia has been demonstrated.[237]

Two *NN*-dimethylarginines from bovine myelin hydrolysates have been identified by g.c.–m.s.[238] An interfering peak in the g.l.c. of TMS derivatives of partially hydrolysed collagen is ascribed to the TMS-diketopiperazine (24) of prolylhydroxyproline.[239] Acids (25a) and (25b), glyoxylic and pyruvic acid

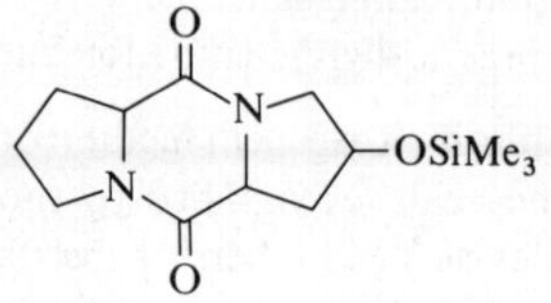

(24) $M^{\ddagger}$ of m/e 282 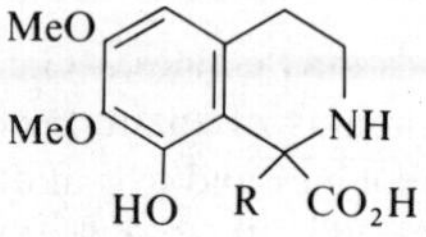   (25)   a; R = H
                                              b; R = Me

condensation products of 3-demethylmescaline, have been identified among the amino-acids of peyote.[240] A group of pyroglutamyl dipeptides has been isolated from mushrooms.[241]

In experiments on the possible role of amino-acetonitrile in chemical evolution, its condensation with amino-acids in basic aqueous solution was found to give low yields of products behaving as dipeptides.[242]

Drugs and Metabolites: (A) Reference Compounds.—A survey[243] of the principal fragments in the mass spectra (recorded in part by g.c.–m.s.) of 58 common drugs has been made, in connection with the development of programs for matching data from 'unknown' samples with reference data stored by a computer. The potential value of CI mass spectrometry in allowing the identification of some drugs more quickly than by g.c.–m.s. has been demonstrated.[244] Of 48 drugs studied, 30 had unique CI mass spectra, while the rest formed nine pairs which

---

[236] W. J. A. VandenHeuvel, J. L. Smith, I. Putter, and J. S. Cohen, *J. Chromatog.*, 1970, **50**, 405.
[237] D. Ring, Y. Wolman, N. Friedmann, and S. L. Miller, *Proc. Nat. Acad. Sci. U.S.A.*, 1972, **69**, 765.
[238] S. W. Brostoff, A. Rosegay, and W. J. A. VandenHeuvel, *Arch. Biochem. Biophys.*, 1972, **148**, 156.
[239] K. Lampiaho, T. Nikkari, J. Pikkarainen, J. Kärkkäinen, and E. Kulonen, *J. Chromatog.*, 1972, **64**, 211.
[240] G. J. Kapadia, G. S. Rao, E. Leete, M. B. E. Fayez, Y. N. Vaishnav, and H. M. Fales, *J. Amer. Chem. Soc.*, 1970, **92**, 6943.
[241] M. R. Altamura, R. E. Andreotti, M. L. Bazinet, and L. Long, jun., *J. Food Sci.*, 1970, **35**, 134.
[242] M. S. Chadha, L. Replogle, J. Flores, and C. Ponnamperuma, *Bio-organic Chem.*, 1971, **1**, 269.
[243] N. C. Law, V. Aandahl, H. M. Fales, and G. W. A. Milne, *Clinica Chim. Acta*, 1971, **32**, 221.
[244] G. W. A. Milne, H. M. Fales, and T. Axenrod, *Analyt. Chem.*, 1971, **43**, 1815.

had the same integral molecular weight but were mostly distinguishable by isotopic composition or by different fragmentations.

Common benzodiazepine-type tranquillizers are readily distinguishable by g.c.–m.s.,[245,246] and studies have been made of barbiturates,[247] sulphonamides,[248] and pyrazoles.[249]

A striking application of g.c.–m.s. has been reported in the determination of erythromycin and its derivatives as TMS ethers (mol. wt. >1000). G.l.c. separation of TMS ethers of erythromycins A, B, and C (26) was achieved on a 1.8 m column at 275 °C (3% polyphenylether-20 phase).[250]

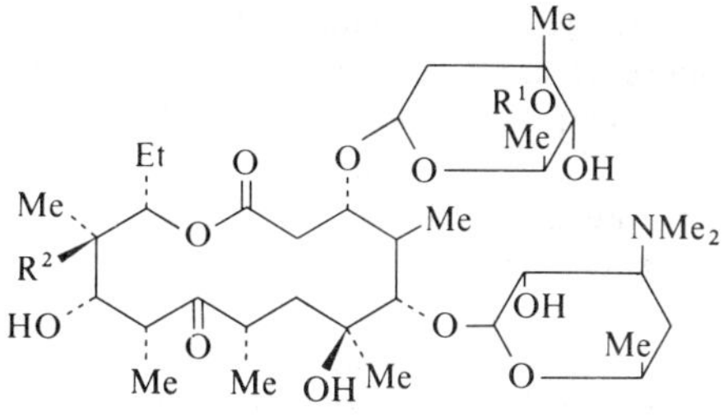

(26)

	No. of TMS	mol. wt.
Erythromycin A; R^1 = Me, R^2 = OH	5	1093
B; R^1 = Me, R^2 = H	4	1005
C; R^1 = H, R^2 = OH	6	1151

The fates of cannabidiol and Δ^9-tetrahydrocannabinol when preparations containing them are used for smoking have been investigated: partial cyclization of cannabidiol appeared to occur.[251]

Drugs and Metabolites: (B) in Biological Material.—Combinations of g.c.–m.s. with a peak-matching accessory[252] and with a computer[253–255] provide efficient techniques for the identification of metabolites of low molecular weight in small amounts of biological material. In conjunction with high-resolution mass spectrometry, such a procedure led rapidly to the identification of

[245] A. Forgione, P. Martelli, F. Marcucci, R. Fanelli, E. Mussini, and G. C. Jommi, *J. Chromatog.*, 1971, **59**, 163.
[246] W. Sadee and E. Van de Kleijn, *J. Pharm. Sci.*, 1971, **60**, 135.
[247] J. N. T. Gilbert, B. J. Millard, and J. W. Powell, *J. Pharm. Pharmacol.*, 1970, **22**, 897.
[248] B. Blessington, *Org. Mass Spectrometry*, 1972, **6**, 347.
[249] U. Rydberg and J. C. Buijten, *J. Chromatog.*, 1972, **64**, 170.
[250] K. Tsuji and J. H. Robertson, *Analyt. Chem.*, 1971, **43**, 818.
[251] F. Mikeš and P. G. Waser, *Science*, 1971, **172**, 1158.
[252] C.-G. Hammar, B. Holmstedt, and R. Kitz, *J. Chromatog.*, 1970, **49**, 402.
[253] F. Hutterer, J. Roboz, L. Sarkozi, A. Ruhig, and P. Bacchin, *Clinical Chem.*, 1971, **17**, 789.
[254] J. R. Althaus, K. Biemann, J. Biller, P. F. Donaghue, D. A. Evans, H.-J. Förster, H. S. Hertz, C. E. Hignite, R. C. Murphy, G. Preti, and V. Reinhold, *Experientia*, 1970, **26**, 714.
[255] E. Jellum, O. Stokke, and L. Eldjarn, *Scand. J. Clin. Lab. Invest.*, 1971, **27**, 273.

propoxyphene (27) in the urine of a comatose patient who, it was inferred, had received an overdose of Darvon [the hydrochloride of (27)].[254] The technique is

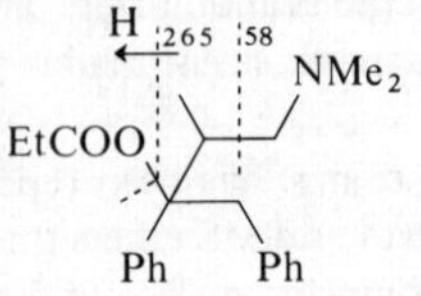

(27) Base peak m/e 58

equally valuable in the diagnosis of metabolic disorders.[255] However, the necessary computer facilities are, as yet, less widely available than the basic equipment for g.c.–m.s., with which most of the studies reported below have been effected.

Phenolic acids in urine of patients with tyrosinemia[256] and of horses[257] have been identified, and a useful compilation has been made[258] of salient m/e values in the mass spectra of derivatives of common urinary acids and related compounds. 'Profiles' of volatile constituents of urine, recorded by capillary g.c., comprised several hundred peaks, of which 40 were identified by m.s.[259]

General procedures have been described for the fractionation of urine (with particular reference to newborn infants) for determination of 'profiles' of urinary acids, bases, sugars, and steroids.[260,261] Studies of drug metabolites in man by g.c.–m.s. have included detection of metabolites of Dianabol (17β-hydroxy-17α-methylandrosta-1,4-dien-3-one) by monitoring the mass spectra of TMS ethers at m/e 143;[39] determination of probenecid as its methyl ester;[262] determination of Rafanoxide [3'-chloro-4'-(4-chlorophenoxy)-3,5-di-iodosalicylanilide] as its di-TMS derivative;[263] and determination of Haloperidol (28) in plasma.[264]

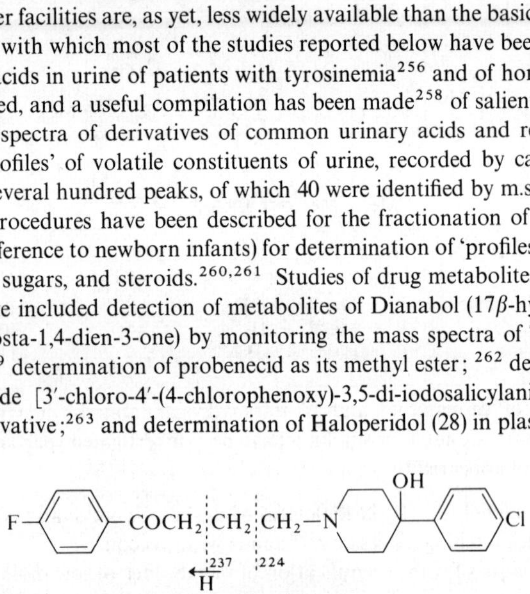

(28) $M^{+\cdot}$ of m/e 375
Base peak 224

[256] J. C. Crawhall, O. Mamer, S. Tjoa, and J. C. Claveau, *Clinica Chim. Acta*, 1971, **34**, 47.
[257] D. I. Chapman, J. R. Chapman, and J. Clark, *Internat. J. Biochem.*, 1970, **1**, 465.
[258] O. A. Mamer, J. C. Crawhall, and Sioe San Tjoa, *Clinica Chim. Acta*, 1971, **32**, 171.
[259] A. Zlatkis and H. M. Liebich, *Clinical Chem.*, 1971, **17**, 592.
[260] M. G. Horning, L. D. Waterbury, E. C. Horning, and R. M. Hill, *Excerpta Med. Int. Congr. Ser.*, No. 183: The Foeto-Placental Unit, Proc. Int. Symp., Milan, 1968, p. 305.
[261] E. C. Horning, P. G. Devaux, A. C. Moffat, C. D. Pfaffenberger, N. Sakauchi, and M. G. Horning, *Clinica Chim. Acta*, 1971, **34**, 135.
[262] K. Sabih, C. D. Klaassen, and K. Sabih, *J. Pharm. Sci.*, 1971, **60**, 745.
[263] C. P. Talley, N. R. Trenner, G. V. Downing, jun., and W. J. A. VandenHeuvel, *Analyt. Chem.*, 1971, **43**, 1379.
[264] F. Marcucci, E. Mussini, L. Airoldi, R. Fanelli, A. Frigerio, F. de Nadai, A. Bizzi, M. Rizzo, P. L. Morselli, and S. Garattini, *Clinica Chim. Acta*, 1971, **34**, 321.

G.c.–m.s. is an excellent method for the detection of 'doping' of racehorses, as illustrated by the work of Momose and Tsuji on antipyrine, sulpyrine, and aminopyrine.[265] Thiamine administered to rats is excreted, in part, as 2-methyl-4-amino-5-hydroxymethylpyrimidine.[266] Urinary metabolites of the anthelmintic Cambendazole in the pig have been characterized.[267]

Analysis of cannabis constituents has received attention in several laboratories. In hashish, the three major cannabinoids with a pentyl side-chain are generally accompanied by their propyl analogues, the cannabivarins, which are especially abundant in some Asian samples: the six compounds are easily separated and identified by g.c.–m.s.[268] Allylic mono- and di-hydroxylation of Δ^9-6a,10a-*trans*-tetrahydrocannabinol (Δ^9-THC) (29a) occurs during its metabolism by rats and rabbits, yielding (29b) and (29c).[269]

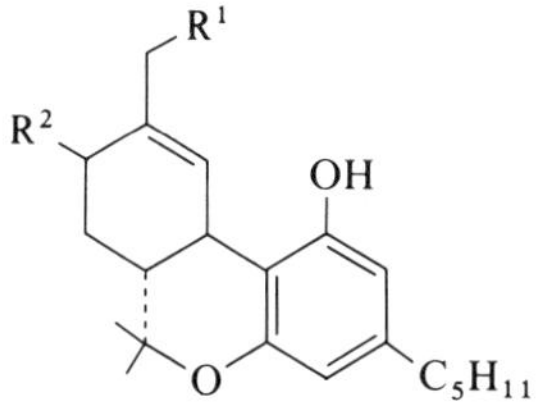

a; $R^1, R^2 = H$
b; $R^1 = OH; R^2 = H$
c; $R^1, R^2 = OH$

(29)

G.c.–m.s. based on a quadrupole mass spectrometer was effective for the forensic identification of procaine and diacetylmorphine in 'street heroin' samples.[270] Metabolism of thiamine by a soil micro-organism,[271] and of cyclopropanecarboxylic acid by a fungus using the acid as its sole carbon source,[272] has been studied.

A preliminary report describes three new metabolites of methadone (in man and rat), *viz.* 4-dimethylamino-2,2-diphenylpentanoic acid arising from oxidative cleavage of the ethyl ketone grouping, and two products of aromatic hydroxylation – a process stated not to have been previously reported in the metabolism of synthetic opiates.[273]

---

[265] A. Momose and T. Tsuji, *J. Pharm. Soc. Japan*, 1972, **92**, 187, 193.
[266] W. W. White, W. H. Amos, and R. A. Neal, *J. Nutrition*, 1970, **100**, 1053.
[267] W. J. A. VandenHeuvel, R. P. Buhs, J. R. Carlin, T. A. Jacob, F. R. Koniuszy, J. L. Smith, N. R. Trenner, R. W. Walker, D. E. Wolf, and F. J. Wolf, *Analyt. Chem.*, 1972, **44**, 14.
[268] T. B. Vree, D. D. Breimer, C. A. M. Van Ginneken, J. M. van Rossum, R. A. de Zeeuw, and A. H. White, *Clinica Chim. Acta*, 1971, **34**, 365; R. A. de Zeeuw, J. Wijsbeek, D. D. Breimer, T. B. Vree, C. A. M. van Ginneken, and J. M. van Rossum, *Science*, 1972, **175**, 778.
[269] F. Mikeš, A. Hofmann, and P. G. Waser, *Biochem. Pharmacol.*, 1971, **20**, 2469.
[270] G. R. Nakamura, T. T. Noguchi, D. Jackson, and D. Banks, *Analyt. Chem.*, 1972, **44**, 408.
[271] W. H. Amos and R. A. Neal, *Analyt. Biochem.*, 1970, **36**, 332.
[272] J. G. Schiller and A. E. Chung, *J. Biol. Chem.*, 1970, **245**, 5857.
[273] H. R. Sullivan, S. L. Due, and R. E. McMahon, *J. Amer. Chem. Soc.*, 1972, **94**, 4050.

Insect Pheromones and Other Secretions.—G.c.–m.s. has been used in two main ways, *viz.* to analyse the complex mixtures of mainly aliphatic compounds found in defensive, signalling, and marking secretions, and to identify compounds showing particular pheromonal activities. In the ant *Lasius niger*, hendecane appeared to act as an alarm pheromone,[274] while the defensive secretion of the daddy-long-legs *Leiobunum vittatum* contained 4-methylheptan-3-one and (*E*)-4,6-dimethyloct-6-en-3-one.[275] Three instructive reports have appeared on a tetradecadienyl acetate found as a sex attractant of the almond moth and the Indian meal moth: (i) 200 μg of active fraction was shown to give tetradecyl acetate by hydrogenation, and 9-oxononyl acetate and acetaldehyde by ozonolysis at −78 °C;[276] (ii) 1 μg was converted by OsO_4 into a tetraol, and g.c.–m.s. of the tetra-TMS ether indicated the original double-bond positions (Scheme 4):[277] (iii) Accurate mass measurement showed the composition to be $C_{16}H_{28}O_2$; hydrogenation and ozonolysis products were identified by g.c.–m.s.; the four possible geometrical isomers were synthesized and the (*Z*)-9,(*E*)-12-isomer was identical with the natural pheromone.[278] Tetradecenyl acetates appear to be sex pheromones of the moth *Adoxophyes orana*.[279]

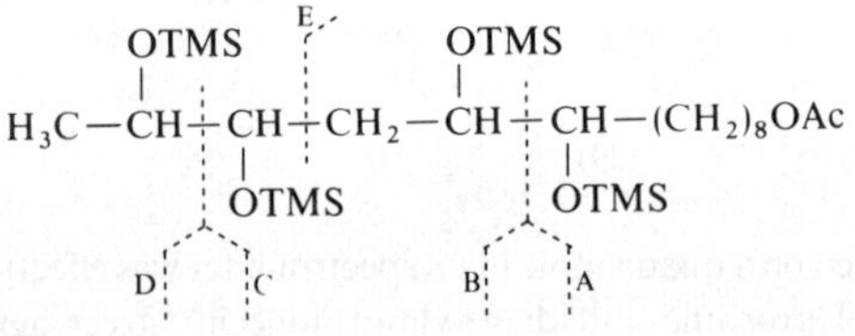

$$H_3C-CH \stackrel{\text{OTMS}}{\underset{}{|}} CH \stackrel{E}{\underset{}{|}} CH_2-CH \stackrel{\text{OTMS}}{\underset{}{|}} CH-(CH_2)_8 OAc$$

$$\underbrace{}_{D} \underbrace{}_{C} \quad \underbrace{}_{B} \underbrace{}_{A}$$

Peaks are at *m/e* 401 (C − 90), 273 (A), 245 (B − 90), 219 (E), 117 (D).

Scheme 4 *Definitive fragmentation of tetraol tetra-TMS ether*

The value of 'reaction gas chromatography', using reagents mounted in series with the column, in the correlation of functional type with pheromone activity, is exemplified in a study of the sex attractant of the female gypsy moth, shown to be *cis*-7,8-epoxy-2-methyloctadecane.[280] G.c.–m.s. was used to check the presence of a terminal isopropyl group in the alkane produced by instantaneous hydrogenation: strong peaks at *m/e* 253 ($[M − 15]^+$) and 225 ($[M − 43]^+$) were consistent with this.

---

[274] G. Bergström and J. Lofqvist, *J. Insect. Physiol.*, 1970, **16**, 2353.
[275] J. Meinwald, A. F. Kluge, J. E. Carrel, and T. Eisner, *Proc. Nat. Acad. Sci. U.S.A.*, 1971, **68**, 1467.
[276] U. E. Brady, J. H. Tumlinson, R. G. Brownlee, and R. M. Silverstein, *Science*, 1971, **171**, 802.
[277] K. H. Dahm, I. Richter, D. Meyer, and H. Röller, *Life Sci., Part II*, 1971, **10**, 531.
[278] Y. Kuwahara, C. Kitamura, S. Takahashi, H. Hara, S. Ishii, and H. Fukami, *Science*, 1971, **171**, 801.
[279] G. M. Meijer, F. J. Ritter, C. J. Persoons, A. K. Minks, and S. Voerman, *Science*, 1972, **175**, 1469.
[280] B. A. Bierl, M. Beroza, and C. W. Collier, *Science*, 1970, **170**, 87.

A survey has been made of volatile components in the cephalic marking secretion of male bumble bees.[281]

Food Flavours and Aromas.—Two useful reviews of applications of g.c.–m.s. in flavour and aroma chemistry have appeared.[282,283] A survey of components identified in vegetable volatiles[284] lists many that were identified by g.c.–m.s. The technique of combined g.c.–m.s. is now firmly established as one of the most powerful available for flavour research.[285]

There has recently been much concern over the presence of carcinogens, particularly nitrosamines, in foods. Many foods contain nitrates, nitrites, and secondary amines, and the possibility exists that they may be converted into nitrosamines during the preparation or digestion of these foods. G.c.–m.s. provides a specific and sensitive method for the detection and estimation of nitrosamines.[21,286] Fazio and co-workers claim to have detected N-nitrosodimethylamine in raw and smoked sea fish.[287] However, they were able to obtain only weak and ill-defined spectra (with high background levels) of the components suspected to be N-nitrosodimethylamine. Williams and co-workers failed to detect NN-dimethyl- or NN-diethyl-nitrosamine in a potable spirit.[288] In 1967, Devik obtained polarographic and chromatographic evidence for the formation of N-nitrosamines from the heat-induced reactions between D-glucose and several L-amino-acids.[289] This work has been repeated with several mixtures of monosaccharides and amino-acids, and the products have been characterized by g.c.–m.s.[290—293] No nitrosamines were found: the major products were alkyl-pyrazines, which also appear to be formed during roasting or frying of most foods. They have been identified by g.c.–m.s. in potato crisps and chips,[294,295] roasted barley,[290] coffee aroma,[296] roasted peanuts,[297] and popcorn.[298]

[281] B. Kullenberg, G. Bergström, and S. Ställberg-Stenhagen, *Acta Chem. Scand.*, 1970, **24**, 1481.
[282] W. H. McFadden and R. G. Buttery, *Adv. Analyt. Chem. Instrum.*, 1970, **8**, 327.
[283] R. Teranishi, P. Issenberg, I. Hornstein, and E. L. Wick, 'Flavor Research: Principles and Techniques', Marcel Dekker, New York, 1970.
[284] A. E. Johnson, H. E. Nursten, and A. A. Williams, *Chem. and Ind.*, 1971, 556, 1212.
[285] E. Honkanen, *Suomen Kem.*, 1970 43, *A*, 75.
[286] K. Heyns and H. Röper, *Z. Lebensm.-Untersuch.*, 1971, **145**, 69.
[287] T. Fazio, J. N. Damico, J. W. Howard, R. H. White, and J. O. Watts, *J. Agric. Food Chem.*, 1971, **19**, 250.
[288] A. A. Williams, C. F. Timberlake, O. G. Tucknott, and R. L. S. Patterson, *J. Sci. Food Agric.*, 1971, **22**, 431.
[289] O. G. Devik, *Acta Chem. Scand.*, 1967, **21**, 2302.
[290] P.-S. Wang, H. Kato, and M. Fujimaki, *Agric. and Biol. Chem. (Japan)*, 1969, **33**, 1775.
[291] K. Heyns and H. Koch, *Tetrahedron Letters*, 1970, 741; *Z. Lebensm.-Untersuch.*, 1971, **145**, 76.
[292] A. Ferretti and V. P. Flanagan, *J. Agric. Food Chem.*, 1971, **19**, 245.
[293] R. A. Scanlan and L. M. Libbey, *J. Agric. Food Chem.*, 1971, **19**, 571.
[294] R. G. Buttery, R. M. Seifert, D. G. Guadagni, and L. C. Ling, *J. Agric. Food Chem.*, 1971, **19**, 969.
[295] G. M. Sapers, S. F. Osman, C. J. Dooley, and O. Panasiuk, *J. Food Sci.*, 1971, **36**, 93.
[296] P. Friedel, V. Krampl, T. Radford, J. A. Renner, F. W. Shephard, and M. A. Gianturco, *J. Agric. Food Chem.*, 1971, **19**, 530.
[297] B. R. Johnson, G. R. Waller, and A. L. Burlingame, *J. Agric. Food Chem.*, 1971, **19**, 1020.
[298] J. P. Walradt, R. C. Lindsay, and L. M. Libbey, *J. Agric. Food Chem.*, 1970, **18**, 926.

Forty-two components of Valencia orange essence and aroma oils have been identified.[299] More detailed studies of long-chain hydrocarbon profiles of orange and tangor juice sacs[300] and carbonyl compounds from orange juice[301] have been carried out. Enzymic oxidative changes of limonene, one of the main components of orange flavour, have been studied during the destruction of cell structure in various fruits.[302] The incorporation of labelled amino-acids into banana aroma substances has been investigated in order to determine their biogenesis.[303] Other studies of fruit aromas include those of bananas,[304] bilberries,[305] blackberries,[306] grapefruit,[307] guava,[308] and pineapple – in which 59 compounds, 35 of them not previously found, were identified.[309]

Heatherbell and co-workers have identified 23 volatile flavour components of raw carrots,[310] and have investigated the effects of canning and freeze-drying on the flavour,[311] and the enzymic regeneration of flavour components.[312] G.c.–m.s. has also been employed in the study of flavours of capsicum,[313] coconut meat,[314] mint,[315] and tomato.[316] Vanilla constituents have been characterized as their O-methyloxime and TMS ether derivatives.[317] Volatile flavour components produced by thermal degradation of thiamine included 2-methylthiophen and its 4,5-dihydro-analogue.[318]

Several studies have been made on the aroma constituents of peanuts, including the basic[296,319] and neutral[320] fractions, and the influences of curing temperature[321] and enzyme activity during maturation[322] on the flavour.

[299] R. L. Coleman and P. E. Shaw, *J. Agric. Food Chem.*, 1971, **19**, 520.
[300] S. Nagy and H. E. Nordby, *Phytochemistry*, 1971, **10**, 2763.
[301] H. L. Dinsmore and S. Nagy, *J. Agric. Food Chem.*, 1971, **19**, 517.
[302] R. Tressl, F. Drawert, W. Heimann, and R. Emberger, *Phytochemistry*, 1970, **9**, 2327.
[303] R. Tressl, R. Emberger, F. Drawert, and W. Heimann, *Z. Naturforsch.*, 1970, **25b**, 704.
[304] R. Tressl, F. Drawert, and W. Heimann, *Z. Lebensm.-Untersuch.*, 1970, **142**, 249.
[305] E. von Sydow, J. Andersson, K. Anjou, and G. Karlsson, *Lebensm.-Wiss. Technol.*, 1970, **3**, 11.
[306] R. A. Scanlan, D. D. Bills, and L. M. Libbey, *J. Agric. Food Chem.*, 1970, **18**, 744.
[307] M. G. Moshonas and P. E. Shaw, *J. Agric. Food Chem.*, 1971, **19**, 119.
[308] K. L. Stevens, J. E. Brekke, and D. J. Stern, *J. Agric. Food Chem.*, 1970, **18**, 598.
[309] R. Näf-Müller and B. Willhalm, *Helv. Chim. Acta*, 1971, **54**, 199.
[310] D. A. Heatherbell, R. E. Wrolstad, and L. M. Libbey, *J. Agric. Food Chem.*, 1971, **19**, 1069.
[311] D. A. Heatherbell, R. E. Wrolstad, and L. M. Libbey, *J. Food Sci.*, 1971, **36**, 219.
[312] D. A. Heatherbell and R. E. Wrolstad, *J. Agric. Food Chem.*, 1971, **19**, 281.
[313] S. Kosuge and M. Furuta, *Agric. and Biol. Chem. (Japan)*, 1970, **34**, 248; A. Müller-Stock, R. K. Joshi, and J. Büchi, *Pharm. Acta Helv.*, 1972, **47**, 7.
[314] F. M. Lin and W. F. Wilkens, *J. Food Sci.*, 1970, **35**, 538.
[315] E. Fedeli and L. Pedrinella, *Riv. Ital. Sostanze Grasse*, 1970, **47**, 782.
[316] R. G. Buttery, R. M. Seifert, D. G. Guadagni, and L. C. Ling, *J. Agric. Food Chem.*, 1971, **19**, 524.
[317] J. C. Lhuguenot, B. F. Maume, and C. Baron, *Chromatographia*, 1971, **4**, 204.
[318] R. G. Arnold, L. M. Libbey, and R. C. Lindsay, *J. Agric. Food Chem.*, 1969, **17**, 390.
[319] C. K. Shu and G. R. Waller, *J. Food Sci.*, 1971, **36**, 579.
[320] B. R. Johnson, G. R. Waller, and R. L. Foltz, *J. Agric. Food Chem.*, 1971, **19**, 1025.
[321] J. A. Singleton, H. E. Pattee, and E. B. Johns, *J. Agric. Food Chem.*, 1971, **19**, 130.
[322] H. E. Pattee, J. A. Singleton, E. B. Johns, and B. C. Mullin, *J. Agric. Food Chem.*, 1970, **18**, 353.

Phenolic wood-smoke components have been identified in smoked meat.[323] A kerosene-like taint in mullet has been investigated,[324] and the identification of meat taints has been discussed.[325] Some flavour volatiles of cooked chicken have been identified.[326] Volatile neutral components of heat-degraded pork fat have been identified, and conversion of undecane and of fatty acids into lactones has been studied.[327] G.c.–m.s. has also been applied to the study of aromatizing substances.[328]

A number of additional compounds have been identified in coffee aroma,[296] and in green[329] and black[330] tea. Among the alcoholic beverages which have been studied by g.c.–m.s. are cider,[331] cognac,[332] rum,[333] sake,[132] whiskey,[333] and wine.[334] It has been demonstrated that ionones in whiskey could be produced by thermal decomposition of β-carotene.[335] Additional volatile hop components have been identified.[336]

Pesticides and Pollutants.—G.c.–m.s. has been used for detection of pesticide residues in many types of sample, e.g. human tissues,[337] foods,[338] plants,[339] and tea.[340] Gas-chromatographic determination of chlorinated pesticide residues is difficult if polychlorobiphenyls (PCB's) are also present in a sample.[341] The main problem of analysis of PCB's is that the commercial products are exceedingly complex mixtures containing many isomers. It has been noted that some components are metabolized more rapidly than others.[342] Nevertheless, PCB's can be adequately characterized by g.c.–m.s.,[343,344] and mixtures of PCB's and chlorinated pesticide residues are conveniently studied with the aid of a small

[323] A. O. Lustre and P. Issenberg, *J. Agric. Food Chem.*, 1970, **18**, 1056; P. Issenberg, M. R. Korureich, and A. O. Lustre, *J. Food Sci.*, 1971, **36**, 107.
[324] J. Shipton, J. H. Last, K. E. Murray, and G. L. Vale, *J. Sci. Food Agric.*, 1970, **21**, 433.
[325] R. L. S. Patterson, *Process Biochem.*, 1970, **5**, 27.
[326] A. Hobson-Frohock, *J. Sci. Food Agric.*, 1970, **21**, 152.
[327] K. Watanabe and Y. Sato, *Agric. and Biol. Chem. (Japan)*, 1970, **34**, 1710; 1971, **35**, 278.
[328] E. Fedeli, *Riv. Ital. Sostanze Grasse*, 1970, **47**, 14.
[329] T. Yamanishi, M. Nose, and Y. Nakatani, *Agric. and Biol. Chem. (Japan)*, 1970, **34**, 599.
[330] S. Sato, S. Sasakura, A. Kobayashi, Y. Nakatani, and T. Yamanishi, *Agric. and Biol. Chem. (Japan)*, 1970, **34**, 1355.
[331] A. A. Williams and O. G. Tucknott, *J. Sci. Food Agric.*, 1971, **22**, 264.
[332] J. Schaefer and R. Timmer, *J. Food Sci.*, 1970, **35**, 10.
[333] R. Timmer, R. ter Heide, H. J. Wobben, and P. J. de Valois, *J. Food Sci.*, 1971, **36**, 462; H. J. Wobben, R. Timmer, R. ter Heide, and P. J. de Valois, *ibid.*, p. 464.
[334] P. J. Hardy and E. H. Ramshaw, *J. Sci. Food Agric.*, 1970, **21**, 39.
[335] E. G. LaRoe and P. A. Shipley, *J. Agric. Food Chem.*, 1970, **18**, 174.
[336] Y. Naya and M. Kotake, *Bull. Chem. Soc. Japan*, 1970, **43**, 3594.
[337] F. J. Biros and A. C. Walker, *J. Agric. Food Chem.*, 1970, **18**, 425.
[338] S. W. Bellman and T. L. Barry, *J. Assoc. Offic. Analyt. Chemists*, 1971, **54**, 499.
[339] M. Zimmer and W. Klein, *Chemosphere*, 1972, **1**, 3.
[340] T. Yagi, S. Takahashi, and T. Murata, *Shimadzu Rev. (Kyoto)*, 1971, **28**, 89.
[341] G. E. Bagley, W. L. Reichel, and E. Cromartie, *J. Assoc. Offic. Analyt. Chemists*, 1970, **53**, 251; R. Edwards, *Chem. and Ind.*, 1971, 1340.
[342] S. Bailey and P. J. Bunyan, *Nature*, 1972, **236**, 34.
[343] D. L. Stalling and J. N. Huckins, *J. Assoc. Offic. Analyt. Chemists*, 1971, **54**, 801.
[344] D. Sissons and D. Welti, *J. Chromatog.*, 1971, **60**, 15.

computer.[345,35] G.c.–m.s. has also been employed in studies of the action of u.v. radiation on PCB's[346] and of highly toxic contaminants in PCB mixtures.[347] A useful collection of mass spectra of chlorinated aromatic fungicides has been published,[348] and a number of impurities in commercial chlorophenols have been identified by g.c.–m.s.[349] The metabolism of Endrin in the rat[350] and the photolysis of Heptachlor[351] and some herbicidal triazin-5(4H)-ones[352] have also been investigated, and a new metabolite of Chlordane, found in milk and cheese, has been identified.[353]

G.c.–m.s. has been used to identify a number of other pollutants, including methylmercury compounds in fish,[354] nitriloacetic acid and related aminopolycarboxylic acids in inland waters,[355] 2,5-di-(benzoxazol-2-yl)thiophen (an optical brightener) in sludge and fish,[356] volatile organic compounds produced during the Kraft pulping process,[357] some organic components of filter papers,[358] and a dioctyl sebacate in 'reagent' grade $CHCl_3$.[359]

It has been estimated that atmospheres within nuclear submarines contain more than 10 000 contaminants. Some of these have been identified by g.c.–m.s., and many derive from cigarette smoke,[360] in which over 1000 components have been identified.[361] K. and G. Grob have applied a refined technique of trace analysis, based on g.c.–m.s. with high-resolution capillary columns, to identify 108 compounds in an urban atmosphere.[362]

Parathion was detected by g.c. and m.s. in autopsy material from two cases of murder by poisoning, though in this instance combined g.c.–m.s. was apparently unsuccessful.[363]

Organic Geochemistry.—Albrecht and Ourisson[364] reviewed the literature concerning biogenic substances in sediments and fossils up to the beginning of

[345] E. J. Bonelli, *Amer. Lab.*, 1971, No. 2, p. 27; ref. *g* (Table, p. 303).
[346] K. Hustert and F. Korte, *Chemosphere*, 1972, **1**, 7.
[347] J. G. Vos, J. H. Koeman, H. L. van der Maas, M. C. ten Noever de Brauw, and R. H. de Vos, *Food Cosmet. Toxicol.*, 1970, **8**, 625.
[348] O. Hutzinger, W. D. Jamieson, and S. Safe, *J. Assoc. Offic. Analyt. Chemists*, 1971, **54**, 178.
[349] D. Firestone, J. Ress, N. L. Brown, R. P. Barron, and J. N. Damico, *J. Assoc. Offic. Analyt. Chemists*, 1972, **55**, 85.
[350] M. K. Baldwin, J. Robinson, and D. V. Parke, *J. Agric. Food Chem.*, 1970, **18**, 1117.
[351] R. R. McGuire, M. J. Zabik, R. D. Schuetz, and R. D. Flotard, *J. Agric. Food Chem.*, 1970, **18**, 319.
[352] B. E. Pape and M. J. Zabik, *J. Agric. Food Chem.*, 1972, **20**, 72.
[353] J. H. Lawrence, R. P. Barron, J.-Y. T. Chen, P. Lombardo, and W. R. Benson, *J. Assoc. Offic. Analyt. Chemists*, 1970, **53**, 261; *cf.* B. Schwemmer, W. Cochrane, and P. B. Polen, *Science*, 1970, **169**, 1087.
[354] B. Johansson, R. Ryhage, and G. Westöö, *Acta Chem. Scand.*, 1970, **24**, 2349.
[355] C. B. Warren and E. J. Malec, *J. Chromatog.*, 1972, **64**, 219.
[356] S. Jensen and O. Pettersson, *Environ. Pollution*, 1971, **2**, 145.
[357] D. F. Wilson and B. F. Hrutfiord, *Tappi*, 1971, **54**, 1094.
[358] G. von Unruh, G. Remberg, and G. Spiteller, *Chem. Ber.*, 1971, **104**, 2071.
[359] R. B. Holtz, P. Swenson, M. Abel, and T. A. Walter, *Lipids*, 1971, **6**, 523.
[360] F. E. Saalfeld, F. W. Williams, and R. A. Saunders, *Amer. Lab.*, 1971, No. 7, p. 8.
[361] K. Grob and J. A. Völlmin, *J. Chromatog. Sci.*, 1970, **8**, 218.
[362] K. Grob and G. Grob, *J. Chromatog.*, 1971, **62**, 1.
[363] G. Bohn, G. Ruecker, and K. H. Luckas, *Z. Rechtsmedizin.*, 1971, **68**, 45.
[364] P. Albrecht and G. Ourisson, *Angew. Chem. Internat. Edn.*, 1971, **10**, 209.

the period of this Report. Since then, work has been continued on terrestrial, meteoric, and lunar geochemistry, largely aided by g.c.–m.s.

Examination of Colorado Green River Shale has yielded further data on steranes,[365,366] terpanes,[365] branched alkanes,[365] a triterpenoid alcohol,[149] various acids,[144,367] and nitrogenous compounds.[368] When [4-^{14}C]cholesterol was incorporated into pulverized Green River Shale and heated, labelled cholest-4-ene and -5-ene were formed.[369] Alkanes have been identified in fluid inclusions in quartz crystals.[370] The sterol contents of samples from a Pleistocene basin and a contemporary lake have been compared.[371] Marine sediments of approximate ages 50 yr and 750 yr both contained carbohydrates.[372]

The first identification of a steroidal carboxylic acid in petroleum has been reported: 23,24-bisnor-5α-cholanoic acid (5α-pregnane-20-carboxylic acid), isolated from a Californian petroleum of Pleistocene age, was characterized by reduction to bisnor-5α-cholane and [22-^{2}H$_1$]bisnor-5α-cholane.[373]

Aromatic hydrocarbons,[374] heterocyclic compounds,[375] and amino-acids, which are apparently of extraterrestrial origin,[376] have been found in the Murchison Meteorite. The possibility of terrestrial contamination of meteoritic samples is high, and it was fortunate that the meteorite which fell at Pueblito de Allende (Mexico) in 1969 could be examined in laboratories which were being prepared for organic analysis of lunar samples. A number of hydrocarbons were found in this meteorite.[377] The possibility that 'organic' material in meteorites may be formed by Fischer–Tropsch-type reactions has been investigated.[378]

The results of g.c.–m.s. analyses of returned lunar samples from the Apollo-11 and -12 missions have been fully documented.[379–386] No clear evidence has been

[365] E. J. Gallegos, *Analyt. Chem.*, 1971, **43**, 1151.
[366] B. Balogh, D. M. Wilson, and A. L. Burlingame, *Nature*, 1971, **233**, 261.
[367] P. Haug, H. K. Schnoes, and A. L. Burlingame, *Chem. Geol.*, 1971, **7**, 213.
[368] B. R. Simoneit, H. K. Schnoes, P. Haug, and A. L. Burlingame, *Nature*, 1970, **226**, 75.
[369] M. M. Rhead, G. Eglinton, and G. H. Draffan, *Chem. Geol.*, 1971, **8**, 277.
[370] K. A. Kvenvolden and E. Roedder, *Geochim. Cosmochim. Acta*, 1971, **35**, 1209.
[371] W. Henderson, W. E. Reed, G. Steel, and M. Calvin, *Nature*, 1971, **231**, 308.
[372] J. E. Modzeleski, W. A. Laurie, and B. Nagy, *Geochim. Cosmochim. Acta*, 1971, **35**, 825.
[373] W. Seifert, E. J. Gallegos, and R. M. Teeter, *Angew. Chem. Internat. Edn.*, 1971, **10**, 747.
[374] K. L. Pering and C. Ponnamperuma, *Science*, 1971, **173**, 237.
[375] C. E. Folsome, J. Lawless, M. Romiez, and C. Ponnamperuma, *Nature*, 1971, **232**, 108.
[376] K. A. Kvenvolden, J. G. Lawless, and C. Ponnamperuma, *Proc. Nat. Acad. Sci. U.S.A.*, 1971, **68**, 486.
[377] R. L. Levy, C. J. Wolf, M. S. Grayson, J. Gilbert, E. Gelpi, W. S. Updegrove, A. Zlatkis, and J. Oró, *Nature*, 1970, **227**, 148.
[378] M. H. Studier, R. Hayatsu, and E. Anders, *Geochim. Cosmochim. Acta*, 1972, **36**, 189.
[379] S. R. Lipsky, R. J. Cushley, C. G. Horvath, and W. J. McMurray, *Science*, 1970, **167**, 778.
[380] P. I. Abell, G. H. Draffan, G. Eglinton, J. M. Hayes, J. R. Maxwell, and C. T. Pillinger, *Geochim. Cosmochim. Acta, Suppl. 1*, 1970, **34**, 1757.
[381] A. L. Burlingame, M. Calvin, J. Han, W. Henderson, W. Reed, and B. R. Simoneit, *Geochim. Cosmochim. Acta, Suppl. 1*, 1970, **34**, 1779.

obtained for the existence of organic compounds of lunar origin except methane, but many artefacts, including contaminants arising from spacecraft materials,[387] were identified. However, it is hoped that g.c.–m.s. will be used in the initial analysis of the Martian atmosphere and soil during the proposed Viking mission.[388,389]

Miscellaneous.—G.c.–m.s. has been applied in analyses of ^{18}O-labelled products of enzymic oxygenation of thymine (coupled to the oxidative decarboxylation of 2-ketoglutarate),[390] and of the bromine oxidation of Synkavit (2-methylnaphthalene-1,4-diol diphosphate).[391] The identification of *meso*-tartaric acid as the product of stereospecific hydration by fumarase of L-*trans*-2,3-epoxysuccinic acid depended on g.c.–m.s. (Scheme 5).[392] Thiol esters have been characterized

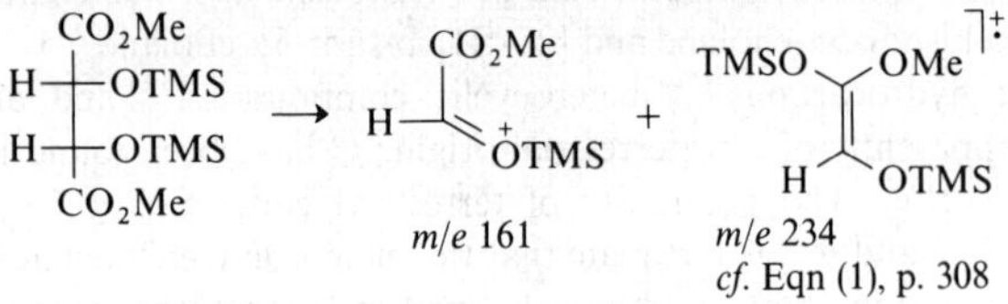

$$\begin{array}{c} \text{CO}_2\text{Me} \\ \text{H}{-}\!\!{-}\text{OTMS} \\ \text{H}{-}\!\!{-}\text{OTMS} \\ \text{CO}_2\text{Me} \end{array} \longrightarrow \begin{array}{c} \text{CO}_2\text{Me} \\ \text{H}{-}\!\!\overset{+}{\underset{\text{OTMS}}{\diagup}} \\ m/e\ 161 \end{array} + \begin{bmatrix} \text{TMSO}\diagdown\!\!\diagup\text{OMe} \\ \text{H}\diagup\!\!\diagdown\text{OTMS} \end{bmatrix}^{\ddagger} \\ m/e\ 234 \\ cf.\ \text{Eqn (1), p. 308}$$

Relative retention data (methyl myristate = 1.00):

3% QF-1, 115 °C: *meso*   0.37
             (+)   0.41
             (−)   0.42

Scheme 5 *Salient g.c.–m.s. data for dimethyl* meso-*tartrate di-TMS ether*

in Galbanum oil.[393] Products of degradation of lignins by saponification[394] and during bleaching[395] have been analysed, as have combustion products of poly(vinyl chloride)[396] and polyphenylene oxide[397] plastics. The structures of

[382] C. W. Gehrke, R. W. Zumwalt, W. A. Aue, D. L. Stalling, A. Duffield, K. A. Kvenvolden, and C. Ponnamperuma, *Geochim. Cosmochim. Acta, Suppl. 1*, 1970, **34**, 1845.
[383] M. E. Murphy, V. E. Modzeleski, B. Nagy, W. M. Scott, M. Young, C. M. Drew, P. B. Hamilton, and H. C. Urey, *Geochim. Cosmochim. Acta, Suppl. 1*, 1970, **34**, 1879.
[384] J. Oró, J. Gibert, W. Updegrove, J. McReynolds, J. Ibanez, E. Gil-Av, D. Flory, and A. Zlatkis, *J. Chromatog. Sci.*, 1970, **8**, 297.
[385] C. W. Gehrke, R. W. Zumwalt, D. L. Stalling, D. Roach, W. A. Aue, C. Ponnamperuma, and K. A. Kvenvolden, *J. Chromatog.*, 1971, **59**, 305.
[386] J. Gibert, D. Flory, and J. Oró, *Nature*, 1971, **229**, 33.
[387] B. R. Simoneit and A. L. Burlingame, *Nature*, 1971, **234**, 210.
[388] P. G. Simmonds, *Amer. Lab.*, 1970, No. 10, p. 8.
[389] L. F. Herzog, *Internat. J. Mass Spectrometry Ion Phys.*, 1970, **4**, 337.
[390] E. Holme, G. Lindstedt, S. Lindstedt, and M. Tofft, *J. Biol. Chem.*, 1971, **246**, 3314.
[391] W. J. A. VandenHeuvel, B. H. Arison, and J. S. Cohen, *J. Chromatog.*, 1971, **59**, 169.
[392] F. Albright and G. J. Schroepfer, *J. Biol. Chem.*, 1971, **246**, 1350.
[393] J. W. K. Burrell, R. A. Lucas, D. M. Michalkiewicz, and G. Riezebos, *Tetrahedron Letters*, 1971, 2837.
[394] R. D. Hartley, *J. Chromatog.*, 1971, **54**, 335.
[395] J. Gierer and L. Sundholm, *Svensk Papperstidn.*, 1971, **74**, 345.
[396] M. M. O'Mara, *J. Polymer Sci., Part A-1, Polymer Chem.*, 1970, **8**, 1887.
[397] G. Ball, B. Weiss, and E. A. Boettner, *Amer. Ind. Hygiene Assoc. J.*, 1970, **31**, 572.

acylphloroglucinol-derived cedrone acetates have been clarified.[398] Studies of mass spectra by g.c.–m.s., not cited above, include surveys (with deuterium labelling) of catechol polymethylene diethers,[399] and dialkyl malonates.[400] The occurrence of doubly charged ions of mass $[M - 30]$ in the spectra of di-TMS derivatives of aromatic dihydroxy-, diamino-, and aminohydroxy-compounds has been explored.[401] Cyclic n-butylboronates of various bifunctional compounds have been examined.[402]

Notable inorganic applications have been reported, e.g. to TMS derivatives of inorganic anions,[403] to the characterization of Ni, Pd, and Pt as their hexafluoromonothioacetylacetonates,[404] and to the determination of Cr and Be at the picogram level as their trifluoroacetylacetonates.[405]

---

[398] J. A. Beisler, J. V. Silverton, A. Penttila, D. H. S. Horn, and H. M. Fales, *J. Amer. Chem. Soc.*, 1971, **93**, 4850.
[399] P. Vouros and K. Biemann, *Org. Mass Spectrometry*, 1970, **3**, 1317.
[400] M. H. Wilson and J. A. McCloskey, *J. Amer. Chem. Soc.*, 1972, **94**, 3865.
[401] J. L. Smith, J. L. Beck, and W. J. A. VandenHeuvel, *Org. Mass Spectrometry*, 1971, **5**, 473 and references there cited.
[402] C. J. W. Brooks and I. Maclean, *J. Chromatog. Sci.*, 1971, **9**, 18.
[403] W. C. Butts and W. T. Rainey, *Analyt. Chem.*, 1971, **43**, 538.
[404] E. Bayer, H. P. Müller, and R. Sievers, *Analyt. Chem.*, 1971, **43**, 2012.
[405] W. R. Wolf, M. L. Taylor, B. M. Hughes, T. O. Tiernan, and R. E. Sievers, *Analyt. Chem.*, 1972, **44**, 616.

Author Index

Aandahl, V., 297, 324
Aaronson, M. J., 208, 213
Aasen, A. J., 145, 146, 310
Abbot, G. G., 144
Abe, E., 92
Abe, H., 127
Abel, E. W., 260
Abel, M., 332
Abell, P. I., 333
Ablov, A. V., 247
Abraham, E. P., 162
Abramovitch, R. A., 78, 117
Achenbach, H., 116, 126
Acheson, R. M., 132
Adam, G., 189
Adamson, J., 100, 170
Addison, C. C., 236
Adelstein, G. W., 114
Adler, A. D., 121, 247
Adlercreutz, H., 176, 319
Aexel, R. T., 318
Aghejanian, Ts. Ye., 113
Agranat, I., 99, 218
Agurell, S., 322
Agwada, A., 187
Ahlgren, G., 197
Ahmann, G., 186
Airoldi, L., 326
Aitzetmüller, K., 182
Akabori, Y., 125
Akermark, B., 197
Akhmedova, F. N., 94
Alakhov, Y. B., 163
Alam, S. N., 315
Alazard, J. P., 183
Albrecht, P., 332
Albright, F., 334
Aldanova, N. A., 156, 163
Alder, R. W., 122
Alexander, E. S., 205
Alexander, R. G., 218
Alexanderson, B., 305
Allcock, M. G., 243
Allendoerfer, R. D., 207
Al-Shamma, A. A., 181
Alt, H., 203
Altamura, M. R., 324
Althaus, J. R., 297, 325
Ambler, R. P., 163
Amiet, R., 231
Amos, W. H., 327
Amoss, M., 323
Amy, J. W., 59, 61, 92, 93, 94, 95, 97, 109, 295
Anders, E., 333
Anders, U., 259

Anderson, D. A., 280
Anderson, J. W., 256
Anderson, R. H., 223
Anderson, R. L., 300
Andersson, J., 330
Andlauer, B., 52
Ando, T., 318
Andreotti, R.-E., 324
Andrus, W. G., 103
Anet, E. F. L. J., 179
Angelici, R. J., 261
Anjou, K., 330
Ansell, M. F., 110
Anthonsen, T., 190
Anthony, G. M., 130, 304, 305
Antosz, F. J., 176
Aoyama, T., 279
Aplin, R.T., 91, 104, 180, 181
ApSimon, J. W., 118
Arai, M., 279
Arhart, R. W., 230
Aringer, L., 317
Arison, B. H., 334
Armarego, W. L. F., 121
Armbruster, R., 124, 127, 133
Arnold, R. G., 330
Arpin, N., 182
Arsenault, G. P., 6, 7, 161, 305, 309
Asahi, Y., 125
Ashe, A. J., 215
Ashton, D. S., 98
Askew, W. B., 60
Aslanov, F. A., 94
Aspinall, G. O., 315
Ast, T., 60, 62, 69, 86, 87, 88, 96, 114, 127
Atallah, A. M., 318
Atkinson, L. K., 236
Atwell, W. H., 208, 257
Audier, H. E., 96, 110, 122, 183
Aue, W. A., 334
Auerbach, R. A., 123
Aumann, R., 260
Auret, B. J., 185
Ausloos, P., 6, 85
Avakyan, N. P., 231
Axelrod, G., 198
Axen, U., 306
Axenrod, T., 7, 9, 10, 152, 167, 324
Ayers, B. O., 306
Aynilian, G. H., 187
Azzaro, M., 133

Babb, D. P., 256
Bacchin, P., 325
Backinowsky, L. V., 314
Bacon, V. A., 293, 310
Bafus, D. A., 38
Bagley, G. E., 331
Bailar, J. C., 105, 241, 244
Bailey, A. J., 159
Bailey, C. A., 279
Bailey, S., 331
Baillie, T. A., 306
Baird, R. B., 257
Baitinger, W. E., 59, 61, 68, 92, 93, 94, 95, 96, 97, 109
Baker, A. D., 2
Baker, C., 2
Baker, G. B., 78, 116
Balaban, A. T., 108, 125
Baldas, J., 91
Baldeschwieler, J. D., 24
Baldwin, M. A., 46, 130
Baldwin, M. K., 332
Ball, G., 334
Ball, R., 259
Ballantine, J. A., 108, 121, 123
Ballio, A., 311
Balogh, B., 333
Bancroft, G. M., 241, 243
Bandi, P. C., 145
Bandurco, V. T., 116
Bandurski, R. S., 315
Banks, D., 327
Barber, M., 93, 165, 267, 276
Barber, R. C., 273
Barcellona, S., 311
Barcelo, G., 259
Barchas, J. D., 322
Barford, A. D., 100, 170
Barker, G., 123, 125
Barlow, M. G., 100
Barnes, C. S., 126, 133
Barnes, W. T., 163
Barnett, J. E. G., 160
Barrett, J. F., 162
Barron, R. P., 19, 305, 332
Barry, J. L., 331
Baron, C., 315, 317, 330
Barthels, M. R., 211
Bartlett, P. A., 186
Bartley, W. J., 115
Barton, L., 197
Barua, A. K., 181
Basak, S. P., 181
Basi, J. S., 251
Basu, K., 173

337

Author Index

Bathala, M. S., 189
Bathelt, W., 262
Batt, R. D., 144
Bauder, W., 248
Bauer, Š., 170
Baum, A. A., 106
Baumeister, W., 201
Baumgärtner, F., 260
Baumgarten, H. E., 194
Bayer, E., 244, 323, 335
Bazinet, M. L., 324
Beachley, D. T., 256
Beal, J. K., 181
Beal, J. L., 189
Beam, C. F., 122
Bear, J. L., 248
Beastall, G. H., 318
Beauchamp, J. L., 24, 26, 27, 28, 31, 83, 142, 223
Beck, J. L., 114, 335
Beck, H.-J., 236
Beck, V., 173
Becker, C., 278
Becker, H. G. O., 124
Beckey, H. D., 19, 20, 21, 22, 23, 51, 52, 53, 56, 154, 169, 174, 179
Beery, K. E., 321
Beggs, D. P., 6, 82, 85
Bégué, R. J., 321
Beisler, J. A., 335
Belcher, R., 243, 244
Belikov, A. B., 127
Bell, D., 273
Bell, L. G., 232
Bellman, S. W., 331
Bellut, H., 256
Belsky, R. C., 300
Benek, J. A., 280
Benezra, C., 134, 218
Benezra, S. A., 32, 50, 68, 82, 106, 108, 110
Benezra, S. E., 54
Benkeser, R. A., 68, 96
Bennett, C. E., 305
Bennett, M. A., 262
Bennett, M. J., 259
Bennett, C. E., 305
Bennett, M. A., 262
Bennett, M. J., 259
Bennett, S. W., 197
Benoit, F., 74, 75, 102, 108, 114, 119
Benson, W. R., 332
Bente, P. F., 24, 86
Bentley, R., 152, 317
Bentley, T. W., 46, 79, 91, 96
Bercaw, J. E., 232
Bergmann, E. D., 99, 134, 215, 216, 218
Bergmann, F., 133
Bergstedt, L., 296, 306, 322
Bergström, G., 328, 329
Bergström, K., 323
Bernath, A., 160
Beroza, M., 302, 309, 328
Berry, A. D., 228
Berry, R. S., 126
Bertino, C. D., 98

Bertoglio, M. R., 278
Bertolini, M., 175
Bertorello, H. E., 98, 118
Bertrand, M., 87
Bessell, E. M., 100, 170
Bethke, H., 129
Betteridge, D., 247
Bettler, C. R., 257
Beugelmans, R., 181
Beuhler, R. J., 10, 168
Beyer, D., 124
Beynon, J. H., 59, 60, 61, 62, 63, 68, 69, 74, 86, 87, 88, 92, 93, 94, 95, 96, 97, 103, 106, 108, 109, 114, 123, 127, 279, 295
Bhakuri, D. S., 192
Bhalerao, U. T., 104, 182
Bhattacharya, A. K., 100, 239
Bichlmeir, B., 207
Bickelhaupt, F., 198, 256
Biddlestone, M., 258
Bidinosti, D. R., 225
Bieber, M. A., 316
Biemann, K., 6, 7, 90, 106, 111, 113, 161, 265, 274, 281, 292, 296, 297, 305, 325, 335
Bierl, B. A., 309, 328
Bigley, D. B., 218
Bild, N., 113
Biller, J., 281, 297, 325
Billets, S., 27
Bills, D. D., 330
Binenbaym, J., 257
Bingham, R. A., 280
Binkley, W., 171
Binks, R., 267
Bir, R., 281
Birchall, T., 230
Bird, C. W., 310
Bird, J. S., 299
Bird, K. E., 299
Birkofer, L., 207, 209, 256
Birnbaum, K. B., 186
Biros, F. J., 331
Bisarya, S. C., 180
Bizzi, A., 326
Björkhem, I., 310
Björndal, H., 171, 315
Black, D. R., 269
Black, D. St. C., 245
Blackburne, I. D., 180
Blackman, A. J., 133, 140
Blackwell, R., 323
Blair, J. A., 116
Blair, L. K., 18, 26, 27
Blais, J.-C., 15
Bland, K. P., 312
Blas, J. C., 138
Blazer, T. A., 101
Blenkinsop, J., 265, 267
Blessington, B., 172, 325
Bloch, D. R., 138
Bloching, S., 19
Blomquist, G. J., 145
Blomstrand, R., 317
Blumenfeld, O. O., 160
Blumenthal, T., 14, 140

Blumer, M., 309, 316
Boal, J. R., 300
Boar, R. B., 189
Bochkarev, V. N., 125, 156
Bock, H., 203
Boeck, L. D., 178
Böhm, R., 113
Boettger, H. G., 137, 205, 209
Boettner, E. A., 334
Bogan, D. J., 42, 117
Bogdanova, I. A., 156
Bogentoft, C., 120, 127
Bogolyubov, G. M., 219, 221, 223
Bohlmann, F., 179, 191
Bohn, G. T., 197
Bojesen, I. N., 132
Boldyreva, O. G., 199
Bolm, G., 115
Bonati, F., 240
Bonelli, E. J., 295, 296, 305, 332
Bonnell, D., 278
Bonnett, R., 248
Bonnichsen, R., 306
Boone, B., 287
Bor, G., 238, 263
Borders, D. B., 179
Borén, H. B., 171
Borga, O., 322
Borisov, G. K., 232
Borowski, E., 176
Borossay, J., 42
Bose, R. J., 136, 199
Bottin, J., 122, 183
Bouchez-Dangye-Caye, M. P., 182
Bouchoux, G., 96, 110
Boulloud, P., 295
Bournot, P., 315
Bowen, D. H., 316
Bowers, C. Y., 162
Bowers, M. T., 29, 30
Bowie, J. H., 14, 15, 65, 80, 90, 91, 96, 97, 99, 107, 113, 123, 130, 133, 135, 136, 138, 139, 140, 191, 192, 214, 218
Boyer, J. H., 258
Boyusch, E., 257
Brabetz, H., 260
Braddley, W. H., 263
Bradford, C. W., 226
Bradley, D. C., 251
Bradshaw, T. K., 65, 96
Brady, L. E., 125
Brady, R. A., 120
Brady, U. E., 328
Brakke, J. W., 145
Brandstrup, N. E., 320
Brauman, J. I., 18, 26, 27, 92
Braun, A. M., 96
Bray, L. S., 259
Brehm, B., 4
Brehm, E., 199, 256
Breimer, D. D., 327
Breitmaier, E., 315
Brekke, J. E., 330

Breland, J. G., 263
Bremer, N. J., 262
Brennan, M. R., 307
Brent, D. A., 111, 137
Breuer, H., 305
Breuer, S. W., 117, 195
Brewer, P., 248
Brewington, C. R., 319
Bridges, J. C., 293
Brieskorn, C. H., 173
Briggs, P. R., 42, 73, 132, 204
Brindley, P. B., 199
Brintzinger, H. H., 232, 260
Brion, C. E., 4
Bristol, D. W., 120
Brittain, E. F. H., 119, 199
Brodie, A. M., 259
Brodskii, E. S., 129
Brokl, O., 144
Bronwen Loder, R., 162
Brooks, C. J. W., 130, 304, 305, 306, 308, 310, 312, 318, 319, 335
Brooks, E. H., 259
Brooks, R., 134
Brostoff, S. W., 324
Broughton, J. M., 318
Broussier, R., 239
Brown, C. A., 104
Brown, C. L., 14, 17, 140, 198
Brown, E. V., 126, 127
Brown, J. P., 179
Brown, L. M., 250
Brown, N. L., 332
Brown, P., 22, 23, 55, 78, 99, 112, 119, 172, 175, 185
Brown, R., 280
Brown, R. A., 262
Brown, R. F. C., 107
Brown, T. L., 194
Browne, E. J., 124
Brownlee, R. G., 328
Brownlee, R. T. C., 27
Brubaker, C. H., 250
Bruce, M. I., 193
Brugel, E., 124, 129
Bruins Slot, J. H. W., 289, 307
Brundle, C. R., 2
Brune, H. A., 260
Bruneteau, M., 158
Brunfeldt, K., 160, 165
Brunner, H., 258
Brüschweiler, F. R., 23, 172, 175, 185
Brutane, D., 105
Bryan, R. F., 180, 256
Bryant, W. F., 195
Bryce, W. A., 43
Bryson, J. G., 220
Buchanan, B. G., 281, 282, 283, 288
Buchardt, O., 122
Buchler, J. W., 248
Buchs, A., 102, 118, 281, 282, 283
Buckeridge, F. A., 173
Budde, W. L., 166, 288

Budylin, V. A., 120, 138
Budzikiewicz, H., 90, 97, 127, 182, 186, 194
Büchi, J., 330
Bühler, R. E., 273
Buell, G. R., 203
Bugge, A., 132
Buhs, R. P., 323, 327
Buijten, J. C., 325
Bukhari, S. T. K., 178
Bu'lock, J. D., 310
Bunton, C. A., 134
Bunyan, P. J., 331
Buonocore, V., 151
Burg, A. B., 258, 260
Burgus, R., 11, 323
Burikov, V. M., 126, 156
Burkholder, P. R., 190
Burks, R., 92
Burlingame, A. L., 33, 90, 91, 92, 120, 125, 159, 186, 265, 273, 300, 329, 333, 334
Burlitch, J. M., 259
Burrell, J. W. K., 334
Bursey, J. M., 125
Bursey, J. T., 83
Bursey, M. M., 25, 31, 32, 50, 51, 54, 66, 68, 82, 83, 106, 107, 108, 110, 112, 194, 233, 240, 246
Burt, R., 260
Busby, M. R., 280
Busch, M. A., 262
Bush, L. P., 318
Bush, P. B., 318
Butcher, M., 107, 121, 323
Buttery, R. G., 329, 330
Buttrill, S. E., 30, 138
Butts, W. C., 335
Buu-Hoi, N. P., 129, 133
Buurmans, H. M. A., 108
Byers, B. A., 145

Cable, J., 111, 114, 116
Cady, G. D., 137
Cady, G. H., 16
Cala, P. C., 323
Calder, I. C., 130
Callot, H. J., 134, 218
Calvert, J. G., 114
Calvin, M., 333
Cambie, R. C., 180, 181
Cambon, A., 98, 133
Cameron, D. W., 191
Campbell, A. J., 276
Campbell, H. F., 186
Campbell, I. M., 322
Campbell, M. M., 130, 131
Cannon, E., 288
Cant, E., 106
Capek, K., 314
Capella, P., 149, 311
Caprioli, R. M., 59, 60, 61, 62, 68, 74, 86, 92, 93, 94, 95, 96, 97, 103, 106, 109, 114, 127
Carbery, E., 211
Cardenas, C. G., 99

Cardillo, B., 190
Cardin, D. J., 227, 261
Caress, E. A., 319
Carey, R. N., 300
Carlin, J. R., 327
Carlson, D. A., 309
Carlson, L. R., 278
Carpenter, I., 192
Carpenter, R. S., 277
Carrel, J. E., 328
Carrick, A., 273, 279
Carrol, C. A., 280
Carroll, S. R., 97
Carter, J. C., 17
Cartledge, F. K., 207
Carty, A. J., 217, 230
Caruso, F., 122
Carver, R. D., 279
Caserio, M. C., 31
Casey, A. C., 129
Casida, J. E., 115
Caspar, A., 113
Caspar, M. L., 258
Cataliotti, R., 239
Cattabeni, F., 306
Cavell, R. G., 220, 258
Cavender, C., 288
Cermak, V., 83
Cetini, G., 226, 230, 259
Cetinkaya, B., 261
Chadha, M. S., 151, 323, 324
Chait, E. M., 20, 60, 104
Chakrabarti, P., 181
Chakraborty, D. P., 181
Challis, J. R. G., 317
Chambers, C. A., 280
Chambers, D. B., 212
Chan, E. K., 257
Chan, T. H., 220, 248
Chan, W. R., 180
Chandler, H. A., 129
Chandra, G., 261
Chaney, E. L., 17
Chang, C., 82
Chang, C.-J., 190
Chang, K. Y., 22, 118
Chapat, J.-P., 105
Chapman, D. I., 326
Chapman, J. R., 267, 276, 285, 326
Charalambous, J., 238, 247
Charles, D., 306
Charlton, T. L., 220
Charpenet, L., 305
Chashcin, V. L., 172
Chasin, D. G., 109
Chatranon, W., 310, 311
Chattoray, S. C., 306
Chaudhuri, R. K., 191
Chen, J.-Y. T., 332
Chen, Y.-P., 189
Cherkez, S., 125
Chervin, I. I., 219
Chiba, T., 127
Chin, M. S., 80
Chisholm, M. H., 251
Chivers, T., 212, 258
Chizhov, O. S., 76, 94, 96, 106, 108, 132, 158, 169, 170, 172, 181, 314

Chochua, K. A., 96
Chondromatidis, G., 129
Chong, S.-L., 84
Chow, Y. L., 260
Christiansen, G. D., 105
Christodoulides, A. A., 17
Christophersen, G., 125
Christophorou, L. G., 17
Chuche, J., 127
Chung, A. E., 327
Chupakhia, M. S., 279
Chupka, W. A., 3, 5, 36
Churchill, M. R., 261
Cicero, T. J., 147
Cimarusti, C. M., 94
Cimirio, G., 180
Clark, J., 127, 128, 129, 326
Clark, R. J., 262
Clarke, D. C., 270
Clarke, J. K. A., 230
Clase, H. J., 197
Clausen, E., 191
Claveau, J. C., 326
Cleaver, R. L., 92, 267
Cleland, A. J., 259
Clemens, D. F., 258
Clemmitt, A. F., 229
Clerc, J. T., 265, 293
Clobes, A. L., 105, 242
Clow, R. P., 28
Coburn, T. C., 293
Cochrane, W., 332
Cochry, R. E., 256
Cohen, J. S., 152, 324, 334
Cohen,S.C.,16,130,196,211
Cohn, K., 258
Cole, R. H., 299
Coleman, K. J., 260
Coleman, R. L., 330
Collier, C. W., 328
Collin,J.E.,42,129,130,137
Collins, P. M., 17
Collins, J. H., 3, 44, 298
Collman, J. P., 226
Colton, R., 246
Comin, J., 187
Comisarow, M. B:, 27
Compernolle, F. C., 115, 123, 156
Compton, R. N., 16, 17, 18
Condé-Caprace, G., 42, 129, 130
Cone, C., 161, 200
Conner, R. L., 144
Connor, J. A., 235, 257
Conrad, H. E., 144
Conrads, R. J., 280
Conroy, A. P., 250
Constantin, V., 126
Conville, J. J., 224, 260
Conway, D. C., 262
Cook, J. M., 187
Cook, M., 61, 100
Cook, R. J., 101
Cooke, D. W., 323
Cooke, M., 260
Cooks, R. G., 37, 45, 47, 57, 62, 63, 69, 87, 88, 96, 105, 116, 131, 136, 169, 172, 197, 239

Cooney, J. D., 118
Cooper, D. J., 173, 178
Cooper, J. L., 121
Cope, B. T., 182
Copperthwaite, R. G., 251
Corbin, J. L., 108
Corey, E. R., 228
Corey, J. Y., 202, 207
Corn, J. E., 68, 96
Cornford, A. B., 2
Cornu, A., 92
Corral, R. A., 121
Cortegiano, H., 154
Cortés, E., 118, 120
Corval, M., 70, 101
Coskran, K. J., 220
Costa, E., 306
Cotter, J. L., 98, 121, 205
Cottin, M., 15, 138
Cottrell, I. W., 315
Couch, M. W., 113
Coulson, C. J., 129
Courchene, W. L., 3, 44
Courrier, W. D., 243
Cousins, R. J., 185
Coutant, J. E., 93, 277, 279
Coutts, R. S. P., 232
Coutts, R. T., 115, 121
Cover, R. E., 154
Cowan, D. O., 106
Cowley, A. H., 220
Cowley, S. W., 110
Cox, M. R., 180
Cozens, R. J., 245
Crable, G. F., 79
Cradwick, P. D., 180
Cragg, R. H., 42, 137, 198, 199, 200, 208, 223
Craig, J., 164
Cranwell, P. A., 154
Crawford, L. R., 285, 286
Crawford, R. W., 293
Crawhall, J. C., 326
Cren, M.-C., 188
Critterden, A. L., 16, 137
Croisy, S., 133
Cromartie, E., 331
Cronholm, T., 319
Cross, B. E., 179
Cross, R. J., 229
Croteau, R., 311
Crow, J. P., 262
Crow, W. D., 187
Crozier, A., 316
Cruse, W. B. T., 181
Csákvári, B., 42
Cuddy, B. D., 258
Cullen, W. R., 129, 262
Cum, G., 122
Cummings, S. C., 245
Cundy, C. S., 259
Curran, D. J., 91
Curtis, J. R., 57, 131
Curtis, R. F., 108
Curtius, H.-C., 303, 321
Cushley, R. J., 333
Cutcliffe, A. B., 262

Dabrowiak, J. C., 323
D'Agostino, M., 192

Dahl, A. R., 257
Dahl, L. F., 229
Dahm, K. H., 328
Dale, A. J., 65
Dalgarno, A., 83
Daly, W. H., 131
Damico, J. N., 19, 305, 329, 332
Damoth, D. C., 305
Danby, C. J., 4, 54
Daniel, A., 179
Daniels, P. J. L., 171, 178
Danielsson, B., 127
Danilenko, G. I., 94
Darensbourg, M. Y., 194
Das, B. C., 144, 160, 181, 188
Das, K. G., 137, 180, 209
Dasgupta, B., 173
Davies, C. S., 260
Davies, F. J., 16
Davies, V. H., 180
Davis, B., 110, 112
Davis, D. L., 318
Davis, H. A., 312
Davis, J. E., 300
Davis, R., 199, 236
Davis, R. E., 259
Davison, A., 257, 261
Dawson, G., 148
Day, R. A., 155
Dayhoff, M. O., 287
Deacon, G. B., 257
Dean, J. R., 322
Dear, R. E. A., 250
Deas, A. H. B., 311
de Beer, J. A., 238
Deberitz, J., 261
de Bertorello, M. M., 98, 118
de Boer, Th. J., 72, 73, 94, 97, 106, 117, 118, 126, 131
de Clerq, M., 211
Decora, A. W., 94
De Corpo, J. J., 38
Deeming, A. J., 263
Defaye, C., 188
DeFilippo, D., 263
Degen, P. H., 322
Dehennin, L., 317
Dehnicke, K., 197
de Jong, F., 73, 132
DeJongh, D. C., 33, 90, 111, 137, 265
Dekeirel, M., 123
Delalieu, F., 97
Delfino, A. B., 281, 282, 283
de Liefde Meijer, H. J., 234, 260, 261
Dellaca, R. J., 227
DeLuca, H. F., 185
del Valle, U. E., 166
de Nadal, F., 326
Denniston, M. L., 256
Derevitskaya, V. A., 158
de Ridder, J. J., 117, 171, 194, 247, 289, 307
Dermirgian, J., 207
de Rosa, M., 310

Author Index

Derrick, P. J., 21
de Schryver, F. C., 115
Desiderio, D. M., 7, 11, 161, 163, 174
Desmarchelier, J. M., 130
DeStefano, S., 180
Deumont, M., 137
Deutsch, J., 105, 109, 133
Dev, S., 180
de Valois, P. J., 331
Devaux, P. G., 318, 320, 326
Devik, O. G., 329
Devillanova, F., 263
Devissagnet, P., 8
de Vos, R. H., 332
Devyatylch, G. G., 232
Dewar, M. J. S., 200
de Witt, H. G. J., 314
Deyrup, J. A., 119
de Zeeuw, R. A., 327
Dhar, M. M., 192
Dias, A. R., 262
Dias, J. R., 55, 104, 113
Diatta, L., 183
Dibeler, V. H., 197
DiCave, L. W. jun., 305
Dickinson, R. J., 94, 125, 162
Dickson, R. S., 259
di Domenico, A., 17
Diekman, J., 60, 104, 106, 137
Dielard, J. G., 92
Dierdorf, D. S., 220
Dillard, J. G., 17
Dillon, J. P., 213
Dimmock, J. R., 78, 116
Dine-Hart, R. A., 121
Dinsmore, H. L., 330
Distefano, G., 2, 79, 97, 197, 203, 225, 238, 240
Dittmer, D. C., 120
Djerassi, C., 55, 59, 60, 74, 90, 94, 103, 104, 105, 106, 111, 112, 113, 114, 116, 117, 121, 125, 137, 185, 186, 210, 281, 282, 283, 288, 318
Dmitriev, B. A., 314
DoAmaral, J. R., 322
Dobbie, R. C., 262
Dobson, G. R., 262
Dolby, R., 259
Dolcetti, G., 226
Dolhun, J. J., 6, 7, 127, 305
Dolphin, D., 186
Dombek, B. D., 211
Donaghue, P. F., 297, 325
Dooley, C. J., 122, 329
Doolittle, F. G., 94
Dornberger, E., 260
Dorokhov, V. A., 199
Doskotch, R. W., 181
Dougherty, R. C., 53, 98, 171
Douglas, A. G., 316
Douraghi-Zadeh, K., 316
Dowie, R. S., 309
Downer, S. W., 273

Downing, G. V. jun., 323, 326
Draffan, G. H., 333
Drake, J. E., 256
Drawert, F., 330
Draxe, K., 92
Dresdner, R. D., 250
Drew, C. M., 334
Dreyfuss, R. M., 257
Drienovsky, P., 260
Dromey, R. G., 297, 299
Dubac, J., 208, 257
Dube, G., 137, 250
Dubois, R. J., 129
Duckworth, H. E., 273
Dudley, F. B., 16, 137
Dudley, K. H., 66, 107
Due, S. L., 327
Dueber, M., 202, 207
Duffield, A., 334
Duffield, A. M., 101, 104, 108, 115, 116, 120, 121, 122, 125, 127, 188, 280, 281, 282, 283, 288
Dugan, J. V., 300
Duholke, W. K., 83
Dunbar, R. C., 17, 26, 28, 29
Duncan, J. H., 146, 314
Dunn, A. D., 131
Dunn, T. F., 11
Dupzyk, R. J., 279
Durden, J. A., 115
Durley, R. C., 316
Dussel, H.-J., 22
Dutta, S. P., 181
Dyatkin, B. L., 250, 255, 257
Dyer, M. C. D., 122
Dyke, S. F., 189
Dzidic, I., 11

Eaborn, C., 197
Eadon, G., 59, 60, 103, 104, 105, 185
Earle, F. R., 311
Earle, N. R., 270
Earley, R. A., 216, 223
Earnshaw, D. G., 94
Ebbinghausen, W., 315, 323
Ebert, H. D., 220, 258
Eck, R. V., 287
Eckhardt, C. J., 275
Edmonson, L. J., 223
Edwards, A., 331
Edwards, T. R., 300
Efraty, A., 217, 231, 232, 258, 260
Ege, G., 92
Egge, H., 310, 311
Eggelte, T. A., 46
Egger, K. W., 229
Eglinton, G., 316, 333
Ehmann, A., 315
Ehntholt, D. J., 260
Ehrl, W., 237
Eilers, N. C., 315
Einhorn, J., 183

Einolf, N., 75, 107
Eisler, W. J., 306
Eisner, T., 316, 328
Eizen, O. I., 96, 97
Eland, J. H. D., 4, 54, 197
Elbein, A. D., 149, 318
Eldjarn, L., 325
Elkin, Y. N., 185
Elleman, D. D., 29, 30
Elliot, R. M., 93
Elliott, W. H., 152, 319
Ellis, G. P., 123, 125
Ellsworth, R. K., 322
Elmes, P. S., 220, 222, 262
Eloranta, J., 138
Elschenbroich, C., 261
Elsken, R. H., 269
Elter, G., 257
Elwood, T. A., 31, 32, 66, 83, 107
Elyakov, G. B., 185
Ernberger, R., 330
Emerson, G. F., 231
Ende, M., 185, 308
Eneroth, P., 317, 319, 320
Engelmore, R. S., 288
Enzell, C. R., 181, 182
Erickson, R., 265
Eriksson, H., 320, 321
Ermakov, A. I., 125
Erni, Von F., 265, 293
Estacio, P., 257
Ettore, R., 263
Eudier, M., 305
Euranto, E. K., 109
Eustache, H., 305
Evans, D. A., 90, 111, 281, 297, 325
Evans, S., 240
Eyem, J., 303
Eyler, J. R., 27

Fackler, J. P., 245
Fagerson, I. S., 311
Fairweather, R. B., 52
Fairwell, T., 163, 164, 288
Falch, E., 133
Fales, H. M., 6, 7, 8, 9, 10, 152, 167, 297, 324, 335
Falick, A. M., 300
Falkenberg, I., 190
Fal'ko, V. S., 133
Falkowski, L., 176
Fallgatter, M. B., 280
Falter, H., 155
Fanelli, R., 325, 326
Farmery, K., 236
Farnham, P., 234
Farnsworth, N. R., 187
Farona, M. F., 262
Farrant, G. C., 13
Farren, J., 300
Farrier, D. S., 186
Fattorusso, E., 180
Faue, W. H., 186
Faul, W. H., 104
Faulkner, D. J., 316
Fay, R. C., 243
Fayez, M. B. E., 324

Fazio, T., 329
Fear, T. E., 195
Feast, W. J., 115
Fechner, K. H., 102
Fedeli, E., 330, 331
Feeney, J., 179
Feher, F., 256
Fehlhaber, H.-W., 115, 173, 184, 187
Fehlner, J. P., 256
Feigenbaum, E. A., 281, 282, 283, 288
Feigina, M. Yu., 156, 163
Feil, V. J., 77, 110
Feinstein, A. I., 118
Felix, R. A., 204, 209, 210
Fellows, R., 323
Feltham, R. D., 221
Fenderl, K., 224, 225, 233, 237
Fenselau, C. C., 22, 101, 103, 106, 129, 146, 314, 317
Fenwick, R. G., 121, 123
Ferguson, E. E., 83
Ferguson, K. A., 144
Ferles, M., 322
Ferngren, H., 320
Ferrari, R. P., 230, 259
Ferraro, G. M., 190
Ferrer-Correia, A. J., 24, 28, 82
Ferretti, A., 329
Ferrier, R. J., 174
Feser, M. F., 256
Fetchin, J. A., 245
Fetizon, M., 96, 110, 122, 183, 188
Fiasson, J.-L., 182
Field, D. S., 228, 263
Field, F. H., 9, 11, 14, 82, 85, 92
Fieldhouse, S. A., 259
Fields, E. K., 97, 118
Fields, R., 230, 258
Filby, W. G., 23, 77, 174
Filho, R. B., 191
Finch, R. W., 304
Finney, C. D., 3, 45
Finney, R. D., 46
Firestone, D., 332
Fischer, E. O., 235, 236, 260, 261, 262
Fischer, R. D., 235
Fishwick, M., 202
Fitzpatrick, J. D., 97
Fjeldstad, P. E., 133
Flanagan, V. P., 329
Flanigan, E., 168
Flesch, G. D., 43, 194
Fletterick, R. J., 260
Flores, I., 291
Flores, J., 324
Flory, D., 334
Flotard, R. D., 332
Förster, H.-J., 297, 325
Foester, R., 258
Foffani, A., 238, 239
Foglia, T. A., 122
Follweiler, J., 133

Folsch, G., 255
Folsome, C. E., 333
Foltz, R. L., 53, 177, 178, 330
Fong, C. W., 208, 257
Fontaine, A. E., 62, 95
Foote, J. L., 313
Forbes, C. E., 246
Forehand, J. B., 22
Forgione, A., 325
Forin, M.-C., 317
Forster, W., 243
Fossey, J., 260
Foster, A. B., 100, 170
Foster, M. S., 28, 83, 223
Foust, A. S., 229
Fowler, L. J., 159
Fox, W. B., 250
Fox, W. G., 299
Fraas, R. E., 62
Francis, G. W., 174, 182
Franklin, J. L., 17, 38, 84, 92, 97
Franklin, S. J., 32
Franz, D. A., 256
Franz, M., 256
Franzen, J., 278
Fraser, D. R., 149
Fraser, J. W., 265
Fraser, P. J., 259
Fraser, R. T. M., 113
Frazer, J. W., 278, 293
Frazer, M. J., 238, 247
Frazer, R. R., 98
Fredericks, R. J., 250
Freeburger, M. E., 203
Freedman, L. D., 258
Freeland, B. H., 259
Frerburger, M. E., 99
Frerman, F. E., 146
Frew, N. M., 273, 284
Freytag, W., 310
Fridlyanskii, G. V., 121, 219
Friedel, P., 329
Friedman, L., 10, 83, 168
Friedman, N., 324
Fries, I., 152
Friesen, R. D., 272
Frigerio, A., 326
Frisch, M. A., 295
Frost, D. C., 2, 129
Fry, A., 15, 137
Frye, C. L., 207
Fujimaki, M., 329
Fujimoto, Y., 127, 133
Fujita, T., 180
Fukanii, H., 328
Fukamiya, N., 181
Fukushimo, S., 125
Funke, H., 181
Furlei, I. I., 138
Furtsch, T. A., 220
Furukawa, Y., 311
Furuta, M., 330
Futrell, J. H., 6, 10, 13, 28, 66, 70, 82, 167

Gadzhiev, M. M., 94
Gäumann, T., 98

Gaffney, T. E., 306
Gaidis, J. M., 42, 204
Gaivoronskii, P. E., 232
Galamova, T. A., 94
Galbraith, M. N., 181
Gallagher, D. A., 42, 200
Gallagher, M. J., 216, 223
Gallegos, E. J., 333
Gallop, P. M., 158, 160
Galy, J.-P., 111
Gambacorta, A., 310
Gambino, O., 226, 230, 259
Gamble, A. A., 77, 81, 110, 114, 131
Gamo, M., 311
Ganguli, G., 186
Gara, A. P., 218
Garattini, S., 326
Garcia-Martinez, N., 118
Gardner, R. J., 116
Garegg, P. J., 171
Garg, H. S., 181, 191
Garner, A. W. B., 259
Garzo, G., 208
Gautheron, B., 239
Gaymard, F., 98
Gdulenwicz, M., 176
Gebelein, C. C., 115
Geddes, A. J., 165
Gehrke, C. W., 334
Geissler, H., 257
Gella, I. M., 120
Gelman, R. A., 145
Gelpi, E., 333
Gennaro, A., 190
George, M. V., 137, 209
George, R. D., 228, 257
George, T. A., 211
George, W. O., 107
Georgobiani, T. P., 297
Germain, M. M., 230
German, V. F., 176
Géro, S. D., 160
Gerrard, A. F., 104, 106
Ghosal, S., 173, 191
Gianturco, M. A., 329
Gibbon, G. A., 257
Gibson, H. W., 131
Gielen, M., 202, 211
Gierer, J., 334
Giessner, B., 277
Giessner, B. G., 3, 44, 297, 298, 299
Giezendanner, H., 76, 114
Gil-Av., E., 109, 334
Gilbert, E. E., 250
Gilbert, E. J., 10, 167
Gilbert, J., 333, 334
Gilbert, J. N. T., 325
Gilbert, J. R., 51, 77, 81, 110, 114, 129, 225
Gilbertson, J. R., 145
Gillis, R. G., 96, 116, 287
Girard, J.-P., 105
Gitton, B., 15, 138
Gladstone, W. A. F., 117
Glaudemans, C. P. J., 175
Gleason, J. G., 152
Glemser, O., 256, 257
Glennie, G. W., 191

Author Index

Glockling, F., 201, 212, 229, 259
Goad, L. J., 318
Godtfredsen, W. O., 176
Gogan, N. J., 260
Gohlke, R. S., 208
Goldberg, P., 39, 93
Goldberg, S. I., 263
Golden, R., 200
Goldenfeld, I. V., 19
Gol'dfarb, Ya. L., 132
Goldman, D. S., 174
Goldsack, R. J., 126, 133
Golik, J., 176
Goode, G. C., 24, 28, 82
Goodwin, S. L., 147, 315
Goodwin, T. W., 318
Gordon, L. P., 300
Gore, J., 114, 125
Goren, M. B., 144
Gorinsky, C., 189
Gorman, M., 178, 187
Gorodetsky, I. G., 108
Gorodetsky, M., 103
Gose, P., 193
Goto, S., 304
Goto, T., 106
Gottlieb, O. R., 191
Gough, T. A., 305
Goulden, J. D. S., 109
Govindachari, T. R., 181, 191
Govind Rao, M. K., 316
Gower, D. B., 321
Grablyauskas, K. V., 127
Graff, G., 146
Graham, G. N., 165
Graham, W. A. G., 258, 259, 262
Grajower, R., 51, 137
Gramain, J.-C., 183
Grandberg, I. I., 120, 121
Granger, R., 105
Granoth, I., 134, 215, 216
Granström, E., 312
Grant, P. M., 270
Grassberger, M. A., 258
Gray, G. A., 24
Gray, M. F., 319
Gray, R. T., 137, 210
Gray, W. R., 10, 166, 167
Grayson, M. A., 280
Grayson, M. S., 333
Gream, G. E., 99
Greatrex, R., 261
Grechkin, N. P., 219
Green, B. N., 267, 276
Green, J. H., 121, 129, 247
Gréen, K., 306
Green, L. J., 10
Green, M., 259, 260
Green, M. L. H., 261, 262
Green, M. M., 69, 101
Greene, F. D., 120
Greene, L. J., 168
Greene, P. T., 256
Greene, R. L., 94
Greenwood, N. N., 261
Gregory, L. M., 122
Greidanus, J. W., 101

Greiss, G., 235
Grigorian, R. T., 113
Grigsby, R. D., 115, 299
Grimes, R. N., 256, 261
Grimm, L. F., 256
Grimmer, R., 238, 259, 263
Grishin, N. N., 219, 221, 223
Grob, G., 332
Grob, K., 302, 332
Grobe, J., 237, 257, 258, 262
Gröger, D., 189
Grønneberg, T., 126
Grose, W. F. A., 46
Gross, K. P., 17, 198
Gross, M. L., 3, 24, 25, 26, 32, 83, 96, 97, 275, 298
Gross, R., 117
Groszek, E., 256
Grotch, S. L., 289
Grote, A., 198
Grützmacher, H.-F., 96, 102, 112, 116, 170, 307
Grundke, H., 256
Grunwald, C., 318
Guadagni, D. G., 329, 330
Guenzler, H., 273
Gürtler, J., 317, 323
Güsten, H., 65, 74, 97
Guggisberg, A., 113
Guidj, A., 98
Guignes, F., 125
Guillard, R. R. L., 309
Guillemin, R., 11, 323
Guillery, R., 256
Guiochon, G., 233
Guion, J., 100
Gunnarsson, L.-E., 182
Gunnell, K. D., 147
Gunning, H. E., 83
Gunstone, F. D., 144, 145
Gunthard, Hs. H., 267
Gupta, A. S., 180
Gusel'nikov, L. E., 137, 206, 209
Gustafsson, J. A., 309, 317, 319, 320, 321
Guthrie, R. D., 178
Gyimesi, J., 179
György, P., 310

Haag, A., 198, 199, 220, 257
Haake, P., 248
Haddad, H., 207
Haddon, W. F., 269
Hafner, F. W., 293
Haga, M., 169
Hage, A., 256
Hagele, K., 7
Hagen, A. P., 228
Hagenmaier, H., 323
Haines, R. J., 238
Hakomori, S.-I., 148
Hala, S., 309
Halasa, A., 257
Halbert, E. J., 126

Hale, R. L., 104
Hall, G. G., 16
Hall, R. H., 315
Halliday, J. S., 276
Halpern, B., 115, 293, 310
Halpern, D., 198
Hamberg, M., 306, 310, 312
Hamill, R. L., 178
Hamilton, P. B., 334
Hammar, C.-G., 305, 306, 325
Hammarström, S., 149, 313
Hammerum, S., 129
Hammett, A., 240
Hammill, W. H., 3
Hammans, G. J., 120
Han, J., 333
Hanaoka, M., 187
Hancox, N. C., 187
Hand, C. W., 42, 117
Hanebeck, H., 260
Hanessian, S., 169
Haney, M. A., 38
Hanessian, S., 169
Haney, M. A., 38
Hanrahan, R. J., 280
Hansen, C. O., 299
Hara, H., 328
Harayama, T., 182
Harborne, J. B., 191
Harbourne, D. A., 262
Harder, V., 237
Harding, J. H., 272
Hardy, A. D. U., 180
Hardy, P. J., 331
Hargreaves, K. J. A., 137, 208
Harkness, A. L., 182
Harland, P. W., 92, 142
Harland, W. A., 310
Harley-Mason, J., 119
Harman, J. S., 258
Harpp, D. N., 116, 152
Harris, M. H., 97
Harrison, A. G., 3, 34, 45, 80, 86, 101, 103, 111, 129
Harrison, D. R., 132
Harrison, F. A., 321
Harrison, P. G., 228
Hart, N. K., 189
Hart-Davis, A. J., 260, 262
Hartley, B. S., 165
Hartley, R. D., 334
Hartwell, G. E., 194
Harvey, D. J., 183, 184, 211, 308, 314, 317, 318
Hasatome, M., 239
Hasegawa, T., 182
Hassall, C. H., 176
Hasselmann, D., 112
Hassid, D. V., 107
Haszeldine, R. N., 100, 205, 230, 257, 258
Hata, T., 154
Hatfield, W. E., 246
Hattori, F., 321
Haug, P., 128, 278, 333
Haumann, J. R., 306
Hauser, C. R., 122
Havlicek, S. C., 307

Hawks, R. L., 186
Haworth, R. D., 154
Hawthorne, M. F., 261
Hayashi, A., 313
Hayashida, K., 128
Hayatsu, R., 151, 333
Hayden, J. L., 93
Hayes, H. W., 109, 310
Hayes, J. M., 333
Hayes, R. G., 231
Hayes, S. E., 259
Haywood Farmer, J. S., 4
Heacock, R. A., 120
Heaney, H., 108
Heap, R. B., 317, 321
Heath, E. C., 146
Heatherbell, D. A., 330
Hecht, S. S., 120
Hedberg, F. L., 239, 263
Heerma, W., 117, 243
Hegarty, M. P., 154
Heikkilä, J., 319
Heil, H. F., 235
Heimann, W., 330
Heindrichs, A., 19
Heller, B., 322
Hellerqvist, C. G., 315
Hellwinkel, D., 98, 215, 222
Hemingway, J. C., 180
Hemingway, R. J., 180
Henderson, W., 317, 333
Henderson, W. G., 27
Henglein, A., 137, 141
Henis, J. M. S., 17
Henneberg, D., 270, 296
Henson, E., 160
Herber, R. H., 257
Herberhold, M., 260, 261
Herberich, G. E., 235, 261
Herlan, A., 93, 225, 265
Herod, A. A., 86
Herring, F. G., 2
Herron, J. T., 92
Hertz, H. S., 90, 281, 292, 297, 325
Hertzberg, M., 42
Herz, J. E., 317
Herzberg, G., 2
Herzog, L. F., 93, 334
Herzog, W., 257
Hesse, G., 198, 199, 256
Hesse, M., 76, 113, 114, 187, 189
Hessling, R., 306
Heurtevant, C. W., 131
Heyman, M. L., 116
Heyns, K., 329
Heywood, A., 131
Hibino, T., 182
Hickling, R. D., 35
Hieber, W., 226
Higgens, C. E., 178
Higgins, F. R., 228
Hignite, C. E., 297, 325
Hill, G. C., 259
Hill, R. M., 320, 326
Hillman, M. J., 189
Hills, L. P., 66, 70
Hinton, B. B., 273
Hipps, P. P., 315

Hirata, S., 121, 124
Hirota, K., 94
Hirst, P. A., 117
Hites, R. A., 274, 292, 296, 299
Hitzke, J., 100
Hjeds, H., 191
Ho, A. C., 15, 138, 139
Ho, C. T., 181
Hobson-Frohock, A., 331
Hodges, R., 144, 180
Höfer, M., 262
Høg, J. H., 132
Hoehn, M. M., 178
Hörlin, D., 306
Hoffman, J. H., 93
Hoffman, L., 262
Hoffman, M. K., 25, 31, 32, 51, 54, 68, 82, 106, 107, 112
Hofmann, A., 327
Hofstetter, H. H., 145, 310
Hogan, J. C., 226
Hogg, A. M., 6
Holick, M. F., 185
Holland, H. L., 185
Holland, J. F., 268
Holland, W. H., 272, 315
Holloway, P. J., 311
Holm, A., 130
Holm, R. H., 246
Holman, R. T., 109, 145, 146, 310
Holme, E., 334
Holmes, J. F., 114
Holmes, J. L., 39, 74, 75, 94, 99, 102, 108, 119
Holmes, W. F., 272
Holmlund, C. E., 145
Holmstedt, B., 305, 306, 322, 325
Holtz, D., 27, 83
Holtz, R. B., 315, 332
Holtzclaw, H. F., 194
Holubova, N., 322
Holzapfel, C. W., 187
Honda, T., 189
Honig, R. E., 44
Honkanen, E., 329
Hopkinson, J. A., 39, 93, 123
Hoppe, H.-J., 173
Hoppen, H.-O., 305
Horlick, G., 300
Horn, D. H. S., 181, 335
Horning, E. C., 313, 317, 318, 320, 321, 326
Horning, M. G., 183, 211, 308, 313, 314, 318, 320, 321, 326
Hornstein, I., 329
Horodniak, J. W., 121
Horri, Z. I., 188
Horton, D., 171
Horton, E. W., 312
Horvath, C. G., 333
Horváth, Gy., 102, 179
Ho Sam Ahn, 322
Hosfield, S. T., 257
Hota, N. K., 217, 256
Hotop, H., 4

Hottmann, K., 257
Houghton, E., 124
Houghton, L. E., 185
Houghton, R. G., 300
Houseman, J., 293
Howard, A. S., 259
Howard, J. W., 261, 329
Howe, I., 33, 63, 67, 69, 71, 86, 96, 113, 114
Hoyano, J., 259
Hoyes, S. D., 144
Hrung, C. P., 262
Hrutfiord, B. F., 332
Hsieh, A. T. T., 201, 227
Hsü, H.-Y., 189
Hu, C. K., 103
Hubbard, R., 100
Huckins, J. N., 331
Hüther, H., 260
Huggins, D. K., 250
Hughes, B. M., 203, 335
Hughes, R. E., 260
Huhtaniemi, I., 320
Hui, W. H., 181
Hulley, G., 259
Hummel, D., 22
Humphrey, S. A., 15, 138
Humski, K., 101
Hunt, D. F., 12, 13
Hunter, T. L., 109
Huntress, W. T., 28, 29, 30, 84
Hurtado, L., 113
Husain, S., 133
Husband, J. P. N., 42, 200
Hussain, M., 126
Hustert, K., 332
Hutchison, R. D., 187
Hutterer, F., 297, 325
Huttner, G., 261
Hutzinger, O., 61, 100, 120, 186, 332
Hvistendahl, G., 40, 43, 113, 116, 118, 126, 127
Hydes, P., 258

Ibanez, J., 334
Ichikawa, H., 127
Ida, Y., 170, 173
Idler, D. R., 317, 318
Igeta, H., 127
Ignat'ev, V. M., 219
Ihara, M., 189
Ihrig, P. J., 109, 118
Iida, H., 121, 128
Iida, I., 119
Ikeda, M., 111, 126, 131, 188
Ikekawa, N., 321
Ikonen, M., 319, 320
Illingworth, S. M., 259
Imanari, T., 315
Inatomi, H., 154
Inatsu, Y., 170, 173
Indictor, N., 121
Innorta, G., 4, 35, 53, 79, 97, 238, 240
Inubushi, Y., 182
Inukai, F., 154

Author Index

Ionin, B. I., 219
Irgolic, K. J., 223
Irving, P., 35, 167
Irwin, E., 126
Isaeva, S. S., 94
Isenhow, T. L., 273, 284, 291
Ishaq, M., 231
Ishibashi, M., 321
Ishigura, S., 170
Ishii, S., 328
Islamov, J. H., 137, 209
Israel, G., 124
Issenberg, P., 329, 331
Itoh, T., 123
Ivanov, V. T., 156
Ivemark, B., 320
Iwataki, I., 187
Iyengar, B. T. R., 145, 310

Jackson, B., 191
Jackson, D., 327
Jackson, L. L., 145
Jackson, R. A., 197
Jacob, J., 311
Jacob, T. A., 327
Jacobs, G., 141
Jacobsberg, F. R., 145
Jacobson, A. E., 129
Jacquier, R., 122
Jacquignon, P., 133
Jacquot, A., 305
Jacquot, P., 305
Jaeger, E., 157
Jäger, H., 173
Jänne, O., 320
Jaffe, H. H., 27
Jager, K., 137
Jain, S. K., 133
Jalonen, J., 40, 120, 127
James, J. D., 309
James, M. N. G., 181
James, T. A., 258, 262
Jamieson, W. D., 61, 100, 120, 186, 332
Janet, R. S., 173
Jankowski, K., 128
Janse van Vuuren, P., 260
Janssen, G., 156
Janssen, M. J., 73, 132
Jansson, P.-A., 297
Janzen, A. F., 137, 209
Jarman, M., 100, 170
Jary, J., 314
Jean, J. St., 108
Jeffs, P. W., 186, 188
Jellum, E., 325
Jen Chin, 265
Jenkins, A. D., 260, 261
Jenkins, C. R., 243
Jennings, K. R., 24, 28, 30, 33, 35, 37, 82, 87, 93, 116, 142
Jennings, P. W., 68
Jensen, G. M., 130
Jensen, S., 332
Jereczek, E., 176
Jernigan, R. T., 262
Jetz, W., 259

Jery, J. B., 134
Jimerez, M., 118
Job, B. E., 108
Joblin, K. N., 180
Joela, H., 138
Johannessen, J. C., 134
Johanson, G, A., 33, 90, 159, 186, 265
Johansson, B., 196, 305, 332
John, D., 247
Johne, S., 189
Johns, E. B., 330
Johns, R. B., 130
Johns, S. R., 187, 189
Johnson, A. E., 329
Johnson, B. F. G., 226, 259, 260, 263
Johnson, B. R., 329, 330
Johnson, F. A., 256
Johnson, G. S., 169, 172
Johnson, L. F., 186
Johnston, R. D., 193
Johnstone, R. A. W., 3, 46, 79, 91, 96, 128, 162, 297
Jolly, W. L., 257
Joly, M., 208, 257
Jemmi, G. C., 325
Jones, C. E., 220
Jones, C. J., 245
Jones, E. G., 87
Jones, E. M., 235
Jones, G., 120
Jones, K. B., 312
Jones, N. A., 272
Jones, R. A., 92
Jones, S., 37, 116
Jonnet, C., 295
Joshi, B. S., 180
Joshi, R. K., 330
Jullien, J., 54, 103
Jung, G., 315
Junk, G. A., 193, 225, 238, 296
Junkes, P., 258
Juraydini, A., 98
Jurs, P. C., 284, 291
Jutzi, P., 202

Kabaara, A. M., 311
Kadentsev, V. I., 94, 108
Kärkkäinen, J., 314, 315, 324
Kaesz, H., 259
Kagal, S. A., 116
Kahrs, K., 115
Kain-Sze Kwok, 285
Kaiser, K. H., 302
Kaiser, R., 304
Kalinovsky, A. I., 181
Kalir, A., 134, 215, 216
Kalyanaraman, P. S., 191
Kalyanaraman, V., 137, 209
Kamano, Y., 185
Kamerling, J. P., 171, 289, 307
Kametani, T., 121, 124, 189
Kan, G., 230

Kanazawa, A., 318
Kaneda, T., 145
Kanellakopulos, B., 260
Kang, J. W., 232
Kapadia, G. J., 324
Kaplan, F., 27
Kapoor, P. N., 214, 260
Karady, S., 169
Karasek, F. W., 296, 304, 305
Karlajaimen, A., 128
Karlander, S.-G., 149
Karlén, B., 120
Karliner, J., 129
Karlsson, G., 330
Karlsson, K.-A., 148, 149, 313
Karpati, A., 111
Karpenko, N. F., 94
Karyalainen, A., 120
Kashman, Y., 125
Kaska, W. C., 260
Kates, M., 146
Kato, H., 329
Kato, K., 170, 173
Kato, S., 131
Kato, T., 127
Katz, J. J., 182
Kaussmann, E. U., 187
Kawaguchi, H., 189
Kawai, M., 16
Kawalek, J. C., 145
Kawasaki, T., 170, 173, 279
Kazaryan, S. A., 163
Kazbulatova, N. A., 96
Kearns, G. L., 79
Keat, R., 256
Kebarle, P., 18
Kelly, R. W., 306, 312
Kelm, J., 117
Kemball, C., 309
Kemlo, J. A., 243
Kempling, J. C., 309
Kepner, R. E., 304
Keppie, S. A., 227
Kerber, R. C., 260
Keskiner, R., 120
Kevan, L., 83, 84
Khajizov, Kh., 120
Kharitmov, K. S., 247
Khmel'nitskii, R. A., 120, 121, 122, 126, 127, 129
Khuong-Huu, F., 183
Khuong-Huu, Q., 8, 183
Khusainova, B. N., 279
Khvostenko, V. I., 133, 138
Kieboom, A. P. G., 108
Kier, L. B., 53
Kiesel, V., 222
Kifer, E. W., 257
Kilner, M., 236, 256, 261
Kilthau, G., 215
Kimigara, M., 133
Kimland, B., 182
Kimura, B. Y., 194
King, A., 234
King, B. M., 111
King, R. B., 193, 214, 217, 231, 232, 237, 258, 260
King, R. W., 179

Kingsbury, C. A., 123
Kingston, B. M., 227
Kingston, D. G. I., 78, 116, 191, 213
Kinneberg, K., 152, 273
Kinneberg, K. F., 115, 147
Kinstle, T. H., 195
Kirkien, A. M., 46, 130
Kirmse, W., 112
Kiryushkin, A. A., 5, 10, 156, 157, 163, 166, 167, 168
Kiser, R. W., 62, 90, 193
Kitamura, C., 328
Kitchens, J., 248
Kitching, W., 208, 257
Kitson, F. G., 20, 104
Kitz, R., 325
Kiyozumi, M., 170, 173
Klaassen, C. D., 326
Klásek, A., 186, 188
Klasinc, L., 65, 74, 97, 101
Klebe, K. J., 69, 122
Kleemann, L. P., 9
Kleiman, R., 311
Klein, H. F., 262
Klein, P. D., 306
Klein, R. A., 147
Klein, W., 331
Klemann, L. P., 259
Klemon, K., 116
Kleschick, W. A., 94
Klimowski, R. J., 272
Kløsen, H., 182
Klosowski, J. M., 207
Kloster-Jensen, E., 100
Klots, C. E., 38, 46
Kluge, A. F., 328
Klynev, N. A., 120, 126, 127
Knabe, B., 215, 222
Knewstubb, P. F., 46
Knights, B. A., 116, 311
Knoche, H. W., 144
Knock, B. A., 291
Knöppel, H., 21
Knof, H., 138
Knox, S. A. R., 259
Knunyants, I. L., 250, 255
Knutti, R., 273
Kobayashi, A., 331
Kobayashi, Y., 256
Kober, F., 237
Koch, C. W., 129
Koch, H., 329
Koch, M., 187
Kochetkov, N. K., 158, 169, 170, 172, 314
Kochler, P., 257
Kodicek, E., 149
Koegler, W., 265
Koeltzow, D. E., 144
Koeman, J. H., 332
König, W. A., 174, 323
Koepp, P., 321
Koerner von Gustorf, E., 226
Kolattukudy, P. E., 144, 311
Kollmeier, H. J., 235, 261
Kolodziejczyk, P., 176

Kolor, M. G., 128
Kominar, R. J., 119
Komori, T., 170, 173, 279
Konakahara, T., 119, 121
Konda, Y., 154
Kongpricha, S., 256
Konishi, T., 297
Koniuszy, F. R., 327
Konno, T., 128
Koob, R. D., 105, 242
Koreeda, M., 181
Korenawski, T. F., 237
Kornfeld, R. A., 60, 86, 104
Korte, F., 332
Korte, W. D., 260
Korureich, M. R., 331
Kosasayama, A., 128
Koshiyama, H., 189
Koslow, S. H., 306
Kossanyi, J., 105, 127
Kost, A. N., 126, 127, 138
Koster, R., 256, 258
Kostyanovskii, R. G., 115, 120, 125, 219
Kosuge, S., 330
Kotake, M., 331
Kotienko, T., 176
Kotz, J. C., 197, 239
Kováč, P., 170
Kováčik, V., 170
Kovacs, J., 154
Kowalski, B. R., 284
Koyama, T., 182, 310
Koyano, I., 5, 85
Kozuka, 5, 115
Krafft, D., 138
Kraihanzel, C. S., 224, 260
Kramer, E. A., 137, 209
Kramer, J. K. G., 146
Kramer, J. M., 126
Krampl, V., 329
Krasnova, S. G., 232
Kray, A. M., 278
Kreiter, C. G., 235, 236, 261
Krichevesky, M. I., 268
Krieg, B., 262
Krieger, H., 128
Kriegsmann, H., 257
Kriemler, R., 138
Krishnappa, S., 180
Kritskaya, I. I., 231
Kronberg, L., 127
Krone, H., 23, 174
Krueger, V. P. M., 137
Kruglaya, O. A., 257
Kruse, A. E., 261
Kubelka, V., 129, 145, 309, 322
Kucherov, B. F., 108
Kucherov, V. F., 76, 106, 181
Kuenzle, C. C., 174
Kugelman, M., 171
Kuhn, W. F., 22
Kuivila, H. G., 213
Kulkarni, P. S., 137, 209
Kullenberg, B., 329
Kulonen, E., 324
Kumar Das, W. G., 213

Kumna, E. A., 127
Kunderd, A. W., 92, 166, 288, 295
Kunze, U., 257
Kupchan, S. M., 180
Kupperman, A., 295
Kurås, M., 309
Kusamran, K., 310
Kuschel, H., 116
Kushwaha, R. P. S., 311
Kushwaha, S. C., 146
Kusunose, M., 311
Kuszmann, J., 102
Kuwahara, Y., 328
Kvenvolden, K. A., 333, 334
Kwart, H., 72, 101
Kyba, E. P., 78, 117

Laake, M., 190
Lacey, M. J., 70, 108
Lachance, A., 202
Lageot, A., 93
Lageot, C., 44, 114, 118
Laine, R., 207
Laine, R. A., 149, 314, 318
Lala, A. K., 183, 317
Lalancette, J. M., 202
Lamberton, J. A., 187, 189
Lammens, H., 231
Lampe, F. W., 203
Lampiaho, K., 324
Lancaster, J. E., 179
Lande, S., 161
Landis, W. R., 103
Lane, M. J., 245
Langenscheid, E., 116
Lanthier, G. F., 100, 130, 212
Lappert, M. F., 197, 202, 213, 227, 259, 260, 261
Larin, N. V., 232
Larkins, J. T., 15, 138
LaRoe, E. G., 331
Larsen, E., 186
Larson, G. L., 137
Larson, J. G., 239
Last, J. H., 331
Latimore, M. C., 72
Lau, K. H., 46
Laubereau, P. G., 260
Lauer, G., 277
Lauer, W. M., 146
Laurie, W. A., 14, 131, 183, 333
Laver, M. L., 311
Law, N. C., 297, 324
Lawesson, S.-O., 15, 113, 131, 136
Lawless, J., 333
Lawless, J. D., 151
Lawless, J. G., 323
Lawrence, J. H., 332
Lawrie, T. D. V., 319
Lawson, A. M., 152, 305, 308
Lawson, D. E. M., 149
Lawson, G., 42, 200
Leach, J. B., 256

Author Index

Leach, W. P., 225
Leber, K. H., 24
Ledercq, P. A., 161
Lederberg, J., 281, 282, 283, 288, 293
Lederer, E., 144, 159, 160
Ledoux, W. A., 256
Lee, A., 129
Lee, A. G., 201, 202
Lee, R. K., 247
Lee, W. K., 181
Leemans, F. A. J. M., 152
Leete, E., 324
Leferink, J. G., 270
Leffler, H., 149
Leftin, J. H., 109
Lehman, T. A., 31, 32, 83
Lehmann, B., 16
Lehrle, R. S., 83
Leipziger, F. D., 280
Leitch, L. C., 94, 102, 103
Leite, R. J., 273
Leitich, J., 188
Lenard, J., 160
Lengyel, I., 120, 208, 213
Lennarz, W. J., 146, 314
Lenoir, D., 184
Leon, N. H., 130
LeQuesne, P. W., 187
Lesclaux, R., 85
Lester, G. R., 62, 123
Leupold, M., 235
Levase, L. J., 305
Levenberg, M. I., 178
Leverett, P., 262
Levine, S. P., 98, 101
Levinthal, E. C., 293
Levkoeva, E. I., 125
Levsen, K., 18, 20, 21, 51, 52, 53
Levy, J. B., 216
Levy, R. L., 333
Lewis, D., 3
Lewis, J., 226, 259, 260, 263
Lhuguenot, J. C., 330
Li, H.-Y., 273
Liaaen-Jensen, S., 174, 182
Liardon, R., 98
Libbey, L. M., 311, 329, 330
Licht, K., 257
Lichtenstein, H. A., 323
Lieb, F., 258
Liebich, H. M., 326
Liedtke, R. J., 104, 106, 116
Liehr, J. G., 125
Liengme, B. V., 262
Lifshitz, C., 35, 51, 137, 218
Lightner, D. A., 40, 78, 96, 105, 113, 126
Likhosherstov, L. M., 158
Lilja, A., 149
Lin, F. M., 330
Lin, P.-H., 32
Lin, S. H., 46
Lind, W., 201
Lindberg, B., 171, 315
Lindgren, J. E., 322
Lindholm, E., 7
Lindley, H., 166

Lindner, E., 220, 238, 257, 258, 259, 263
Lindsay, P. H., 195
Lindsay, R. C., 329, 330
Lindsay-Smith, J. R., 322
Lindstedt, G., 334
Lindstedt, S., 334
Ling, L. C., 329, 330
Ling, N., 323
Linnarsson, A., 322
Lintvedt, R. L., 194
Lipkin, V. M., 163
Lipsky, S. R., 121, 135, 161, 275, 333
Lisboa, B. P., 321
Little, W. F., 68, 233
Littler, J. S., 92, 267
Litvinov, V. P., 132
Litzow, M. R., 197, 227
Liu, C. S., 256
Livingstone, S. E., 245
Llewellyn, P. M., 28
Lloyd, D. R., 2, 240
Lloyd, J. P., 236
Lloyd, T., 322
Locht, R., 16, 137
Lock, C. J. L., 243
Locksley, H. D., 191
Locock, R. A., 115, 121
Loder, J. W., 189
Lodge, B. A., 308
Löhle, U., 66
Loewenstein, R. M. J., 99
Loffredo, R. E., 256
Lofqvist, J., 328
Logan, N., 236
Lombardo, P., 332
Long, L., jun., 324
Long, T. V., 228
Longevialle, P., 8, 183, 190
Lorberth, J., 196, 197, 213
Los, M., 312
Loser, R. W., 280
Loskot, S., 258
Lossing, F. P., 40, 43, 119
Loudon, A. G., 46, 91, 96, 97, 130
Loughran, E. D., 120
Loveland, P. M., 311
Lovelock, J. E., 305
Lovins, R. E., 163, 164, 288
Lowry, R. N., 243
Lowry, W. T., 173
Lucas, F., 165
Lucas, R. A., 334
Lucero, D. P., 305
Luchte, A. J., jun., 305
Luckas, K. H., 332
Luftmann, H., 94
Luhan, P. A., 188
Lukacs, C., 190
Lukashenko, I. M., 96, 97
Lumpkin, H. E., 272
Lundström, J., 322
Lusinchi, X., 183, 190
Lustre, A. O., 331
Luukkainen, T., 319, 320
Lyall, R. J., 126
Lyle, R. E., 125
Lynch, J., 236

Lynch, J. M., 310
Lythgoe, S., 257
Lytle, F. E., 291

Maarse, H., 304
Mabie, C. A., 17
McAdoo, D. J., 24, 60, 104
McAllister, T., 14, 82, 141
Macasek, F., 260
McCall, M. T., 54, 106
McCamish, M., 115, 129
McCleverty, J. A., 245, 258, 262
McCloskey, J. A., 11, 137, 152, 174, 308, 335
Maccoll, A., 46, 91, 96, 130
McCoy, R. D., 306
McDaniel, E. W., 83
MacDiarmid, A. G., 193, 228
MacDonald, C. G., 70, 108, 156
McDonald, R. N., 57, 131
McDowell, C. A., 2
McDowell, M. V., 193, 228
McEwen, C. N., 12
McFadden, W. H., 304, 329
McGarry, E. J., 191
McGillivray, D., 39, 94, 99
McGrew, J. G., 69
McGuire, R. R., 332
McIntyre, N. S., 225
McIver, R. T., 26, 28
Mackay, K. M., 228, 257
McKenzie, D. A., 131
MacKenzie, S. M., 129
McKinley, I. R., 136, 199
McLafferty, F. W., 24, 35, 52, 60, 67, 69, 83, 86, 93, 96, 104, 108, 167, 265, 272, 277, 279, 285
MacLean, D. B., 126, 189
MacLean, I., 335
MacLeod, J. K., 94, 116
McMahon, E., 230
McMahon, R. E., 306, 327
McMahon, T. B., 26
MacMillan, A., 267, 316
McMurray, W. J., 121, 135, 161, 275, 333
MacNeil, K. A. G., 16, 42, 92, 137, 142
Macomber, R. S., 221
McPhail, A. T., 188
McReynolds, J., 334
Madaiah, M., 155
Madden, D. P., 230
Madsen, J. O., 15, 113
Maehly, A. C., 306
Märkl, G., 206, 258
Magee, J. L., 300
Maheshwari, M. L., 176
Maitlis, P. M., 232
Majer, J. R., 243
Majeti, S., 40, 96
Major, H. W., 93
Makina, I., 319
Malaidza, M., 202
Malec, E. J., 332

Malik, Z. A., 322
Mallams, A. K., 171
Mallory, C. W., 144
Mallory, F. B., 144
Malorni, A., 303
Mal'tsev, A. K., 196
Malysheva, N. N., 172
Mamayev, V. F., 127
Mamer, O., 326
Mamer, O. A., 119, 326
Mancini, V., 79, 97
Mandel, L. R., 322
Mandelbaum, A., 105, 106, 109, 111
Mann, C. D. M., 259
Mannan, C. A., 115
Mannering, D. G., 299
Manni, P. E., 103
Manning, D. J., 109
Manojlovic-Muir, L., 261
Manske, R. H. F., 189
Manson, M. E., 321
Manville, J. F., 97
Maquestiau, A., 97
Maragnon, J., 295
Marchelli, R., 120
Marcotte, R. E., 83
Marcucci, F., 325, 326
Marde, Y., 306
Marigliano, H. M., 178
Marino, G., 151, 162, 303
Mark, R. V., 120
Markey, S., 322
Markey, S. P., 274, 309
Markgraf, J. H., 129
Marks, T. J., 259
Markwell, R. E., 179
Marsel, J., 65, 74, 97, 124
Marshall, C. J., 257, 259
Martani, A., 133
Martelli, P., 325
Martin, A. J. P., 302
Martin, C., 258
Martin, J. C., 138
Martin, N. H., 188
Martynov, B. I., 250, 255, 257
Maruca, R., 104
Maruyama, M., 180
Marvich, R. H., 232
Mascaretti, O. A., 190
Masdet, P., 70, 101
Maseles, F., 200
Mason, C. J., 273
Mason, P. R., 262
Mason Hughes, B., 99
Masot, R., 265
Massey, A. G., 130, 230, 260
Massey, R. I., 28, 82
Massot, R., 92
Massy-Westrop, R. A., 218
Mastafanova, L. I., 125
Mastin, S. H., 248
Masuda, Y., 94
Matejcek, K.-M., 263
Mathai, K. P., 181
Matheson, T. W., 227
Mathias, A., 39, 88, 93, 131
Mathur, H. H., 176

Matin, S. B., 321
Matsubara, T., 313
Matsumoto, S., 132
Matsuo, A., 179
Matsuoka, S., 151
Matteson, D. S., 256
Matthews, R. F., 311
Maume, B. F., 315, 317, 321, 330
Maume, G. M., 321
Mautner, M., 121, 129, 247
Mawby, R. J., 260
Maxwell, J. R., 333
Mayer, K. K., 94
Mayer, M. S., 309
Mayers, G. L., 154
Maynard, R., 128
Mays, M. J., 227
Mazdiyasni, K. S., 250
Mazengo, R. Z., 97
Mazerolles, P., 208, 257
Mead, T. E., 305
Mead, T. J., 65
Mead, W. L., 77, 174
Mechanic, G., 158
Meier, H., 116, 133
Meijer, G. M., 328
Meinwald, J., 260, 316, 328
Meisels, G. G., 3, 44, 82, 297, 298, 299
Meissner, B., 198
Meldrum, R. D., 265
Melkersson, S., 297
Meller, A., 199
Mellon, F. A., 3, 46, 297
Melton, C. E., 91
Memel, J., 129
Meredith, J. O., 273
Merkuza, V. M., 190
Merlini, L., 190
Mermison, V. G., 42
Merren, T. O., 91, 265, 267, 276
Merrimson, V. G., 184
Mertschenk, B., 234, 237, 261
Merz, P. L., 206
Mestres, R., 206
Metzger, H. G., 221
Meyer, D., 328
Meyer, R. A., 277
Meyerson, S., 94, 97, 102, 103, 108, 109, 118
Michalkiewicz, D. M., 334
Michel, G., 158
Michie, E. A., 320
Michnowicz, J., 11, 12, 76
Midcalf, C., 236, 261
Middleditch, B. S., 116, 304, 305, 306, 310, 311, 318
Middleton, S., 126, 222
Midelfart, A., 190
Midgley, J. M., 183
Midwinter, G. M., 165
Migahed, M. D., 19, 20
Mihkelson, A. E., 245
Mikaelyan, R. G., 196
Mikeš, F., 325, 327

Mikhailov, B. M., 199
Mikhailov, J. A., 129
Mikhlina, E. E., 125
Mikulaj, V., 260
Milivojević, D., 65, 74, 97
Millard, B. J., 129, 133, 183, 325
Miller, A. H., 133
Miller, C. D., 256
Miller, J. A., 134, 258
Miller, J. J., 256
Miller, J. M., 100, 130, 212, 214
Miller, J. R., 225
Miller, N. E., 256
Miller, R. W., 311
Miller, S. L., 324
Miller, V. R., 256
Milliet, P., 183
Millington, D. S., 297
Mills, L., 273
Milne, G. W. A., 6, 7, 9, 10, 152, 167, 297, 324
Minale, L., 180, 310
Ming Chu, I., 145
Minghetti, G., 240
Minks, A. K., 328
Miroshnikov, A. I., 163
Mishra, I. B., 258
Misuraca, G., 192
Mital, R. L., 133
Mitchard, L. C., 261
Mitchell, D. K., 260
Mitchell, D. L., 265, 267
Mitchell, E. D., 115
Mitchell, F. L., 321
Mitchell, J., 265
Mitchell, S. R., 127
Mitchum, R. K., 147, 287
Mitera, J., 129
Mitra, C. R., 181, 191
Mitscher, L. A., 177, 178, 189
Mitsuhashi, K., 126
Mittelbach, H., 127
Miyazaki, H., 321
Mochalin, V. B., 125
Modzeleski, V. E., 333, 334
Møller, J., 132, 191
Möller, U., 257, 258, 262
Moelwyn-Hughes,J.T., 259
Moers, F. G., 249
Moffat, A. C., 321, 326
Mogto, J. K., 105
Mohamed, P. A., 181
Mohammedi-Tabrizi, F., 162
Mokotoff, M., 129
Moldowan, J. M., 69
Molenaar-Langeveld,T. A., 72, 102, 117
Moller, J., 129
Mollère, P. D., 207
Momigny, J., 16
Momose, A., 327
Momose, T., 188
Monahan, M., 323
Mondon, A., 186
Monneret, C., 183
Moon, B. J., 189

Moore, L. P., 151
Moorhouse, S., 260
Moreland, C. G., 220
Mori, C., 321
Morizur, J.-P., 112
Morris, H. R., 149, 162, 163, 165
Morris, M. L., 105, 242
Morrison, J. D., 285, 286, 297, 299
Morselli, P. L., 326
Mortensen, T., 190
Morton, R. B., 176
Moseley, K., 232
Moser, G. A., 233
Moser, R. J., 126
Mosesman, M. M., 30
Mosharrafa, M., 300
Moshier, R. W., 243
Moshonas, M. G., 330
Moss, C. P., 93
Moss, J. R., 258
Most, B. H., 316
Moyer, C. L., 119
Moynahan, E. B., 263
Müller, E., 126, 133
Müller, H., 261
Müller, H. P., 244, 335
Müller, J., 193, 223, 224, 225, 233, 234, 235, 236, 237, 261, 262
Müller-Stock, A., 330
Muhlenbruch, B., 255
Muir, K. W., 261
Mukerjee, S. K., 126
Mukherjee, D. K., 181
Mular, M., 99
Muller, R., 124
Mullin, B. C., 330
Mullin, M. M., 309
Munawar, Z., 127
Munson, B., 7, 11, 12, 75, 76, 107, 305
Murakami, S., 313
Murakami, T., 154
Murata, T., 331
Murawski, U., 310, 311
Murphy, M. E., 334
Murphy, R. C., 297, 325
Murray, I. G., 191
Murray, J. C. F., 258
Murray, K. E., 331
Murray, K. S., 245
Musker, W. K., 137
Mussini, E., 325, 326
Muthukumaraswamry, N., 191
Mysov, E. I., 255

Naccarato, W. F., 145
Näf-Müller, R., 330
Naga, S., 106
Nagarajan, R., 178
Nagy, B., 333, 334
Nagy, S., 330
Nahm, F. C., 256
Nair, R. M. G., 162
Nakadaira, Y., 256
Nakagawa, T., 322

Nakagawa, Y., 132, 172
Nakamura, A., 229, 231
Nakamura, G. R., 327
Nakanishi, K., 181
Nakata, H., 82, 106, 109, 110, 125
Nakatani, Y., 331
Nakayama, N., 5, 85
Namethan, N. S., 137, 206, 209
Narasimhachari, N., 321, 322
Narayanan, C. R., 183, 317
Nasielski, J., 202, 211
Nasini, G., 190
Nasta, M. A., 228
Natile, G., 238, 263
Natvig, T., 133
Nawar, W. W., 103
Naya, Y., 331
Nayar, M. S. B., 180
Nazarenko, V. A., 19
Neal, R. A., 327
Nearn, R. H., 189
Neeter, N., 126
Neeter, R., 69, 126
Nefedov, O. M., 196
Neiman, Z., 133
Nekrasov, Yu. S., 113, 118, 231
Nelson, D. R., 16, 146
Nesmeyanov, A. N., 231
Neumann, P., 126
Newlands, M. J., 205, 228, 263
Newton, W. E., 238, 257
Ney, K. H., 310
Ng, K. K., 181
Ng, T. W., 217
Niazi, G. A., 122
Nibbering, N. M. M., 46, 69, 72, 73, 94, 97, 102, 106, 117, 118, 126
Nicholas, H. J., 318
Nicholas, K., 259
Nicholson, G., 323
Nicholson, J. M., 15, 138
Nicolaides, N., 311
Nicoletti, R., 126
Niedenze, K., 256
Nichaus, A., 4
Nielsen, J. G., 191
Nielsen, J. T., 132
Niemeyer, D. H., 114
Nier, A. O., 93
Nikitina, S. B., 120
Nikkari, T., 324
Nilsson, M., 197
Nilsson, N. J., 281
Nishikawa, Y., 176
Nishishita, T., 297
Nishiwaki, T., 54, 122
Nishiyama, T., 127, 133
Noble, A. C., 103
Nørgaard, P., 191
Nöth, H., 261
Noguchi, J. T., 327
Nolde, C., 15, 113
Noltes, J. G., 214
Nonhebel, D. C., 122

Nordby, H. E., 330
Norman, A. D., 256, 257
Noro, K., 248
Noro, T., 125, 248
Norris, F. A., 173
Norris, K., 160
North, B., 231
Norton, J. R., 226
Nose, M., 331
Noshire, M., 304
Nota, G., 303
Novikov, S. S., 94
Novikova, O. S., 158
Novotny, M., 302, 304
Nowak, A. V., 266
Núñez-Alarcón, J., 191
Nuretidinov, I. A., 219
Nursten, H. E., 329
Nussey, B., 15, 80, 123, 140, 214
Nyholm, R. S., 226

Oae, S., 115
Obermann, H., 184, 185, 308
Obermeier, R., 164
O'Brien, H., 223
O'Brien, R. J., 259
Ochterbeck, E., 19
O'Connor, T. J., 217
Odell, G. V., 147
Öfele, K., 261
Oertel, M., 206
Ogihara, Y., 176
Ogura, H., 123, 127
Ogura, K., 182, 310
Ohashi, M., 54, 123
Ohkuma, H., 189
Ohlaff, G., 105
Ohlsson, A., 322
Okada, S., 128
Okotore, R. O., 155
Oldfield, D., 126
Olfky, R. S., 42
Oliver, A. J., 259
Oliver, J. G., 249
Oliver, R. W. A., 107
Olsen, R. W., 91, 265, 273
O'Malley, R. M., 24, 28, 30, 82, 142
O'Mara, M. M., 334
Omura, S., 154
Onak, T., 256
Onak, T. P., 17, 198
Onda, M., 154
Opheim, K., 316
Op het Veld, P. H., 249
Orazi, O. O., 121
Orchard, A. F., 240
Orchard, D. G., 258
Orcutt, D. M., 318
Orlov, V. M., 5, 168
Orlov, V. Y., 137, 206, 209
Oró, J., 323, 333, 334
Orr, J. C., 318
Oshima, S., 297
Osman, S., 122, 329
Ossko, A., 199
Ostfeld, D., 262

O'Sullivan, W. I., 120
Otsuka, S., 229, 231
Ottinger, C., 52, 66
Ouchida, A., 125
Ouellette, T. J., 256
Ourisson, G., 332
Ovchinnikov, Yu. A., 156, 163, 166
Owen, E. C., 138
Owen, P. J., 191
Ozolins, G., 105

Pace-Asciak, C., 150, 312
Padieu, P., 321
Paetzold, P., 256
Page, R. P., 266
Pai, B. R., 191
Pailer, M., 188
Pain, D. L., 133
Paisley, H. M., 119, 279
Pallotta, U., 311
Palmer, L., 322
Palmer, M. H., 129
Panasiuk, O., 329
Pancirov, R. J., 133
Pande, B. S., 180
Pandey, G. N., 181
Pandey, R. C., 176
Pandler, W. W., 15
Pape, B. E., 332
Papetti, S., 256
Parello, J., 188
Park, J. D., 98
Park, J. Y., 297
Park, R. J., 180
Parke, D. V., 332
Parker, J. A., 272
Parker, J. E., 83
Parker, R. B., 189
Parker, R. G., 194
Parr, W., 323
Parrott, J. C., 257
Pasanen, P., 130
Pascual, C., 100
Patel, H. A., 217
Patel, K. S., 105, 241, 244
Patel, M. B., 187
Patt, S. L., 83
Pattabhiraman, T., 185
Pattee, H. E., 330
Patterson, R. L. S., 329, 331
Patton, S., 321
Patton, W., 310
Pauden, W. W., 138
Paudler, W. W., 124
Paukstelis, J. V., 105
Paul, D. G., 305
Paul, N. C., 113
Pauson, P. L., 263
Pavanaram, S. K., 173
Pavia, A. A., 179
Pavlenko, V. A., 219
Payling, D. W., 128
Paz, M. A., 160
Pazos, J. F., 120
Peach, C. M., 159
Peacock, R. D., 259
Péchiné, J. M., 54, 57, 98, 103

Peciar, C., 170
Pedersen, A. O., 131
Pedersen, C. Th., 132
Pedley, J. B., 197, 202
Pedrinella, L., 330
Peers, A. M., 51
Pehrsson, K., 208
Pelah, Z., 134, 215, 216
Pelham, R. W., 322
Pellet, J., 279
Peluso, R., 258
Penfold, B. R., 227
Pennington, R. H., 269
Penttila, A., 335
Perche, J.-C., 129
Pereira, W., 115, 310
Perichant, A., 132
Pering, K. L., 333
Perkins, E. G., 109, 316
Perone, S. P., 265, 278
Perrine, C., 288
Perry, R., 243
Perry, R. A., 260
Perry, W. O., 94
Persoons, C. J., 328
Peter, F., 100
Peterman, Z. E., 265
Peters, M. D., 136, 199
Petersen, I. B., 132
Petersen, M. R., 316
Peterson, E., 186
Peterson, G., 170
Petersson, G., 137, 308
Petrov, A. A., 219, 221, 223
Petrov, B. I., 257
Petrus, C., 122
Petrus, F., 122
Pettersson, O., 332
Pettit, G. R., 23, 172, 175, 185
Pettit, R., 231, 259
Petty, H. E., 57, 131
Pfoffenberger, C. D., 326
Pfiefer, J. P., 300
Phillips, B., 278
Phillips, D. A. S., 176
Phillips, G. O., 23, 77, 174
Phillips, J., 107
Phillips, L., 113
Picot, A., 190
Pignataro, S., 4, 53, 79, 97, 238, 239, 240
Pihlaja, K., 40, 120, 127, 130
Pikkarainen, J., 324
Pillinger, C. T., 333
Pilotti, Å., 171
Pinar, M., 187
Pincock, R. E., 4
Pines, S. H., 169
Pinkerton, A. A., 220, 258
Pinkney, J. N., 236, 256
Pinkus, J. L., 306
Pirc, V., 124
Pirelahi, H., 133
Piretti, M., 311
Pirt, S. J., 310
Pitts, J. N., 114
Plasz, A. C., 127
Plat, M., 187

Plattner, J. R., 274
Plichta, P., 256
Plöger, E., 127
Plotnikov, V. F., 219
Pobo, L. G., 29, 83
Podczasy, M. A., 189
Podina, C., 126
Poeth, T. P., 228
Pohl, R. L., 261
Poite, J. C., 132
Polan, M. L., 161
Poland, J. S., 213
Polen, P. B., 332
Polgar, N., 310
Polito, A. J., 313
Polivka, Z., 322
Pollock, G. E., 304, 310
Polonia, J., 173
Polviander, K., 120
Polyakova, A. A., 96, 97, 129
Pommier, C., 233
Ponchon, G., 185
Ponnamperuma, C., 324, 333, 334
Ponpipom, M. M., 174
Poole, A. J., 132
Popanova, R. V., 96, 97
Popov, S., 185
Popp, F. D., 121, 127, 129, 263
Porchet, J. P., 267
Porter, G. T., 306
Porter, Q. N., 91, 111, 115, 116, 119
Porter, R. F., 12, 13
Porter, R. J., 262
Post, E. W., 239
Potapov, V. K., 5, 6
Potier, P., 173, 179, 188
Potts, K. T., 122, 124, 127, 129, 133
Potzinger, P., 203
Pound, A. W., 154
Powell, H. M., 280
Powell, J. W., 325
Powell, P., 199
Powers, P., 273, 280
Poyser, N. L., 312
Praet, M.-Th., 2, 39, 94
Pressley, G. A., 295
Preston, F. J., 238
Preston, P. N., 128, 129
Preti, C., 263
Preti, G., 297, 325
Pribytkova, I. M., 259
Price, A. P., 108
Price, C. C., 133
Prinzbach, H., 110
Prota, G., 192
Prouty, W. F., 176
Pryce, R. J., 316
Puchkov, V. A., 113, 118, 156
Pun, M. T., 129
Punch, W. E., 54, 106
Puppe, L., 248
Puschmann, M., 96
Put, J., 115
Putter, I., 152, 324

Author Index

Pyles, M., 129
Pyper, J. W., 272

Quayle, A., 135, 179
Quilliam, J. P., 312
Quin, L. D., 220, 221
Quistad, G. B., 40, 96, 126
Qureshi, A. H., 247

Rabet, F., 257
Rabinowitz, M., 218
Rabjohn, N., 122
Rademaker, W. J., 256
Radford, T., 329
Roesky, H. W., 256
Raimondi, D. L., 270
Rainey, W. T., 335
Rake, A. T., 214
Rakita, P. E., 257
Ralph, C. L., 322
Ramsay, C. C. R., 111, 115
Ramsey, R. B., 318
Ramshaw, E. H., 331
Randall, G. L. P., 260
Rang, S. A., 96, 97
Rankin, P. C., 137, 145
Rao, D. V., 173
Rao, E. V., 173
Rao, G. S., 324
Rao, R. P., 133
Rapoport, H., 104, 182
Rapp, U., 123
Rashman, R. M., 107
Rasmussen, R. A., 179
Rausch, M. D., 233, 255, 259
Ray, B. D., 268
Razumova, N. A., 219
Reagan, W. J., 250, 260
Rechter, M. A., 169
Redwood, M. E., 220
Reed, R. I., 94, 98, 131, 183, 285
Reed, W. E., 333
Rees, H. H., 318
Reger, D. L., 261
Regulski, T. G., 96
Reichel, W. L., 331
Reichert, C., 62, 241, 243
Reichstein, T., 23, 172, 173, 175, 188, 189
Reid, D. M., 316
Reid, N. W., 4
Reid, W. W., 310
Reiff, G., 98
Reiffsteck, A., 317
Reimann, H., 173
Reimendal, R., 296, 306
Reinhardt, R. W., 18
Reinhold, V., 297, 325
Reinhoudt, D. N., 111
Reisch, J., 114
Reistad, K. R., 133
Remberg, G., 93, 332
Renaud, R. N., 94
Rennekamp, M. E., 105
Renner, J. A., 329
Replogle, L., 324

Ress, J., 332
Rettig, M. F., 234
Reynolds, W. E., 280, 293
Rhead, M. M., 333
Rhee, I., 261
Rhyne, T. C., 17
Ricci, A., 133
Rich, S., 305
Richards, F. F., 163
Richardson, B., 318
Richardson, J. W., 62
Richter, I., 328
Richter, W., 161
Richter, W. J., 20, 56, 120, 125
Riddock, A., 276
Ridge, D. P., 142
Ridley, R. G., 289, 291
Riediger, W., 124
Riepe, W., 133
Riezebos, G., 334
Riley, P. N. K., 197
Rinehart, K. L., 105, 175, 176, 241, 244, 297
Ring, D., 324
Ring, M. A., 257
Ritter, F. J., 328
Ritter, G., 257
Riveros, J. M., 27
Rivier, J., 323
Rix, M. J., 132
Rizzo, D. J., 128
Rizzo, M., 326
Roach, D., 334
Robbins, J. M., 207
Robbins, W. E., 318
Robertson, A. J. B., 21
Robertson, A. V., 282
Robertson, D. H., 285
Robertson, G. B., 262
Robertson, J. G., 144
Robertson, J. H., 325
Robertson, J. S., 126
Robinson, B. H., 227, 259
Robinson, C. H., 101, 129, 317
Robinson, C. J., 299
Robinson, G. E., 119
Robinson, J., 332
Robinson, J. W., 91
Robinson, P. J., 257
Robinson, P. W., 226
Robinson, R. J., 208
Robinson, W. T., 226, 227
Roboz, J., 297, 325
Rochow, E. G., 257
Rocher, P., 305
Rodeheaver, G. T., 13
Rodia, R. M., 47, 130
Rodrigo, R. G. A., 189
Rodriguez-Gonzalez, B., 181
Roe, D. M., 230, 260
Roedder, E., 333
Röller, H., 328
Röper, H., 329
Roepstorff, P., 160, 165
Röschenthaler, G.-V., 259
Roesky, H. W., 257
Rogerson, P. F., 66, 107, 194, 240

Roggero, J., 132
Rollgen, F. W., 19, 20
Romero, M. C., 118
Romiez, M., 333
Rona, P., 200
Ronteix, P., 273
Roques, B., 233
Ros, A., 303
Roscoe, J. S., 256
Rose, F. L., 132
Rose, G. G., 246
Rose, P. D., 257
Rosegay, A., 324
Roseman, L., 206
Rosenberg, E., 207
Rosenberg, H., 263
Rosenblum, M., 231
Rosenstock, H. M., 92
Rosenthal, D., 107, 258
Rosinov, B. V., 163, 181
Ross, E. P., 262
Rosset, J.-P., 133
Rostovtseva, L. I., 157
Roth, H. J., 255
Rowland, C. G., 54, 62, 63, 97
Rowland, M., 321
Rowlands, R. J., 166
Roy, R. B., 101
Royer, M., 313
Rozynov, B. N., 156
Rubenacker, K., 22
Rubesch, M., 105
Rubinstein, I., 318
Rubio-Lightbourn, J., 321
Ruby, E. D., 280
Rudolph, C. J., 147
Rudolph, R. W., 256, 258
Rücher, G., 115
Ruecker, G., 332
Ruhig, A., 297, 325
Ruliffson, W. S., 172
Rullkötter, J., 97
Runyan, J., 278
Ruotsalainen, H., 120
Russell, D. R., 259
Russell, D. W., 155, 156
Russell, M. E., 5
Russell, R. D., 265
Russo, P. J., 228
Ruveda, E. A., 190
Ryan, J. F., 12
Ryang, M., 261
Rydberg, U., 325
Ryhage, R., 152, 196, 297, 305, 306, 310, 319, 332

Saalfeld, F. E., 15, 42, 138, 193, 228, 332
Sabih, K., 326
Sach, E., 160
Sadée, W., 129, 325
Sadovskaya, V. L., 42, 184
Safe, L. M., 317, 318
Safe, S., 61, 97, 100, 120, 332
Safe, S. H., 120, 317
Saha, N. C., 152
Sahini, V. E., 126

Saiki, Y., 125
Saillant, R., 259
Sainsbury, M., 189
Saint-Ruf, G., 129
Saito, K., 311
Sakauchi, N., 326
Sakurai, H., 13, 106, 256
Salmón, M., 120
Salmona, G., 111
Samoilova, Z. E., 115
Sample Woodgate, S. D., 125
Sams, J. R., 262
Samuel, E., 255
Samuelsson, B., 149, 306, 312, 313
Samuelsson, B. E., 148, 149, 313
Samuelsson, K., 313
Sander, D. L., 129
Sandermann, W., 181
Sanger, A. R., 259
Šantavý, F., 186, 188, 189
Santoyo, Y., 317
Sapers, G. M., 329
Sappa, E., 226
Saran, H., 256
Sarkozi, L., 297, 325
Sartori, G., 231
Sasaki, K., 176
Sasaki, S., 127
Sasakura, S., 331
Sasse, J. M., 181
Sastry, S. D., 152
Sato, S., 331
Sato, Y., 331
Satoh, D., 172
Sauers, R. R., 103
Saunders, J. K., 189
Saunders, R., 273
Saunders, R. A., 279, 332
Saxby, M. J., 110
Scanlan, R. A., 311, 329, 330
Schacht, R. J., 176
Schaefer, J., 331
Schaeffer, R., 257
Schally, A. V., 162
Schaper, B. J., 250
Schaumburg, K., 132
Scheinmann, F., 191
Scheppele, S. E., 115, 147, 287
Scherer, O. J., 258
Schermer, E. D., 263
Scheuer, P. J., 307
Schiebel, H. M., 97
Schiller, H. W., 258
Schiller, J. G., 327
Schmid, G., 199, 261
Schmid, H., 76, 114, 187
Schmid, H. H. O., 145
Schmidbaur, H., 229, 262
Schmidt, K., 174
Schmidtberg, G., 209
Schmir, G. L., 135
Schmitz, F. J., 185
Schneehage, H. H., 248
Schneider, H. H., 151
Schneider, R. J. J., 260

Schnitzler, M., 262
Schnoes, H. K., 176, 185, 333
Schnorr, H., 257
Schoengold, D. M., 7, 304, 305
Scholer, F. R., 261
Scholler, R., 317
Schomberg, G., 270, 296
Schrader, S. R., 91
Schreiber, K., 186
Schroepfer, G. J., 334
Schroll, G., 113, 131, 281, 282
Schubert, B., 306
Schuetz, R. D., 332
Schulte-Elte, K. H., 105
Schulten, H. R., 19
Schulze, P., 20, 56, 302
Schwab, J. M., 101
Schwab, W., 260
Schwarberg, J. E., 243
Schwartz, D. P., 319
Schwarz, R. A., 122
Schweinler, H. C., 17
Schwemmer, B., 332
Schweren, T., 22
Scott, W. M., 97, 289, 334
Scriven, E. F. V., 78, 117
Searle, R. H., 281
Searles, S., 6, 85
Sedmera, P., 186, 189
Sedova, V. F., 127
Seel, F., 219, 259
Sefick, M. D., 257
Segard, C., 233
Seibl, J., 116
Seidel, P.-R., 186
Seif, A. E., 116
Seifert, R. M., 329, 330
Seifert, W., 333
Seiler, N., 151
Sekija, M., 170
Selbin, J., 246
Selke, E., 103
Seltzer, R., 129
Selvarajan, R., 258
Sen, B., 201
Sendra, J. C., 257
Sen Sharma, D. K., 17
Serebryakov, E. P., 181
Servi, S., 190
Seto, S., 182, 310
Setser, D. W., 37, 116
Sévenet, T., 173, 179, 188
Severinsen, S., 160
Shabarov, Yu. S., 96
Shackleton, C. H. L., 320, 321
Shakhidayatov, Kh., 108
Shambhu, M. B., 126
Shamma, M., 189
Shannon, J. S., 70, 108, 156
Shannon, P. V. R., 102
Shannon, T. W., 42, 59, 73, 106, 132, 204
Shapiro, D., 218
Shapiro, R. H., 52, 56, 74, 98, 100, 101, 109
Sharkey, A. G., 305

Sharma, G. M., 190
Sharma, S. C., 121
Sharp, D. W. A., 258
Shaw, M. A., 65, 98, 258
Shaw, P. E., 330
Shaw, R. A., 258
Shaw, S. J., 102, 174
Sheehan, M., 111, 113
Sheikh, Y. M., 104, 185, 282
Sheinker, Yu. N., 125
Sheldrick, B., 166
Shemyakin, M. M., 156, 163
Shenin, Y., 176
Shephard, F. W., 329
Shepherd, T. M., 243
Sherk, J. A., 3, 45
Sherman, W. R., 147, 315
Sherry, A. E., 300
Sherwood, P. J., 199
Shibuya, S., 121, 124, 180
Shildcrout, S. M., 115
Shima, T., 125
Shino, M., 279
Shio, M., 121, 124
Shiotani, A., 229
Shiotani, S., 126
Shipley, P. A., 331
Shipton, J., 331
Shively, J. M., 144
Shiyonok, A. I., 172
Shizhmamedbekova, A. Z., 94
Shoemake, G. R., 305
Shonfield, W., 278
Showalter, H. D. H., 177
Shreeve, J. M., 256
Shu, C. K., 330
Shu, P., 179
Shukla, Y. N., 192
Shvachkin, U. P., 127
Siebert, W., 198
Sieck, L. W., 6, 85
Siegel, A. L., 319
Siegel, A. S., 68, 75, 95, 101, 115, 131, 132
Siegert, F. W., 260
Siekmann, L., 305
Sievers, R., 244, 335
Sievers, R. E., 243, 245, 335
Siggia, S., 91
Silhacek, D. L., 309
Silverstein, R. M., 328
Silverthorn, W. E., 261
Silverton, J. V., 335
Sim, G. A., 180
Sim, W., 220
Simic, D., 258
Simmie, J. M., 42, 98
Simmonds, P. G., 305, 334
Simolin, A. V., 181
Simoneit, B. R., 333, 334
Simons, B. K., 123, 136
Simov, D., 133
Simpson, D., 273
Simpson, J., 202
Singh, P., 105
Singh, U. P., 122, 124
Singleton, J. A., 330
Singy, G. A., 102

Author Index

Sink, J. D., 321
Sinnige, H. J. M., 73, 132
Sioe San Tjoa, 326
Sisido, K., 102
Siskin, M., 133
Sissons, D., 331
Sizoy, V. F., 42, 184
Sjöqvist, F., 305
Sjövall, J., 296, 306, 319, 320, 321
Skurat, V. E., 108
Slade, M., 115
Sloan, K. B., 122
Sloan, S., 183, 211, 317
Slywka, G. W. A., 121
Smakman, R., 131
Smirnova, N. A., 94
Smith, A. G., 318
Smith, C., 128
Smith, D. C., 236
Smith, D. H., 91, 273, 285, 288
Smith, D. M., 191
Smith, G. A., 93
Smith, G. C., 110, 137
Smith, G. R., 273
Smith, I. C., 291
Smith, J. A., 245
Smith, J. L., 114, 152, 323, 324, 327, 335
Smith, J. S., 60, 104, 108
Smith, L. C., 161
Smith, P. J., 69, 97
Smithson, L. D., 100, 239
Smolina, Z. I., 125
Smyth, K. C., 26, 27
Smythe, R. J., 305
Snaith, R., 250
Snatzke, G., 173, 186, 189
Sneddon, W., 189
Snieckus, V., 230
Snyckers, F. O., 188
Snyder, J. P., 116
Sodini, G. C., 162
Sokolov, S. D., 122
Soliday, C. L., 145
Solomon, J. J., 12, 13
Solomon, M. D., 115
Somanathan, R., 182
Sommerville, I. F., 303
Sonnenberg, K.-D., 151
Soothill, R. J., 131
Sorokin, V. V., 5, 6
Spaide, J., 322
Spalding, T. R., 194, 197, 202, 227
Spangler, R. J., 111, 113
Spear, M. J., 119
Spear, R. K., 165
Spencer, G. F., 311
Spencer, J. L., 227
Spencer, R. B., 92, 166, 288, 295
Spialter, L., 99, 203
Spilners, I. J., 239
Spiteller, G., 91, 92, 93, 94, 184, 185, 293, 308, 309, 332
Spiteller-Friedmann, M., 184, 185, 293, 308

Srivastava, R. C., 260, 261
Srivastava, T. S., 262
Sroka, G., 82
Staab, H., 198
Staab, H. A., 123
Stace, A. J., 51, 129
Stacey, J. S., 265
Ställberg-Stenhagen, S., 329
Stafford, C., 99
Stafford, F. E., 295
Stagg, H. E., 92
Stahl, D., 265
Stalling, D. L., 331, 334
Stanner, F., 226
Stanovnik, B., 124
Stanton, M., 230
Stanyer, J., 120
Stapfer, C. H., 257
Stark, W. M., 178
Starkovskii, N. I., 297
Steed, E., 293
Steedman, W., 129
Steel, G., 310, 317, 318, 333
Steelink, C., 97
Steen, G. O., 148, 149, 313
Steiger, W., 127
Steinberg, F. S., 78, 96
Steiner, G. W., 173
Steinfelder, K., 91, 138
Stenberg, A., 309, 320
Stepanov, F. N., 94
Stephen, W. I., 243, 244
Stephens, M., 199
Sterlin, S. R., 250, 255, 257
Stermitz, F. R., 173
Stern, D. J., 330
Stevens, H. N. E., 128
Stevens, K. L., 330
Stevens, M. F. G., 128, 129
Stevenson, D. P., 44
Stevenson, G. M., 134
Stewart, C. P., 179
Stewart, R. B., 273
Stewart, R. P., 259
Stewart, W. B., 4
Stewart, W. H., 304
Stickley, D. J., 119
Still, G. G., 115
Still, I. W. J., 129
Stillman, R. A., 280
Stillwell, R. N., 270
Stillwell, W. G., 305
Stobart, S. R., 228, 257, 259
Stockdale, J. A., 16
Stockdale, J. A. D., 17, 18
Stokke, O., 325
Stoll, M., 105
Stollings, H. W., 115
Stone, F. G. A., 259
Storr, A., 201
Story, M. S., 6
Stout, C. D., 234
Strafford, R. G., 201
Strain, H. H., 182
Strakov, A., 105
Stransky, K., 145, 309
Strausz, O. P., 83
Streibl, M., 145, 309
Streicher, W. 188

Strelow, F., 188
Strigina, L. I., 185
Strobel, G. A., 173
Strocchi, A., 311
Strong, F. M., 176
Struck, A. H., 93
Struckmeyer, K., 173
Studier, M. H., 151, 333
Stüwer, D., 278
Stump, R. K., 293
Sturm, J. E., 224
Su, T., 83, 84
Sucrow, W., 129
Suda, T., 185
Suggs, J. L., 258
Sugihara, J. M., 77, 110
Sugimoto, S., 123, 127
Sugiyama, Y., 297
Sukhoverkhov, V. D., 94
Sukiasian, A. N., 132
Sukkestad, D. R., 146
Sullivan, H. R., 327
Sullivan, R. E., 90
Sumida, Y., 131
Summerford, C., 198, 250
Sun, K. K., 109, 310
Sun, T., 164
Sunderman, F. W., 113
Sundholm, L., 334
Surapaneni, C. R., 124
Sutherland, G. L., 281, 282
Sutherland, M. D., 180
Sutton, E. J., 19
Suzuki, A., 156
Suzuki, Y., 128
Svec, H. J., 43, 193, 194, 225, 238
Svec, W. A., 182
Svensson, S., 171, 315
Svensson, U., 322
Sviridov, A. F., 170
Svoboda, J. A., 318
Sweeley, C. C., 148, 152, 306, 313, 314, 316
Sweeley, C. G., 268
Sweetman, B. J., 117, 312
Swenson, P., 332
Swinkell, R. F., 256
Szepes, L., 42

Taagepera, M., 27
Tabet, J.-C., 122
Taft, R. W., 27
Tailby, G. R., 259
Takahashi, H., 115
Takahashi, N., 156
Takahashi, S., 328, 331
Takamizawa, A., 132
Takamuku, S., 13
Takano, M., 169
Takayama, T., 174
Takeda, K., 186
Talley, C. P., 326
Talrose, V. L., 108
Tamás, J., 187
Tamborski, C., 100
Tamura, S., 156
Tamura, Y., 126, 131
Tamura, Z., 315

Tanaka, I., 5, 85
Tanaka, Y., 185
Tandon, J. S., 192
Tani, K., 231
Tannenbaum, H. P., 78, 116
Tanouti, M., 102
Tanzer, M. L., 158
Tassel, M., 183
Tatematsu, A., 82, 92, 106, 109, 110, 125, 265
Tattershall, B. W., 113
Taulov, I. G., 133
Tauveron, P., 295
Taylor, D. R., 180
Taylor, F. B., 247
Taylor, M. L., 335
Taylor, S. R., 279
Taylor, W. G., 78, 116
Tebby, J. C., 258
Tedder, J. M., 98
Teeter, R. M., 333
Teh-Liang Chang, 305
Tejima, S., 169
Teller, G., 113
Tempe, H.-J., 124
Tennent, A., 98
ten Noever de Brauw, M. C., 332
Tenschert, G., 22, 52
Teranishi, R., 329
Terent'ev, P. B., 126
ter Heide, R., 331
Terlain, B., 176
Terlouw, J. K., 194, 243, 247
Terrazas, R., 265
Tesarek, J. M., 31, 66, 107, 120
Teshima, S.-I., 318
Teuber, H.-J., 121
Thal, C., 173, 179
Tham, W. S., 227
Thawley, A. R., 305
Thenot, J. P., 317
Thiele, W., 197
Thomas, A. F., 105
Thomas, B. S., 201
Thomas, C. B., 119, 123
Thomas, D. W., 159, 160, 176
Thomas, J. L., 231, 260
Thomas, J.-P., 176
Thomas, M. T., 129
Thomas, R. A., 279
Thommen, E., 113
Thompson, C. J., 312
Thompson, D. T., 238
Thompson, J. C., 256
Thompson, M. J., 318
Thompson, M. L., 256
Thompson, R. M., 176
Thomson, I. J., 243, 244
Thomson, J. B., 230
Thomson, R. H., 192
Thornstad, O., 116
Thorpe, F. G., 195
Throck Watson, J., 117
Thyes, M., 110
Thynne, J. C. J., 16, 42, 92, 137, 142

Tibbetts, F. E., 68, 233
Tickner, A. W., 43
Tiernan, T. O., 83, 84, 99, 203, 335
Tighe, L. E., 295
Tillett, J. G., 77, 81, 110, 114, 131
Timberlake, C. F., 329
Timmer, R., 331
Timms, A. W., 127
Timms, P. L., 262
Tin-Wa, M., 187
Tipping, A. E., 205
Tišler, M., 124
Tjoa, S., 326
Tkach, P. W., 171
Tkachuk, R., 154
Todd, J. F. J., 112, 137, 198, 199, 200, 208, 218
Tökés, L., 217
Toft, M., 334
Toft, P., 308
Togashi, M., 279
Toma, S., 263
Tomes, K., 109
Tomer, K. B., 56, 74, 100, 109
Tomkins, I. B., 262
Toren, E. C., 300
Tornabene, T. G., 309
Torroni, S., 4, 53
Tou, J. C., 19, 20, 22, 47, 67, 102, 118, 130, 216, 222
Toube, T. P., 119, 221
Toubiana, R., 160
Traeger, J. C., 297
Traubel, T., 178
Trenner, N. R., 323, 326, 327
Treppendahl, S., 125
Tressl, R., 330
Trieber, L. R., 320
Trinh-Toan, 229
Trogu, E. F., 263
Tronchet, J., 174
Tronchet, J. M. J., 174
Troy, F. A., 146
Trudell, J. R., 125
Tsai, K.-H., 213
Tsai, S.-C., 86
Tsai, T., 257
Tsang, C. W., 101, 111
Tschesche, R., 173, 187
Tschuikow-Roux, E., 42, 98
Tsuboyama, K., 174
Tsuboyama, S., 174
Tsuchiya, T., 92, 127
Tsuda, S., 15, 16, 138
Tsuji, K., 325
Tsuji, T., 327
Tsujimoto, K., 54, 123
Tsujimoto, N., 126
Tsunoda, M., 128
Tsurkanova, L. P., 122
Tsutsui, M., 262
Tsutsumi, S., 261
Tucker, R. B., 293
Tucker, W. P., 176
Tucknott, O. G., 329, 331

Tümmler, R., 91, 138
Tumlinson, J. H., 328
Turk, J., 52, 56, 100
Turner, D. W., 2
Turner, G., 243
Turner, R. B., 137, 198, 208
Tursch, B. M., 185
Tveita, P. O., 133
Tweedale, A., 197
Twine, C. E., 82, 107
Tzschach, A., 222

Uccella, N. A., 63, 71, 113, 114, 122
Uchimaru, F., 128
Uden, P. C., 243, 244
Udre, V. E., 133
Ueda, M., 315
Ueno, A., 125
Uhlenbrock, W., 257
Uliss, D. B., 120
Ulmann, R., 174
Ulrick, J., 265
Umaba, T., 15, 138
Umirzakov, B., 76, 106
Undheim, K., 40, 43, 113, 116, 118, 126, 127, 133
Ungar, G., 163
Ungermann, C., 256
Unkovsky, B. V., 125
Unruk, G. V., 293
Upadhyay, R. R., 182
Updegrove, W. S., 333, 334
Upham, R. A., 12
Urey, H. C., 334
Urry, G., 257
Urshibara, T., 128
Usdin, E., 306
Ushakova, R. L., 206
Usov, A. I., 169
Ustynyuk, Y. A., 259
Utimoto, K., 102

Vaglio, G. A., 230, 259
Vahrenkamp, H., 237, 256
Vaisbein, Z. Y., 247
Vaishnav, Y. N., 324
Vale, G. L., 331
Vale, W., 11, 323
Valente, L., 162
Valentin, M., 122
Vallon, P., 265
Vance, J., 317
van de Graaf, B., 108, 111
Van de Kleijn, E., 325
Vandendunghen, G., 202
Van den Heuvel, W. J. A., 114, 152, 322, 323, 324, 326, 327, 334, 335
Vanderhaeghe, H., 156
Van der Maas, H. L., 332
Vanderwalle, M., 106
van der Wielen, A. J., 257
Vanderzanden, R. J., 197
Van Dyke, C. H., 257
van Fossen, R. Y., 137
Van Ginneken, C. A. M., 327
Vanhaelen, M., 173

Author Index

van Havenbeke, Y., 97
van Houte, J. J., 69, 122
van Lear, G., 175
Van Oven, H. O., 234, 261
van Rossurn, J. M., 327
van Thuijl, J., 69, 122
Vant Klooster, H. A., 265
van Veen, R., 198, 256
Varfalvy, L., 128
Varshavsky, Y. M., 5, 168
Varvoglis, A. G., 45, 116
Vecchi, M., 114, 183, 322
Vegar, M., 94
Veibel, S., 121
Veith, H. J., 113
Velleman, K.-D., 219
Venema, A., 73, 94, 97, 106
Venkataraghavan, R., 167, 272, 277, 285
Vergedal, S., 176
Vermeer, J. H., 322
Vermeglio, A., 98
Vernay, H. F., 171
Vernon, J. M., 119
Vestal, M., 35
Vestal, M. L., 6, 70
Vetter, W.,114,161,183,322
Vetter-Diechtl, H., 161
Vidal, J.-P., 105
Viglino, P., 263
Vihko, R., 320
Vilkas, E., 159
Villa, J. F., 246
Vincent, E.-J., 111
Viney, B. W., 67
Vink, J., 171, 289, 307
Vinogradov, B. A., 219
Vinogradova, E. I., 156, 163
Virtanen, P. O. I., 120
Visotskii, V. I., 120, 121
Viswanathan, N., 181
Vitzhum, G., 257
Vliegenthart, J. F. G., 171, 289, 307
Vöumin, J. A., 302, 303, 321, 332
Voelter, W., 174, 315
Voerman, S., 328
Voetsch, W., 323
Vogel, L., 121
Vogt, J., 100
Voiland, A., 158
von Ardenne, M., 91
von Euw, J., 173
von Halasz, S. P., 256
von Minden, D. ., 123
von Putkammer, E., 4
von Sydow, E., 330
von Unruh, G., 309, 332
Voorhees, R. L., 256
Vorobjova, V. Ya., 125
Voronkov, M. G., 133
Vos, J. G., 332
Vouros, P., 73, 113, 132, 183, 184, 211, 308, 314, 317, 321, 335
Voyevodskaya, T. I., 259
Vredenberg, S., 13, 82
Vree, T. B., 327
Vyazankin, N. S., 257

Wachs, M. E., 97
Wachs, T., 35
Wade, K., 198, 250
Wagner, P. J., 96, 114
Wagner, R., 226
Wahl, G. H., 94, 130
Wahlberg, I., 181
Waight, E. S., 317
Wailes, P. C., 232
Waite, N. E., 258
Wakabayashi, K., 128
Walker, A. C., 331
Walker, B. J., 258
Walker, R. W., 322, 327
Wallbridge, M. G. H., 202
Wallenfals, K., 126
Waller, G. R., 115, 152, 273, 329, 330
Wallington, M. J., 273
Walls, E. C., 273
Walradt, J. P., 329
Walsh, I., 140
Walter, T. A., 332
Walther, W., 114, 183, 322
Walton, D. R. M., 206
Walton, J. C., 98
Walton, T. J., 144, 311
Wan, A. S., 318
Wander, J. D., 171
Wang, C.-S., 216, 222
Wang, P. -S., 329
Wang, S. Y., 22
Wangen, L. E., 284, 291
Wannagat, U., 257
Ward, D. N., 11
Ward, J. E. H., 262
Ward, R. S., 258
Ward, S. D., 3, 297
Wardle, R., 229
Warner, C. M., 214
Warnhoff, E. W., 157
Warren, C. B., 332
Warren, J., 43
Warren, J. D., 262
Warren, L. E., 246
Warren, L. F., 261
Warren, R. F., 300
Wascho, L. S., 270
Waser, P. G., 325, 327
Watanabe, K., 331
Waterbury, L. D., 326
Watson, J., 312
Watson, J. T., 312
Watson, T. R., 126
Watts, J. O., 329
Watts, L., 231
Waugh, F., 206
Wayland, B. B., 228
Webb, K. S., 305
Weber, H., 263
Weber, L., 199
Weber, W. B., 209
Weber, W. P., 14, 137, 140, 204, 205, 207, 210
Webster, B. R., 132, 161
Wegener, J., 257
Wegener, J. A., 305
Wehner, H. W., 261
Weidlein, J., 201
Weigel, H., 136, 199

Weinberg, D. S., 99
Weinberg, K. G., 220
Weiner, N., 322
Weinstein, J., 171
Weiss, B., 334
Weiss, M., 51
Weller, F., 197
Wells, C. H. J., 119
Wells, J. N., 103
Wells, P. R., 213
Wells, R. J., 94
Welti, D., 331
Welzel, P., 115, 184
Wenkert, E., 190
Wentrup, C., 131
Weringa, W. D., 65, 105, 132, 179
Werner, H., 237
West, B. O., 220, 222, 245, 262
West, D. W., 138
Westmore, J. B., 241, 243
Westöö, G., 196, 305, 332
Westover, L. B., 19
Westwood, J. H., 100, 170
Westwood, R., 65, 66, 98, 119
Wewerka, E. M., 120
Wexler, S., 29, 83
Weyenberg, D. R., 207
Weygand, F., 164
Whalley, W. B., 183
Whan, D. A., 309
Wheeler, M. A., 145
Whelan, D. J., 134
Whimp, P. O., 262
Whipple, E. B., 220
Whistler, R. L., 170
White, A. H., 327
White, C., 260
White, E., 125, 137, 308
White, G., 42, 201
White, J. B., 246
White, J. D., 115, 129
White, M. J., 27
White, P. A., 161, 163
White, P. Y., 15, 65, 96, 97, 130, 135, 140
White, R. H., 329
White, W. W., 327
Whitesides, T. H., 230
Whitfield, G. F., 110
Whiting, M. C., 122
Whitney, J. G., 178
Whitney, T. A., 9
Whittaker, D., 262
Whitten, D. G., 54, 106
Whittle, J. A., 103
Wiberg, N., 201, 257
Wick, A. E., 186
Wick, E. L., 329
Widdowson, D. A., 189
Widmark, G., 296, 306, 322
Wiebers, J. L., 127
Wiechers, A., 188
Wies, R., 258
Wiezer, H., 257
Wiggans, P. W., 230
Wijsbeek, J., 327
Wikstrom, S., 297